Principle 11

A sound marketing plan enables a new firm to identify the target customer, set its marketing objectives, and implement the steps necessary to sell the product and build solid customer relationships.

Principle 12

Effective leaders coupled with a good organizational plan, a collaborative performance-based culture, and a sound compensation scheme can help align every participant with the goals and objectives of the new firm.

Principle 13

Effective new ventures use their persuasion skills and credibility to secure the required resources for their firm in order to build a well coordinated mix of outsourced and internal functions.

Principle 14

All new technology business ventures should formulate a clear acquisition and global strategy.

Principle 15

The design and management of an efficient, real-time set of production, logistical, and business processes can become a sustainable competitive advantage for a new enterprise.

Principle 16

A new firm with a powerful revenue and profit engine and a reputation for ethical dealings can achieve strong but manageable growth leading to a favorable harvest of the wealth for the owners.

Principle 17

A sound financial plan demonstrates the potential for growth and profitability for a new venture and is based on the most accurate and reliable assumptions available.

Principle 18

Many kinds of sources for investment capital for a new enterprise exist and should be compared and managed carefully.

Principle 19

The creation and communication of a compelling story about a venture and the resulting skillful negotiations to close a deal with investors are critical to all new enterprises.

Principle 20

The ability to continuously and ethically execute a business plan and adapt to changing conditions provides a firm with a sustainable competitive advantage.

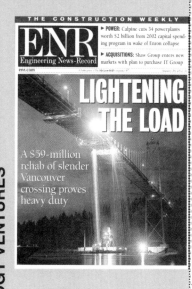
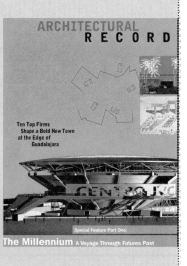

Technology Ventures
From Idea to Enterprise

Technology Ventures
From Idea to Enterprise

Richard C. Dorf
University of California, Davis

Thomas H. Byers
Stanford University

Mc Graw Hill **Higher Education**

Boston   Burr Ridge, IL   Dubuque, IA   Madison, WI   New York   San Francisco   St. Louis
Bangkok   Bogotá   Caracas   Kuala Lumpur   Lisbon   London   Madrid   Mexico City
Milan   Montreal   New Delhi   Santiago   Seoul   Singapore   Sydney   Taipei   Toronto

The McGraw·Hill Companies

 **Higher Education**

TECHNOLOGY VENTURES: FROM IDEA TO ENTERPRISE

Published by McGraw-Hill, a business unit of The McGraw-Hill Companies, Inc., 1221 Avenue of the Americas, New York, NY 10020. Copyright © 2005 by The McGraw-Hill Companies, Inc. All rights reserved. No part of this publication may be reproduced or distributed in any form or by any means, or stored in a database or retrieval system, without the prior written consent of The McGraw-Hill Companies, Inc., including, but not limited to, in any network or other electronic storage or transmission, or broadcast for distance learning.

Some ancillaries, including electronic and print components, may not be available to customers outside the United States.

This book is printed on acid-free paper.

2 3 4 5 6 7 8 9 0 QPF/QPF 0 9 8 7 6 5 4

ISBN 0-07-285353-0

Publisher: *Elizabeth A. Jones*
Senior Sponsoring Editor: *Suzanne Jeans*
Developmental Editor: *Kathleen L. White*
Marketing Manager: *Dawn R. Bercier*
Senior Project Manager: *Kay J. Brimeyer*
Senior Production Supervisor: *Sherry L. Kane*
Lead Media Project Manager: *Audrey A. Reiter*
Media Technology Producer: *Eric A. Weber*
Senior Coordinator of Freelance Design: *Michelle D. Whitaker*
Cover Designer: *Joanne Schopler/Graphic Visions*
(USE) Cover Images: left: © *Corbis;* middle: © *Samuel Ashfield/Getty Images;* right: © *Mario Beauregard/Corbis*
Compositor: *ElectraGraphics, Inc.*
Typeface: *10.5/12 Times Roman*
Printer: *Quebecor World Fairfield, PA*

Library of Congress Cataloging-in-Publication Data

Dorf, Richard C.
   Technology ventures : from idea to enterprise / Richard Dorf, Thomas Byers. — 1st ed.
      p. cm.
   Includes bibliographical references and index.
   ISBN 0-07-285353-0
   1. Information technology. 2. Entrepreneurship. 3. New business enterprises. I. Byers, Thomas (Thomas H.).
II. Title.

HC79.I55D674   2005
658.1'1-dc22   2004040288
                                                                                             CIP

www.mhhe.com

DEDICATION

We wish to dedicate this book to our spouses: Joy M. Dorf and Michele L. Mandell. We recognize their love and commitment to this publication that will help others create important technology ventures for the benefit of all.

RICHARD C. DORF, THOMAS H. BYERS

ABOUT THE AUTHORS

Richard C. Dorf   is Professor of Electrical and Computer Engineering and Professor of Management at the University of California, Davis. He is a Fellow of the American Society for Engineering Education (ASEE) in recognition of his outstanding contributions to the society, as well as a Fellow of the Institute of Electrical and Electronic Engineering (IEEE). The best-selling author of Introduction to Electric Circuits (6th Ed.), Modern Control Systems (10th Ed.), Handbook of Electrical Engineering (3rd Ed.), Handbook of Engineering (2nd Ed.), and Handbook of Technology Management, Dr. Dorf is co-founder of six technology firms.

Thomas H. Byers   is Professor of Management Science and Engineering at Stanford University and founder of the Stanford Technology Ventures Program, which is dedicated to accelerating high-technology entrepreneurship education around the globe. After receiving his MBA and Ph.D. from the University of California, Berkeley, Dr. Byers spent over a decade in leadership positions in technology ventures including Symantec Corporation.

BRIEF CONTENTS

SECTION IV

CONTENTS

FOREWORD

by John L. Hennessy, President of Stanford University

I am delighted to see this new book on technology entrepreneurship by Dorf and Byers. High-technology companies are both an important part of our world's economic growth story as well as the place where many young entrepreneurs realize their dreams.

Unfortunately, there have been relatively few complete and analytical books on high-technology entrepreneurship. Dorf and Byers bring their years of experience in teaching to this book, and it shows. Their personal experiences as entrepreneurs are also clear throughout the book. Their connections and involvement with startups—ranging from now established companies like Sun Microsystems and Yahoo to new ventures just delivering their first products—add a tremendous amount of real-world insight and relevance.

One of the most impressive aspects of this book is its broad coverage of the challenges involved in high-technology entrepreneurship. Part I talks about the core issues involved in deciding to pursue an entrepreneurial vision and what characteristics are vital to success from the very beginning. I am pleased to see that building and maintaining a competitive advantage and the critical issue of market timing are key topics. During the Internet boom, while several great new companies were built, too many entrepreneurs and investors forgot several key principles: have a sustainable advantage, create a significant barrier to entry, and be a leader when the market and the technology are both ready. Hopefully, the material in these chapters will help prevent future irrational behavior by both entrepreneurs and investors.

Part II examines the major strategic decisions that any group of entrepreneurs must grapple with: how to balance risk and return, what entrepreneurial structure to pursue, how to find and cultivate the best employees and help make them productive, and the critical issues of intellectual property. Indeed, these are problems that every company faces, and ones that must be continuously examined by the leadership in any organization.

Part III discusses the operational and organizational challenges that all entrepreneurs must tackle. Virtually every start-up led by a technologist that I have been close to inevitably wonders whether they need sales and marketing. Sometimes in such companies, you hear a remark like: "We have great technology and that will bring us customers, nothing else matters!" I remind them that without sales, there is no revenue, and without marketing, sales will be diminished. Understanding how to approach these vital aspects of any successful business is crucial. The related topics of building the organization, thinking about acquisitions, and managing operations are also important. If you fail to address these aspects of your company, it will not matter how good your technology is.

The final part of this book talks about putting together a solid financial plan for the company, including exit and funding strategies. Of course, such topics are crucial, and they are often the sole or dominant topics of "how-to" books on entrepreneurship. Certainly, the financing and the choice of investors are key, but unless the challenges discussed in the preceding sections are overcome, it is unlikely that a new venture, even if well financed, will be successful.

In looking through this sage and comprehensive treatment, my overwhelming reaction was, "I wish I had read a book like this, before I started my first company (MIPS Technologies in 1984)." Unfortunately, I had to learn many of the topics covered here in real-time and often by making a mistake on the first attempt. In my experience, it is the challenges discussed in the earlier sections that really proved to be the minefields. Yes, it is helpful to know how to negotiate a good deal and to structure the right mix of financing sources, especially so that as much equity as possible can be retained by employees. If, however, you fail to create a sustainable advantage or have a sales or marketing plan that is solid, the employee's equity will not be worth much.

Those of us who work at Stanford and live near Silicon Valley are in the heart of the land of high-technology entrepreneurship. With this new book, many others will get to share the extensive and deep insights of Dorf and Byers on this wonderful process that builds tomorrow's companies and business leaders.

Entrepreneurship is a vital source of change in all facets of society, empowering individuals to seek opportunity where others see insurmountable problems. For much of the past century, entrepreneurs have created many great enterprises that subsequently led to job creation, improved productivity, increased prosperity, and a higher quality of life. With one-third of the world's population lacking access to basic energy needs and two-thirds with annual incomes of less than $2,000, entrepreneurship can play an important role in finding solutions to these challenges facing civilization.

Many books have been written to help educate others about entrepreneurship. Our textbook is the first to thoroughly examine a global phenomenon known as "technology entrepreneurship." Technology entrepreneurship is a style of business leadership that involves identifying high-potential, technology-intensive commercial opportunities, gathering resources such as talent and capital, and managing rapid growth and significant risks using principled decision-making skills. Technology ventures exploit breakthrough advancements in science and engineering to develop better products and services for customers. The leaders of technology ventures demonstrate focus, passion, and an unrelenting will to succeed.

Why is technology so important? The technology sector represents a significant portion of the economy of every industrialized nation. In the United States, more than one third of the gross national product and about half of private-sector spending on capital goods are related to technology. It is clear that economic growth depends on the health and contributions of technology businesses.

Technology has also become ubiquitous in modern society. Note the proliferation of cell phones, personal computers, and the Internet in the past decade and their subsequent integration into everyday commerce and our personal lives. When we refer to "high technology," we include information technology and electronics companies, life science and biotechnology businesses, and those service firms where technology is critical to their missions (e.g., Fidelity Investments and Schwab in the financial industry). At the dawn of the 21st century, many technologies show tremendous promise, including photonics and Internet advancements, medical devices and drug discovery, nanotechnology, and materials technologies related to energy and the environment.

The drive to understand technology venturing has frequently been associated with boom times. Certainly, the often-dramatic fluctuations of economic cycles can foster periods of extreme optimism as well as fear with respect to entrepreneurship. However, some of the most important technology companies have been founded during recessions, including Intel, Cisco, and Amgen. This book's principles endure regardless of the state of the economy.

APPROACH

Just as entrepreneurs combine things to create innovations, we integrate the most valuable entrepreneurship and technology management theories from some of the world's leading scholars, educators, and authors. We also provide an action-oriented approach to the subject through the use of examples, exercises, and lists. By striking a balance between theory and practice, we hope our readers will benefit from both perspectives.

Our comprehensive collection of concepts and applications provides the tools necessary for success in starting and growing a technology enterprise. We show the critical differences between scientific ideas and true business opportunities. Readers will benefit from the book's integrated set of cases, examples, business plans, and recommended sources for more information.

To illustrate the book's concepts, our examples and exercises include a blend of traditional high-technology firms (e.g., Microsoft, eBay, and Genentech) and other companies that use technology strategically (e.g., Starbucks, Southwest Airlines, and Wal-Mart). How do they develop enterprises that have such positive impact, sustainable performance, and realistic potential for longevity? In fact, the book's major principles are applicable to any high-growth, high-potential venture. This includes non-profit (often called "social") enterprises such as Conservation International and the Kauffman Foundation.

AUDIENCE

This book is designed for students in colleges and universities, as well as others in industry and government, who seek to learn the essentials of technology entrepreneurship. No prerequisite knowledge is necessary, although an understanding of basic accounting and finance principles will prove useful.

Colleges and universities have traditionally taught entrepreneurship exclusively to business majors. Because entrepreneurship education opportunities now span the entire campus, we wrote this book to be approachable for students of all majors. Our primary focus is on science and engineering majors enrolled in entrepreneurship and innovation courses, but the book is also valuable to business and other students with a particular interest in technology ventures.

For example, the courses at Stanford University and the University of California, Davis based on this textbook regularly attract students from majors as diverse as computer science, product design, political science, economics, premed, electrical engineering, history, biology, and management. Although the focus is on technology entrepreneurship, these students find this material applicable to the pursuit of a wide variety of endeavors. Entrepreneurship education is a wonderful way to teach universal leadership skills, which include being comfortable with constant change, contributing to an innovative team, and always demonstrating passion in their effort. We particularly encourage instructors to design courses where the students form study teams early in the term and learn to work together effectively on group assignments.

FEATURES

The book is organized in a modular format to allow for both systematic learning and random access of the material to suit the needs of any reader. It is a reference and companion tool to keep on hand for future use. We deploy the following wide variety of methods and features to achieve this goal, and we welcome feedback and comments to our e-mail addresses provided below.

Preview—Each chapter opens with a preview that outlines its content and objectives.

Principles—A set of twenty fundamental principles are developed and defined throughout the book.

Examples—Examples of the concepts are provided in a shaded box format.

Sequential Case—A case about an actual biotechnology firm, AgraQuest, runs from one chapter to the next.

Exercises—Exercises are offered at the end of each chapter to test comprehension of the concepts.

Business Plans and Cases—Two full business plans and seven complete cases are included in the back of the book.

References—References are indicated in a box format [Smith, 2001] and may be found as a complete set in the back of the book, along with valuable sources for additional information.

Chapter Sequence—The chapter sequence represents our best effort to organize the material in a format that can be used in various entrepreneurship courses. The chapters follow the four-section layout shown in Figure P1.

Website—Search for this book at www.mcgrawhillengineeringcs.com for additional information applicable to educators, students and professionals. For example, a complete syllabus for an introductory course on high-technology entrepreneurship is provided to assist instructors.

ACKNOWLEDGEMENTS

Many people have made this book possible. Our editors at McGraw-Hill were Ryan Blankenship, Suzanne Jeans, Betsy Jones, Jonathan Plant, Marianne Rutter, and Katie White. We thank all of them for their insights and dedication. We also thank Kay Brimeyer and her production team at McGraw-Hill for their diligent efforts as well as Josephine Chu for her skillful preparation of the manuscript.

Our colleagues at Stanford University and the University of California, Davis were helpful in numerous ways. We are indebted to them for all of their great ideas and support. At Stanford, they include Scott Cannon, Kathleen Eisenhardt, Kailash Gopalakrishnan, Yvonne Hankins, John Hennessy, Randy Komisar, Thomas Kosnik, Kelley Porter, James Plummer, Elisabeth Pate-Cornell,

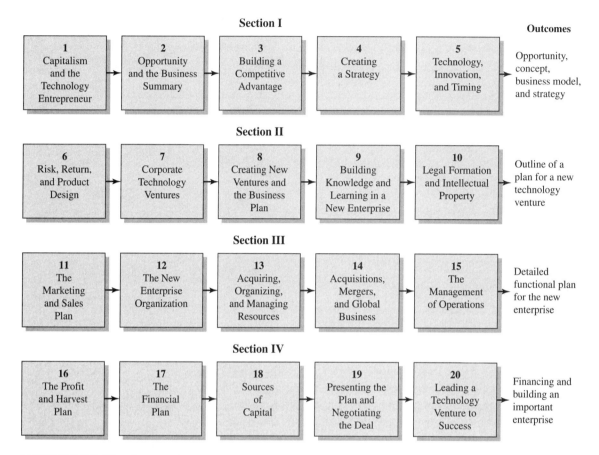

FIGURE P.1 The four sections.

Emily Ma, Asha Nayak, Tina Seelig, and Robert Sutton. At UC Davis, they include Robert Smiley, Andrew Hargadon, Nicole Biggart, Jerome Suran, Ben Finkelor, and Kurt Heisinger.

Practitioners and industry leaders who provided key input included Brook Byers of Kleiner Perkins, Ken Byers of Byers Engineering Company, Gordon Eubanks of Oblix, Bob Fung of Fair Isaac, Jeremy Jaech, Pamela Marrone of AgraQuest, and John Walter of MarkeTech Group. We also express sincere appreciation to all of the manuscript's reviewers: Tom Mason of Rose-Hulman, Tom Miller of North Carolina State University, David Barbe of the University of Maryland, Ed Zschau of Princeton University, John Ochs of Lehigh University, Steve Nichols of the University of Texas, Austin, Elizabeth (Liz) Kisenwether of Penn State University, and Andrew Isaacs of the University of California, Berkeley.

Richard C. Dorf, University of California, Davis, rcdorf@ucdavis.edu
Thomas H. Byers, Stanford University, tbyers@stanford.edu

CHAPTER 1

Capitalism and the Technology Entrepreneur

Our aspirations are our possibilities.
Robert Browning

The entrepreneur provides the creative force capitalism needs to work. Entrepreneurs strive to make a difference in our world and contribute to its betterment. They are also motivated by achievement, independence, and the accumulation of wealth. In this chapter, we describe the characteristics of the people called entrepreneurs and the process they use to create new enterprises. We describe the four types of entrepreneurship used to respond to opportunity: incremental, innovative, imitative, and rent-seeking. Engineers and scientists often respond to the challenge to build important new enterprises by combining their knowledge of new technologies with sound business practices. The technology entrepreneur's role in the improvement of an economy and the role of knowledge in the creation and growth of new enterprises are described. Finally, the firm or organization as the key structure for a new enterprise and the system of innovation used by new ventures are depicted. ■

1.1 THE ENTREPRENEUR AND THE CHALLENGE

Wealth and social change are created by people who strike out on their own and are devoted to worthy tasks and enterprises that make a difference in the world. An **entrepreneur** is a person who undertakes the creation of an enterprise or business that has the chance of profit (or success). Entrepreneurs distinguish themselves through their ability to accumulate and manage knowledge, as well as their ability to mobilize resources to achieve a specified business or social goal [Kuemmerle, 2002].

The entrepreneur is a bold, imaginative deviator from established business methods and practices who constantly seeks the opportunity to commercialize new products, technologies, processes, and arrangements [Baumol, 2002]. Entrepreneurs are skilled in applied creativity, thrive in response to challenge, and look for unconventional solutions. They experience challenges, create visions for solutions, build stories that explain their visions, and then act to be part of the solution. They forge new paths and risk failure, but persistently seek success.

The Horatio Alger myth describes the rise of a young man from rags to riches through entrepreneurism. In this myth, the entrepreneurial hero personifies freedom and creativity. A century ago, this was a common possibility, although actually limited to a few success stories such as those of John D. Rockefeller and Andrew Carnegie. The key virtue was self-reliance and diligence. While this possibility remains for a few today, almost all entrepreneurs are educated, experienced, and skilled. Furthermore, entrepreneurship is an attitude and capability that diffuses beyond the founding team to all members of its organization. For most, collective entrepreneurship represents the path toward a promising economic future. Most growing firms strive to infuse the culture of the entire company with the entrepreneurial spirit. Thomas Edison created an enterprise that became General Electric. Steve Jobs and Steve Wozniak founded Apple Computer, one of the first personal computer companies. These entrepreneurs combined their knowledge of valuable new technologies with sound business practices to build important new enterprises that continued to maintain their entrepreneurial spirit for years after founding.

Entrepreneurship is more than the creation of a business and the wealth associated with it. It is focused on the creation of a new enterprise that serves society and makes a positive change. Entrepreneurs can create great firms that exhibit performance, leadership, reputation, and longevity. Examples of new enterprises that have made a significant contribution to life in our day are provided in Table 1.1. What organization would you add to the list?

Entrepreneurs seek to achieve a certain goal by starting an organization that will address the needs of society and the marketplace. Entrepreneurs are prepared to respond to a challenge to overcome obstacles and build a business. When faced with difficult situations, they are prepared to make the extra effort to overcome these obstacles and succeed. As Martin Luther King, Jr. [1963] said:

> The ultimate measure of a man is not where he stands in moments of comfort and convenience, but where he stands at times of challenge and controversy.

TABLE 1.1 Important new enterprises that started or emerged from 1973 to 2003.

■ Amazon.com	■ Intel
■ Amgen	■ Microsoft
■ Apple Computer	■ Nature Conservancy
■ Cisco	■ Nokia Corporation
■ Conservation International	■ Qualcomm
■ Dell Computer	■ Southwest Airlines
■ Doctors Without Borders / Medecins Sans Frontières	■ Starbucks
■ eBay	■ Virgin Group
■ Federal Express (FedEx)	■ Wal-Mart
■ Genentech	

For an entrepreneur, a **challenge** is a call to respond to a difficult task and the commitment to undertake the required enterprise.

Richard Branson, the creator of Virgin Group, reported [Garrett, 1992]:

Ever since I was a teenager, if something was a challenge, I did it and learned it. That's what interests me about life—setting myself tests and trying to prove that I can do it.

Entrepreneurs are resilient people who pounce on problems, determined to find a solution. The elements of the ability to overcome a challenge are summarized in Table 1.2.

Over nearly a decade, Fred Smith worked on perfecting a solution to what he viewed as a growing problem of organizations to find ways to rapidly ship products to customers. To address this challenge, Smith saw an opportunity to build a freight-only airline that would fly packages to a huge airport and then sort, transfer, and fly them onto their destinations overnight. He turned in his paper describing this plan to his Yale University professor, who gave it an average grade, said to be a C. After he graduated, Smith served four years as a U.S. Marine Corps officer and pilot. Following his military service, he spent a few years in the aviation industry building up his experience and knowledge of the industry. Then, he prepared a fully developed business plan for an overnight freight service. By

TABLE 1.2 Elements of the ability to overcome a challenge.

■ Able to deal with a series of tough issues.	■ Resilient in the face of setbacks.
■ Able to create solutions and work to perfect them.	■ Willing to work hard and not expect easy solutions.
	■ Well-developed problem-solving skills.
■ Able to handle many tasks simultaneously.	■ Able to learn and acquire the skills needed for the tasks at hand.

1972, he had secured financial backing, and Federal Express took to the air in 1973. Federal Express became a new way of shipping goods that revolutionized the cargo shipping business worldwide.

Smith and other entrepreneurs recognize a change in society and its needs, and then, based on their knowledge and skill, they respond with a new way of doing things. Typically, entrepreneurs create a novel response to an opportunity by recombining people, concepts, and technologies into an original solution. Smith saw that the combination of dedicated cargo airplanes, computer-assisted tracking systems, and overnight delivery would serve a new market that required just-in-time delivery of critically important parts, documents, and other valuable items. Smith adapted computer technology to manage the complex task of tracking and moving packages.

When does a person know that he or she is ready to assume the mantle of entrepreneur? When a person is ready to assume the risk and effort and is truly motivated to organize an enterprise to meet the entrepreneurial challenge, they will most likely know it since they will be unable to think of anything but the challenge. A straightforward test of a potential entrepreneur is provided in Table 1.3. Take the test and see if you are ready. Perhaps the right opportunity has not yet emerged, but when it does, be prepared to seize it.

An **opportunity** is a favorable juncture of circumstances with a good chance for success or progress. It is the job of the entrepreneur to locate new ideas and put them into action. Entrepreneurs respond to opportunities by exploiting changes, needs, or new skills or knowledge within the context of their industry. Thus, **entrepreneurship** may be described as the identification and exploitation of previously unexploited opportunities [Hitt, 2001]. Fortunately for the reader, it is a systematic, repeatable discipline that can be learned.

Entrepreneurship can consist of innovation or the introduction of creative change. Change is generally considered as part of the entrepreneurial expectation. In that sense, the entrepreneur is a change agent. Change agents thrive in a land of opportunity in which people can rise from nothing to greatness, depending on their talents and their hard work. Entrepreneurs act as change agents for progress when, as Abraham Lincoln said in 1864, they are offered "an open field and a fair chance for industry, enterprise and intelligence."

Since only about one-third or fewer new ventures survive their first three years, entrepreneurial ventures can be viewed as experiments or probes into a market. This approach is consistent with the goals of change agents who are willing to accept failure as a potential outcome of their venture.

Regardless of whether the right opportunity has emerged, a person can learn to act as an entrepreneur by trying the activity in a low-cost manner. The would-be entrepreneur should, if possible, engage in this sequence: do it, then reflect on it. To avoid the realm of daydreams and fantasy, a person needs to start the practice of experimenting, testing, and learning about their entrepreneurial self [Ibarra, 2002]. The first step is to craft small experiments in new activities with entrepreneurial teams or small ventures. Through these small experiments, the entrepreneur develops new contacts and mentors. They may also find a challenge that serves as a catalyst for a new venture.

TABLE 1.3 Entrepreneur test.

Are you an entrepreneur? Answer each question by checking yes or no.

		Yes	No
1.	When I am faced with a challenge, I am confident that I can work through it.	—	—
2.	I want to be financially independent and be rewarded for my accomplishments.	—	—
3.	Trying something new is attractive, even if I know the risk of failure is significant.	—	—
4.	I would prefer to gain independence and control my destiny.	—	—
5.	Building a new enterprise is important to me.	—	—
6.	My experiences during my youth and early career have shown me the benefits of starting a new enterprise.	—	—
7.	Starting a new business some day soon is always in my thoughts.	—	—
8.	I like working with others and can provide leadership when called upon.	—	—
9.	Our society and my family provide a strong, supportive base for my initiatives.	—	—
10.	I possess strong technical and relationship skills in the industry I wish to enter.	—	—
	Add your total score for yes and no:	—	—

Seven or more yes answers indicate that you may be ready to act as an entrepreneur in the near future.

1.2 ENTREPRENEURIAL ACTIVITY BASED ON INNOVATION AND TECHNOLOGY

Capitalism and enterprise are about having a dynamic economy and innovation, but ultimately, they rest on the actions of businesspeople who assume and accept the benefits and risks of an initiative. It is people acting as leaders, organizers, and motivators who are the central figures of modern economic activity. Most entrepreneurs strive to make a productive, useful contribution to their society while creating wealth for the shareholders and themselves. Profit maximization, however, is not the only goal of these creative businesspeople, who also value independence and leadership challenges.

Three factors comprise entrepreneurial action: 1) a person or group who is responsible for the enterprise, 2) the purposeful enterprise, and 3) initiation and growth of the enterprise. The individuals responsible for the organization were described in Section 1.1. The purposeful enterprise may be a new firm organized for a suitable and attractive purpose or a new unit within or separated from an existing business corporation. Furthermore, the organization may be based on incremental changes, innovation, imitation, or rent-seeking behavior.

TABLE 1.4 Four types of entrepreneurship.

1. **Incremental venture:** The founding and management of a routine business exhibiting modest novelty.
2. **Innovative venture:** The initiation and operation of a business based on an innovation.
3. **Imitative venture:** The identification and imitation of a novel business or venture.
4. **Rent-seeking venture:** The founding of a business that utilizes standards, regulations, and laws to share in some of the value of an existing enterprise.

The first type of enterprise emphasizes the founding and management of a business that entails moderate novelty, such as a new restaurant in town. The founder may still be thought of as an entrepreneur. In the second form, the entrepreneur engages in an innovative activity that results in novel methods, processes, and products. The third, imitative venture, is founded by an entrepreneur who is involved in the rapid dissemination of an innovative idea or process. This person or group finds a novel innovation and transfers it to another region or country. The final means of entrepreneurship is called rent-seeking or profit-seeking and focuses on the use of regulation, standards, or laws to appropriate some of the value of a monopoly that is generated somewhere in the economy. These four types of entrepreneurship are summarized in Table 1.4.

In this book, we emphasize the creation of the innovative venture that will have a significant impact on a region, nation, or the world. A minor or significant incremental innovation may afford the entrepreneur a new opportunity. Alternatively, a radical or transforming innovation may provide an entrepreneur an important and very significant opportunity to make a productive contribution.

The third factor of entrepreneurial action is the initiation and growth of the enterprise. To start an innovative venture, the entrepreneurial team identifies an attractive opportunity that also matches their skills. The opportunity offers the entrepreneurial team a favorable chance to solve a problem or meet a need by creating or applying a technology.

Technology includes devices, artifacts, processes, tools, methods, and materials that can be applied to industrial and commercial purposes. Intel was formed to apply semiconductor technology to the design and manufacture of semiconductor circuits. Microsoft was formed to create and distribute computer software products for applications in industry and the home. The four steps that an entrepreneur typically follows to start a business are summarized in Table 1.5.

Perhaps the most critical aspect of enterprise formation is narrowing in on the best opportunity. Most people see many opportunities but find it difficult to know when to select a specific one and act on it. One useful method of selecting an opportunity is to look for the sweet spot that matches opportunity with interests and capabilities, as shown in Figure 1.1. Most entrepreneurs-to-be will experience a set of good opportunities that flow by over time. They also will have interests, activities, and tasks they like to do. Furthermore, they have capabilities or skills and knowledge that qualify them for certain tasks.

TABLE 1.5 Four steps to starting a business.

1. The founding team or individual has the necessary skills or acquires them.

2. The team identifies the opportunity that attracts them and matches their skills. They create a solution to match the opportunity.

3. They acquire (or possess) the financial and physical resources necessary to launch the business by locating investors and partners.

4. They complete an arrangement or contract with their partners, investors, and within the founder team to launch the business and share the ownership and wealth created.

Good opportunities display the characteristics of a potential to solve important problems within economic constraints. Usually, they will look attractive because they can be profitable to the new venture as well as valuable to the customers. An attractive opportunity displays the five characteristics as listed in Table 1.6. The entrepreneur seeks a timely, solvable, important problem with a favorable context that can lead to profitability.

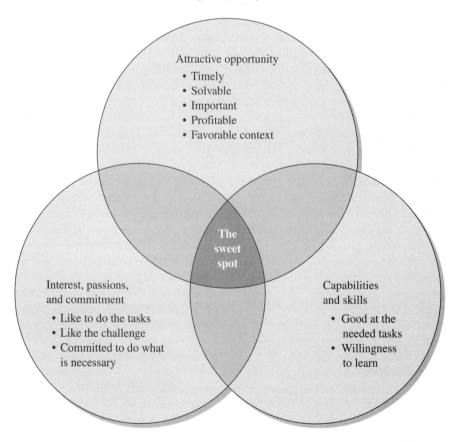

FIGURE 1.1 Selecting the right opportunity by finding the sweet spot.

TABLE 1.6 Five characteristics of an attractive opportunity.

■ Timely—a current need or problem.	■ Profitable—the customer will pay for the solution and allow the enterprise to profit.
■ Solvable—a problem that can be solved in the near future with accessible resources.	
	■ Context—a favorable regulatory and industry situation.
■ Important—the customer deems their problem or need important.	

It is the entrepreneur who adds value to the opportunity by creating a response to a good opportunity. The opportunity, and a general response to it, is not unique—many recognize but few possess the relevant passion to solve the problem as well as the capability to do so. It is really the passion and capabilities that distinguish the entrepreneurial team. The selection process consists of looking for the best match of opportunity, capabilities, and interest (passion).

Jeremy Jaech attended the University of Washington, receiving a BA in mathematics in 1977. He joined the computer science graduate program after graduation, completing his master's degree in 1980, while working at Boeing on computer graphics. In 1983, he joined Atex, a maker of computer systems for newspapers. After nine months, Atex closed the facility where he worked, and Jaech needed to find an opportunity for himself. His capabilities were computer programming for graphics, and his interest was to achieve independence and success. His passion was for developing software for desktop computer graphics. His former boss at Atex suggested they form their own company that would create software for desktop computer graphics. Jaech was a good technical leader, and his boss was a good manager; together, they made a solid team. In 1984, the two men founded Seattle-based Aldus Corporation, which created the software called PageMaker that launched desktop publishing on personal computers.

By 1989, while Aldus had grown, Jaech was faced with a new challenge. He wanted to broaden the product line, while his partner/CEO wanted to remain focused on desktop publishing. Jaech saw an opportunity to create a Windows-based software product for general-purpose drawing. He matched his capabilities with his interests and in 1990 started a new firm that was later called Visio Corporation. When the company's first product was shipped in 1992, it had 14 employees. It went public as a 200-person company in 1995 and was eventually purchased in January 2000 by Microsoft Corporation for $1.5 billion in stock. Jaech served as a vice president of Microsoft for another year. Jaech had exploited two successive opportunities: Aldus and Visio both used his ability to design software while matching his capabilities and skills with his passions and interest to create two important companies.

To summarize, entrepreneurship is centrally focused on the identification and exploitation of previously unexploited opportunities. An opportunity is a favorable juncture of circumstances with a good chance for success or progress. Fortunately for the reader, successful entrepreneurs do not possess a rare entrepreneurial gene. Entrepreneurship is a systematic, organized rigorous discipline

TABLE 1.7 Eight elements of entrepreneurship.

■ Initiate and operate a purposeful enterprise.	■ Ability to assess and mitigate uncertainty and risk associated with the initiation of the enterprise.
■ Operate within the context and industrial environment at the time of initiation.	■ Ability to provide an innovative contribution or at least a contribution that encompasses novelty or originality.
■ Identify and screen timely opportunities.	■ Enable and encourage a collaborative team of people who have the capabilities and knowledge necessary for success.
■ Ability to accumulate and manage knowledge and technology.	
■ Ability to mobilize resources—financial, physical, and human.	

that can be learned and mastered [Drucker, 2002]. The eight elements of entrepreneurship, which also include mobilizing resources, mitigating uncertainty, and building a collaborative team, are listed in Table 1.7.

1.3 ENTREPRENEURIAL CAPITAL AND THE VALUE OF A VENTURE

The quality of entrepreneurial capability is expressed as its ability to generate future income and wealth. One measure of the quality of entrepreneurial capability is entrepreneurial capital (EC), which can be formulated as a combination of entrepreneurial competence and entrepreneurial commitment [Erikson, 2002]. Thus, we say that

Entrepreneurial Capital = entrepreneurial competence
$$\times \text{ entrepreneurial commitment}$$

or more succinctly,

$$EC = Ecomp \times Ecomm \tag{1.1}$$

where Ecomp is entrepreneurial competence and Ecomm is entrepreneurial commitment. Note that the symbol $\times$ is a multiplication sign, but it must be recognized that this equation is qualitative in nature.

The presence of competence without any commitment creates little entrepreneurial capital. The presence of commitment without competence may waste both time and resources. Both commitment and competence are required to provide significant entrepreneurial capital (hence, we use the $\times$ sign). **Entrepreneurial competence** is the ability 1) to recognize and envision taking advantage of opportunity and 2) to access and manage the necessary resources to actually take advantage of the opportunity. **Entrepreneurial commitment** is a dedication of the time and energy necessary to bring the enterprise to initiation and fruition. Entrepreneurial capital reflects the aggregation of competence and commitment of the entrepreneurial team.

The accretion of knowledge and experience over time leads to increased competence as people mature. However, commitment (energy and time) may tend to decline when people become less interested in or available for the necessary entrepreneurial competence activities. No firm rules should be assumed about the appropriate age, but most entrepreneurs emerge by the age of forty. Both commitment and competence are qualities of the leadership team, and they may be complementary qualities shared among the team members.

An opportunity is an auspicious chance of an action occurring at a favorable time. Thus, the entrepreneur identifies a propitious enterprise at a time that appears to be right for success.

We can then propose that the economic value of a venture is

$$\text{Economic Value} = \text{Opportunity} \times \text{Entrepreneurial Capital}$$

or

$$EV = Opp \times Ecomp \times Ecomm \tag{1.2}$$

where Opp = opportunity. The economic value of a venture, EV, may grow eventually to a market value (MV) after a period of T years from initiation.

This economic growth is a complex system dependent on all management and leadership decisions as well as the forces of competition, market evolution, and intellectual capital developed or attained over the period T. The allocation of entrepreneurship between productive and unproductive activities can greatly influence the innovativeness of the firm and its intellectual capital and competencies as well as husband or squander resources. Productive entrepreneurial activities can be summarized as a result of effective and efficient management, M. In addition, we will represent changes in the contextual situation, such as a recession or new government regulations, by context, C. Then, an approximate qualitative model for market value is:

$$MV = M \times C \times EV$$

Substituting EV from Equation 1.2, we have

$$MV = M \times C \times Opp \times Ecomp \times Ecomm \tag{1.3}$$

In words: the expected market value of an enterprise after a period will be the result of the cumulative value of management, context, opportunity, competence, and commitment. All of these factors must be strongly present to achieve success. The enterprises that promise the greatest returns lead to entrepreneurs being attracted to opportunities that seem to be brighter. They evaluate their management skills, the context, the opportunity, their team competencies, and their commitment leading to a choice of a new venture.

The firm Google was founded in 1999 by two 26-year-olds who developed a search engine. As a search gateway to the Internet and over 3 billion Web pages, Google is an attractive and useful website. A daily tool for millions of users, it can be queried in 36 languages. It is an excellent example of a powerful combination of entrepreneurial capital, competence, and commitment. The opportunity for Google to become a very useful worldwide search engine is very big and important to many users. The management and leadership of Google are excellent [Hardy, 2003]. As a result, the implied market value of Google is significant.

1.4 BUILDING AN ENTERPRISE

Sun Microsystems was founded in 1982 by a team of four in their twenties. The concept of a workstation—a high-performance desktop computer—linked to a network was developed at Xerox Palo Alto Research Center (PARC) in 1980. By 1982, Xerox PARC had a network of workstations running sophisticated applications. Workstations were leading-edge devices for computer-aided design.

From the age of 15, Vinod Khosla had wanted to start a company. After receiving his bachelor's degree in electrical engineering at the Indian Institute of Technology, he went to Carnegie-Mellon University to study for his master's in biomedical engineering. Khosla then entered the Stanford University MBA program. At graduation, he joined a small start-up called Daisy Systems, a firm in the computer-aided engineering (CAE) industry.

At Daisy, he saw the need for a workstation to support CAE software. After a year at Daisy, Khosla decided to start his own firm to build workstations. Khosla's skills were in the design of computers and knowledge of the CAE industry. In 1981, he drew up a specification for a workstation that was influenced by Xerox PARC knowledge. At that point, Khosla looked for a partner. There was a project at Stanford called the Stanford University Network (SUN), and there he found a talented graduate student, Andy Bechtolsheim, who agreed to join Khosla in January 1982 to form a company. They wrote a business plan and within a month attracted several million dollars of venture capital. With the funds in hand, they got Scott McNealy, an MBA classmate, to join the team. By May 1982, they had a prototype computer and had made their first sale.

To create the software for their computer, they recruited Bill Joy in June 1982 from the University of California, Berkeley. Joy had led the project to create Berkeley UNIX. By June 1982, the company was led by four engineers and MBAs in their twenties who then built it into the world-renowned Sun Microsystems. Khosla had marketing, design, and leadership skills; Joy was a leading software designer of UNIX; McNealy possessed manufacturing and management skills; and Bechtolsheim had strong skills for designing the hardware workstation. This powerful team of young men created a company that revolutionized the computer industry.

Khosla and his team possessed a great measure of commitment and competency. The opportunity was very attractive, and the industrial context of the new firm was very supportive of their venture. The team's management skills were good for their few years of experience. When Khosla and his team examined the opportunity and their new venture, they saw that the potential market value of their enterprise could be very significant, if they executed their plan effectively. Equation 1.3 succinctly summarizes the potential market value as

$$MV = M \times C \times Opp \times Ecomp \times Ecomm$$

Any entrepreneurial team can estimate the qualitative market value of their enterprise by reviewing five factors of the qualitative equation: management, context, opportunity, competences, and commitment.

1.5 ECONOMICS, THE ENTREPRENEUR, AND PRODUCTIVITY

All entrepreneurs are workers in the world of economics and business. **Economics** is the study of humans in the ordinary business of life [Mankiw, 2000]. Economics can also be defined as the study of how society manages its scarce resources. Society, operating at its best, works through entrepreneurs to effectively manage its material, environmental, and human resources to achieve widespread prosperity. An abundance of material and social goods equitably distributed is the goal of most social systems. Entrepreneurs are the people who arrange novel organizations or solutions to social and economic problems. They are the people who make our economic system thrive.

Entrepreneurs flourish in nations that provide legal and social incentives for their activities. A business environment with sound infrastructive and legal protections will encourage entrepreneurs. A culture that supports and protects intellectual capital such as patents will provide the necessary context for risk-taking ventures. The ranking of 10 selected nations by measures of economic activity is provided in Table 1.8. Nations large and small can provide the context for en-

TABLE 1.8 Ranking of 10 selected nations by measures of economic activity.

Country	Business environment[1]	Start-up index[2]	Patents[3]	Nobel Prize winners[4]
United States	1	1	2	1
United Kingdom	3	3	5	2
Netherlands	2	4	6	7
Germany	8	8	3	3
Switzerland	4	7	8	5
Finland	5	2	9	9
Sweden	7	5	7	6
France	9	9	4	4
Japan	10	10	1	8
Ireland	6	6	10	10

[1] Ease of doing business, infrastructure, and policies.
[2] Ease of starting new enterprises.
[3] Number of patents granted to residents, 1998.
[4] Nobel Prize winners in physics and chemistry, 1901–2000.
Source: *Pocket World in Figures,* The Economist Books, London, 2002.

trepreneurial activity. In the United States, 10.5 percent of the working population was engaged in entrepreneurial activity in 2002. The figure for 2000 was 16.6 percent [Breeden, 2002]. New ventures and start-ups have been the source of an estimated one-half to two-thirds of the new jobs created in the United States over the past decade. The entrepreneur turns a social problem into an opportunity, a productive organization, and new, well-paid jobs.

An economic system is a system for the production and distribution of goods and services. Given the limitations of nature and the unlimited desires of humans, economic systems are schemes for 1) administering the scarcities and 2) improving the system to increase the abundance of the goods and services. For a nation as a whole, its wealth is its food, housing, transportation, health care, and other goods and services. A nation is wealthier when it has more of these goods and services. Nations strive to secure more prosperity by organizing to achieve a more effective and efficient economic system. It is entrepreneurs who organize and initiate that change.

Almost all variation in living standards among countries is explained by **productivity,** which is the quantity of goods and services produced from the sum of all inputs, such as hours worked and fuels used. A model of the economy is shown in Figure 1.2. The inputs to the economy are the natural capital, intellectual capital, and financial capital. The outputs are the desired benefits or outcomes and the undesired waste. An appropriate goal is to maximize the beneficial outputs and minimize the undesired waste [Dorf, 2001].

Natural capital refers to those features of nature, such as minerals, fuels, energy, biological yield, or pollution absorption capacity, that are directly or indirectly utilized or are potentially utilizable in human social and economic systems. Because of the nature of ecologies, natural capital may be subject to irreversible change at certain thresholds of use or impact.

Financial capital refers to financial assets, such as money, bonds, securities, land, patents, and trademarks. The **intellectual capital** of an organization is the talents of its people, the efficacy of its management systems, the effectiveness of its customer and supplier relations, and the technological knowledge employed and shared among its people and processes. Intellectual capital is knowledge that has been formalized, captured, and used to produce a process that provides a significant value-added product or service. Intellectual capital is useful knowledge

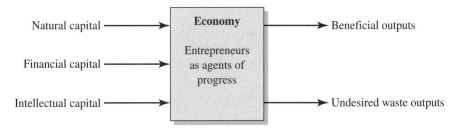

FIGURE 1.2 A model of the economy.

that has been recorded, explained, and disseminated, and is accessible within the firm [Stewart, 2001]. The sources of intellectual capital are threefold: human capital, organizational capital, and relationship capital. Human capital (HC) is the combined knowledge, skill, and ability of the company's employees. Organizational capital (OC) is the hardware, software, databases, methods, patents, and management methods of the organization that support the human capital. Relationship capital (RC) is the quality of relationships with a firm's suppliers, allies, partners, and customers. Relationship capital is often called social capital.

The economy as portrayed in Figure 1.2 consists of the summation of all organizations, for-profit as well as nonprofit and governmental, that provide the beneficial outputs for society. These are the organizations that we study and will label as enterprises or firms. Entrepreneurs constantly form new organizations or enterprises to meet social and economic needs.

Productivity growth is important since it provides all the increases in people's standard of living. Over the past half-century, the U.S. workforce (including immigration) has grown at about 1.7 percent annually, and productivity per worker has risen at 2.2 percent, generating real economic growth (excluding inflation) averaging 3.9 percent. This is an excellent record, due in great part to the impact of technology entrepreneurship.

Rising output per worker comes from two sources: 1) new technology, and 2) smarter ways of doing work. Both paths have been followed throughout human history, and they became faster tracks with the coming of the Industrial Revolution. The twentieth century started with new techniques of management and many new inventions. The century ended with smarter management techniques and dramatic advances in electronic technology, which helped revive productivity growth after limited gains through much of the 1970s and 1980s.

The business system works to drive out inefficiency and forces business process renewal. During the past 25 years, the forces of entrepreneurship, competition, and deregulation have encouraged new technologies and business methods that raise efficiency and efficacy. In recent years, due to competition, much of the benefits of strong productivity have flowed to consumers in the form of lower prices. Innovation, entrepreneurship, and competition are important sources of productivity growth.

1.6 THE KNOWLEDGE ECONOMY

Ideas are many, but knowledge is rare. Ideas are filtered and transformed into knowledge, which can be used to guide the actions of entrepreneurs. Ideas are the raw material from which knowledge is produced. We start with an idea and pass it through an authentication process in which it may be verified, refuted, or transformed using additional information. Business ideas are filtered through an authentication process, and if commercialized, they will be validated or invalidated by the market.

The flow of knowledge from science and technology leads to the application of this knowledge in products, processes, and services—the essence of business.

TABLE 1.9 Three elements of the intellectual capital of an organization.

Human capital (HC): The skills, capabilities, and knowledge of the firm's people.

Organizational capital (OC): The patents, technologies, processes, databases, and networks.

Social capital (SC): The quality of the relationships with customers, suppliers, and partners.

$$IC = HC + OC + SC$$

The complexity of science and technology, appropriately applied, can lead to simple and easily understood products. In many ways, we may think of products as embedded knowledge. Knowledge involves transformed expertise and information. **Knowledge** can be defined as the awareness and possession of information, facts, ideas, truths, and principles in an area of expertise. Thus, a person may be said to be knowledgeable of finance but less knowledgeable of product design or manufacturing methods.

Knowledge can be used for wise actions by entrepreneurs. Intellectual capital is the sum of knowledge assets of an organization. This knowledge is embodied in the talent, know-how, and skills of the members of an organization. The intellectual capital of a firm is used to transform raw material into something more valuable. Mondavi Winery succeeds because of the human capital of its grape growers and wine makers. McDonald's relies on the organizational capital of its recipes and processes. A local coffee shop where the waiter recognizes you and knows your favorite latté relies on its social capital. Social capital is based on strong, positive relationships. The elements of intellectual capital are summarized in Table 1.9. We state that intellectual capital (IC) is a summation:

$$IC = HC + OC + SC \tag{1.4}$$

where OC = organizational capital, HC = human capital, and SC = social capital.

For many, if not most, firms, intellectual capital is the organization's most important asset. It is more valuable than its other physical and financial assets. Many firms depend on their patents, copyrights, software, and the capabilities and relationships of their people. This intellectual capital, appropriately applied, will determine success or failure. Knowledge has become the most important factor of production.

The role of a firm* is to transform inputs into desirable outputs that serve the needs of customers. A firm exists as a group of people because it can operate more effectively and efficiently than a set of individuals acting separately. Furthermore, a firm creates conditions under which people can work more effectively than they could on their own. A firm is more effective because 1) it has lower transaction costs and 2) the necessary skills and talent are gathered together in effective, collaborative work. A model of the firm as a transformation entity is

---

* Henceforth, we use *firm* to represent organizations, enterprises, and corporations.

Inputs   The Firm   Output

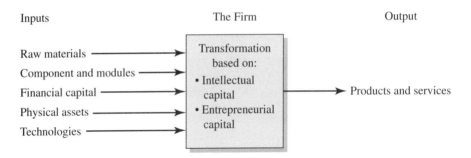

FIGURE 1.3 The firm as transforming available inputs into desired outputs.

shown in Figure 1.3. The transformation of inputs into desired outputs is based on the intellectual capital and the entrepreneurial capital of the firm. As an example, consider Microsoft, a powerful software firm. It creates and purchases technologies, develops new software, and builds a client base. The transformation of its inputs into outputs is based on its formidable stock of intellectual capital and entrepreneurial capital.

One hundred years ago, successful companies such as U.S. Steel were primarily managing physical assets. Today's successful firms such as Microsoft manage knowledge and intellectual capital. Intellectual capital, along with physical assets, transforms raw material to valuable products. The growth of knowledge-based innovation enables economic progress to continue to spur social progress. Acting through an organization, the entrepreneur works for new ideas and change in the face of go-slow opponents [Mokyr, 2003].

Human capital, embodied in people, has mobility—it goes where it is well treated. Thus, a firm needs to attract and retain the best people for its requirements in the same way that it seeks the best technologies or physical assets. Many talented people leave their jobs to join start-up firms because they seek achievement, independence, and opportunity.

Two characteristics of intellectual capital give it power to add value [Stewart, 2001]. Firms can use intellectual capital to reduce the expense of physical assets or maximize the return on those assets. Financial companies, for example, can expand their reach to more customers using software and websites to enable online banking as an alternative to building more branches. Another way to expand their reach efficiently through organizational capital is to establish mini-branches in grocery stores. No longer do a firm's physical assets limit its reach.

The intellectual capital of a new firm encompasses its people's cognitive knowledge, skills, system understanding, creativity, synthesis, and trained intuition [Quinn, 1997]. Fortunately, knowledge is one of the few assets that grows when shared. By organizing around intellectual capital, a new firm strives to leverage it, usually through collaboration, development, and sharing.

A causal diagram can help to portray causal links in a system. Variables are related by causal links, shown by arrows. The link of Figure 1.4a implies that if x increases, then y increases. The link shown in Figure 1.4b implies that if x

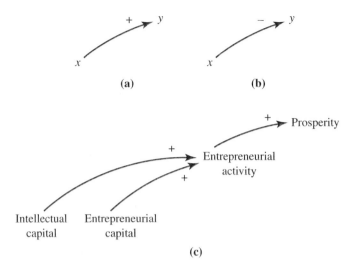

FIGURE 1.4 Causal diagram. **(a)** *y* increases as *x* increases, **(b)** *y* decreases as *x* increases, **(c)** Entrepreneurial activity leads to increasing prosperity

increases, then *y* decreases. For example, Figure 1.4c implies that as a firm's intellectual capital and its entrepreneurial capital increase, it will increase its entrepreneurial activity and prosperity [Stedman, 2000].

1.7 THE FIRM

The purpose of a firm is to establish an objective and mission and carry it out for the benefit of the customer. Thus, the purpose of Merck Corporation is to create pharmaceuticals that protect and enhance its customers' health. To do so, a firm acts to develop, attract, and retain intellectual capital. The firm develops and uses intellectual capital to build the strengths of the firm and to provide the desired products.* The firm provides a place where people can collaborate, learn, and grow.

The firm's actions are based on its knowledge of its customer, its product, and its markets. The firm must identify and understand its customers, competitors, and their values and behavior. Knowledge of organizations, design, and technologies is filtered through a firm's strengths and weaknesses. The firm acts on all this knowledge.

First, a firm is clear about its mission and purpose. Second, the firm must know and understand its customers, suppliers, and competitors. Third, a firm's intellectual capital is understood, renewed, and enhanced as feasible. Finally, the firm must understand its environment or context, which is set by society, the market, and the technology available to it. We can call this the **theory of its business,** or how it understands it total activities, resources, and relationships. Figure 1.5 depicts the business theory of a firm. One hundred years ago, firms were

_____
* Henceforth, we use *products* to refer to products and services.

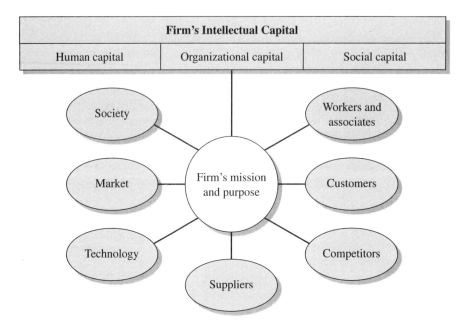

FIGURE 1.5 A firm's theory of business depicts how it understands its total resources, activities, and relationships.

hierarchical and bureaucratic with a theory of business that emphasized making long runs of standardized products. They regularly introduced "new and improved" varieties and provided lifetime employment. Today, firms compete globally with high-value, customized products. They use flattened organizations and base their future on intellectual capital. Firms look to brands and images to cut through the clutter of messages. In the future, a firm's human capital—talent—will become more important.

One way to look at the future of a firm is as a competition among its stakeholders. Flexibility and leanness mostly benefit the firm's shareowners. Placing a high valuation on talent gives more power to the workers. A good reputation means the firm needs to look after its community and society. Customers stand to gain power as competitors vie for their attention. The entrepreneur in the new firm strives to build a firm that serves all its stakeholders well.

1.8 DYNAMIC CAPITALISM AND CREATIVE DESTRUCTION

One view of economic activity describes a world of routine in which little changes. In this static model, all decisions have been made, and all alternatives are known and explored. Clearly, no economy is static, and change appears to be certain. In a world of change, entrepreneurs seek to embrace it. Entrepreneurs match ideas for change with opportunity. These changes include the adoption of new and better (or cheaper) sources of input supplies, the opening of new markets, and the introduction of more profitable forms of business organization.

Economic progress can be described as the generation of new types of goods and services that can be produced efficiently. Progress occurs because individuals engage in creating insights that change the nature of both economic inputs and outputs. Entrepreneurial insight is the recognition of a profit opportunity that was previously unnoticed [Holcombe, 2001].

The profit of the new firm is the key to economic growth and progress. By introducing a new and valuable product, the innovator obtains temporary monopoly power. Lower costs may give the innovative firm profits higher than those of its rivals, which must continue to sell at higher prices to cover their higher expenses. Alternatively, a superior product may permit a price above that charged by other firms. The same concept clearly fits all forms of successful change. The free spirit of entrepreneurs provides the vital energy that propels the capitalist system.

Dynamic capitalism is the process of wealth creation characterized by the dynamics of new, creative firms forming and growing and old, large firms declining and failing. In this model, it is disequilibrium—the disruption of existing markets by new entries—that makes capitalism lead to wealth creation [Kirchhoff, 1994]. New firms are formed by entrepreneurs to exploit and commercialize new products or services, thus creating new demand and wealth. This renewal and revitalization of industry leads to a life cycle of formation, growth, and decline of firms.

Joseph Schumpeter (1883–1950) described this process of new entrepreneurial firms and waves of change as **creative destruction.** Born and educated in Austria, Schumpeter taught at Harvard University from 1932 until his death in 1950. His most famous book, *Capitalism, Socialism and Democracy,* which appeared in 1942 [Schumpeter, 1984], argued that the economy is in a perpetual state of **dynamic disequilibrium.** Entrepreneurs upend the established order, unleashing a gale of creative destruction that forces incumbents to adapt or die. Schumpeter argued that the concept of perfect competition is irrelevant because it focused entirely on market (price) competition, when the focus should be on technological competition. Creative destruction incessantly revolutionizes the economic structure from within, destroying the old structure and creating a new one. The average life span of a company in the Standard and Poors 500 declined from 35 years in 1975 to less than 20 years today. Less than four of the top 25 technology companies 25 years ago are leaders today—perhaps only IBM and Hewlett-Packard.

Schumpeter's theory was based on disruptive (radical) innovations. He depicted the innovator as creative and nonhedonistic with a goal of major change and improvement. He described firms that faced uncertainty, change, and competition and were unable to rationally develop profit-maximizing strategies. Little doubt now exists that the economy is driven by firms that capitalize on change, technology, and challenge. This book is focused on helping the reader to purposefully become an agent for creative destruction by creating his or her own firm. An example of an agent for creative disruption is Bill Gates, who established Microsoft and introduced DOS, Windows, and Office. Gates saw a discontinuity from mainframe computers to personal computers. A recent disruptive innovation, the digital video disk (DVD), has also created a new wave of creative destruction in the movie rental business as DVDs replace videotapes.

With his partners, Jeremy Jaech (see Section 1.2) founded Aldus Corporation when he saw a disequilibrium or **discontinuity** in the newspaper graphics industry. The old system was based on larger workstation computers, and he saw a transition to desktop personal computers using Microsoft Windows. Jaech saw the opportunity and matched it with his interests and capabilities in a new firm.

Most entrepreneurs should look for opportunities based on discontinuities since they can lead to important results, creative destruction, and significant wealth creation. Discontinuities can occur through a new technology, a big cultural change, or a new threat to society such as severe acute respiratory syndrome (SARS) and terrorism.

The entrepreneur of the creative destruction receives a temporary monopoly until rivals figure out how to mimic the innovation. The high profits of the original new product will attract imitators quickly. They imitate the original monopoly and help to disseminate the new product or service.

1.9 THE SEQUENTIAL CASE: AGRAQUEST

> The AgraQuest case illustrates and illuminates the issues raised in each chapter. It focuses on a real-life emerging firm in the life science industry that illustrates each factor described in a chapter. AgraQuest (www. agraquest.com) is a valuable, well-led entrepreneurial firm that may significantly contribute to improved environmental and social conditions and agricultural industries around the world. Read the segment on the case at the end of each chapter and learn of a real-life effort that could make a big difference to the world.

Every seven years in the woodsy town of Killingworth, Connecticut, where she grew up, Pamela Marrone would feel the droppings of gypsy moth caterpillars raining down on her head as the cyclical pests gorged on maples and oaks. Desperate to save a heavily infested dogwood, her father once ignored his own organic gardening tenets and blasted the tree with a chemical called a carbamate.

By the next morning, every bee, every ladybird beetle, every lacewing—all the "good" bugs that fed on plant pests—lay dead on the ground. In her youth, Marrone knew that she wanted to keep the good bugs while deterring bad pests. She recognized a great opportunity that, if solved, could help farmers prosper while using natural pest control agents (not chemicals). Furthermore, as a youth, Marrone had tried, with her parents' encouragement, several modest entrepreneurial ventures at craft fairs and state fairs.

Marrone studied entomology (the study of the forms and behavior of insects) at Cornell University, going on to North Carolina State University, from which she received her doctorate in 1983. She then spent seven years as the leader of the new pest control unit at Monsanto in St. Louis, where she acted on her dedication to the natural control of pests. At Monsanto, Marrone built her technical and entrepreneurial skills. As a result, in 1990 she was recruited by Novo Nordisk, a Danish company, to create a biopesticide subsidiary called Entotech Inc. in Davis, California.

Entotech's goal was to hunt for natural products that can defeat plant scourges without wreaking havoc on human beings, animals, helpful insects, or soil. But in 1995, Entotech was sold to Abbott Laboratories, prompting Marrone to start her own firm to meet the challenge of building a successful company that would use a new search process for identifying natural products for pest control. Thus was born AgraQuest. Marrone possessed the interest and passion, the capabilities and skills, and saw an attractive opportunity in the sweet spot of Figure 1.1.

1.10 SUMMARY

The entrepreneur is the creative force that allows free enterprise to flourish. Entrepreneurship is the process through which individuals and teams bring together the necessary resources to exploit opportunities and in doing so create wealth, social benefits, and prosperity.

The critical ideas of this chapter are:

- The entrepreneur as creator of a great enterprise.
- The entrepreneur responds to an attractive opportunity.
- A person can learn to be an entrepreneur.
- The entrepreneur knows how to use knowledge to create innovation and new firms.
- Positive entrepreneurship activity flows from a combination of entrepreneurial capital and intellectual capital that leads to productivity and prosperity.
- The entrepreneur uses an appropriate organizational structure to achieve his or her goals.

Principle 1
The entrepreneur develops an enterprise with the purpose of creating wealth and prosperity for all participants—investors, customers, suppliers, employees, and themselves—using a combination of intellectual and entrepreneurial capital.

1.11 EXERCISES

1.1 Consider the opportunities that occurred to you over the past month and list them in a column. Then, describe your strong interests and passions, and list them in a second column. Finally, create a list of your capabilities in a third column. Is there a natural match of opportunity, interests, and capabilities? If so, does this opportunity appear to offer a good chance to build a business? What would you need to do to make this opportunity an attractive chance to build a business?

1.2   Steve Jobs left college after one year and joined Atari, a video games company in San Jose, California. He renewed his friendship with a high school friend, Steve Wozniak, who introduced Jobs to the Homebrew Computer Club. Jobs and Wozniak had met while working at Hewlett-Packard as summer interns. A dropout from the University of California at Berkeley, Wozniak possessed a passion for creating electronic devices. Jobs persuaded Wozniak to work with him toward building a personal computer. Jobs saw the opportunity, and Wozniak had the electronics skills.

Jobs and Wozniak put together their first computer, called the Apple I. They sold it in 1976 at a price of $666. The Apple I was the first single-board computer. Jobs was marketing the Apple I to hobbyists like members of the Homebrew Computer Club who would be able to perform operations on their personal computers. Jobs and Wozniak created $774,000 in sales of the Apple I. The following year, Jobs and Wozniak developed the Apple II.

Consider the founding of Apple Computer using the format of Figure 1.1. Describe the attractiveness of the opportunity, the capabilities of the team, and the interests and passions of the two founders. Was this a timely match of opportunity and the two partners? Discuss the factors that led to the rapid success of this firm founded by two 20-year-olds. What industry could you enter and do as well in today?

1.3   Complete the following tasks to select a favorable opportunity for yourself or your team.

   1.   Describe an opportunity that attracts you.
   2.   Describe the competencies and skills you and your team members possess.
   3.   Describe the passion and commitment you have for the opportunity.
   4.   Is this a good opportunity for you?

1.4   The quantity of undesired junk e-mail called spam has risen to exceed the number of desired e-mails. Using Figure 1.4, determine the entrepreneurial capital and the intellectual capital needed to provide the necessary entrepreneurial activity to reduce the impact of spam.

1.5   Building and airport worker security and access are important needs worldwide. Worker badges can be used to control access but also can be easily passed from one person to another. Suggest a means of reliable personal security, such as fingerprints, and describe the opportunity using the model of Table 1.6.

1.6   Bette Nesmith, working as a secretary, wondered why artists could paint over their mistakes, but typists couldn't. The solution was "liquid paper." Determine how Nesmith created an enterprise based on the opportunity (see inventors.about.com).

CHAPTER 2

Opportunity and the Business Summary

In the field of observation, chance only favors minds which are prepared.
Louis Pasteur

Entrepreneurs identify and evaluate opportunities while striving to find one that fits their capabilities, interests, and resources. Examining social, technological, and economic trends can lead to the identification of important emerging needs. Entrepreneurs seek to build new ventures and act on a good opportunity when it has the potential to provide independence and a good economic return while requiring a reasonable work effort and modest risk. The entrepreneur must make a difficult decision to act or not act on a good potential opportunity. The choice of an opportunity and the decision to act is a critical juncture in the life of an entrepreneur. With the decision to act, the entrepreneur prepares a business summary for the venture that is used to test the new venture with potential investors, employees, and customers. The six steps to action as an entrepreneur are shown in Figure 2.1, which summarizes the tasks described in this chapter. ■

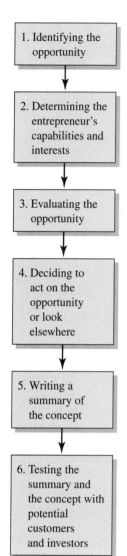

FIGURE 2.1 Six steps to acting as an entrepreneur.

2.1 OPPORTUNITY IDENTIFICATION

Ideas for new ventures are easy to find but difficult to evaluate. An opportunity is a timely and favorable juncture of circumstances providing a good chance for a successful venture. Entrepreneurship is the identification and exploitation of previously unexploited opportunities. Good opportunities are usually disguised, so most people don't easily recognize them. New opportunities open up because customers' needs change or new technologies lead to new ways of accomplishing tasks. Good opportunities also emerge from circumstances of employment or experience. Often they emerge from the personal experience of a need or problem that cries out for a solution. An example is the need for a pharmaceutical that can mitigate or cure the effects of AIDS. This type of opportunity can be called *opportunity pull,* since the size of the opportunity draws opportunity seekers to attempt to exploit it [Vesper, 2003].

Another type of opportunity occurs from the discovery of a capability or resource that can be applied to a problem or need. An example of this type of opportunity is the discovery of a new technology, such as digital television. This type of opportunity can be called a *capability push,* since it flows from a capability or resource availability. Often being in the right line of business at the right place and time is the source of good opportunity.

The founders of new industries capitalize on opportunity pull to create disruptive innovations that lead to new products that solve significant problems. New organizational firms and industries are founded by individuals who recognize big opportunities as a result of technological change. For example, Steve Jobs and Steve Wozniak, founders of Apple, recognized in 1976 the opportunity to build and sell personal microcomputers. Apple's sales jumped from $7.8 million in 1978 to $117 million in 1980 and to over $5 billion by 2001. Industry creating innovations opens up huge new markets. The replacement of the vacuum tube by the transistor is a good example. These innovations provide superior performance and lower costs. Another example is the introduction of the Polaroid instant camera in 1947. These two innovations created paradigm shifts: 1) the mainframe computer to the personal computer, and 2) the film camera to the Polaroid instant photography camera.

The capability-push opportunity can be equally attractive. Cisco Systems was formed in 1984 to exploit the capabilities of the founders and their associates at Stanford University. The firm was founded by Sandra Lerner and Leonard Bosack, who discovered the capability to enable a router to transmit and translate data to and from disparate computers [Bunnell, 2000]. By 2001, Cisco had sales of $22 billion.

A good opportunity has the potential to create significant value for the customer. Another way of describing a good opportunity is to describe the customer's *pain,* which represents the extent of need for the solution to a problem. The pain of need is the converse of value. A high-value solution is sought by a customer who feels significant pain of need. For example, today's airline customer often experiences a fear of flying. The solution to that problem should lead to the improved value of security of airline travel. Both the customer and the airline want a security solution. Some would-be entrepreneurs have a new technology and often mistake it for a solution. Customers want a solution to their problem and usually do not care what technology is employed. Unfortunately, some believe that entrepreneurship is having a great technological idea. Entrepreneurship is really about creating a new business that solves a problem.

Successful new ventures are often initiated by people who have experienced significant painful problems as customers or employees. By 1830, agriculture was a major industry in the United States. The short harvest season was a painful limitation to any farmer's major expansion in wheat output. Harvesting was an arduous manual process involving relatively uncertain labor costs and availability. The answer to the farmers' problem was the reaper, developed by Cyrus Hall McCormick in 1831 and patented as a commercial model in 1834. By 1848, McCormick had moved to Chicago to be close to his markets. McCormick added an innovative marketing strategy—sales on credit—and built a powerful business that enabled farmers to increase their productivity and profitability. McCormick recognized the problem and created a technical and business solution that was effective.

Other new successful ventures occur due to shifts in regulatory policies. The opening of the wireless radio spectrum for mobile devices and cell phones is an example of a big shift in opportunity. We only need look around us to see the proliferation of the wireless revolution.

We can summarize the nine categories of opportunity as shown in Table 2.1. We use these nine categories of opportunity to describe a way of identifying opportunities. The first, and perhaps most common, is to increase the value of the product or service. This can include improved performance, better quality or experience, and improved accessibility or other values unique to the product. For example, a winemaker might strive to provide a high-quality, good-tasting, low-priced wine.

The second method seeks new applications of existing means or technologies. Credit cards with magnetic stripes were available in the 1960s, but a thoughtful innovator recognized the application of this technology to hotel door cards and created a wholly new application and industry.

TABLE 2.1 Nine categories of opportunity.

1. Increasing the value of a product or service.

2. New applications of existing means or technologies.

3. Creating mass markets.

4. Customization for individuals.

5. Increasing reach.

6. Managing the supply chain.

7. Convergence of change.

8. Process innovation.

9. Increasing the scale of the firm.

The third method concentrates on creating mass markets for existing products. A good example is the introduction of the disposable 35-millimeter camera, which is often used at weddings or parties.

Customization of products for individuals affords a new opportunity for an existing product or technology. Examples of customization can be found in the personal computer industry, with Dell Computer as a good one.

Expanding geographic reach or online reach allows a new venture to increase its number of customers. Schwab is an example of a company that has extended its connection to more customers by using the Internet.

Managing the supply chain is a powerful force for improvement. Wal-Mart used its distribution system and large stores connected to an inventory information system to reap the economic benefits of inventory management. Dell Computer is another example of a firm with a well-developed inventory and supply chain management systems.

Convergence of change and industries affords potential benefits to innovative teams. For example, E-Trade now offers securities brokerage as well as banking and insurance services. The boundaries between industries are constantly breaking down.

Innovation of the business and manufacturing processes is a source of opportunity. For example, the shipping of goods has been greatly changed by the introduction of FedEx and other airborne shipping systems.

Finally, the ninth category of opportunity is the increasing scale or consolidation of industry. Consolidation examples include the waste removal and video rental industries. Through mergers and acquisitions, an industry can be consolidated with attendant cost savings and value for the customer.

Railroad companies have built and operated transportation systems on their own rights-of-way in the United States since the 1850s. The first era of railroads worked to build the value of their network by improving on-time rates and providing higher speed service. Then, they strived to expand the reach of their service by increasing the length of their lines and connecting more towns. From 1840 to the 1860s, the technology of railroad transportation was rapidly perfected [Beatty, 2001]. Thus, the first efforts increased value and reach. By increasing reach, new markets were opened by connecting geographically and opening to new cargos such as coal and timber. By 1860, 9,000 miles of railroad track had replaced rivers, lakes, and canals as the primary means of most transportation. The primary value of the railroad was reliable, scheduled, all-weather transportation of goods. By 1870, with the completion of the transcontinental railroad, over 70,000 miles of track were in operation.

Customization for goods shippers and passenger travel became available by 1880, with special cars for coal, liquids, and other goods, as well as passenger cars with classes of carriage and sleeping accommodations. The railroads became the first modern business enterprise to use centralized management and improved process innovation and supply chain management. By the 1880s, the railroad construction and operation industry employed 450,000 workers. By the turn of the century, consolidation of railroads began, and today, there are five major railroad companies, down from the thousands of companies in the late 1890s. Henry Huntington, Cornelius Vanderbilt, and Leland Stanford were some of the leading creators of the American railroads.

Great opportunities are often disguised as difficult problems. However, as Charles Kettering stated, "a problem well stated is a problem half solved." Scott Cook saw a problem experienced by individuals who wanted to easily and reliably keep their own home budget records and pay the bills. He thought that problem could be solved by using a personal computer. The software program his new firm developed was intuitive and easy to use so that most people can use it without resorting to the manual—thus, the name of the firm: Intuit (www.intuit.com). Scott Cook solved a big problem with an easy-to-use solution.

Once a problem has been identified, a solution can be deduced by first asking how would an unconstrained person solve the problem. Starting without constraints such as price and physical limits opens up many possibilities. Once a good unconstrained solution appears attractive, it can often be rearranged to accommodate reasonable constraints [Nalebuff, 2003].

The identification of opportunity depends on preparation, experience, competence, and a keen sense of observation. Curiosity and an observant nature are needed. Railroads and semiconductor manufacture exhibit economies of scale. Today, many innovators have recognized these scale economies and created enterprises such as Intel, Analog Devices, and Xilinx.

The power of *serendipity*—making useful discoveries by accident—can lead to good opportunities. Working in a microwave lab, Percy Spencer observed a chocolate bar melting by microwave power—thus, leading to the microwave oven. Clarence Birdseye was a fur trader in Canada when he noticed a phenomenon while ice fishing. At 50 degrees below zero, fish froze rock-hard almost instantly, yet when thawed, they were fresh and tender. After some experimentation, he learned that the key was the speed at which foods were frozen. That observation led to a quick freezing process that created a multibillion-dollar industry and made Birdseye a wealthy man.

One means of finding a good opportunity is to look for a discontinuity in culture, society, or markets. Table 2.2 describes examples of discontinuities that lead to a big opportunity. An example of a big opportunity is addressing the need for creation of new pharmaceuticals to help prevent the increasing incidence of Parkinson's and Alzheimer's diseases among older people. Any specific business opportunity may be portrayed in the three-dimensional cube of Figure

TABLE 2.2 Sources of discontinuities.

Society	Technology	Markets
■ Aging society	■ Innovation	■ Deregulation
■ Lifelong education	■ Disruptive technologies	■ Supply chain disruption
■ Food and population	■ New knowledge	
■ Regulation		

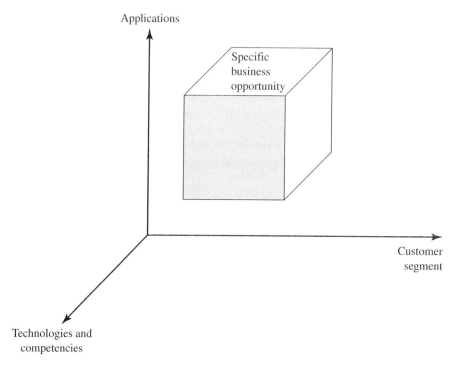

FIGURE 2.2 Finding a specific business opportunity with a combination of customer segment, technology and competencies, and applications.

2.2. The entrepreneur identifies the customer, the required technology, and the application of this technology to create a solution. Several websites for identifying new ideas are listed in Appendix C.

2.2 TRENDS, CONVERGENCE, AND OPPORTUNITIES

Trends in demographics and technologies can lead to large opportunities. Opportunities abound in medicine, agriculture, materials, energy, transportation, housing, computers, and education, to mention a few industries.

The world's food supply and nutritional sources will get a big boost from biotechnology in the decades ahead. Biotech's benefits will include more environmentally friendly agriculture. Farmers will have more tools to combat pests, overcome difficult conditions, and grow more food from fewer acres and resources.

Crops may be designed with built-in resistance to diseases and pests, boosting yields worldwide. This is already being done with corn, cotton, and soybeans. Plants will also be endowed with new tolerance to weather, greatly expanding the land area where grains and vegetables can be profitably grown. Genetic engineering will also produce trees that grow faster or resist disease.

As prosperity grows and spreads worldwide, many opportunities occur from lifestyle changes. Starbucks, for example, offers a quality, customized coffee product and makes it widely available. Other premium product segments such as wine, gifts, and flowers will grow with prosperity. Premium product companies include Häagen-Dazs, Mondavi winery, and Borders bookstores and cafes. Entrepreneurs should try to identify destabilizing influences. These come about through technological change, as well as through changes in taste. Entire industries can be made or broken by a shift in fashion. It took nylon stockings replacing silk stockings to popularize synthetic fabrics in clothing.

The creation of a nationwide air taxi industry is a solid opportunity. Business people would value the convenience, cost, and speed of a jet air taxi that can accommodate their needs. Adam Aircraft Industries of Englewood, Colorado, is building a jet that may bring the total cost per mile down to an economic level. Using carbon fiber design instead of aluminum and a double tailfin (twin boom), the jet may revolutionize the business air travel industry (see www.adamaircraft.com).

Table 2.3 lists some important technology trends and opportunities for the future. Perhaps the most important advances will come from the life sciences. The boundaries between many once-distinct businesses, from agribusiness and chemicals to pharmaceuticals and health care to energy and computing, will blur, and out of that *convergence* will emerge what promises to be the largest industry in the world: life science [Enriquez, 2000].

TABLE 2.3 Trends and opportunities.

- Life sciences: Genetic engineering, genomics, biometrics
- Information technology: Internet, wireless devices
- Food preservation: Improved distribution of food
- Video gaming: Learning, entertainment
- Speech recognition: Interface between computers and people
- Security devices and systems: Identification devices, baggage checkers, protective clothes
- Nanotechnology: Devices 100 nanometers or less for drug delivery, biosensors (see www.nano.gov)
- Fuel cells: Electrochemical conversion of hydrogen or hydrocarbon fuels into electric current
- Superconductivity: Energy savings on utility power lines
- Designer enzymes: Protein catalysts that accelerate chemical reactions in living cells for consumers and health products
- Smart cards: Wallet-sized personal information cards
- Software security: Blocking unsolicited e-mail (spam)
- Robots: Teams of small, coordinated robots for monitoring and safety functions

TABLE 2.4 **Finding a large-impact discontinuity in the functions of life.**

■ Mobility	■ Communications
■ Food	■ Entertainment and leisure
■ Shelter	■ Natural environment
■ Energy substitute for human effort	■ Community
■ Health	■ Learning

One can often find a large-impact discontinuity lodged within the functions of life, as listed in Table 2.4. Consider that an estimated 1 billion people in the developing world have uncorrected vision. They are nearsighted, farsighted, or in need of reading glasses, and they cannot afford or don't have access to eyeglasses. Joshua Silver has developed cheap, adjustable eyeglasses and formed Adaptive Eyecare Ltd. to meet this need (see www.adaptive-eyecare.com) [Shulman, 2003]. Another company, Low Cost Eyeglasses, is also developing self-adjustable glasses (see www.lowcosteyeglasses.com).

The **convergence** of technologies or industries is the coming together or merging of several technologies or industries thought to be different or separate. Genetic engineering is the convergence of electron microscopy, micromanipulation, and supercomputing. Convergence is the unification of functions evidenced as the coming together of previously distinct products or technologies. Often they emerge from creative combinations that build on complementary technologies [Chandler, 2003]. One example of industry convergence is that of computing and communications, which merged into the field of networks. Another example is convergence of a handheld computer and a cell phone. The idea is to let users carry just one device instead of two or three and still stay connected via voice, e-mail, or data. Observers say the devices could be a boon to the wireless industry.

A simple example of convergence is the emergence of the outdoor cinema (drive-in movie theater) in 1933. With the convergence of the widespread use of autos and the talking movie, many people migrated to the drive-in movie theater. By 1948, there were 800 drive-ins across the United States. By 1958, there were 4,000 such theaters, acting as hallmarks of the auto- and movie-centric society. However, by the 1980s, the multiplex theater eclipsed the drive-in. Today, only about 400 drive-ins remain in the United States (see www.drive-ins.com). Many industries have a limited life, but drive-ins experienced a solid 40 years of popularity. The construction and operation of a chain of drive-ins afforded many entrepreneurs a good and profitable life.

Demographic and cultural trends offer many examples of convergence and opportunity. Several social and cultural trends are listed in Table 2.5. The biggest current trend in America is the aging of the baby-boom generation—those born between 1946 and 1973. During those years, 107.5 million Americans were born, making up 50 percent of everyone alive in 1973 [Hoover, 2001]. Those

TABLE 2.5 Social and cultural trends that will create opportunities.

■ Aging of the baby-boom generation.	■ Changing role of religions organizations.
■ Increasing diversity of the people of the United States.	■ Changing role of women in society.
■ Two-working-parent families.	■ Pervasive influence of media—television, DVDs, Internet.
■ Rising middle class of developing nations.	■ Growth of the Latino population in the United States.

born in the peak-birth year of 1961 will be 45 in 2006 and acting as wealthy consumers of goods and services such as new homes, furniture, travel, and retirement plans.

Other critical trends include the rise of diversity as massive waves of immigrants arrive in the United States and the changing role of women in many societies and nations in the world. Any new enterprise must fit the social and cultural context of its service area. One of the most promising areas of science and engineering is based on several breakthroughs that enable the manipulation of matter at the molecular level. Mass production of products with these molecular adjustments now offers a world of possibilities. Nanotechnology will make materials lighter, more durable, and more stain-resistant. One nanometer is one-billionth of a meter. Soon we may also get a host of miniaturized products, from semiconductors to pumps, that work more efficiently and accurately. The areas of application range from medical and industrial to the home [Brenner, 2002].

Tiny robots acting in coordinated teams may be used in events such as fires, toxic spills, and bomb threats. This activity, like that depicted in the movie *Minority Report,* can be used for safety and reconnaissance activities [Grabowski, 2003].

An example of a current trend that results in a big problem is the unsolicited e-mail (spam) that is received by all e-mail users. The number of unsolicited mass mailing attacks rose from 2 million in 2001 to 5.5 million in 2002 [Mangalindan, 2002]. Any new firm that can sell a foolproof spam blocker would solve a problem for most e-mail users.

An excellent example of the convergence of two technologies leading to an opportunity is the development of global positioning systems (GPS) and their wide use by hikers, travelers, surveyors, and farmers. Satellite imaging and data and the handheld computer converged into the GPS device, which is a widely used, inexpensive device that addresses the need for accurate locational data.

With the need for security and safety of personal information, the emergence of personal identification cards, or smart cards, may be the next trend in America. Cards for pay telephones and money transfer are one application. Another important use of smart cards would be a common approach for driver's licenses and personal information. A s*mart card* is a plastic card incorporating an integrated circuit chip and memory that stores and transfers data such as personal data and identification information such as finger or palm point or facial

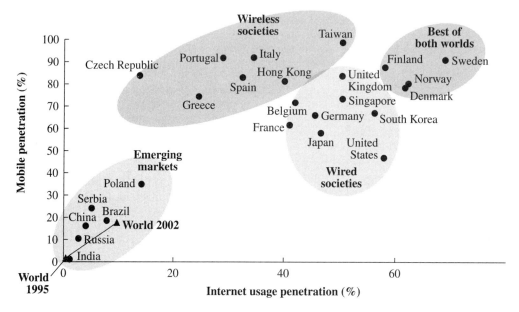

FIGURE 2.3 Convergence of the Internet and mobile phone usage in the world.
Source: Applied Materials Corporation.

scans. These cards have been adopted in several European and Asian countries and could spread worldwide.

Think creatively about possible convergences. How about the convergence of scanners, computers, and security systems that enables shoppers to bag their own groceries in a self-checkout system (see www.optimal-robotics.com)? Another example is the new world of medicine driven by innovation and the needs of aging Americans for ever-more-intense levels of care. One big trend is the convergence of the computer and communications and the trend toward wireless phones and devices, as illustrated in Figure 2.3. Cell phones become more like computers, and handheld computers transform to phones [Lohr, 2003]. People are excited by the opportunities as wireless start-ups proliferate. Wildseed Ltd. designed specialized software and cell phone "skins" that change the look and operation of a smart phone so that it can combine a phone with a gaming machine, music player, or video platform—all tailored to a person's favorite games, rock stars, sports teams, or fashion tastes (see www.wildseed.com). Look for a new technology in one industry that can be applied in another. For example, a gene chip uses semiconductor technology to speed up gene lab analysis.

Already, the United States spends $1.4 trillion, or 14 percent of its gross domestic product, on medical care. The health care transformation could be as big as the computer revolution.

2.3 THE ENTREPRENEUR AND OPPORTUNITY

The first role of the entrepreneur*—an individual or a group of people—is to identify and select an appropriate opportunity. Entrepreneurship begins with an idea that upon reflection is a valuable opportunity. Effective entrepreneurs recognize and pursue opportunities that are based on meeting a need in the marketplace, solving a problem, or filling a niche within a reasonable time. There is timeliness to every opportunity. Peter Drucker [1993] describes the role of the entrepreneur as innovator:

> Innovation is the specific tool of entrepreneurs, the means by which they exploit change as an opportunity for a different business or a different service.

Effective entrepreneurs often find opportunity a creative process that relates a need to one or more associated methods, means, or services that combine to solve the problem. This association process is related to the concept of *convergence,* which is two or more means, methods, or technologies that solve the problem and overcome the pain.

Often the opportunity will be reviewed by a team or group of creative individuals working together to select a good opportunity. Entrepreneurs are often dreamers, visionaries, or just good thinkers. Often one or more of the team have strong intuitive skills and can generate ideas easily. It is then up to the team to help select the opportunity that best fits the capabilities and resources that the team has or can secure.

Entrepreneurs are opportunity-driven and work to find a strategy that can reasonably be expected to bring that opportunity to fruitful success. They seek new means or methods and are willing to commit to solving a social or business problem that will result in success. Entrepreneurs work toward needing shorter time periods to decide on an appropriate strategy and seize opportunities. Typically, entrepreneurs have a passion to build an enterprise that will solve an important problem. They seek ways to express themselves and validate their ideas. They are creative, internally motivated, and attracted to new, big ideas or opportunities.

One or more of the entrepreneur team usually have some experience in the industry in which the new venture will be operating. Some say a successful entrepreneur has good luck, but we need to remember the old saying that luck is where preparation and opportunity meet. The team has to have the necessary competencies and be committed to the project.

Good entrepreneurs seek to be flexible so they can adapt to changing conditions and reduce the risks of the venture. They create an overarching vision of the venture and use it to motivate employees, allies, and financiers. Perhaps the most important qualities or characteristics of an entrepreneur are the abilities to accomplish the necessary tasks, meet goals, inspire others to help with these tasks, and do so with the sustained effort required.

---

* Henceforth, the word *entrepreneur* can refer to an individual or a team of individuals.

TABLE 2.6 Required capabilities of the entrepreneurial team.

- Has talent, knowledge, and experience within the industry where the opportunity occurs.

- Seeks important opportunities with sizable challenges and valuable potential returns.

- Able to select an opportunity in a short period: timely.

- Creatively explores a process that results in the concept of a valuable solution for the problem or need.

- Able to convert an opportunity into a workable and marketable enterprise.

- Achievement-oriented: wants to succeed.

- Able to accommodate uncertainty and ambiguity.

- Flexibly adapts to changing circumstances and competitors.

- Seeks to evaluate and mitigate the risks of the venture.

- Creates a vision of the venture to communicate the opportunity to staff and allies.

- Attracts, trains, and retains talented, educated people capable of multidisciplinary insights.

Members of the entrepreneurial team must exhibit leadership qualities. **Leadership** is the ability to create change or transform organizations. Leadership within an organization enables the organization to adapt and change as circumstances require. A real measure of leadership is the ability to acquire needed new skills as the situation changes. Leadership competencies are attained from experience and native talent. Talent consists of brilliance, charm, and ambition. Entrepreneurship is about people who collectively act within or through organizations to identify and exploit opportunities and ultimately create value (wealth) for society. Entrepreneurs are innovators or change agents, or know a good opportunity when they see one. Entrepreneurs exhibit robust confidence, sometimes bordering on overconfidence. As Ted Turner, cable TV pioneer, once said: "This is America. You can do anything here."

The required capabilities of the entrepreneurial team are shown in Table 2.6. In general, all entrepreneurs should have most of these qualities to participate in a new venture. Of course, not all team members will have the same blend of capabilities, but it is useful if all team members have some measure of each characteristic (or appreciate and respect others who do).

Successful entrepreneurial teams depend on the core capabilities of the team members and the ability to accommodate and manage risk. They can attract, train, and retain intellectually brilliant and educated people capable of multidisciplinary insights [van Praag, 2001]. Using the Meyers Briggs Type Indicator, entrepreneurs are most often profiled as ENTP. They have an external (extrovert) orientation, E; are innovative or intuitive, N; are responsive to change and ideas, T; and are perceptive, P (see www.mbti.com for the Meyers Briggs test).

Members of an entrepreneurial team decide whether to act as entrepreneurs based on the seven factors listed in Table 2.7 [Gatewood, 2001]. Good entrepreneurs tend to seek independence, financial success, self-realization, validation of achievement, and innovation, while fulfilling leadership roles. At the same time, potential entrepreneurs evaluate the risk and work efforts associated with an op-

TABLE 2.7 Factors people use to determine whether to act as entrepreneurs.

Positive factors or benefits	
■ Independence: Freedom to adapt and use their own approach to work and flexibility of work, autonomy	■ Self-realization: Recognition, achievement, status
■ Financial success: Income, financial security	■ Innovation: Creating something new
	■ Roles: Fulfilling family tradition, acting as leader

Negative factors	
■ Risk: Potential for loss of income and wealth	■ Work effort and stress: Level of work effort required, long hours, constant anxiety

portunity and balance them with the benefits. Helpful characteristics of entrepreneurs are good emotional balance, self-confidence, a realistic attitude toward the vagaries of life, and the ability to manage stress. David Packard, co-founder of Hewlett-Packard, said that he didn't start his firm to make a lot of money. He wanted to be his own boss, and he could not do that working at General Electric. Packard knew he was a talented, hard worker who sought achievement.

About 16 percent of the male population and about 8 percent of the female population of the United States engages in entrepreneurial activity [see www.entreworld.org/GEM2000]. A list of some successful entrepreneurs is provided in Table 2.8. The age of these people when they launched their enterprise ranges from 19 to 45. Most entrepreneurs start their first business by age 50.

People act as self-employed entrepreneurs when that career path is felt to be better than employment by an existing firm. Consider the satisfaction (utility) derived from an employment arrangement. A utility function, U, is [Douglas, 1999]:

$$U = f(Y, I, W, R, O)$$

where Y = income, I = independence, W = work effort, R = risk, and O = other working conditions. Also, it may be assumed that income depends in turn on ability. People will have an incentive to be entrepreneurs when the most satisfaction (utility) is obtained from the entrepreneurial activity. In other words, entrepreneurship pays off due to higher expected income and independence when reasonable levels of risk and work efforts are required.

For new entrepreneurial activities, the results of the venture are less known, and expected returns, independence, work effort, and risk can only be estimated. It is the combination of all four of these characteristics that influences the decision. The actual utility is achieved over a period of N years. Neglecting the factor O, we may postulate a utility index that we will call the Entrepreneurial Attractiveness (EA) index. The index may be calculated as a linearly weighted index:

$$EA = \int_0^N (w_1 Y + w_2 I - w_3 W - w_4 R) dt$$

TABLE 2.8 Successful entrepreneurs.

Entrepreneur	Company founded	Age of entrepreneur at time of start	Year of company start
Bezos, Jeff	Amazon.com	31	1995
Carpenter, Jake Burton	Burton Snowboards	23	1977
Cohen, Ben, and Jerry Greenfield	Ben & Jerry's Ice Cream	27	1978
Cook, Scott	Intuit	31	1983
Dell, Michael	Dell Computer	19	1984
Dubinsky, Donna	Palm Computing	37	1992
Gates, William	Microsoft	20	1976
Hewlett, William	Hewlett-Packard	27	1939
Johnson, Robert L.	Black Entertainment TV	33	1980
Blank, Arthur	Home Depot	36	1978
Rowland, Pleasant	Pleasant Company	45	1986
Schultz, Howard	Starbucks	34	1987
Smith, Fred	Federal Express	29	1973
Stemberg, Tom	Staples	36	1985

where w_i = weighting factors and $w_1 + w_2 + w_3 + w_4 = 1$. Clearly, a very risk-averse person might use $w_4 = 0.8$ and avoid almost all entrepreneurial ventures. Both w_3 and w_4 have negative signs since they are weighted as disutility factors [Levesque, 2001]. In addition, we expect that a very confident, talented, experienced person may accommodate more risk.

As a simple example, consider a five-year venture for a person who has equal weights $w_i = \frac{1}{4}$, and where Y = 3, I = 4, W = 2, and low risk, R = 1. We use a scale of 1 to 5 with 1 = low, 3 = medium, and 5 = high for Y, I, W, and R. In this case, income will be good, independence high, work moderate, and risk low. Then, we have

$$EA = \frac{1}{4}[(Y + I) - (W + R)] \times N \qquad (2.1)$$

$$= \frac{1}{4}[7 - 3]5 = 5$$

a very positive attractiveness index that could lead a person to choose this entrepreneurial path.

In summary, a person will choose to pursue an entrepreneurial path when the benefits of independence and income outweigh the work effort required and the risk of the venture. The expected income, independence, work effort, and risk of a venture depend on the quality of the opportunity, the proposed venture strategy, and the team of people who will execute the venture strategy. Regrettably, many entrepreneurs overweigh the benefits of independence and income, and underestimate the work effort required.

George Bernard Shaw [1903] summarized it well:

> The reasonable man adapts himself to the world: the unreasonable one persists in trying to adapt the world to himself. Therefore all progress depends on the unreasonable man.

2.4 EVALUATING THE OPPORTUNITY

Choosing the right opportunity is a difficult and important task. We select the opportunity that affords the best chance of success within the context of the marketplace. This choice is analogous to the selection of an equity investment in a company. Entrepreneurs will invest time, effort, and money in the venture they choose in a manner similar to how people invest in the stock of a company. Some sound investment principles that can be used for selecting opportunities are listed in Table 2.9.

The entrepreneur finds and thoroughly analyzes the best opportunities, since for many people, only one or two are needed to make a good life of entrepreneurial activity. One goal is to invest in a firm for which you pay less than it is worth; this provides some cushion for unforeseen challenges. Also, the entrepreneur tries to find an opportunity with solid long-term potential in an industry they understand. They also put together a good management team that can execute the strategy for this opportunity. They also ensure that the customer will allow their firm to profit from the venture. Thus, they avoid industries that sell commodities where price is the only differentiation unless they have a new, innovative business process that enables their firm to be the low-cost provider.

Tom Stemberg, the founder of Staples, conceived the idea of a supermarket store for office supplies in the mid-1980s. He didn't like the politics of big companies and sought independence. He started with a single store in Brighton,

TABLE 2.9 Guiding principles for selecting good opportunities.

■ Only one or two very good opportunities are needed in a lifetime.	■ If the opportunity is selected and turns out unfavorably, can you exit with minor losses?
■ Invest less time, money, and effort in the venture than it will be worth in one or two years. Calculate the probability of a large return in four years.	■ Does this opportunity provide a potential for a long-term success, or is it a fad? Go to where the potential future gains are significant. Can the management team execute the strategy selected for this opportunity?
■ Do not count on making a high-priced sale of your firm to the public or another company.	■ Will the customer enable your firm to profit from this venture?
■ Carry out a solid analysis of the current and expected conditions of the industry where the opportunity resides.	

Massachusetts, and built 1,500 outlets. He carried out a complete analysis of the opportunity and determined it was a $100 billion market growing at 15 percent per year with large profit margins.

The review of opportunities will always include the evaluation of alternatives. The **opportunity cost** of an action is the value (cost) of the forgone alternative action. Selecting one opportunity will involve rejecting others. Consider a 28-year-old woman who has just received a $20,000 bonus. Should she put the money in a mutual fund, make a down payment on a condo, use the money to pay the tuition for a graduate degree, or leave her job and start a software design firm? A start on the analysis is to use the expected annual return on the mutual fund—10 percent—as the opportunity cost. The expected return on the graduate degree might be 12 percent, the return on new software firm is estimated to be 11 percent, and the return on the condo is 9 percent. Factoring in the psychic return of the pleasures of learning accorded by the graduate degree, that would be the alternative selected by some. Others might value independence highly and choose to start the firm.

For most potential entrepreneurs, the new opportunity is contrasted with their existing employment and the potential it affords. Consider the straightforward alternatives for a successful marketing manager in the electronics industry. He can earn $60,000 annually in his existing job (Y in Equation 2.1). However, he values the independence of the new venture highly (I). The work effort for the new venture is estimated to be the same as for his current work. However, the risk is higher for the new independent venture. The potential entrepreneur estimates that he can obtain the same income over the next two years, although he will need a four-month period with a lower income at the start. The entrepreneur can make a sample calculation, as shown in Table 2.10. In this case, over the first two years, the benefits of the new venture are $Y + I = 8$, and the costs of the venture are $W + R = 7$. The benefits of the existing job are equal to 5, and the costs are 6. Therefore,

$$\text{New venture:} \quad (Y + I) - (W + R) = 8 - 7 = +1$$
$$\text{Existing job:} \quad (Y + I) - (W + R) = 5 - 6 = -1$$

TABLE 2.10 Summary of the entrepreneur's analysis of a new opportunity and the opportunity cost using a two-year period.

Factor	New venture	Existing job
Income over two years (Y)	$120,000 $Y = 3$	$120,000 $Y = 3$
Independence (I)	$I = 5$	$I = 2$
Work effort (W)	$W = 4$	$W = 4$
Risk (R)	$R = 3$	$R = 2$

The new opportunity looks more favorable due to this entrepreneur's desire for independence. Using the guiding principles of Table 2.9, this opportunity appears to qualify as a good opportunity, warranting in-depth analysis.

This example is relatively easy to evaluate. In more complex industries, all the issues need to be considered carefully. One of greatest risks for any person considering a new venture is the general tendency for human beings to be over-confident and expecting things to turn out better than they actually do. Most people expect their own performance to be above average [Szulanski, 2002].

The considerations in this example are involved with the opportunity from the entrepreneur's viewpoint. It is equally important to review the quality of the opportunity in terms of a market assessment, feasibility of implementation, and differentiation of the product. Much of this analysis requires additional information. Judgment regarding the qualities of the opportunity can be made by the entrepreneurial team as they consider all the aspects of the opportunity. A comprehensive analytical approach to evaluation of the opportunity does not suit most start-ups. Entrepreneurs often lack the time and money to interview a representative cross section of potential customers, analyze substitutes, reconstruct competitors' cost structures, or project alternative planning scenarios.

Most entrepreneurial teams follow a basic five-step process, as outlined in Table 2.11. The goal is to quickly weed out unpromising ventures and conserve energy and time for the promising ones. In general, it is best to reject ventures in industries or markets in which the entrepreneurs have little experience or knowledge. Standard checklists or approaches don't work for most entrepreneurs. The appropriate analytical effort and the issues that are most worthy of research and analysis depend on the characteristics of each venture.

In general, however, an entrepreneur works through the five steps of Table 2.11 and eliminates the opportunities that do not pass muster. Those that do pass a quick review are worth looking into further.

It is difficult for many to ascertain their dreams and goals. Often it helps to write them down. Keep revising them until you are sure of them. John James Audubon was a taxidermist who had a passion for painting the birds of America. By 1829, he published the first of his famous four volumes entitled *The Birds of*

TABLE 2.11 Basic five-step process of evaluating an opportunity.

1. **Capabilities:** Is the venture opportunity consistent with the capabilities, knowledge, and experience of the team members?

2. **Novelty:** Does the product or service have significant novel, proprietary, or differentiating qualities? Does it create significant value for the customer—enough so that the customer wants the product and will pay a premium for it?

3. **Resources:** Can the venture team attract the necessary financial, physical, and human resources consistent with the magnitude of the venture?

4. **Return:** Can the product be produced at a cost so that a profit can be obtained? Is the expected return of the venture consistent with the risk of the venture?

5. **Commitment:** Do the entrepreneurial team members feel compelled to commit to this venture? Are they passionate about the venture?

America. Most of the paintings were life-size. His opportunity became a life's work and a legacy that we value today.

The entrepreneur has to live with critical uncertainties, such as the relative competence of rivals or the preferences of customers, which are not easy to analyze. Who could have forecast, for example, that IBM would turn to Microsoft for an operating system for its personal computer, allow Microsoft to retain the rights to this operating system, and thus gain monopolistic dominance of the operating system marketplace? Entering a race requires faith in one's ability to finish ahead of whoever else might participate.

A new product has to offer customers exceptional value at an attractive price, and the company must be able to deliver it at a good profit. The initial opportunity review can be based on the five characteristics of the opportunity and its assessment by the team: capabilities, novelty, resources, return, and commitment, as depicted in Table 2.11.

Scott Jordan, 39, founded Scott eVest in 2001 to design, produce, and sell vests and jackets with 16 to 22 pockets that hold phones, pagers, PDAs, and assorted gear. The patent-pending product has great appeal to the tech gadget-burdened person. Jordan found an opportunity that afforded a good return for a new product that passed the test of Table 2.11. Selling for $80 to $150, these garments with lots of pockets meet the needs of people on the go (see www.scottevest.com).

Consider the following opportunity: people like to use guides to help them select movies for viewing in theaters or at home. John Olson loves movies and has significant skills in developing Internet-based survey instruments. John and his wife, Alice, are interested in starting a nationwide movie online survey using a simple, four-question survey form. They expect to be able to sell advertising space on their website to local movie theaters as well as movie distribution companies. Does this opportunity pass the evaluation process of Table 2.11?

Kemmons Wilson, a Memphis home builder and owner of movie theaters, drove with his wife and five children across country in 1951 and was disappointed in the quality and price of accommodations. He was particularly disturbed by the surcharge for each child. He was an experienced entrepreneur and was confident that he could do better for families who travel. He told his wife he was going to build a chain of motels and never charge for children as long as they stayed in the same room as their parents. Using a tape measure, he measured the motel rooms he encountered on the trip and sketched his first room and site plans. When he returned to Memphis, his draftsman formulated a motel design according to his sketches. The draftsman, Eddie Bluestein, had watched a rerun of a Bing Crosby movie the preceding night and wrote the movie title, *Holiday Inn,* on the plans. Wilson obtained a $325,000 bank loan and built the first Holiday Inn in Memphis at a cost of $280,000 [Jakie, 1996]. The first Holiday Inn included a swimming pool, air conditioning, and in-room telephones, and allowed children to stay free in their parent's room. Twenty-eight years later, when Wilson retired, there were

1,759 Holiday Inns around the world. This new opportunity in 1951 easily passed the five-step evaluation process. Wilson and his building company possessed the necessary capabilities and experience. He designed a novel, original motel for travelers and attracted the financial capital required from a bank that trusted him. The expected return was consistent with the risk, and Wilson was strongly committed to the venture. He was an observant, skillful, intelligent entrepreneur who recognized the customers' need by experiencing it himself.

When evaluating an opportunity, the entrepreneur considers whether it fits or matches the contextual conditions, the team's capabilities and characteristics, and their ability to secure the necessary resources to initiate a new venture based on the opportunity. Figure 2.4 shows a diagram of fit or congruence that can be

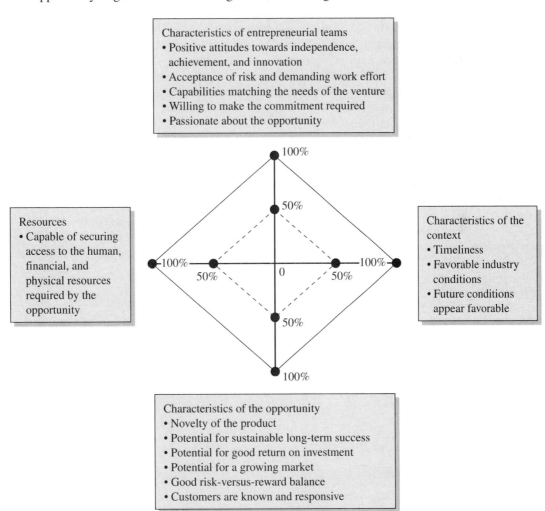

FIGURE 2.4 Diagram of the fit of an opportunity, the context, the entrepreneurial team, and the resources required. Rate each factor on a scale of 0 to 100 percent.

used to review an opportunity. A big diamond with high grades of fit are best. Let us consider the opportunity discovered by Kemmons Wilson. Wilson's plan for the Holiday Inn was highly congruent with all four factors of fit, and we should not be surprised at the success of the Holiday Inn. We would probably rate each factor at the 100 percent level (with the advantage of hindsight).

Let us consider another opportunity that has existed for over 100 years—the electric automobile. We will assume that a capable set of engineers is available and the entrepreneurial team has the attitudes and capabilities required. However, the team is insecure about the risky nature of the venture, given the numerous failures over the past century. We will rate the entrepreneurial team at 70 percent on the team scale. The characteristics of the context are very mixed since regulations and support for electric cars are continually changing as potential customers and government organizations adjust their assessment of the benefits and costs of these vehicles. We will rate this opportunity as only 50 percent on the context scale. Next we turn to the opportunity, which is challenged by costs, limited life batteries, and short ranges before a recharge is required. The characteristics of the opportunity call for a rating of only 50 percent on the opportunity scale. Given these limited ratings, most teams would be severely limited in their ability to secure the tens of millions of dollars required to launch this venture. Thus, we rate it only 40 percent on the resource scale. Clearly, this opportunity is a poor one. Without a technical breakthrough in battery performance and cost, electric autos have a very limited future, and most rational entrepreneurs would quickly dismiss this opportunity—valuable as an electric car might be to the environmental conditions of auto-impacted regions such as Los Angeles.

Another way of envisioning the concept of a fit with an opportunity is shown graphically in Figure 2.5. An opportunity exists within its industry and market context, as shown in Figure 2.5a. With the appropriate competencies, resources, strategy, and execution, a new venture can achieve a good fit, as shown in Figure 2.5b.

(a) 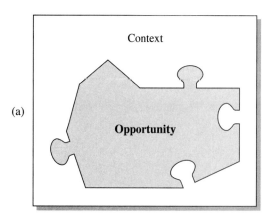   (b)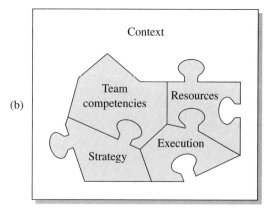

FIGURE 2.5 (a) The opportunity within its context; (b) the fit of resources, competencies, strategy, and execution to match the opportunity.

Pleasant Rowland started her own enterprise at age 45. She wanted to give her nieces, aged 8 and 10, dolls that did not celebrate a teen queen or a mommy. She realized she had a real unserved need for a doll with an associated book that didn't force girls to grow up too fast. An author of children's books, Rowland conceived a series of books about 9-year-old girls growing up in different periods of history and corresponding dolls with accurate clothes and accessories with which girls could play out stories. When she explained the idea to a focus group, they were unimpressed. Then, she showed them a sample book and doll, and they loved it. American Girl sold $1.7 million in its first year, 1986. In the second year, sales grew to $7.6 million. Rowland sold American Girl to Mattel for $700 million in 1998.

Source: www.fortune.com/fortune/smallbusiness and www.americangirl.com.

2.5 THE DECISION TO ACT OR CONTINUE LOOKING ELSEWHERE

After evaluating an opportunity by using the factors in Table 2.11, the entrepreneurs should decide whether to act. With the knowledge generated by using the five-step process in Table 2.11, the entrepreneurs will tend to act on their estimate of the potential benefits and gains, B, while accounting for the total costs of the venture, C. Within the total cost accounting, there will need to be a recognition of their security needs and loss aversion. An individual will tend to act if the ratio B/C is greater than 1. The lucrative opportunity (high benefits and low losses) will tend to cause higher intention to act [McMullen, 2002]. If one acts and it is a false choice, the cost of that choice is important. Opportunities that can be attempted with low initial financial and time commitment costs may offer the chance for lucrative returns at a low initial cost.

The matrix in Figure 2.6 shows the decision to act or not act. Then, the actual resulting quality of the opportunity is shown (this can only be determined after the decision). Life is about choices, and the best case is when we choose to act and it turns out we are right!

We can portray the decision matrix actions of Figure 2.6 in the form of a decision tree, as shown in Figure 2.7. Our goal is improve our decision-making skills so that the probability, p_1, of selecting a good opportunity is high.

The entrepreneur attempts to make a rational decision based on 1) his or her current psychological and financial assets, and 2) the possible consequences of the choice [Hastie, 2001]. The decision challenge is the task of turning incomplete knowledge of an opportunity into an action consistent with that knowledge. Competitive advantage comes from actually doing something that others cannot do. Analysis and reports cannot substitute for action. Reworking a plan is no substitute for acting to get things done. In the end, an opportunity can only be evaluated so much [Pfeffer, 2000]. Ambiguity remains, and the entrepreneur

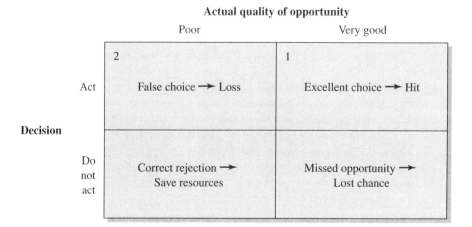

FIGURE 2.6 Decision matrix.

needs to act on or reject the opportunity. Fear of failure may overwhelm all but the best opportunities.

Perhaps the best way to find a really good opportunity is to examine it by estimating fit in the matrix of Figure 2.5 and then act on the best opportunity, trying it out in the marketplace of ideas and investors. This can lead to a refinement of the opportunity. Tom Peters and Robert Waterman called this approach "ready, fire, aim" [Peters, 1982]. Banishing fear of failure and learning from a series of small failures can lead to a good new venture.

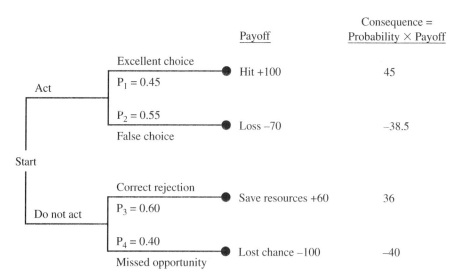

FIGURE 2.7 Decision tree for acting or not acting.

TABLE 2.12 Critical assets of an entrepreneur.

■ Knowledge of the marketplace and industry.	■ Reasonable acceptance of risk while acting prudently.
■ Competencies required.	■ Willingness to act on the best opportunities and learn to adjust them as required.
■ Willingness to accept and mitigate small failures and setbacks.	

Schumpeter wrote in his seminal book [1984]:

To undertake such new things is difficult and constitutes a distinct economic function, first, because they lie outside of the routine tasks which everybody understands and secondly, because the environment resists in many ways that vary according to social conditions from simple refusal either to finance or to buy a new thing, to physical attack on the man who tries to produce it. To act with confidence beyond the range of familiar beacons and to over-come that resistance requires aptitudes that are prevalent in only a small fraction of the population and that define the entrepreneurial type as well as the entrepreneurial function. This function does not essentially consist in ei-ther inventing anything or otherwise creating the conditions which the en-terprise exploits. It consists in getting things done.

Few ideas are unique. Many may have the idea, but few have the will, the passion, and the competencies to pursue it. For example, many propose to ex-ploit the new science of nanotechnology to solve various problems, but few will act. The true entrepreneur finds the best opportunity that matches his or her in-terests, skills, knowledge, and acts to get it done. Investors are looking for the entrepreneur who has the critical assets listed in Table 2.12. The act-review-fix cycle, as shown in Figure 2.8, summarizes the critical ability to act, review, and learn from the results, and then, fix and adjust the business scheme as required. As John Stuart Mill stated:

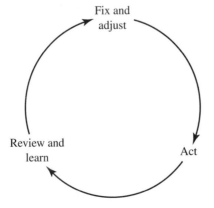

FIGURE 2.8 Act-learn-fix cycle of building a new venture.

TABLE 2.13 Five questions for the potential entrepreneur.

■ Are you comfortable stretching the rules and conventional wisdom?	■ Are you willing and able to shift strategies quickly?
■ Are you prepared to take on powerful competitors?	■ Are you a good closer and decision-maker?
■ Do you have the perseverance to start small and grow slowly?	

There are many truths of which the full meaning cannot be realized until personal experience has brought it home.

Successful entrepreneurs are able to answer positively the five questions listed in Table 2.13 [Kuemmerle, 2002b]. Stretching the rules without breaking them is a challenge the fledging entrepreneur will encounter. Entrepreneurs will need to take on powerful, entrenched competitors in a new market. A new venture requires the patience and perseverance to start small and grow slowly since most investors and customers will want the entrepreneurs to prove themselves step by step.

Most entrepreneurs need to adjust their strategy in response to market change or demands. Finally, entrepreneurs know how to negotiate and close a deal. They can finalize deals under pressure. Furthermore, they must be able to make decisions and deals with incomplete information.

You can learn to excel in all these five skills by first working in the industry in which you intend to develop a start-up. Also, you can find a mentor who can help you gain the skills and confidence required. Finally, find a partner so that between the two of you, you have all the skills.

Entrepreneurial innovators tend to exhibit high self-efficacy—the belief that they can organize and effectively execute actions to produce desired attainments [Markman, 2002]. They believe they possess the capabilities and insights required for the entrepreneurial task.

2.6 THE NEW VENTURE STORY AND SUMMARY

Once a business opportunity has been selected for action, it is important to prepare a **concept summary** of the new venture. This can be a simple statement of the problem the new venture is solving and how the new venture will act to solve it. This statement of the business concept is a short description of the new business. The elements of a concept summary are given in Table 2.14. For ex-

TABLE 2.14 Elements of a concept summary.

1. Explain the problem or need and identify the customer.

2. Explain the proposed solution and the uniqueness of the solution.

3. Tell why the customer will pay for the solution.

TABLE 2.15 Elements of the business story.

1. **Background:** Describe the current situation, characters, and problem.

2. **Challenge:** Describe the challenges and conflicts that impede a coherent plan to solve the problem.

3. **Resolution:** Portray a solution to the challenges and the problem and how the venture will succeed by resolving the problem.

ample, the original business concept for Amazon.com might be summarized as "an Internet-based retail service that allows customers to search for and purchase at a discounted price books that will be delivered quickly."

A **story** is a narrative of factual or imagined events. The new business story depicts a business problem responded to with a new means to solve the problem. The story tells the goal of the venture, the challenge, and the response of the new firm. The creation of the story is used to communicate verbally the business idea and the profitable solution of the problem. The investor or new team member will be drawn to a good story. The three elements of a story summarized in Table 2.15 are 1) background, 2) challenge, and 3) resolution.

As an illustration of an important story, consider the world's energy challenge. Energy is the lifeblood of industrial civilization and necessary for lifting the world's poor out of poverty. However, current methods of mobilizing energy are highly disruptive of local and global environmental conditions and processes. Thus, the challenge is to develop a new, more favorable energy system and its associated sources. The resolution of this challenge will, it is hoped, be the discovery of an energy technology that can economically convert solar energy to a locally useful form. One possibility is a solar conversion system yielding hydrogen to be used in fuel cells. Many technology ventures could exploit this opportunity favorably. It can become a great story with an important outcome.

In today's fast-paced, dynamic world, a business concept and associated story are all one needs to start working on building a business. For the first steps, the entrepreneur 1) builds a concept to solve the business challenge, 2) fashions a story that conveys the meaning of the new venture, and 3) prepares a presentation of a few slides that tell the story and explain the concept. The elements of a presentation are given in Table 2.16. After testing the concept summary and the story, the entrepreneur may go on to draft a complete business plan, as described in Section 8.9.

TABLE 2.16 Elements of the presentation.

1. Explain the concept and give the story. Emphasize the customer and their pain.

2. Clearly explain the problem and the solution. Tell why the customer cares.

3. Describe the competencies of the team. Tell about the passion and skills of the team.

4. Provide a picture of the competition. Name a few competitors and tell how you are different and better.

The story can be told to all would-be investors or employees. The concept summary can be left with them for later review. The presentation is for more formal occasions with investors or allies. The story, concept summary, and presentation should be professional, novel, provocative, creative, and, where possible, customized. *Novelty* refers to newness and freshness. *Provocative* and *creative* mean it provokes interest and is creative in layout or format.

For many entrepreneurs, the executive summary of the business plan is what the investors and potential team members will want to review. If the entrepreneurs can piece together the initial elements of a business plan, they can write a reasonable summary without actually completing the plan. In a sense, the executive summary is the essence of the business plan. As such, for many new ventures, it stands alone as a short business plan. The executive summary succeeds by capturing the readers' attention and imagination, causing them to want to learn more. When readers finish the executive summary, they should have a good sense of what the entrepreneurs are trying to do in their business. The executive summary should be no longer than three pages. Most professional investors will ask you to e-mail it to them. An executive summary states the problem, the solution, the customer, the competitive advantage, and who will lead the effort. This summary is intended to convey the core of the business and draw the reader into a follow-up conversation.

The executive summary portrays the content and purpose of your business. The elements of an executive summary are listed in Table 2.17. Not all these elements will be necessary for all ventures. An example of a business summary is provided in Table 2.18. Also see Appendix A for examples of executive summaries. The executive summary for AgraQuest is provided in Table 2.19.

TABLE 2.17 Elements of an executive summary.

1. Business concept: The problem and the solution.
2. Market, customer, and industry.
3. Marketing and sales strategy.
4. Organization and key leaders.
5. Financial plan: Four years of summary results.
6. Financing and key allies required.

TABLE 2.18 Example of a Business Summary.

Security Robots Inc. (SRI) was formed in 2004 to design and build mobile robots for clearing and cleaning up facilities that have or may have experienced security breaches. Office buildings, factories, schools, and laboratories may be subject to intrusion by terrorists who plant biological, chemical, or explosive devices. The SRI robots are capable of remote operation by security and police organizations and can be used to examine and clear or destroy terrorists' weapons.

TABLE 2.18 (continued)

SRI is a Subchapter S corporation seeking an initial set of investors to bring its new products to market. Founded in 2004 by Dr. Henry Morgan and Ms. Angela Wolfe, the firm has designed a mobile robot platform that can be customized for many high danger security tasks. SRI has filed for a design patent on the robot platform system.

Dr. Morgan holds a MS in electrical engineering and a PhD in mechanical engineering from the University of Texas. Dr. Morgan served as chief technology officer of FMA Corporation of Dallas, Texas, from 1994–2003. Ms. Wolfe, CPA, holds an MBA from Duke University and was formerly CFO of Moore Systems, Austin, Texas.

SRI has secured 3000 sq. ft. of industrial space in Austin Technology Park. The current staff of six has an operating robot under review and certification by the U.S. Department of Homeland Security and the Texas State Police. Manufacture of the robot product line is to be provided by SelectTech Systems, a national contractor, at its Huntsville, Alabama, facility. The marketing plan calls for a regional strategy in the first year, 2005, with expansion throughout the southern and eastern United States in 2006. A highly trained direct sales force will sell the SRI robots to police and security organizations.

The funds requested for commencement of manufacturing and marketing operations are $400,000. The co-founders of the firm have already invested $120,000 of their funds. The funds will be used for facilities, equipment, contracting, and marketing communications. The founders will not receive a salary until January 1, 2006, or at cash-flow breakeven, whichever occurs first.

Financial projections show revenues of $1.3 million in 2005 and $7.4 million in 2006. The company intends to go public (IPO) within five years of beginning sales operations. Investors may purchase units of 10,000 shares for $20,000. After issuing the 200,000 shares to the investors for $400,000, there will be total of 2 million shares issued. All individual investors should contact the firm for an investment prospectus and further information.

Contacts:	Henry Morgan, CEO; Angela Wolfe, CFO	
	Security Robots Inc.	(512) 555-0121
	Austin Technology Park	www.securityrobots.net
	Austin, Texas 78712	

TABLE 2.19 AgraQuest executive summary.

Mission: AgraQuest's mission is to be the best and most efficient at discovery and development of environmentally friendly natural products for pest management.

The business: AgraQuest discovers, develops, and markets environmentally friendly natural product pesticides from microorganisms. It has three sources of revenues: 1) sales of natural product pesticides to farmers and consumers, 2) sales of lead molecules that do not fit our development criteria to large pesticide companies, and 3) contract testing for pesticide companies.

Market need and market opportunity: Twenty-five billion dollars are spent each year on chemical pesticides. Consumers have increasing expectations that their food is free of pesticide residues. Society has increasing concerns about how chemicals affect the environment, including fish, wildlife, groundwater, and air quality. The regulatory agencies are responding by establishing stricter criteria for registration of new chemical pesticide products and reregistration of older ones. The cost and time for registering a new chemical pesticide have ballooned to $40–70 million and 7–10 years. As a result, few new products are being registered, and many older products are being taken off the market or are so tightly regulated that their use is limited.

(continued on next page)

TABLE 2.19 (continued)

Technology: Natural products are substances produced by microbes, plants, and other organisms that can kill pests. Unlike natural products, currently marketed biopesticides, such as *Bacillus thuringiensis* (Bt), insect viruses, and insect-killing fungi, use living organisms as pesticides. As a result, they are negatively affected by heat, wind, rain, and sunlight. Therefore, they do not have efficacy as good as chemical pesticides and have not significantly penetrated chemical pesticide markets. Natural products can have efficacy against the targeted pest that is as good as chemical pesticides. This is not speculation. We have found them. We know they are there. Unlike many chemical pesticides, natural products are biodegradable and specific to the pest, without harmful effects on fish, wildlife, and beneficial insects.

   Microbial natural products can be registered with the U.S. Environmental Protection Agency (EPA) as "biochemicals." This means that bringing a specific natural product to the market takes considerably less time and money (approximately 3–5 years and less than $5 million) than chemical pesticides.

Competition: If microbial natural product pesticides are so ideal, why aren't they the target of large companies? Pharmaceutical companies have the technical expertise for discovery of microbial natural products with pesticidal activity, but they are often not set up to assess agricultural applications of the molecules and lack the knowledge and experience to commercialize them. There is currently no independent company dedicated to screening for microbial natural product insecticides, fungicides, nematicides, and herbicides.

Company's competitive advantage: AgraQuest can find a higher number of novel pesticidal natural products more quickly than anyone else. We have unique knowledge of the groups and sources of microorganisms that yield the highest number of novel pesticidal natural products. Our proprietary isolation and fermentation media generate higher numbers of "hits." We find more novel natural products because of our focus on difficult chemistry that very few in the industry attempt. We have proprietary automated, high-throughput in-vivo and in-vitro pesticidal and extraction assays. At a very early stage, we can rapidly recognize pesticidal molecules with product potential and activity as good as synthetic chemicals. We know how to develop and market bio-based pesticides in specialty markets; we have extensive and unique knowledge of the market and competition. We are experienced at creating a company culture that results in exceptional and sustained productivity, creativity, motivation, and commitment by employees.

Management team: AgraQuest has assembled a management team experienced in pesticide, biopesticide, and agricultural biotechnology business, research and development, marketing, management, and finance.

Pamela G. Marrone, PhD, President/CEO. Dr. Marrone left Novo Nordisk in January 1995 to start up AgraQuest. Under her tenure as president of Novo's subsidiary, Entotech, Inc., which she built from the ground up, the company extended its Bt product line into three new crop segments, brought a Bt product, two new Bt product formulations, and a new gypsy moth virus product formulation to the market. In addition, Entotech found six novel pesticidal natural products, including a novel Bt enhancer (now on the commercial track), and has filed or pending 20 patent applications. Dr. Marrone wrote and implemented marketing plans and developed a new approach to generating revenue from biopesticides, which is now the flagship strategy of the division. Prior to Novo Nordisk, Dr. Marrone worked for Monsanto Agricultural Company (1983–1990). Her Insect Biology group led pioneering projects in natural product and genetically engineered microbial pesticides and Bt transgenic crops (to be on the market in 1996).

Ralph Sinibaldi, Vice President of Research and Development. Dr. Sinibaldi worked for Sandoz Agro, Inc., from 1982–1994 and was most recently Associate Research Director and Project Manager, where he coordinated two major products on crop transformation and regu-

(concluded on next page)

TABLE 2.19 (concluded)

lation of gene expression. Dr. Sinibaldi has received or filed several patents, and he turned over three major pieces of technology to Sandoz Seeds for development.

Duane Ewing, Vice President of New Business and Product Development. Duane Ewing has 13 years of management experience and a total of 17 years of experience in agriculture-related industries. As one of the first employees of Pan-Ag Labs, Inc. (1981), Mr. Ewing played a crucial role in Pan-Ag's growth from $120,000 to almost $6 million annually in roles as Director of Field Research, Director of Business Development, Vice President, and de facto President during the owner's absence.

Bruce Holm CPA, Chief Financial Officer. Bruce Holm has over 30 years of accounting experience. For 16 years (1971–1987), he was Corporate Controller for Zoecon Corporation, where he was responsible for financial reporting in six SEC filings. Following Zoecon (1987–1991), he was employed by California Energy Company, Inc., as Corporate Controller and Joint Venture Controller.

Financial summary and amount and structure of proposed financing: AgraQuest requires start-up funding in the first full year of approximately $1.1 million for equipment, $2.5 million for operations, and $2.9 million for cash reserves. *First-round financing of $2.5 million will allow us to identify our first commercial product candidate from our own R and D and to develop an externally acquired product.* The following two years are expected to require approximately $11.4 million for operating expenses, $5.8 million funded by sales of research services and molecules and product sales, leaving a net operations requirement of $5.1 million. Also, approximately $1.1 million is projected for equipment and improvements purchases, and $0.8 million is required for cash reserves.

A public offering is projected to occur early in year five, with a target of $20 million. We are confident that AgraQuest can, by year three, develop a pipeline that subsequently generates 5–10 new natural products per year. Our novel natural product portfolio will specifically include two for corn rootworm (to be sold), one for sucking insects, one fungicide, one nematicide, and one herbicide (to be sold).

The business projects a profit in year five and approximately $40 million in sales of molecules, services, and products in year seven.

Projected capital requirements:

Year ended June 30:	1996	1997	1998	1999
	($thousands)			
Operating expenditures and interest	$2,500	$4,700	$6,700	$9,800
Equipment and furniture	1,100	500	600	400
Cash reserves buildup	2,900		1,300	
Total	$6,500	$5,200	$8,600	$10,200

Projected funding:

	1996	1997	1998	1999
Revenue from contract screening, molecule and product sales, and government grants	$200	$1,800	$4,000	$7,500
Equity financing	5,200	2,600	4,300	
Capital lease and/or bank financing of equipment, net of repayment	1,100	300	300	
Cash reserves usage		500		2,700
Total	$6,500	$5,200	$8,600	$10,200

Status of the company: AgraQuest was incorporated in the state of Delaware in January 1995. The company is in the process of completing seed financing (approximately $200,000), which is being used for starting the microbial library and pest colonies. Also, we expect to obtain one product candidate from outside the company and secure at least one corporate collaboration.

2.7 AGRAQUEST: THE OPPORTUNITY AND THE BUSINESS SUMMARY

Pam Marrone and her colleagues at Entotech were informed by Novo Nordisk that the firm was being sold to Abbott Laboratories. Marrone believed that natural biological controls could protect crops—an old idea that environmental enterprises are making fresh again. Driving the quest is pressure from government and consumer activists to reduce the use of synthetic chemicals on the nation's farms and ranches. The challenge for such companies is to develop reliable biopesticides at a price with chemicals.

Marrone contemplated leaving Entotech and starting a new venture based on developing an innovation (item 8 of Table 2.1) in the biotechnology industry: finding naturally occurring micro-organisms that can serve as biopesticides and developing a process for producing them for reliable use on farms. With the growing trend toward natural, environmentally friendly products and processes, the opportunity looked favorable.

Marrone had already worked in corporate new ventures with Monsanto and Novo Nordisk, and she was confident of her technical and leadership competencies. She examined the opportunity using the principles of Table 2.9 and determined that this opportunity was very good. The agricultural pesticide industry showed a tendency to be slow to adopt risky innovations such as natural pesticides. However, she was convinced she could overcome the risk-averse nature of the farmer and the long regulatory review by government. The fit of her proposed company (Figure 2.4) with the opportunity seemed to be high. Therefore, she decided to act by finding the key members of her team and founding the company, to be called AgraQuest. She was convinced that the opportunity was of very high quality (see Figure 2.6).

Pam Marrone and her fellow founders wrote an executive summary dated May 5, 1995, provided in edited form in Table 2.19.

2.8 SUMMARY

The entrepreneur identifies numerous ideas and needs that may point to good opportunities that can be made into great companies. However, he or she searches for the one that best fits the capabilities of the team, the characteristics of the business context, the characteristics of the opportunity, and the team's capability to secure the necessary resources. Then, the entrepreneur decides whether to act or not act on that best-fit opportunity.

The important ideas of the chapter are:

- A great enterprise displays leadership in its industry, profitability, reputation, and longevity.
- Great opportunities are often disguised as problems that are difficult to describe.
- An important problem well stated is a problem on its way to solution.
- The entrepreneurial team should cumulatively possess all the necessary capabilities.

■ Entrepreneurs should, if possible, act on favorable opportunities in a timely manner.

■ Entrepreneurs should prepare a story and summary of the venture and use it to test the venture with potential customers, employees, and investors.

Principle 2
The capable entrepreneur knows how to identify, select, describe, and communicate an opportunity that has good potential to become a successful venture.

2.9 EXERCISES

2.1 One big trend is toward a more active, outdoor life. This trend extends through all ages and regions of the world. As an example, visit Johnson Outdoors (www.johnsonoutdoors.com) or REI (www.rei.com) and summarize the potential opportunity for a new outdoor recreational equipment and clothing enterprise.

2.2 Apple Computer's Steve Jobs announced in 2003 the opening of an online music store with 99-cent downloads of popular music. The iTunes Music Store is limited to MacIntosh computers. Examine the system and describe new opportunities opening up in the online music business. Write a short one- or two-sentence description of the opportunity.

2.3 Wi-Fi is short for "wireless fidelity," which is a new way to connect to the Internet. Companies, hotels, and cafés are installing wireless networks. Combine Wi-Fi with Internet-based telephony, and you have an intriguing potential competitor for existing cellular networks. Consider the opportunities offered by Wi-Fi networks. Write a short description of the opportunity.

2.4 Great companies often create tools that solve people's everyday problems. People like to chat and say hello often. What are the opportunities for instant messaging via cell phones or other devices?

2.5 View the movie *Startup.com* about two young entrepreneurs who start a company called govWorks. Test their experience against the questions of Table 2.11. Were they wise to have started their firm? Explain your answer.

2.6 The next big wave of innovation may be the convergence of bio-, info-, and nano- technologies. Each holds promise in its own right, but together, they could give rise to many important products. Describe one opportunity in this new field, and write a story about the opportunity.

2.7 Glass can now be made with an inner layer that makes it strong enough to withstand strong winds and bomb blasts, and support the weight of people [Herrick, 2003]. Examples of this new technology are PVB laminate and ionoplast made by Dupont and Solutia. Consider this opportunity for new building materials, and develop a story for this opportunity.

2.8   The convergence of biology with computers and nanotechnology may lead to safer and more effective medicines [Jonietz, 2003]. Visit www.research.cornell.edu/anmt and examine the potential for nanomedical technologies. Write a brief concept summary for a nanotechnology start-up.

2.9   Avis and Hertz rent cars 24 hours a day, so why not rent hotel rooms at any hour? Think of the traveler who arrives on an overnight flight at 6 A.M. and wants to rent a room. Around-the-clock check-in could generate a new opportunity for a hotel. Review this opportunity and evaluate it using the process of Table 2.11.

2.10   The trend of performance of two electronic technologies is given in Figure E2.10. Determine the performance trend of another technology, and prepare a chart of its performance over time.

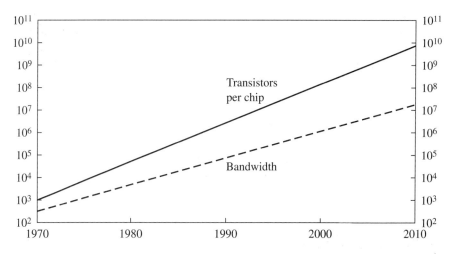

FIGURE E 2.10 Technology trends: (a) transistors per chip; (b) bandwidth per household (bits/second).
Source: Dorf, 2004.

CHAPTER 3

Building a Competitive Advantage

Success in any enterprise requires the right product, methods, and workers,
and each must complement the others.
Joseph Burger

A new business is defined by the wants or needs customers satisfy when they buy a product or service. To create a theory of a new business, the entrepreneur must cogently and clearly describe the customers and their needs and how the new venture will satisfy those needs. To describe the business, the entrepreneur prepares a series of statements and propositions that clearly outline the business. These are ultimately summarized in a model of the business activities and goals. Based on the core competencies of the organization coupled with the business model and the key resources available, the firm acts to attempt to create and retain a sustainable competitive advantage. The six steps of designing and creating a theory of the new business are summarized in Figure 3.1. ■

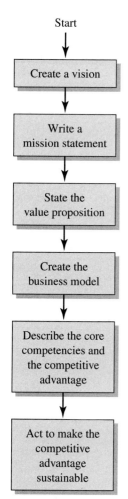

FIGURE 3.1
Creating a business theory of a new venture.

3.1 THE VISION

Once the entrepreneur identifies a good opportunity and decides to pursue it, the next step is to formulate a vision. A **vision** is an informed and forward-looking statement of purpose that defines the long-term destiny of the firm. Thus, if the entrepreneur recognizes a good opportunity to meet a real customer need, he or she describes a vision of a future venture that will respond effectively to that opportunity. The vision is a statement of insight, intention, ambition, and purpose. It reflects clearly the novelty of the solution and uniqueness of the entrepreneur's commitment. Successful entrepreneurs are able to communicate their vision and their enthusiasm about that vision to others. A vision often constitutes a novel idea about serving its market. McDonald's vision is: low-priced, fast food in a clean restaurant for people short on time. Google's vision is: online search that reliably provides fast and relevant results.

A solid vision provides direction and shows a path forward. A vision also motivates and influences decisions that are made by the team members. A clear vision can bind and inspire the entire community of a firm. A good vision is clear, consistent, unique, and purposeful [Hoover, 2001]. A clear vision is also easily understood. A consistent vision is one that does not change in response to daily challenges and fads. Any sound vision clearly explains the purpose of the firm. Remember, a good opportunity embodies a response to a big problem and calls forth a clear picture of your response. The four elements of a vision are summarized in Table 3.1.

The purpose of the firm defines the enduring character of the organization, consistently held to and understood through the life of the firm. The purpose, or core ideology, of Hewlett-Packard has been a respect for the individual, a dedication to innovation, and a commitment to service to society. This core ideology provides the glue that holds an organization together [Collins, 1996]. The vision provides a clear picture of the future for all concerned. The core ideology is based on the core values of the organization, such as respect for the individual.

A vision describes a specific desired outcome and promotes action and change. It serves as a picture of its destiny as the firm moves through challenge and change. It also provides the basis for a strategy. A vision is an imaginable picture of the future. It is like a rudder on a boat in a turbulent sea. An example of a simple, clear vision is given in Table 3.2. This vision statement provides the reader with a clear mental model of where the firm is going and how it will get

TABLE 3.1 Elements of a vision.

- **Clarity:** Easily understood, focused.
- **Consistency:** Holds constant over a time period, but is adjustable as conditions warrant.
- **Uniqueness:** Special to this organization.
- **Purposeful:** Provides reason for being.

TABLE 3.2 Example of a vision for an innovative firm.

We strive to preserve and improve life through the innovation of biomedical devices while supporting, training, and inspiring our employees so that individual ability and creativity is released and rewarded. Our goal is to be the leader in our industry by 2006 and be widely known throughout the world.

there. Notice that the vision statement of Table 3.2 states the values and goals of the firm, and it inspires and motivates people. The vision statement of eBay is:

We help people trade practically anything on earth through an online system.

Entrepreneurs need to create a shared vision or meaning for their venture. A dialogue of meaning and commitment will help bring a shared sense of urgency and importance for the venture. The vision can be written as a statement and verbally expressed as a story. The vision is used as a part of the business plan and described often to potential team members and investors. Stories play an important role in the processes that enable new businesses to emerge [Lounsbury, 2001]. These stories serve to legitimate new ventures. A story is a narrative version of the vision, told in an engaging way. It helps to make the unfamiliar new enterprise more familiar, understandable, acceptable, and thus more legitimate to key constituencies.

Jim Clark started three companies: Silicon Graphics, Netscape, and Healtheon (now WebMD). As recounted in *The New New Thing* [Lewis, 1999], Clark stated:

"The only thing I can do is start 'em." His role in the Valley was suddenly clear: he was the author of the story. He was the man with the nerve to invent the tale in which all the characters—the engineers, the VCs, the managers, the bankers—agreed to play the role he assigned them. And if he was going to retain the privilege of telling the stories, he had to make sure the stories had happy endings.

Clark had a vision for eliminating waste in the $300 billion costs of the U.S. health system by using the Internet to enable all the parties of any health transaction to connect via an online network—no paper forms, no hassle. Clark sketched a diamond depicting the players, as shown in Figure 3.2, and placed his proposed company, Healtheon, in the middle as intermediary. This was the way Clark told his ambitious stories—graphically, using sketches. Tales told by the entrepreneur aim to show plausibility and build confidence that the enterprise can succeed. To construct an identity that legitimates a new venture, entrepreneurial stories must have narrative clarity and resonate with the expectations, interests, and agendas of potential stakeholders [Lounsburg, 2001].

Entrepreneurs need to learn how to tell their story about their team and venture, and to explain how their products will solve a problem. Their vision of the future can capture the interest of investors and team members.

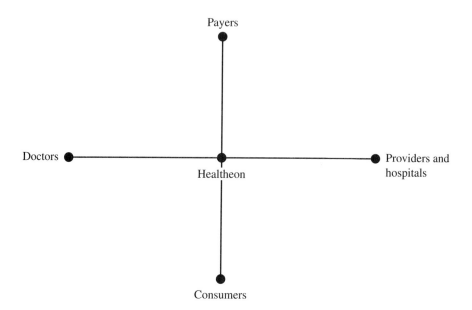

FIGURE 3.2 Vision for Healtheon (now WebMD). WebMD's three main businesses are providing electronic transaction services to doctors and hospitals; marketing software to help doctors run their practices; and providing online health information to doctors and consumers (see www.webMD.com).

In late 1966, Rollin King visited Herb Kelleher in San Antonio with an idea for a new airline to serve San Antonio, Dallas, and Houston. King described the three-city route, the Golden Triangle of Texas, as perfect for a point-to-point airline. He sketched the three-city route on a cocktail napkin and said, "Herb, let's start an airline. We could offer fares so low people would fly instead of drive." This compelling story led to the formation of Southwest Airlines [Montague, 2001].

A vision told as a story helps people to see the situation and visualize the solution. The vision can also help people respond to the emotionally charged idea and want to help bring about useful change to the situation. The vision can be told as a story describing the potential outcome.

Henry Ford had a vision in 1910 of an automobile that could be available to all [Hounshell, 1985]:

> "[The] greatest need today is a light, low-priced car with an up-to-date engine of ample horsepower, and built of the very best material. . . . It must be powerful enough for American roads and capable of carrying its passengers anywhere that a horse-drawn vehicle will go without the driver being afraid of ruining his car."

3.2 THE MISSION STATEMENT

The mission statement for a new venture will more completely describe the company's goals and customers, while incorporating the basic tenets of the vision statement. A vision is an imaginative picture of the future, while a mission is a description of the course of action to implement the vision. The mission of an organization is lofty and visionary: it provides for a theory of change.

The potential elements of a mission statement are shown in Table 3.3. Most mission statements include only some of these elements. For example, the mission statement of eBay is given in Table 3.4.

Most mission statements are short—fewer than 100 words. The mission statement should be a concise, clear explanation of the purpose, values, product, and customer. The eBay statement clearly describes its mission. An example of a concise yet clear mission statement for an electronics firm is:

> Our mission is to design and manufacture electronic devices that serve the needs of the aerospace industry on a timely basis and at reasonable prices.

A good mission statement can help align all the stakeholders and provide a rationale for allocating resources. If possible, the mission statement should be developed by the entrepreneurial team with other employees. The mission statement for Starbucks is shown in Figure 3.3. This statement is very complete and describes its commitment to all its stakeholders—customers, employees, and community. The mission statement of Symantec Corporation is provided in Table 3.5. This statement speaks clearly to the customers—Symantec breeds confidence.

TABLE 3.3 Possible elements of a mission statement.

■ Core values	■ Competitive advantage
■ Customers and/or stakeholders	■ Values provided to customer
■ Products	■ Markets or industry

TABLE 3.4 Mission statement of eBay.

We help people trade practically anything on earth. eBay was founded with the belief that people are basically good. We believe that each of our customers, whether a buyer or a seller, is an individual who deserves to be treated with respect.

We will continue to enhance the online trading experiences of all—collectors, hobbyists, dealers, small business, unique item seekers, bargain hunters, opportunistic sellers, and browsers. The growth of the eBay community comes from meeting and exceeding the expectations these special people.

Establish Starbucks as the premier purveyor of the finest coffee in the world while maintaining our uncompromising principles while we grow. The following six guiding principles will help us measure the appropriateness of our decisions:

• Provide a great work environment and treat each other with respect and dignity.

• Embrace diversity as an essential component in the way we do business.

• Apply the highest standards of excellence to the purchasing, roasting, and fresh delivery of our coffee.

• Develop enthusiastically satisfied customers all of the time.

• Contribute positively to our communities and our environment.

• Recognize that profitability is essential to our future success.

FIGURE 3.3 Starbucks mission statement.

TABLE 3.5 Mission statement of Symantec Corporation.

At Symantec, we know what happens when people have the confidence to achieve their best; we help make it possible.

    We're the global leader in Internet security—solely dedicated to making the connected world a safer place. The more connected the world becomes, the more pivotal our role in making it an environment where commerce, culture, and ideas can flourish.

    Symantec breeds confidence.

Source: *Symantec 2003 Annual Report.*

3.3 THE VALUE PROPOSITION

Value delivered to the customer results in a satisfied customer who will pay a reasonable price in return for the product or service. **Value** is the worth, importance, or usefulness to the customer. In business terms, value is the worth in monetary terms of the social and economic benefits a customer pays for a product or service. To be successful, firms must offer products that meet the needs and values of the customer. The values of the customer often include ease of locating or accessing the product as well as its qualities and features.

    The five key values held by a customer can be summarized as product, price, access, service, and experience. These five values are listed in Table 3.6, along with specific descriptors for each value. Price, for example, can have high value to the customer when it is fair, visible, and consistent. A product may have value if it has high performance and quality, and is easy to find and use.

    The value proposition defines the company to the customer. Most value propositions can be described using the five key values. Crawford and Matthews

TABLE 3.6 Five values offered to a customer.

1.	**Product:**	Performance, quality, features, brand, selection, search, easy to use, safe.
2.	**Price:**	Fair, visible, consistent, reasonable.
3.	**Access:**	Convenient, location, nearby, at-hand, easy to find, in a reasonable time.
4.	**Service:**	Ordering, delivery, return, check-out.
5.	**Experience:**	Emotional, respect, ambiance, fun, intimacy, relationships, community.

have shown that one value is selected to dominate the value proposition offered to the customer. A second value differentiates the offering, and the remaining three values must meet the industry norm [Crawford, 2001]. Consider a performance rating on a 1-to-5 scale where 5 is world-class, 1 is unacceptable, and 3 is industry par. Crawford states that a venture should plan a good product offering to have a value score of 5, 4, 3, 3, 3 for its five value proposition attributes in the following order: dominant, differentiating, norm, norm, norm.

Consider Wal-Mart, where price is the dominant value of its offering. Wal-Mart differentiates itself on product in terms of selection and quality. The values offered by Target are dominated by product and differentiated by price. Good service is about human interaction. For example, Honda has great service as its dominant value, and its secondary, differentiating, value is product.

Access can be described by ease of locating, connecting to, and then navigating the physical or virtual facility of a business. Very good accessibility is offered by Amazon.com, and its website is relatively easy to navigate. Accessibility can also be described as convenience or expedience. For a customer with a high demand for time, convenience is very important. A readily accessible website can be very valuable to a time-starved customer.

The experience of a business is the entertainment factor as well as how a customer feels about the transaction. Starbucks provides a very good experience (it pays attention to your order) and good accessibility (its stores seem to be everywhere).

By adding an experience to a regular purchase, one can add value for the customer. Consider the Pike Place Fish Market in Seattle (www.pikeplacefish.com). When you order a fish for dinner, the fishmongers start hollering back and forth, and one of them pitches the fish to the other, who catches it in a newspaper fashioned into a catcher's mitt and wraps it for you. You get the fun of the experience and the value of the fish. Starbucks and Dave & Busters are also excellent examples of a good experience added to a good product.

Most customers seek a provider of a product or service who saves them time, charges a reasonable price, makes it easy to find exactly what they want, delivers where they ask, pays attention to them, lets them shop when they want to, and makes it a pleasurable experience. Any firm that fashions a value proposition to that set of customer values and actually delivers on that promise should do well.

TABLE 3.7 Primary and secondary values for leading firms.

		Primary Value				
		Product	**Price**	**Access**	**Service**	**Experience**
Secondary Value	**Product**	——	Wal-Mart	Amazon.com	Honda	Harley-Davidson Disney World
	Price	Target	——	Holiday Inn	Land's End	Olive Garden
	Access	Google Barnes & Noble	Priceline Visa	——	Dell Computer	Starbucks
	Service	Toyota Home Depot Intel	Southwest Airlines	McDonald's	——	Carnival Cruise Line
	Experience	Mercedes	Virgin Atlantic Best Buy	AOL	Nordstrom	——

The product value is described by its performance, range of selection, ability to search for it, and quality. Volvo built its business on the idea of product safety. Volvo became the first car company to offer three-point, lap and shoulder, seat belts. Home Depot focuses on providing a very wide selection of quality products. The differentiating (secondary) value for Home Depot is its service. The primary and secondary values for selected leading firms are shown in Table 3.7.

Remember, a firm must meet at par the three remaining variables. Consider the plight of today's department stores. Their primary value is product selection. However, they are struggling to be accessible to today's shopper and just be at par on service, price, and experience.

> What are the primary and secondary values for Google? It offers product as its primary value with fast, relevant results for the most ill-described inquiry. Its secondary value is access, which is embodied in the easy online connection right to the search page without annoying pages or advertisements obscuring the search box.

McDonald's is very accessible, being almost everywhere, but sometimes its service fails to differentiate its offering.

The **value proposition** states who the customer is and describes the values offered to this customer. The value proposition for Amazon.com could be described as:

> An easily accessible Internet site that is convenient all of the time to provide a wide selection of books, CDs, and videos at a fair price to the busy, computer-literate customer.

The value proposition for Starbucks could be described as:

> We provide a friendly, comfortable, well-located place offering a wide range of fresh, customized quality coffees, teas, and other beverages for the person who enjoys a good experience and a good beverage.

Home Depot and Lowe's stores are the two large home-improvement chains in the United States. Home Depot's dominant value is product selection, and its secondary value is service. Lowe's has a dominant value of accessibility and a secondary value of product selection. These value differences lead to clearly separate value propositions offered to the customers of these two competitors.

The **unique selling proposition** (USP) is a short version of a firm's value proposition and is often used as a slogan or summary phrase to explain the key benefits of the firm's offering versus that of a key competitor. For example, Hewlett-Packard uses a USP as follows:

> Excellent technical products with reliable service at a fair price.

The clear, simple USP for FedEx is:

> Positively, absolutely overnight.

USPs are useful for succinctly describing a new venture to would-be investors, customers, or team members. In the jargon of investors, it is often called "the elevator pitch." This is a short description of your venture that can be told during the brief ride on an elevator between getting on and getting off. The USP is widely used in Hollywood for screenwriters to "pitch" their movie idea in a single sentence. For example: "A black cop from Detroit goes to Beverly Hills to find his partner's killer" is the pitch for *Beverly Hills Cop*.

New ventures can use their value proposition and unique selling proposition to clarify the business values offered to the customer. This will help all stakeholders understand the purpose of the firm's business concept.

3.4 THE BUSINESS MODEL

The design of a business is the means for delivering value to customers and earning a profit from that activity. The **business design** incorporates the selection of customers, its offerings, the tasks it will do itself and those it will outsource, and how it will capture profits. Business design is often called business concept. A successful business model represents a better way than existing alternatives. Creating a new business model is like writing a new story. It is this story that attracts investors, customers, and team members.

A good business design involves what your firm will and will not do and how the firm will create a sound value proposition. The business design answers three key questions: who is the customer; how are the needs of the customer satisfied; and how are the profits captured and profitability protected. The resulting outcome of the business design process is the **business model,** which is the description of the business and how it will work in economic terms. A business

TABLE 3.8 Elements of a business model.

■ **Customer selection:**	Who is the customer?
	Is our offering relevant to this customer?
■ **Value proposition:**	What are the unique benefits?
■ **Differentiation and control:**	How do we protect our cash flow and relationships?
	Do we have a sustainable competitive advantage?
■ **Scope of product and activities:**	What is the scope of our product activities?
	What activities do we do, and what do we outsource?
■ **Organizational design:**	What is the organizational architecture of the firm?
■ **Value capture for profit:**	How does the firm capture some of the total value for profit?
	How does the firm protect this profitability?
■ **Value for talent:**	Why will good people choose to work here?
	How will we leverage their talent?

model is a set of planned assumptions about how a firm will create value for all its stakeholders [Magretta, 2002].

The business model answers questions about the customer, profit, and value. The elements of a business model are shown in Table 3.8. The business model for Dell Computer is given in Table 3.9 [Slywotzky, 2000]. The first critical element of a business model is the selection of the customer. The business design aims to specify the customers with unmet or latent needs, which will then define the target market. Dell uses four market segments to describe its customers and then prepares separate offerings for each segment. It is important to choose customers who will permit you to profit and spurn customers who want great value but are difficult or unfairly demanding. Instead of making all customers very happy, focus on the right customer and create an offering that allows good value to the customer and a reasonable profit to your firm. If possible, the price of your offering should include a reasonable profit margin, as well as good value for the customer. Once you know who your selected customer is, start saying no to those who don't fit the model.

The second step is to clearly state a unique value proposition that will provide differentiation for your firm. Show how the value proposition will address

TABLE 3.9 Dell Computer business model.

■ Customer selection: High relevance	Four segments: Corporate, government, education, consumer
■ Value proposition: Unique benefits	A customized computer at a good price with great service readily accessible via phone or Internet
■ Differentiation and control: Sustainable competitive advantage	Customized products via a direct sales channel via phone or Internet with strong service and customer relationships
■ Scope of product and activities	Desktops, laptops, and servers Strong supply-chain management
■ Organizational design	Divisional organization for each customer segment
■ Value capture for profit	Opportunities for cross-sell and up-sell. Avoid price as the key value and focus on service and accessibility.
■ Value for talent: Learn, grow, prosper	Training, learning, and career opportunities

the market segment you have identified. Then, explain the scope of product and activities and organizational design that will enable you to implement the value proposition. Dell sells customized computers at a good price with great service. Dell differentiates itself by relying on a direct sales model via the phone, mail, or the Internet and providing offerings suitable for each separate market segment.

A clear path to profitability is critical. Furthermore, it is important that you can retain good profit margins so that you can invest for the future. In general, it is best to avoid competing solely on price and making price the dominant value of the value proposition. Wal-Mart, Costco, Dollar General, and Family Dollar stores are examples of firms that use price discounting as their dominant value. Dollar General and Family Dollar use smaller stores in well-located strip malls so that accessibility is their differentiating value. Wal-Mart and Costco compete on the product quality and selection as their secondary value. The business model of Wal-Mart is successful because of its use of technology to achieve strong supply-chain management and store inventory control.

Southwest Airlines is another example of a business with price as the dominant value in its value proposition. Its secondary value is service: on-time arrival and departure, online ticket ordering, and a customer-friendly attitude. It captures profit from the valued service by controlling costs. It uses one type of aircraft, which keeps its costs of maintenance and training lower than its competitors'. It also heavily promotes the online sales of tickets. As a result, Southwest has been profitable every year since 1973 [Freiberg, 1997]. The business model of Southwest Airlines is compared with the business model of American Airlines in Table 3.10.

TABLE 3.10 Business model of two airlines.

	American Airlines	**Southwest Airlines**
Customer	Traveler who needs to fly many places throughout the world	Traveler who desires to fly routes served point to point in the U.S.

Value proposition:
    Dominant value
    Differentiating value
	Product
	Accessibility
		Price
		Service

Differentiation	Wide scope of product: goes almost anywhere	Limited point-to-point flights Easy maintenance and training for low cost
Scope of products and activities	Very broad: connects everywhere	Narrow: only flies to selected cities (point to point)
Organizational design and implementation	Hub-and-spoke High fixed cost	Point-to-point Lower, flexible costs Control costs
Value capture for profit	Dominate hub city Requires high occupancy	Requires high occupancy
Value for talent	High pilot salaries Good career	Participation in stock options and camaraderie

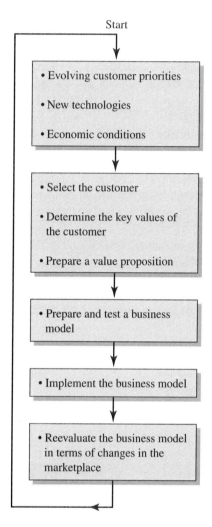

Start

- Evolving customer priorities

- New technologies

- Economic conditions

- Select the customer

- Determine the key values of the customer

- Prepare a value proposition

- Prepare and test a business model

- Implement the business model

- Reevaluate the business model in terms of changes in the marketplace

FIGURE 3.4 Business design process.

Customers influence changes in sound business models as their priorities change. Many business models fit a context that eventually evolves and necessitates changes in the models. The obsolescence of an outmoded business model and the necessity for a redesign of the business model is called value migration [Slywotzky, 1996]. For example, the appropriate business model for Hewlett-Packard Corporation has changed significantly since the company's founding in 1938. In recent years, many of the firm's manufacturing activities have been outsourced as the company migrated toward a computer company competing on the primary value of product and the secondary value of price.

The business design process is summarized in Figure 3.4. The dynamic firm continuously tests for changing conditions and redesigns its value proposition to meet the values of its customers.

3.5 BUSINESS MODEL INNOVATION IN CHALLENGING MARKETS

The collapse of stability in the marketplace challenges any business team to keep their place in the list of successful companies. Business model innovation is the capacity to reconceive existing business models in new ways that create new value for customers [Hamel, 2000]. For example, Hewlett-Packard changed its model to become America's premier brand for computer printers. The key to Hewlett-Packard's business model is the consistency of product and service. Southwest Airlines has taken the lead for profit and return on investment in the airline industry. IKEA has designed a high-volume business model for selling affordable, well-designed home furnishings. Searching for a new job in another state can be accomplished using Monster.com, which lists hundreds of thousands of jobs across the country. All these firms have effectively reconceived their business models over time in response to market change.

While competitors will always attempt to imitate the best practices of a market leader, a unique, difficult-to-imitate business model is often based on a unique competency or technology—or both. Dell's business model is based on a direct sales, customized product capability, and an information system that enables Dell to manage its supply chain efficiently. As a result, Dell has an average inventory of four days, while a typical competitor would have 40 days' inventory.

Markets are dynamic, and companies and nations respond slowly. The Forth and Clyde canals were built to connect Edinburgh and Glasgow, Scotland, over 100 years ago. The purpose of the canals was to move goods between the east and west coasts. The new Falkirk Wheel represents an innovation to move large ships quickly through locks (see www.thefalkirkwheel.co.uk). The economic benefit of this innovation is significant and could result in the renewed utility of the canals.

Amazon.com started as a bookseller, but it now looks like the Wal-Mart of the Internet. Gas stations have evolved into convenience stores selling beverages, food, newspapers, and fuel. The entrepreneur within an existing company can help to build new value for customers and new profit for the company by reconfiguring the firm's business model before its competitors do theirs. One powerful way to find a new business model is to look for the customers' latent, unstated dissatisfactions with existing business practices.

ChildrenFirst provides backup child care to employees of over 260 corporations and consortia throughout the United States and Canada. Corporate clients pay a yearly fee in exchange for a specific number of days of backup child care. The clients' employees are then able to bring their children to a ChildrenFirst center whenever primary child care plans fall through. In 2001, the company had 35,869 registered children and provided 73,172 child visits. ChildrenFirst currently employs over 220 people across the United States and Canada. Not only do backup child care providers increase

(continued on next page)

(continued from page 67)

worker productivity, they also give working parents peace of mind from knowing that their children are being well cared for in a safe environment. Such centers give parents the opportunity to bring their families with them on business trips, encouraging greater family togetherness.

The customers are families with working parents. ChildrenFirst offers a valuable product with safety and good care as necessary elements of the children's care. Will ChildrenFirst be able to consistently capture profit from the value provided? (See www.childrenfirst.com.)

3.6 CORE COMPETENCIES AND COMPETITIVE ADVANTAGE

The **core competencies** of a firm are its unique skills and capabilities. A capability is the capacity of the firm, or a team within the firm, to perform some task or activity. Firms with core competencies that match those necessary to effectively implement their business model have the best chance to succeed. It is very important that the core competencies of your firm match the requirements of your business. The core competency of Honda is the ability to design and build internal combustion engines of all sizes. The core competency of Intel is the ability to design and manufacture integrated circuits for computers and communication systems.

We care about competencies since they are the roots of competitive advantage. The real sources of advantage are found in the competencies of a firm. Core competencies include the collective learning in the organization, the skills of its people, and its capabilities to coordinate and integrate know-how and proprietary knowledge. Unlike physical assets, core competencies do not deteriorate as they are applied and shared. They can grow as a firm learns to build its competencies. Physical assets wear out, but intellectual assets such as core competencies can improve over time.

The core competencies of 3M are in designing and manufacturing materials, coatings, and adhesives, and devising various ways of combining them for new, valuable products. Honda's core competencies in engines and power trains have enabled it to provide distinctive products for lawnmowers, motorcycles, automobiles, and electric generators. Core competencies provide potential access to a wide variety of markets. Core competencies are the wellspring of new business ventures.

A successful firm's core competencies are valuable, unique capabilities that enable the firm to implement its business model and thus deliver a valuable product or service to its customers. These unique capabilities will be rare, difficult to imitate, and difficult to substitute.

Core competencies are dynamic by nature and an integral part of organizational learning and competence building. These distinctive capabilities are those activities that a firm does better than its competitors. These competencies are the critical asset of a technology venture.

> The core competency of Google is the design and operation of a software search engine. It is the dominant online search engine. While starting as a search tool for finding information on diverse subjects, it has also become a tool for people searching for an online site selling a product they want to buy.

3.7 SUSTAINABLE COMPETITIVE ADVANTAGE

The **competitive advantage** of a firm is its distinctive factors that give it a superior or favorable position in relation to its competitors. Competitive advantage is measured relative to a firm's competitors. A **sustainable competitive advantage** is a competitive advantage that can be maintained over a period of time—hopefully, measured in years. The duration, D, of a competitive advantage, CA, leads to the estimate of the market value, MV, of a firm as

$$MV = CA \times D \qquad (3.1)$$

That is, the market value of the firm is proportional to the size or magnitude of the competitive advantage and dependent on the expected duration of that advantage. A pharmaceutical firm with a 17-year patent and a strong competitive advantage will be highly valued, indeed!

The competitive advantage of a firm is directly dependent on its core competencies, its assets, and its organization architecture. A firm such as General Electric is said to have a sustainable competitive advantage in the electric power industry. It has higher profit margins than all its competitors in this field.

In general, the more value, V, customers place on a firm's products, the higher price, P, the company can charge for these products. The cost of producing the products is C, and the profit margin is P – C. The company profits as long as P > C. The value created is V – C, and the net value to the customer is V – P. These relationships can be portrayed as shown in Figure 3.5. The profit margin, P – C, is vanishingly small in some very competitive industries where the competitive advantage is small or nonexistent for all the firms.

American Express invented the traveler's check as a means of getting money while traveling abroad. The value to the customer was high. The cost to issue the checks was low, and (V – P) remained high to the customer. All parties have been pleased with this business model for more than 100 years [Magretta, 2002].

Many profitable firms are built on differentiation: offering customers something they value that competitors don't have. This unique offering can be in the product, service, or sales, delivery, or installation of the product. While the basic product may be a commodity, the differentiation can be obtained somewhere in the various interactions or services for the customer. Firms selling personal

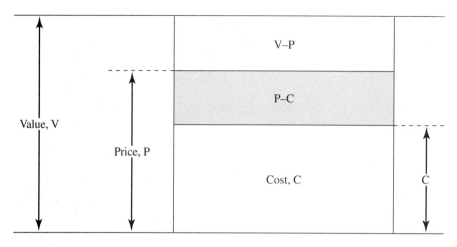

FIGURE 3.5. Value and return to the customer and the firm.

computers attempt to differentiate themselves by offering high-quality service. Harley-Davidson lends money to people to buy its motorcycles.

A competitive advantage is a significant difference in a product or service that meets a customer's key buying criteria. The sustainability of a firm's competitive advantage is a function of the competitors' difficulty in imitating or innovating around the incumbent's unique product or service attributes. One hospital service company successfully differentiates saline, a commodity product, by delivering it premeasured and frozen in plastic bags directly to hospital wards, thereby saving hospitals handling costs.

All firms seek to erode competitors' advantages by acting to *imitate* their product or service attributes or innovation.

Competitive advantage can be based on lower costs or differentiation of product or both. Most firms try to improve the efficiency of their operations to lower costs. They also strive to innovate or provide superior quality to outdo their competitors. Another point of differentiation can be in customer relationships. These approaches are summarized in Table 3.11.

TABLE 3.11 Sources of competitive advantage.

Source	Example
Efficiency, low costs	Alcoa
Product innovation	Intel
Quality, reliability	Mercedes
Customer responsiveness	Dell
Manufacturing innovation	Toyota

FIGURE 3.6 Pyramid of value creation.

A firm can create new value and thus establish sustainable competitive advantage. The pyramid of value creation is shown in Figure 3.6. From a solid base of assets, a firm builds its capabilities, which lead to its core competencies. With its core competencies and knowledge, it develops new products, processes, and other activities to build a competitive advantage. The sustainability of a firm's competitive advantage depends on its ability to continually innovate.

The duration, D, of a competitive advantage is longer when it is difficult to imitate. This difficulty is present when unique skills and assets are required and hard for a competitor to replicate or obtain.

Charles Tiffany founded Tiffany & Company in 1837, an upscale chain of retail stores selling silverware and jewelry. Tiffany's has always been known for its artistic displays and product design and selection. Tiffany watched for the chance to acquire unusual jewelry and items so he could offer customers the top of the line. In 1887, when the French government sold the crown jewels, Tiffany made his largest investment yet, buying roughly a third of the whole lot for about $500,000. The core competency of Tiffany's is the ability to select and display high-value, artistic items. This remains its core competency today. Tiffany's has retained a sustainable competitive advantage over the past century.

Selling the vision of the sustainable venture requires a passionate commitment to the venture. Candy Lightner started a nonprofit organization called Mothers Against Drunk Driving (MADD) after her 13-year-old daughter was killed by a drunk driver in a hit-and-run accident. She communicated the vision of the organization with a passion born of loss and injustice.

TABLE 3.12 Ten types of sustainable competitive advantage.

Type	Example
■ High quality	Toyota
■ Customer service	Starbucks
■ Low-cost production or operation	Wal-Mart
■ Product design and functionality	Hewlett-Packard
■ Market segmentation	Dell Computer
■ Product-line breadth	Amazon.com
■ Product innovation	Medtronic
■ Effective sales methods	Pfizer
■ Product selection	Tiffany's
■ Intellectual property	Microsoft

The most powerful new venture provides a great sense of value at a reasonable price resulting in a high ratio of value to price for the customer. With an added sense of emotion or importance, the potential success of a venture can be seen as the ratio:

$$\text{Potential success} = \frac{\text{Value} + \text{Emotion}}{\text{Price}}$$

Clearly, organizations such as MADD or Doctors Without Borders incorporate the powerful emotion of a cause or importance.

A business model is the result of a firm's decision about how a business should be structured. The securities brokerage industry, in the past, operated on a theory of high commission fees and personal service. In the 1990s, the model changed to low commission fees and reduced personal service. Few business models are unchallenged.

Core competencies built by a firm that have the potential generate value can be a source of competitive advantage in the marketplace. For airlines, one important capability is providing a memorably pleasant experience to passengers during flight. In the software services business, a dominant capability lies in the combination of high quality and low cost. Having capabilities that are distinctive and difficult for others to imitate can give your firm sustainable competitive advantage.

Ten types of sustainable competitive advantage are given in Table 3.12.

3.8 AGRAQUEST

Like the hungry microbes in its natural products, AgraQuest is eating its way into the $28 billion global pesticide market—a field dominated by chemical giants Dow, DuPont, and Monsanto. Steering the biotech start-up into the fray is company president and CEO Pam Marrone, an international expert in agricultural biotechnology and biopesticide science. Marrone has led AgraQuest in its

TABLE 3.13 AgraQuest's vision statement.

Agriculture badly needs safer, biodegradable pesticides that fit well in pest management systems in order to create an environmentally sustainable agricultural system. The goal is to reduce the use of synthetic chemicals on the nation's farms and ranches. AgraQuest develops its own natural product pesticides that meet its criteria for in-house development and aggressively licenses or acquires natural products from outside the company to reduce the time line until market entry. AgraQuest discovers new pesticidal natural products from microorganisms and sells these natural compounds to agrochemical companies for non-core markets. AgraQuest plans to build a natural pesticide and herbicide business that will make a difference in world agricultural practices and environmental impacts. We will be the premier source of pest management knowledge and technology, and be accountable to our customers, our shareholders, our families, our community and ourselves.

vision to research and develop safe and environmentally friendly alternatives for farm, home, and public health pest management.

Natural products are nonliving substances such as proteins and enzymes that are produced by organisms such as microbes and plants. Everyday, scientists at AgraQuest make their rounds, carefully checking and rechecking various biological experiments and meticulously recording their findings. They check out new samples of soils, plant roots, or lichen arriving from across the globe, hoping that one will lead to the next breakthrough natural, environmentally safe pesticide or fungicide.

The vision statement for AgraQuest is given in Table 3.13. The vision of this company is to make a difference in the worldwide agricultural industry by providing pesticides and herbicides that do not cause environmental problems.

The mission statement, provided in Table 3.14, is clearly stated and motivational in character. The vision and mission statements are useful to provide information about the company to employees, investors, and other stakeholders.

The value proposition of AgraQuest must clearly state the key values for the firm, while also identifying those values that will match the competition. The five core values for the firm are provided in Table 3.15. The dominant value is

TABLE 3.14 AgraQuest's mission statement.

AgraQuest discovers, develops, manufactures, and markets effective, safe, and environmentally friendly natural products for farm, home, and public health pest management.

TABLE 3.15 Five values for AgraQuest's products.

Dominant value: Product—The efficacy of the product is equivalent to that of chemicals, but it also can be used right up to harvest time. Furthermore, natural products are less susceptible to pest resistance buildup. It also is safer, reliable, and easy to use.

Differentiating value: Experience—A "green," natural product that is environmentally friendly and can lead to more sustainable agriculture and healthy conditions worldwide

Norm value: Price

Norm value: Service

Norm value: Access

TABLE 3.16 AgraQuest's value proposition.

AgraQuest discovers, develops, manufactures, and markets effective, safe, and environmentally friendly natural products for pest management that serve worldwide agriculture and make it more environmentally sustainable.

TABLE 3.17 Business model for AgraQuest.

■ Customer selection	Farmers with higher-value products who want a safer "green," natural pesticide solution
■ Value proposition	"Green" products for herbicide and pesticide use at comparable efficacy and price
■ Differentiation	Natural products that can be used right up to harvest time
■ Scope of product	A moderate range of natural products
■ Organizational design	A flat organization with good communication
■ Value capture	Reasonable costs and growing revenues allowing for net positive income
■ Value for talent: Scientists and staff	Opportunity to work in an organization with a "green" mission

to ensure that the efficacy of the product is as good as that for chemicals while providing safer products that can be used right up to harvest time. The differentiating value is that of a "green," natural product than can lead to a sustainable agricultural system. AgraQuest matches its competitors on price, service, and access. The value proposition is provided in Table 3.16.

The unique selling proposition for AgraQuest is:

Innovative natural product solutions for pest management

The business model of AgraQuest is given in Table 3.17. Note that AgraQuest is well positioned to grow in the global pesticide market. Its challenge is to exploit its differentiation as a natural, "green," and safe product in a somewhat skeptical agricultural industry.

3.9 SUMMARY

The theory of a business is a description of the elements required for the entrepreneur to act to build a business that satisfies the customers' needs. Coupled with the firm's core competencies and resources, the firm uses the elements of its business design to build a sustainable competitive advantage. The elements of a firm's theory of its business include: vision, mission, value proposition, business model, competitive advantage, and how it acts to retain a sustainable competitive advantage.

■ Great vision is a statement of purpose (or story) in response to an opportunity.

- The mission describes the firm's goals, products, and customers, providing a theory of change for all to see.
- The value proposition describes the customers' needs that will be satisfied.
- The business model describes the elements economics, and activities of the new business.
- The firm strives to create a competitive advantage and make it sustainable.

Principle 3

The vision, mission, and value proposition embodied within the business model of a firm and powered by a sustainable competitive advantage can lead to compelling results.

3.10 EXERCISES

3.1   HeadBlade Company makes a patented razor fashioned to use a man's hand as the handle and readily shave his head. A mission statement for HeadBlade might be: "HeadBlade provides a safe, effective way for men to shave their heads." HeadBlade is delivered direct from www.headblade.com. What are the dominant and differentiating values for this product?

3.2   Frequent-flyer businesspeople are always looking for carry-on luggage geared for their needs. SkyRoll is a new kind of carry-on luggage that fits easily into overhead bins of airplanes and helps minimize wrinkles to clothing.

   Don Chernoff, the founder and designer of SkyRoll, worked as an engineer at Intel and developed SkyRoll after traveling constantly for work. He had the idea that if clothes could be rolled instead of folded, they would wrinkle less and be easier to carry on the plane.

   Key elements of the luggage include a detachable garment bag that has space for suits and clothing and a hollow cylinder into which can fit items such as shoes and toiletries. SkyRoll can hold enough clothing for one to three days. Retailing at about $150, SkyRoll can be found at retailers around the United States and Canada (see www.skyroll.com).

   Is this a good growth business?

   What is the business model of SkyRoll?

3.3   Every house needs some improvement—that is the essence of Home Depot's business concept. Home Depot was based on the radical idea of stocking merchandise directly from the manufacturer. Most other hardware stores were stocked by one distributor or cooperative that handled a small range of products. Home Depot offered as wide a selection of products as possible selling at a lower cost than their competitors'. Consider Home Depot today in terms of the pyramid of value creation shown in Figure 3.6.

   Describe the elements of the pyramid for Home Depot.

3.4   Online matchmaking is a serious business alternative to people finding mates in singles' bars and personal ads. Match.com is in the online matchmaking business to help single adults find compatible relationships. Yahoo Personals and American Singles are competing for the same market [Orenstein, 2003]. Describe the business model for a reliable, safe online dating and matching service. Can this firm create a sustainable competitive advantage?

3.5   State the vision and value proposition for Yahoo. What is the business model for Yahoo?

3.6   Using the elements of Table 3.1, describe the vision of First Data Corporation (www.firstdatacorp.com). Describe the competitive advantage for FDC. Is this competitive advantage sustainable?

3.7   Secure wireless fleet management and logistics are in demand today. Other applications for wireless services are in health care, retail, manufacturing, and utilities. Select an application and prepare a value proposition for that application.

3.8   Hot Topic is a mall-based chain of retail stores specializing in apparel, accessories, gifts, and music for teenagers. Describe the business model of Hot Topic in the format of Table 3.8.

3.9   Purchasing a used car is one of the least desirable experiences for most people. EBay Motors offers fraud protection, a warranty, and a title history (www.ebaymotors.com). What is the value proposition for eBay Motors? Would you buy a car using eBay?

3.10   E-Trade has a value proposition based on the concept of an online banking and securities firm. It competes on price as the primary value attribute. E-Trade offers stock brokering, refinancing, home equity loans, and credit cards for people with modest assets. Describe the strengths and weaknesses of this value proposition (see www.etrade.com).

3.11   Major league baseball is trying a new venture selling abridged game recordings that provide a game in 20 minutes or less. Visit www.mlb.com to see the results. What is the value proposition for this business?

3.12   Think of a box with a clock, alarm, radio, and CD player all controlled by a keypad like that on a telephone. The clock and function display would be digital. All functions would be intuitively and easily controlled. Write a value proposition and business model for a firm called Easy Electronics [Nalebuff, 2003].

CHAPTER 4

Creating a Strategy

*Praise competitors. Learn from them. There are times when you can cooperate
with them to their advantage and to yours.*
George Mathew Adams

Every new venture has a strategy or approach to achieve its goals. This strategy is in response to its plan to implement a solution to an important problem or opportunity. The process for creating a strategy for a new firm is shown in Table 4.1. Steps 1 and 2 were described in Chapter 3. With sound vision and mission statements and an initial business model, the entrepreneur examines the political and economic context of the industry, along with its growth rate and typical profit margins (step 3). Once the industry is understood, steps 4 and 5 are used to describe the firm's strengths and weaknesses and its opportunities and threats (SWOT). In step 6, the entrepreneur integrates his or her knowledge of the industry and competitors with his or her own SWOT to identify key success factors. Based on the information gathered in the preceding steps, the entrepreneur refines his or her vision, mission, and business model and creates a strategy to achieve a sustainable competitive advantage. ∎

TABLE 4.1 Management process for developing a strategy.

1. Develop the vision and mission statements, and the business model.

2. Describe the firm's core competencies, its customers, and its competitive advantage.

3. Describe the industry and context for the firm and its competitors.

4. Determine the firm's strengths and weaknesses in the context of the industry and environment.

5. Describe the opportunities and threats for the venture.

6. Identify the key factors for success using the six forces model.

7. Formulate strategic options and select the appropriate strategy.

8. Translate the strategy into action plans with suitable measures and controls.

4.1 VENTURE STRATEGY

A **strategy** is a plan or road map of the actions that a firm or organization will take to achieve its mission and goals, but it is not static. Imagine the difficulties of navigating through most towns with a map from 1900. In other words, a strategy is a firm's theory about how to compete successfully within the current realities of its industry. To be useful, the plan must be action-oriented and based on the firm's opportunities, strengths, and competencies. For example, the most efficient route for a cyclist to move from point A to point B may be different for a motorist. A corporate or organizational strategy is an integrated plan for the whole organization [Hill, 2001]: a firm's way of doing things, or a theory of business [Drucker, 1995]. The desired outcome of a strategy is a sustainable competitive performance. Because of the dynamic nature of the competitive business world, a strategy has to be simple and clear. This allows everyone to work on a commonly understood strategy.

Strategies help to set a firm on a course and focus effort on that course. Often, a strategy emerges as actions are taken and tested, eventually converging toward a pattern [Mintzberg, 1998]. With a strategy, the firm can differentiate its offerings and activities. For some, the essence of strategy is choosing what not to do [Magretta, 2002]. The process for developing a strategy is summarized in Table 4.1.

A strategy is a response to opportunity. The word *opportunity* is derived from the Latin expression "toward the port." The builder of value is like a merchant sea captain who secures the right payloads from the best customers, manages his crew, and adjusts his mix of established ports and new ports with high potential [McGrath, 2001]. The formulation of a sound strategy is based on deep knowledge of the opportunity, the industry, and its context. In describing the opportunity as a vision, a sense of drama and vitality emerges. With this vitality, the entrepreneur motivates the team and the investors to share the vision, embrace the strategy, and act on it. In this case, the strategy emerges as the details unfold.

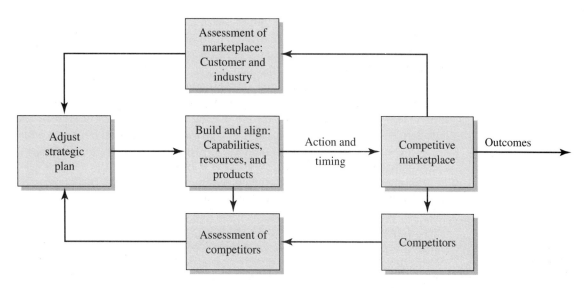

FIGURE 4.1 Framework for a firm operating in a dynamic marketplace.

Long-term planning is very difficult due to the dynamic nature of the competitive marketplace. Industries are not in equilibrium, and industry analysis is difficult. It is hard to define where an industry begins and ends. Also, it is difficult to distinguish competitors from collaborators from suppliers. Thus, all strategies are subject to change and reemergence as conditions, alliances, and competition change.

Entrepreneurs start in the center of Figure 4.1 by building and aligning their capabilities, resources, and products. They then act on their initial strategy or business plan. Entry into the competitive marketplace will force a reassessment of the marketplace and industry as well as their competitor analysis. This leads strategic managers to redeploy and adjust the capabilities, resources, products, and actions to effectively compete in the dynamic market. These managers strive to attain a competitive advantage by securing and managing the assets of the firm. How the internal management responds to a changing customer, industry, and competition is crucial in the reestablishment of the strategic plan and the firm's assets to act competitively. Venture leaders strive to identify the fundamental forces for creating and capturing customer value. Those who focus on continuously adjusting and aligning a firm's strategy and capabilities will constantly evolve from one strategic position to the next strategic position in response to changing conditions.

GE Aircraft Engines (GEAE) provides an example of an adjustment in a strategic plan as a result of changes in the market. GEAE had a product strategy to develop engines with more power, efficiency, and better reliability. Because of relentless competition and shorter product cycles, sustaining profitability was difficult. GEAE shifted to operating as an engine production and services provider, generating significant profits in the after-market services business [Demos, 2002]. Faced with a dynamic marketplace, the strategic leader develops a strategic response and adapts to the changes in the market.

To summarize Figure 4.1, the first step is to determine the basic driving forces in the industry: the economic, demographic, technological, or competitive factors that either constitute threats or create opportunities. The second step is to formulate a strategy that addresses the driving forces identified in step 1. The third step is to create a plan to implement the new strategy. Finally, the new strategy is implemented by building and realigning the firm's capabilities, resources, and products.

Entrepreneurs define their strategy within their perception of opportunity. They are not constrained by the present resources or capabilities but seek to acquire the necessary resources and capabilities. The theory of resource dependence states that a company's freedom of action is limited to satisfying the needs of customers and investors that give it the resources to survive [Christensen, 1999]. Investors and customers dictate how money will be spent because companies that do not satisfy them will be unable to survive.

A good strategy answers the questions asked by Kipling [1902]:

I keep six honest serving-men (They taught me all I knew);

Their names are What and Why and When and How and Where and Who.

The six questions for creating a sound, dynamic strategy are summarized in Figure 4.2. With solid, effective answers to these six questions, a firm will have formed a strategy that has the potential to lead to profitability.

A strategy can be viewed as a plan that integrates a firm's goals and actions into a cohesive whole that draws effectively on its resources and capabilities. The essence of strategy is choosing the priorities and deciding what to do and what not to do. The strategic priorities determine how a business is positioned

Profitability		
Why are we pursuing this objective? • The vision • The mission	Where will we be active? • The customer • The market	How will we achieve our objective? • Innovation • Acquisitions
When will we act and at what speed? • Timing • Execution	What will differentiate our product? • Positioning	With whom will we compete and cooperate? • Competition • Alliances

FIGURE 4.2 The six questions for creating a dynamic strategy. Profitability rests on six solid answers to these questions.

relative to the alternatives. As the competitive conditions change, the new venture adjusts its strategy to meet the new conditions.

4.2 THE INDUSTRY AND CONTEXT FOR A FIRM

The eight steps for developing a strategic plan are outlined in Table 4.1. In the remaining sections of the chapter, we will discuss steps 3 through 8 since steps 1 and 2 were described in Chapter 3. In this section, we address step 3 of Table 4.1. Also, multiple methods exist for understanding the activities of Figure 4.1. We will highlight some of them in this and later sections.

A full description of the customer and the industry will help the entrepreneur build a sound strategic plan. The main elements of an industry analysis are given in Table 4.2. The first step is to accurately name and describe the industry in which the firm is or will be operating. The definition should be narrow and focused. An **industry** is a group of firms producing products that are close substitutes for each other and serve the same customers. Thus, selecting the telecommunications industry may be too broad. The definition of the industry should be more focused, such as "the Internet service provider industry serving homes and businesses in Ohio and Indiana." If data is not available for the targeted area of the market, the closest proxy should be used. For example, if statistics are not available for Ohio and Indiana, they may be available for the Midwest or the United States. Then, define this market and describe the customer. The second step is to describe the regulatory and legal issues within the industry. Both national as well as state and local regulations should be considered.

The third step of Table 4.2 suggests describing the growth rate and state of evolution of the industry. Most industries tend to emerge through an initial period of slow growth with limited sales and few competitors. Then, they expand through a period of rapid growth as sales take off and many firms enter the industry. This is followed by a third period of maturation marked by slower growth and stability. Eventually, the number of firms in the industry declines [Low and Abrahamson, 1997]. We depict in Table 4.3 these four stages as 1) emergence, 2) growth, 3) maturation, and 4) decline. It is important to know where your industry is in the evolution cycle. In the emerging phase, significant product and market uncertainty exists. Producers are unsure of what features are required for the product. Customers may be unsure of the elements of the product they need. Many technology ventures begin in the emerging phase of an industry.

TABLE 4.2 Five elements of an industry analysis.

1. Name and describe the industry.

2. Describe the regulatory, political, and legal issues in this industry.

3. Describe the growth rate of the industry and the state of the evolution of the industry.

4. Describe the profit potential and the typical return on capital in the industry.

5. Describe the competitors in the industry and the rivalry among them.

TABLE 4.3 Four stages of an industry life cycle.

Stage	Examples
1. Emergence	Artificial organs
	Nanotechnology
	Genomics
2. Growth	Medical technology
	Software
	Electronics
3. Maturation	Electric appliances
	Automobiles
	Movie theaters
4. Decline	Steel
	Roller-skating rinks
	Bowling alleys

The growth stage emerges when the necessary features and performance become clear and a dominant design emerges. A **dominant design** is one whose major components and core concepts do not substantially vary from one product offering to another. With the emergence of a dominant design, the number of competitors stabilizes.

Eventually, an industry enters its mature phase as the number of competitors stabilizes and profit margins slowly decline as price becomes the primary competitive weapon. Finally, an industry enters a declining phase as the number of firms decline and profit margins erode. These four phases are described in Table 4.3.

The personal computer market began in 1978, with a number of small, emerging firms such as Apple Computer. IBM entered the personal computer market in 1982, and its PC quickly emerged as the dominant design. Many other firms entered after IBM made the design open to all, and the PC industry experienced a growth phase between 1984 and 1998. Eventually, the market reached a period of maturity, with only a few dominant firms having standardized or slightly differentiated products and relatively stable sales and market shares.

Table 4.2 shows that the next step in the industry analysis is a statement of the profit potential and the typical return on investment capital in the industry. The **return on capital** is defined as the ratio of profit to the total invested capital of a firm. The average return on capital in the computer software industry is about 16 percent, while the return on capital in the steel industry is about 6 percent. The steel industry is less attractive, while the computer software industry is attractive. One of the most effective ways to identify realistic profit opportunities for a new venture is to look at the Securities Exchange Commission filings of a young representative firm in the industry (www.sec.gov).

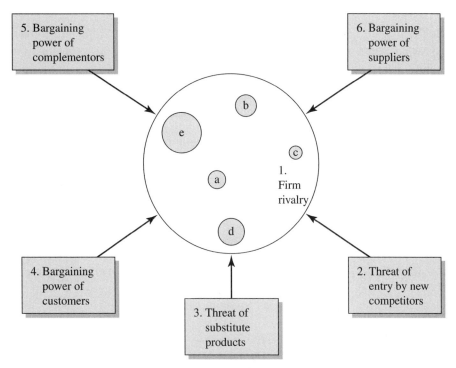

Note: Firms are represented by a circle; for example, (a) represents firm a. The size of the circle indicates the size of revenues of the firm. The six forces are numbered for clarity. The rivalry of the firms is shown as a vortex of competition.

FIGURE 4.3 Six forces model.

The **six forces model,** shown in Figure 4.3, is one popular method for evaluating the competitive forces in an industry. The six forces are: 1) firm rivalry, 2) threat of entry by new competitors, 3) threat of substitute products, 4) bargaining power of customers, 5) bargaining power of complementors, and 6) bargaining power of suppliers. This framework is an extension of the five forces model [Porter, 1998]. The six forces model enables the analyst to consider all the issues facing a new entrant by describing the key industry factors. The rivalry among the industry competitors may be intense or modest. In some industries, the bargaining power of the customer may be modest.

Consider the automobile industry, which has about 10 competitors. The rivalry is extremely intense. The bargaining power of customers regarding a new vehicle is very high since they have access to broad information on the relative performance and price of the products of the competitive companies and their dealers. The bargaining power of the suppliers in the industry is modest. Furthermore, the threat of a substitute product is small. The threat of new entrants is very small, due to the costs of developing a new product and dealer network. Thus, the auto industry experiences intense competition with the buyer wielding significant power.

Consider the online bookselling industry: Amazon.com and Barnesand-Noble.com are the two large, national online booksellers, but there are many regional competitors such as Powells.com. The rivalry among these competitors is high. Their suppliers have low bargaining power, and the barriers to entry are moderate. The bargaining power of the customer is large, resulting in low prices, and profitability is modest. The threat of substitute products is low, but e-books could undermine the printed book industry if a suitable e-book reader emerged.

By contrast, many new firms enter the computer software industry each year. The bargaining power of customers is moderate, and the threat of substitute products is low. As a result, profitability in the industry is high. However, the rivalry of the firms is intense.

A competitive analysis explains how you will do better than your rivals. And doing better, by definition, means being different. Organizations achieve superior performance when they are unique, when they do something no other business does in ways that no other business can duplicate. In military competition, strategy refers to the large-scale plan for how the generals intend to fight and win a war. The word *tactics,* in contrast, refers to small-scale operations, such as the conduct of a single battle [Clemons and Santamaria, 2002]. Very few strategic plans survive the first contact with competitors. Competitors respond and change the situation.

Complementors are companies that sell complements to the enterprise's own product offerings. A **complement** is a product that improves or perfects another product. For example, the complementors to the Sony PlayStation are the companies that produce the video games that run on the PlayStation. Without an adequate supply of complementary products, demand for the player product would be modest. The complementary product to the automobile is the interstate road system that enables automobiles to safely and rapidly travel long distances. Without suitable, widely located electric recharge stations, the future of electric vehicles is very limited.

The entrepreneurial firm is likely to be a new entrant to the industry. Thus, the new venture should describe the barriers to entry, the threat of substitutes, and the bargaining power of the suppliers, customers, and complementors. One of the main factors that drives traditional analyses of the determinants of market structure involves comparing the size that a firm must be to compete efficiently to the overall size of the market in which it competes. If the industry has few firms, a new firm may be able to readily enter and gain market share. Using the six forces model, a new technology venture is likely to perform better when it operates in an industry with high barriers to entry, low rivalry, low threat of substitutes, low buyer power, low supplier power, and low bargaining power of complementors.

In Figure 4.3, examine the bargaining power of the suppliers. When the supplier industry is composed of many small companies and the buyers are few and large, the buyers tend to dominate the supply companies. An example is the automotive component supply industry in which the buyers are few and large and dominate the many small suppliers.

To complete the industry analysis, it will be necessary to name the competitors and describe the profitability of the industry. One method is to use *Stan-*

dard and Poors Reports or the *Value Line Investment Survey.* For example, if the new firm is entering the biomedical devices industry, the leading competitors are Medtronics, Stryker, and Guidant. Using Value Line, we note that the average return on total invested capital for these companies is 15 percent. Value Line projects a 13 percent future growth rate of sales for this industry. With these attractive measures, the industry appears to be very attractive to new entrants with well-differentiated, fairly priced products.

4.3 STRENGTHS AND OPPORTUNITIES— SWOT ANALYSIS

Steps 4 and 5 of the management process for developing a strategic plan (Table 4.1) suggest that a strategy is based on the firm's strengths and opportunities, while avoiding or mitigating its weaknesses and managing threats. As discussed in Chapters 2 and 3, a new firm is focused on securing the capabilities and resources necessary to succeed in its industry. Furthermore, the new firm concentrates on an attractive opportunity that was selected using Table 2.9. Thus, a strategy addresses the four aspects of the setting in which a firm operates: 1) a firm's strengths, 2) its weakness, 3) the opportunities, and 4) the threats in its competitive environment. This analysis is often called a SWOT analysis, which allows a firm to match its strengths and weaknesses with opportunities and threats and find the purpose for which it is best suited.

A firm's strengths are its resources and capabilities. Its weaknesses are its limitations of organization or lack of capabilities or resources. A firm's opportunities are its chances for success in a new entry or product in its industry. The threats are actions or events outside its control in the competitive environment.

A basic SWOT analysis for Southwest Airlines is given in Table 4.4. The SWOT analysis provides the questions for a strategic response and helps a firm exploit its strengths, avoid or fix its weaknesses, seize its good opportunities, and mitigate its threats. Examples of threats are market shifts, regulatory changes, and delays in product development. Positive opportunities include increasing demand, repeated use, and willingness to pay.

TABLE 4.4 SWOT analysis for Southwest Airlines.

Organizational (internal)	Environmental (external)
1. Strengths:	1. Opportunities:
■ Highly productive pilots, and ground and flight crews	■ Ability to add scheduled flights to new cities
■ Low costs	■ Low prices enable market-share growth
2. Weaknesses:	2. Threats:
■ Inability to provide nonstop, long-distance travel	■ Inability to secure new gates at airports
	■ Competition from a low-cost rival such as JetBlue

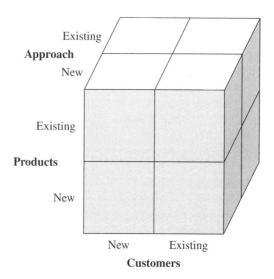

FIGURE 4.4 Three dimensions for examining opportunities.

We can examine opportunities in three dimensions, as shown in Figure 4.4. Perhaps the safest strategy is to take new products to existing customers via existing distribution channels using existing approaches. We can call the three dimensions: products, customers, and approach [Black and Gregersen, 2002]. Approach is the method or means of taking the product to the customer. The most risky strategy would be a new product taken to new customers via a new approach. Amazon.com started selling books (existing products) to book buyers (existing customers) via a new approach—online.

4.4 BARRIERS TO ENTRY

Barriers to entry are factors that make it costly for companies to enter an industry. The greater the costs that potential competitors must bear to enter an industry, the greater are the barriers to entry. The six potential barriers to entry are listed in Table 4.5. Economies of scale can be a barrier in industries where the costs of production are low for a narrow range of volume or occur only for

TABLE 4.5 Potential barriers to entry into an industry.

- Economies of scale
- Cost advantages independent of scale
- Product differentiation
- Contrived deterrence
- Government regulation
- Switching costs

higher volumes. An example is the aircraft design and production industry. It is difficult to enter that industry since a low volume of production of aircraft is most likely uneconomic for the new entrant [Barney, 2002].

Cost advantages independent of scale may be held by existing companies and will deter a new company from entering. For example, incumbent firms may have proprietary technology, know-how, favorable geographic locations, and learning-curve advantages. These can all be barriers to a new entrant.

Product differentiation means that incumbent firms possess brand identification and customer loyalty that serve as barriers to new entrants. For example, Dell, Hewlett-Packard, and Gateway have brand and customer loyalty, making it difficult for a new personal computer company to enter the industry on a large scale. Of course, this barrier may be less important to a specialty manufacturer that seeks a small niche in the personal computer market. A formidable barrier to entry is the reputation or brand equity of the incumbents. Providing ratings for bonds is an attractive industry since it isn't asset-intensive and the profit margins are very good. If a new firm tries to enter this market, it would have to compete with Moody's and Standard and Poors, both competitors with strong reputations.

Contrived deterrence as a barrier occurs when incumbent firms strive to throw up unnatural barriers at a cost to them. They can use lower prices, newer products, or brand building to send a signal to potential entrants that intense responses will result if they try to enter. For example, a potential entrant to television broadcasting is deterred by government allocation of regular broadcast channels. A response to this limitation is for the new entrant to choose another means such as cable as the distribution channel—for example, the Fox Channel.

Two kinds of economic markets exist: substitutable and nonsubstitutable. Substitutable products are commodities such as groceries, cola drinks, and gasoline. In a nonsubstitutable market such as semiconductor manufacturing equipment, the required associated infrastructure means that once purchasers choose a system, they are not inclined to switch due to high switching costs.

Switching costs are the costs to the customer to switch from the product of an incumbent company to the product of the new entrant. When these costs are high, customers can be locked into the product of the incumbents even if new entrants offer a better product. An example is the cost of switching from Microsoft to the Apple computer operating system. Users would need to purchase a new set of software to use on the Apple computer as well as train their employees to use the new software.

The home building industry is attractive to many entrepreneurs since it is relatively easy to enter. It is a growth industry, but localized and fragmented. Thousands of properties are sold in a region annually. Furthermore, most of the money to purchase a property can be borrowed. It offers an opportunity to an entrepreneur to add value where resources are scarce and knowledge and experience are important. No wonder so many entrepreneurs enter the home building business every year.

4.5 ACHIEVING A SUSTAINABLE COMPETITIVE ADVANTAGE

A distinctive or unique competency is a matchless strength that a firm can use to achieve superior operating conditions that lead to a strong competitive advantage. A SWOT analysis helps the entrepreneur identify this unique competency. The unique competency of a firm arises from its capabilities and resources, as shown in Figure 4.6. Resources are financial, human, physical, and organizational, and include patents, brand, know-how, plants and equipment, and financial capital. The capabilities of a firm include skills, methods, and process management. It is the usefulness of both capabilities and resources in a coordinated way that leads to distinctive competencies. A firm must have: 1) a valuable set of resources and the capability to exploit those resources, or 2) a unique capability to manage common resources. Intel possesses unique patent and know-how resources and the capabilities to exploit that knowledge and intellectual property. Southwest Airlines possesses common resources—aircraft and aircraft equipment—but has a unique capability to manage these resources. Disney has unique resources in its film library, brand, and theme parks but a mixed record of managing them well.

If a new technology venture possesses a particular valuable resource, then that firm can gain a competitive advantage and thus improve its efficiency and effectiveness in ways that competing firms cannot [Barney, 2001].

As shown in Figure 4.5, a firm uses its unique competencies to manage its innovation, efficiency, product quality, customer relations, and supplier relations to differentiate its product and manage its costs. A technology venture works to design and produce at a low cost the highest-quality product that possesses unique differentiating factors. Four common ways in which a firm will distinguish itself from its competitors are: differentiation, cost, combined differentiation and cost, and niche, as summarized in Table 4.6.

The goal of a differentiation strategy is to create a unique product based on a firm's unique competencies. The low-cost strategy is based on unique competen-

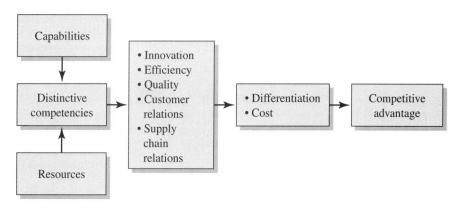

FIGURE 4.5 Distinctive competencies lead to a competitive advantage.

TABLE 4.6 Four common types of strategies and their characteristics.

Factor	Type of Strategy			
	Differentiation	**Low cost**	**Differentiation-cost**	**Niche**
Distinctive competencies	Innovation and relationships	Processes, logistics	Innovation and processes	Relationships
Product differentiation	High	Low	Medium	Medium
Market segmentation	Many segments	Mass market	Many segments	One or two segments
Examples	Intel	RadioShack	Dell	Getty Images
	Microsoft	Wal-Mart	Southwest Airlines	Incyte

cies that enable the efficient management of processes. Many firms can achieve a combined differentiation–low cost strategy that blends the best of low cost and differentiation. The niche strategy is directed toward one or two smaller segments of a larger market. This niche can be geographic or a product or price segment.

Since the founding of Intel, its strategy was focused on technology leadership, first-mover advantage, and the dominance of important new markets. Intel emerged as the dominant supplier of microprocessors, which are used in 90 percent of personal computers. Intel is also a leading manufacturer of flash memory, embedded control chips, and communication chips. A unique competency is Intel's ability to build, manage, and exploit the world's best semiconductor manufacturing facilities. As an example of its technology leadership strategy, Intel announced a new material that will replace silicon, enabling Intel to build more density (transistors per area) while reducing heating and current leakage. Intel has a highly successful differentiation strategy.

Niche ventures often require less capital and achieve financial success rather quickly. Typically, a niche business is too small for the mass-market supplier, and thus, competition is low. A niche can be geographic or a product or price segment. Niche businesses typically are started in one market segment and based on a focused core competency and good customer and supplier relationships.

Getty Images sells and licenses still and moving images and related services that are distributed digitally to its customers (www.gettyimages.com). This firm is a good example of satisfying a niche market while achieving a good profit margin.

Southwest Airlines is an example of an airline that started as a niche, low-cost business operating only in Texas. It served three cities—Dallas, Houston, and El Paso—and operated using standardized Boeing 737 aircraft. It used highly productive crews, frequent, reliable departures, and a no-frills (low-cost), short-haul, point-to-point system. Eventually, Southwest moved to other western states and many locations across the nation. Thus, its strategy evolved from a niche strategy to a differentiation-cost strategy.

Fastenal Company of Winona, Minnesota, is the largest distributor of nuts and bolts in the United States. It uses a low-cost strategy for its manufacturing and distribution business with about 1,000 warehouse stores achieving total sales exceeding $800 million. Each store has at least one delivery truck. Customers talk to the local store and receive personal service. Fastenal sees itself as an inventory and delivery manager offering excellent customer service. It currently has a fleet of 1,200 pickup trucks that respond to customer orders on an expedited basis (www.fastenal.com).

A differentiation strategy is commonly based on an innovation or capability that others do not possess. For example, Tom Siebel, founder of Siebel Systems, addressed the problems of sales, marketing, and customer service, and built a proprietary software architecture to enable firms to apply information and communication technology to the problems.

Paychex is an example of a company with a differentiation-cost strategy. It provides payroll-processing services and began by targeting small- and medium-sized businesses that needed this service. The company offers customer service and payroll accuracy at a reasonable price, leading to wide acceptance. Once it has a satisfied customer, the switching costs for this customer are sizable. Paychex's revenues have grown to over $1 billion, and it serves more than 350,000 businesses. Paychex's annual revenue growth rate has been greater than 18 percent for over 20 years.

IKEA provides furniture to customers who are young, not wealthy, likely to have children, and work for a living. These customers are willing to forgo service to obtain low-cost furniture. IKEA designs its own low-cost, modular, and ready-to-assemble furniture. In large stores, it displays a wide range of products. While IKEA is a low-cost provider, it also offers several differentiated factors, such as extended hours and in-store childcare. Its strategy is a differentiation-cost strategy.

4.6 MATCHING TACTICS TO MARKETS

A company can be said to be successful if it outperforms its competitors over time. Another view of how to formulate the best strategy for a venture is to match the firm's approach to the pace of the market. Table 4.7 summarizes three competitive approaches [Eisenhardt and Sull, 2001]. The first approach is based on establishing a *position* in an industry and defending it. The goal is to position the company so that its capabilities provide the best defense against the competitive forces of Figure 4.3 [Porter, 1998]. Furthermore, the positioning approach can be defended by anticipating shifts in the six forces of Figure 4.3 and responding to them.

The second method focuses on *resources,* such as patents and brand, and attempts to leverage those resources against the resources of the competitors. For example, the powerful brand of Southwest Airlines has enabled the firm to issue its own Visa card to many of its customers.

TABLE 4.7 Three types of competitive tactics.

	Position	Resources	Emergent
Approach	Establish a position and defend it	Leverage resources such as brands, patents, or assets	Pursue emerging opportunities
Firm's basic question	Where should we be?	What should we be?	How should we take our next step?
Basic steps	Identify an attractive market Locate a defensible position, and fortify and defend it	Acquire unique, valuable resources	Choose one or two core strategic processes and use them to guide to the next step
Works best in	Slowly changing, well-understood markets	Moderately changing, well-understood market	Rapidly changing, uncharted markets
Duration of competitive advantage	Relatively long (3–6 years)	Relatively long (3–6 years)	Short period (1–3 years)
Risk or difficulty	Difficult to change position	Difficult to build new resources, if needed	Difficult to choose best opportunities
Performance goal	Profitability	Long-term dominance	Growth and profitability

The third approach may be called *emergent* and is based on flexible and simple rules [Eisenhardt and Sull, 2001]. Firms using this method to develop a strategy select a few significant strategic processes and build simple rules to guide them through the ever-changing marketplace. The strategic processes could be innovation, alliances, or customer relationships. Dell, for example, has chosen its customer relationships and customized products as its basic strategy. It then adjusts this strategy as conditions require.

Cisco Systems used an innovation strategy to guide it through emergent opportunities for its first years of operation. Later, it changed to a basic strategy of acquisitions to respond to rapidly changing markets. These basic tenets for guidance in emerging markets may be called simple rules and are summarized in Table 4.8 [Eisenhardt and Sull, 2001]. These rules allow a firm to compete in a fast-moving marketplace such as the emergent markets that many technology ventures start in.

TABLE 4.8 Simple rules for emergent markets.

Rules	Purpose	Example
Boundary	State which opportunities can be pursued	Cisco acquisition rule: No more than 75 employees in an acquired company
Priority	Rank the possible opportunities	Expected return on investment
Timing	Synchronize the selection of opportunities and the conditions of the firm	When product must be delivered
Exit	Know when to pull out of opportunities	Key team member leaves

A good way to understand strategic planning in emerging industries is to imagine an American football team trailing by a touchdown with only two minutes left to play, and it has the ball. The team refuses to panic. It has well-established rules of play for this situation. It switches to the "no-huddle" offense, with the quarterback calling the plays at the line of scrimmage as he surveys the defense.

Uncertainty is endemic in strategy formulation. Thus, the quality of a strategy cannot be fully assessed until it is tried. Strategy making can be thought of as an organizational capability, where different approaches are generated and considered, and where past successful approaches are just options for the future among many.

Sam Walton started with a strategy based on low-cost retail discount stores. He gained differentiation by locating many of these stores in relatively rural cities that were only large enough to support one large discount retail operation. His second differentiating factor was his organizational culture, which inspired his employees. As competition emerged, he developed one of the most cost-efficient distribution networks based on information technology systems. Walton's simple rules of strategy and operation were part of Wal-Mart's success.

In addition to matching the approach to the market pace, Table 4.9 summarizes two key factors for determining a successful strategy: 1) specific industry-related competence, and 2) the existing level of competitive rivalry in the industry [Shepard, Ettenson and Crouch, 2000]. The venture capitalists who participated in Shepard's study stated in summary: The most attractive strategy is led by a team that has strong competence in an industry that has not yet built up intense rivalries. The timing of entry may be favorable in these circumstances.

The success of a new venture arises, in part, from a fit between the distinctive competencies of the venture team and the major success factor requirements of the industry. The better the fit, the greater is the competitive advantage. A competitive advantage is sustainable if the competencies of the venture quickly track and match the changing requirements of the industry. In addition, it is important to build alliances with critical stakeholders such as suppliers or distributors, thus erecting barriers to new entrants.

Every business strategy is unique since it is a unique mix of resources, context, goals, competencies, and organizational values. The potential for differen-

TABLE 4.9 Factors for determining a successful strategy, in priority order.

1. Industry-related competencies: Distinctive competencies

2. Competitive rivalry: Low rivalry in the industry

3. Time of entry: Enters industry early at appropriate time

4. Educational capability: Able to obtain the skills, knowledge, and resources required to overcome market ignorance

5. Lead time: Significant time between the pioneer's entry and the appearance of the first follower

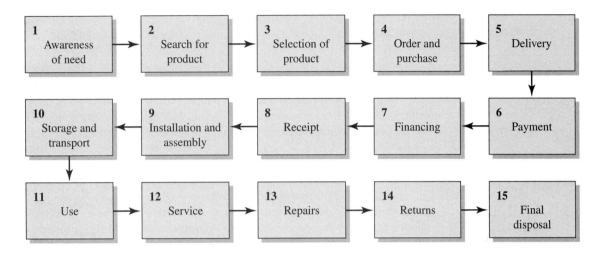

FIGURE 4.6 Consumption sequence.

tiation of a firm's strategy can also occur along a selected part of the consumption sequence shown in Figure 4.6 [McGrath, 2001]. Unique methods, tools, or arrangements can be used at each step in the sequence. Every new technology venture should look at the consumption sequence and decide where it can differentiate its product or service.

The power of Dell Computer is its direct sales model offered to three different customer segments. The Dell direct sales model incorporates all 15 steps of the consumption sequence. On the other hand, CDW Computer Centers (www.cdw.com) acts as a middleman reseller for Hewlett-Packard and offers excellent customer service for the purchaser who needs help in choosing a computer. Its 1,320-person sales force helps customers choose a total system that fits them, and a single salesperson is assigned to each customer for follow-up and later purchases. The CDW sales model incorporates steps 3 through 10 of the consumption sequence.

4.7 AGRAQUEST

AgraQuest has a business model, as given in Table 3.17. The basis of AgraQuest's strategy is differentiation of its product. Its natural products have no environmental impacts, they can be used right up to harvest, and pests do not build up resistance to them as they do to chemicals. Thus, the differentiation is the efficacy of the product compared to chemical pesticides and herbicides.

The industry that AgraQuest participates in is the agricultural pesticide and herbicide industry. The goal of using pesticides is to increase the yield per acre of the crop. The industry is heavily regulated by the U.S. Environmental Protection Agency (EPA), as well as state agencies such as the California EPA. The global pesticide market (2001) is about $28 billion. The largest portion (26 percent) of the market consists of fruits, nuts, and vegetables. AgraQuest's target markets are grapes, tomatoes, peppers, bananas, lettuce, apples, cherries, and home gardens.

AgraQuest's biologically based products fight plant pests and diseases with as much success as synthetic chemical pesticides and compete favorably on cost, pest resistance, shelf life, ease of use, food and worker safety, and environmental impact.

The first useful microorganism was found by one of the AgraQuest scientists in a handful of dirt from a Fresno farmyard. Lab tests showed that the bacterium had an appetite for the fungus that causes bunch rot and mildew in grapes. At that point, AgraQuest went to work on finding a formula to grow the beneficial bug in industrial quantities.

Although the exact formula is proprietary and secret, this is how AgraQuest goes about creating it. A flask of bacteria is dumped into a 10,000-gallon tank filled with a special food source. Forty-eight hours later, the gooey slime in the tank is harvested. To create a usable product, the bacterial concentrate is dried so that it becomes something resembling powdered milk. The powder is put into 24-pound bags and shipped to the farmer, who dumps it into a spray tank, mixes it with water, and applies it just like a chemical fertilizer.

AgraQuest's competitors include many firms worldwide such as Valent Biosciences, Chicago; Dow Agrasciences, Indianapolis; BASF, Germany; Syngenta, Switzerland; and Bayer, Germany.

The barriers to entry are significant since an entrant must have the technical capabilities as well as recognition and reputation in the natural pesticide industry. The biggest strength of AgraQuest is its scientific capability to identify, develop, and manufacture microorganisms for agricultural pesticide control. This capability is the strength of AgraQuest and critical to the firm's success in the industry. The weakness of AgraQuest is its limited ability to build a large product line of products for various crops in a timely way.

The differentiated product strategy based on a strong scientific capability is sound, but its weakness is in the delay of creating new products. The economics of the natural pesticide market requires a product line in place that creates a positive cash flow.

4.8 SUMMARY

The strategy of a new business venture is its plan to act to achieve its goals. Given the challenge of an important problem (opportunity), the strategy provides a road map for the new firm to act to achieve a profitable solution to the problem. The strategy is designed to solve the problem by creating a unique and sustainable way of acting that, it is hoped, will lead to a profitable and valuable outcome for the customer and the firm. A solid strategy is based on:

■ Sound knowledge of the industry and the context for the venture.
■ A deep understanding of the firm's strengths and weaknesses as well as its opportunities and threats.

- A solid competitor analysis and review of the six forces encountered by firms in a rival market.
- A strategic design that can lead to a sustainable competitive advantage.
- A choice of a differentiation, low cost, differentiation and low cost, or niche strategy that provides unique value to the customer.

Principle 4
A clear road map (strategy) for a new venture states how it will act to achieve its goals and attain a sustainable competitive advantage.

4.9 EXERCISES

4.1 The development of book superstores and Web-based booksellers such as Amazon.com has put intense pressure on small, independent booksellers, threatening to drive them out of existence. The movie *You've Got Mail* illustrates the difficulties of the independent stores. Well-planned, well-executed strategies, however, can give these independent bookstores the ability to retain a base of loyal customers and increase sales and profits. For instance, chain stores often have a more difficult time adapting their inventories to suit local tastes and culture, something that independent bookstores have always done. Because they are locally owned and operated, small bookstores are better at stocking books that are unique to a particular community or region. Furthermore, a bookstore can differentiate itself by offering better customer service.

Two examples of bookstores with strong differentiators and a loyal customer base are Book Passage in Corte Madera, California (www.bookpassage.com), and Powells' Books in Portland, Oregon (www.powells.com).

What are the differentiators offered by these two stores? What strategy would you recommend to your local independent bookstore owners?

4.2 The beauty industry is a $160 billion arena encompassing makeup, skin and hair care, fragrances, cosmetic surgery, health clubs, and diet pills [*Economist,* 2003b]. Select one of these subindustries and describe the competitive situation using the six forces model of Figure 4.3.

4.3 Consider the specialty chemicals industry. Select Airgas or Ecolab, and complete a SWOT analysis for the firm.

4.4 Cypress Semiconductor is an integrated circuit chip company in a very competitive industry. Identify the firm's core industry. Briefly describe the firm's strengths and weakness, and its threats and opportunities.

4.5 Zipcar offers a sophisticated form of car sharing (see www.zipcar.com). The firm opened for business in Boston in late 2000. Describe the strategy of Zipcar using the six questions of Figure 4.2. Is the Zipcar strategy sustainable, and will it lead to profitability?

4.6 Trader Joe's sells upscale foods at discount prices in 200 stores in 18 states. The firm uses a "less is more" philosophy. It focuses on products that it can buy and sell at a good price. Trader Joe's cuts costs by having its 18 expert buyers go directly to its suppliers [Wu, 2003]. Describe the strategy of Trader Joe's. Is its competitive advantage sustainable?

4.7 Commerce Bancorp of Cherry Hill, New Jersey, has had huge success following the creative strategy of treating its customers well. Its growth has largely come at the expense of its bigger rivals, where customers often are ignored. Few banks have ever grown as fast for as long as has Commerce Bancorp [Brown, 2003]. The bank is open long hours and on weekends, and has 300 branches. Determine and describe the strategy used by Commerce Bancorp using Tables 4.6 and 4.7.

4.8 Flexplay Technologies offers EZ-D DVDs that have the same sound and picture quality as conventional DVDs. A customer purchases a movie for about $5 and plays it as many times as desired for 48 hours, after which point it stops working. If the package is unopened, it has a shelf life of one year. (See www.flexplay.com and video.movies.go.com/ez-d/). Use Table 4.8 to determine the strategy used by Flexplay and comment on the sustainability of its competitive advantage.

4.9 An opportunity exists to design and sell a black box (event recorder) for automobiles. It would record speed, braking, and other driving data. Prepare a SWOT analysis of a new venture to design and sell such a device.

4.10 Two firms are competing for the sea rescue service for small recreational boats [Smillie, 2003]. Boat US (www.boatus.com) and Sea Tow Services (www.4seatow.com) are the two largest marine assistance firms operating as the sea equivalent of AAA. Describe the competitive advantage for each firm, and outline their competitive strategy using Figure 4.3.

CHAPTER 5

Technology, Innovation, and Timing

Imagination is more important than knowledge.
Albert Einstein

Many people believe that those quick to act will win the race while the slow and deliberate will trail behind. The decision to be the first mover needs to be addressed by all entrepreneurs. Using an idealized model of a window of opportunity, the entrepreneur can decide when to act. The entrepreneur needs to maintain a sense of urgency but avoid being too early or too late to market.

Entrepreneurs establish and build a network of partners who work with them to achieve the new venture's goals. This partnership or set of alliances will include suppliers, customers, complementors, and often competitors. Entrepreneurs also seek to build an innovation strategy that involves new technologies, ideas, and creativity that lead to invention and ultimately commercialization. An innovation strategy is part of most new firms' road map to success. A firm that encourages creativity and inventiveness can create the ingredients of sustained innovation. ■

5.1 FIRST MOVER VERSUS FOLLOWER DECISION

Many entrepreneurs believe that the quick survive while the slow struggle. The firm that leads the way with a new product or into a new market expects to lock in a competitive advantage that ensures superior profits over the long run. In this section, we consider the circumstances in which a pioneer may benefit from being a first mover. A **first-mover advantage** is the gain that a firm attains when it is first to market a new product or enter a new market.

We will describe the industries that a new venture enters as mature, growing, or emergent, as noted in Table 5.1. **Emergent industries** are newly created or newly recreated industries formed by product, customer, or context changes [Barney, 2002]. **Mature industries** have slow revenue growth, high stability, and intense competitiveness. **Growing industries** exhibit moderate revenue growth and have moderate stability and uncertainty. New technology ventures often start in uncertain, emergent industries.

The pioneering, first-mover firm has to bear the costs of promoting and establishing a product, including the potentially high costs of educating customers and suppliers. Furthermore, due to the high uncertainty of emergent markets, it is subject to potential mistakes in product, strategy, and execution. The follower firm can learn from the pioneer's mistakes and exploit the market potential created by the pioneer. Some firms successfully exploit a **follower strategy.**

Early entrants (second or third movers) into an emergent industry can also benefit from the additional time to develop, commercialize, and exploit new products if they possess the resources to wait for the opportunity to materialize [Agarwal, Sakar, and Echambadi, 2002]. Many examples exist of new start-ups that arrived early but didn't stay long. Pets.com, Webvan, and eToys all burned through their investment capital before attracting enough customers to sustain a business. For most start-ups, it is more like a marathon, where how fast you get out of the starting block is irrelevant.

In many cases, pioneer entrants tend to make a large and lasting impression on customers, obtaining strong brand recognition, and buyers often face high switching costs in moving their business to a later entrant. The simplest reason in favor of a first-mover strategy is the ease of recalling the first brand name in

TABLE 5.1 Three types of industries and their characteristics.

Characteristics	Type of industry		
	Mature	**Growing**	**Emergent**
Revenue growth	Slow	Moderate	Potentially fast
Stability	High	Moderate	Low
Uncertainty	Low	Moderate	High
Industry rules	Fixed	Fluid	Unestablished
Competitiveness	High	Moderate	Low or none

a category. However, one study found that pioneers gained significant sales advantages but incurred large cost disadvantages relative to a fast follower entrant [Boulding and Christen, 2001]. The return on investment for pioneers was less than that for followers.

Of course, many conditions exist in which a first-mover advantage may be clear and compelling. Consider a mature industry such as restaurants or grocery stores. The attainment of a strategic resource such as a superior location may warrant acting as a first mover in a geographic market segment. Starbucks, for example, wants a store on the busiest corner in a city and acts when it finds an available site.

If a market is insufficiently ordered or unstable, the first entry may be too early. A market is said to be stable if the requirements necessary for success will not change substantially during the period of industry development. Amazon.com entered the online bookstore market and created intellectual property and the standard for this market. However, it incurred high development costs and had its advantages challenged by a later entry, BarnesandNoble.com. Nevertheless, Amazon stayed the course in the marathon and became the leader in the race.

Pioneers are often said to gain a low-cost advantage from having a head start down the experience curve, which describes improvements in productivity as workers gain experience. Often these lower costs are an advantage over later entrants [Shepard and Shanley, 1998]. New technology ventures often act as pioneers in a new or emerging industry to gain brand, cost, and switching cost advantages. The potential advantages and disadvantages of first-mover action are summarized in Table 5.2.

Numerous examples exist of later entrants overtaking first movers and eventually bypassing them in profitability. Superior performance comes from distinctive competencies combined with an appropriate strategy leading to a competitive advantage (see Figure 4.5). Unfortunately, the first mover can develop a strategy based on uncertain or inaccurate assumptions about the six forces (see

TABLE 5.2 First-mover potential advantages and disadvantages.

Possible advantages	Possible disadvantages
■ Create the standard and the rules	■ Short-lived advantages disappear with competition
■ Low-cost position	■ Higher development costs
■ Create and protect intellectual property	■ Established firms circumvent or violate patents and intellectual property rights.
■ Tie up strategic resources	■ Cost of attaining the resources
■ Increase switching costs for the producer	■ High uncertainty of designing the right product. If vision is wrong, then costs to switch are large.
■ Increase switching costs for the customer	■ Customer is reluctant to buy when a large cost to switch may be incurred.

Figure 4.3). A follower who learns from the first mover's mistakes can move quickly to catch up or pass the first mover. The first mover also suffers from uncertainty about the customer, the organizational capabilities needed, and the industry context.

However, pioneering ventures can use their lead time to build relationships among suppliers, customers, and even competitors. These relationships can build trust and brand that a follower may not easily reproduce. A first-mover advantage can usually be attained under conditions of low market and internal firm uncertainty. Regrettably, most new ventures encounter large measures of uncertainty and must weigh carefully when to enter the market [Kessler and Bierly, 2002]. Entrepreneurs should emphasize speed to market in predictable markets. In an uncertain market, the new venture can probe or test the market by trying product tests, focus groups, and other means of market probes.

A commitment is an action taken in the present that binds an organization to a future course of action. A decision to act as a first mover is usually binding and should not be taken lightly. Preemptive actions can deter potential rivals from entering but may also result in heavy, irreversible investments [Sull, 2003].

The entrepreneur considers entering a market during an estimated period of opportunity often called a window of opportunity. The first mover envisions a greater cash flow as a result of early entry, as shown in Figure 5.1. Uncertainty about the period of opportunity can erode the actual results. If the first mover misestimates the timing of the window, a less attractive cash flow curve will result.

An entrepreneur's objective is to decide when to stop searching for additional information and enter the new market so as to maximize the expected profit. With insufficient information, a firm can enter too early and incur a large cost. However, if it takes too long to gather sufficient information, the firm may lose the first-mover advantage. Entrepreneurs should stop searching for information and enter a market when they estimate that the marginal benefit of additional knowledge is less than the payoff of entry [Shepard and Levesque, 2002].

An example of when a first-mover strategy is clearly called for is that of a fad product. A fad is a briefly popular fashion or product, such as a Beanie Baby, with a limited period of opportunity. The Hula-Hoop fad swept through the world in 1958. Because no one had a patent on the hoop, almost any plastics manufacturing firm could enter the business, but they had to move fast. In a matter of weeks, 20 million hoops had been sold. The first mover was Wham-O Manufacturing Company of San Gabriel, California, which gained 50 percent market share within two months.

An example of another successful first-mover strategy was Wal-Mart's strategy that Sam Walton described as: "put good-sized stores into little one-horse towns which everybody else was ignoring." Walton sought out isolated rural towns with populations of between 5,000 and 10,000 that could only support one discount store, and he got there first.

History is replete with companies that were first movers that did not succeed. The CPM operating system preceded Apple, which preceded DOS, which eventually became the early dominant operating system for the PC. Safety razors

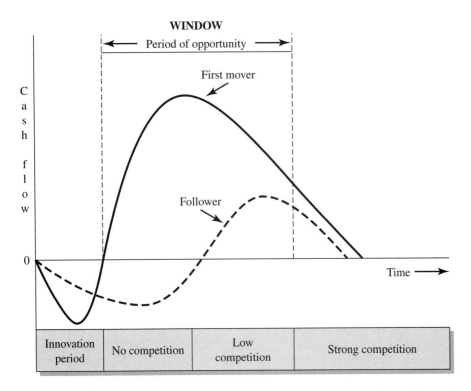

FIGURE 5.1 Expected first mover advantage and the concept of a window of opportunity.

were introduced a decade before Gillette introduced its successful safety razor. The product must have the right mix of attributes and features, and must be understood as well as demanded by the customer. Early movers don't always have all the requisite characteristics. Prodigy was the first commercial e-mail system, but it received poor acceptance. The second entrant, CompuServe, was equally unsuccessful. Only later did AOL and MSN put together the right mix of attributes to succeed.

Many new ventures set a fast pace as they and their competitors enter a window of opportunity. Many start-ups exhibit a torrid pace due to a high sense of urgency, as illustrated in Figure 5.2. As the firm experiences a sense of urgency due to a shortfall of customers, it acts to build capacity to design, build, and sell its products. However, inevitable delays, D, slow down the buildup of capacity. As capacity increases, the firm expects customers to buy, but again it may experience delay as customers consider the purchase carefully. A slowdown in the growth of customer buildup results in a sales shortfall and an increasing sense of urgency [Perlow, Okhuysen and Repenning, 2002]. One way to decrease this unfortunate urgency cycle is to reduce the delay in capacity building and the time delay to customer purchase.

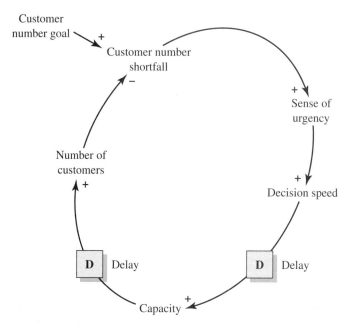

FIGURE 5.2 The sense-of-urgency cycle that can be experienced by new enterprises.

An encouraging case of good timing and entry into a marketplace is that of Estée Lauder. Estée Lauder's start in the cosmetics industry began in the 1920s in the United States. The company's founder was born in 1908 in New York and developed an interest in cosmetics in her youth. Cosmetics are specialty chemical products developed on the basis of chemical technologies. She worked in her father's department store and later, in the mid-1920s, for her uncle in the cosmetic chemistry business. After marrying Joseph Lauder, she registered her own cosmetics business [Koehn, 2001]. With a family, she worked part time on her cosmetics business, naming it Estée Lauder, after herself. During World War II, her products were popular in New York beauty shops. By 1946, she and her husband opened an expanded business called Estée Lauder Cosmetics and envisioned entering the national mass market. They expected a significant surge of women purchasing cosmetics in that postwar period of opportunity. With new packaging for their products, they decided they were selling "pure glamour." They moved the products into the department store channels with such stores as Saks and Neiman Marcus. By 1958, their sales were $1 million. Sales continued to increase by a compound annual rate of 45 percent in the 10 years after 1958. By 2002, sales had grown to over $4.6 billion. Estée Lauder had identified the great potential for personal cosmetics and managed a timely entry with a distinctive competence. Her marketing and product development skills led to a successful entry into a window of opportunity.

Machiavelli wrote in *The Prince* (XVII): "The prince ought to be slow to believe and to act, nor should he himself show fear, but proceed in a temperate manner with prudence and humanity, so that too much confidence may not make him incautious." An entrepreneur will be temperate and patient to move. On the other hand, an entrepreneur has a propensity to act. If a window of opportunity appears to be in the distant future, the entrepreneur may be wise to abandon the distant opportunity and seek one that is available and active now. On the other hand, if the window is about to occur, action may be prudent.

> The formation and launch of Silicon Valley Bank (SVB) is a sound example of good timing. In the early 1980s, the deregulation of the banking industry led to a need for new innovative banks. In the same period, the Bank of America, which served the San Francisco area, was discontinuing its lending to high-tech companies. At that time, one of the founders of SVB had a series of meetings with bankers who were interested in participating in this opportunity. These factors converged, and the lead founder acted on his intuition, all of which led to the formation of Silicon Valley Bank in 1983 (see www.svb.com).

5.2 THE VALUE NET AND ALLIANCES

Many businesses use competitive strategies to shape their business strategies but often ignore *cooperative* strategies. Business is a complex mix of both competition and cooperation. A new venture possesses valuable novelty and innovation that will attract the attention of suppliers, customers, competitors, and complementors, acting as a value network, as shown in Figure 5.3. Recall from

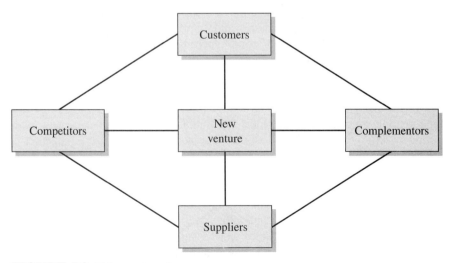

FIGURE 5.3 Value network.

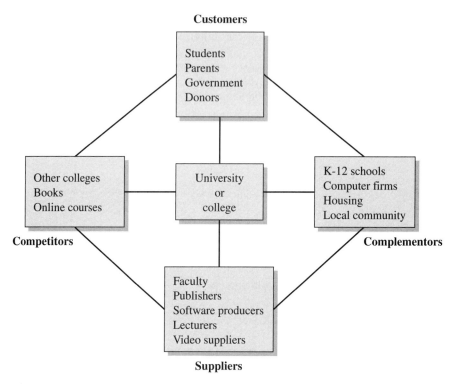

FIGURE 5.4 Value network for a university or college.

Chapter 4 that a *complement* to a product is any other product that makes the first one more attractive to the customer. Hot dogs and mustard, and cars and auto loans are good examples. Microsoft and Intel, sometimes called Win-Tel, is a good example of the power of complementors working together. A complementor is a firm that provides complements to another firm. As depicted in Figure 5.3, all the participants are connected and participate in this network of activity. Consider the value network for a university, shown in Figure 5.4 [Brandenburger and Naleboff, 1997]. The complementors to a university include kindergarten–grade 12 schools, local housing, community activities, and computing systems. All the members of the value network are connected together in the higher education value network. The university, to succeed, must cooperate with its suppliers, customers, competitors, and complementors. Competitors can be seen as rivals but also will be, in many instances, collaborators.

The value network is important to entrepreneurial ventures as they strive to accumulate the resources and capabilities required for success. The value of exploiting complementary resources can be significant [Hitt et al., 2001]. For example, a smaller, new biotechnology firm and a large pharmaceutical firm can both benefit from an alliance. The biotech firm provides new technologies and innovation, while the pharmaceutical firm provides the distribution networks and marketing capabilities to successfully commercialize the new products. The larger established pharmaceutical firm also gains value through access to its partner's innovation. Thus, firms usually search for partners with complementary assets or capabilities. An excellent example of complementary partners is Starbucks and Barnes and Noble bookstores. Starbucks provides a cafe and lounge area in Barnes and Noble superstores that enhance the value of each partner. Firms that are proactive in forming partnerships are likely to enjoy higher performance [Sarker, Echambadi and Harruson, 2001].

A **partnership** or **alliance** is an association of two or more firms that agree to cooperate with one another to achieve mutually compatible goals that would be difficult for each to accomplish alone [Spekman and Isabella, 2000]. Proactive firms take the initiative rather than react to events. Proactive formation of strategic alliances is an important dimension of entrepreneurial activity that enables a new firm to acquire access to unowned but required strategic assets. All alliances are based on some exchange of knowledge in addition to a flow of products, capital, or technology.

The configuration of alliances of a start-up impacts its early performance. External alliances can substitute for internal resources. A firm's decision to enter an alliance can be motivated by a desire to exploit an existing capability or technology, or to explore for new opportunities and new technologies [Rothaermel and Deeds, 2004].

The new firm should consider developing an alliance when it lacks the necessary assets that a complementor can provide. To select a partner, it must be clear which missing capabilities or resources are required. Then, it must determine which firms possess those assets and look at their characteristics. It will be necessary to build a relationship of trust with the potential partner and craft an agreement that will yield benefits for both partners [Doz and Hamel, 1998]. For example, the alliance that Calyx & Corolla, the florist, has with independent floral growers and FedEx enables it to deliver flowers from grower to customer in a few days (see www.calyxandcorolla.com).

Alliances have a variety of structures and are usually governed by a contract that delineates the roles and responsibilities of each partner. Complementor firms may also be potential competitors. Many a well-conceived alliance has fallen apart due to the tension between cooperative and competitive forces. These can be culture clashes, poor conflict management, and lack of effective coordination mechanisms. Furthermore, the entrepreneurial firm may be seeking access to needed assets but may, as a result, be exposed to the risk of losing its own vital internal knowledge. An example of this occurred during the development of the Apple Macintosh. Apple partnered with Microsoft to develop spreadsheet, database, and graphical applications for the Mac. As a result, Microsoft acquired critical knowledge about Apple's graphical user interface products, which eventually enabled its engineers to develop the Windows operating system [Norman, 2001]. Knowledge transfer occurs in conversation and association and is difficult to control. Starbucks wants to share its expertise with its partners. When it places its coffee shops in Barnes and Noble stores, it makes sure the bookstore employees at its counters are well versed in the Starbucks way of doing things.

> Webvan.com tried to quickly build an online grocery store with home de-livery. Webvan.com burned through almost a billion dollars in just over two years, partly by building a series of extremely expensive, highly auto-mated warehouses around the United States and trying to do it all by itself. The survivors, which include MyWebGrocer.com, Safeway.com, and YourGrocer.com, have kept their expenses low and target affluent neigh-borhoods. Building an online grocery business slowly and picking densely populated cities is a way to build returning customers.
>
>   MyWebGrocer.com enables the purchaser to buy groceries online from one of 200 stores or chains nationwide. This series of alliances lets MyWebGrocer set up the network of activity and operate within a value net while avoiding the costs of warehouses and inventory.

The benefits of alliances can be significant. Both firms learn and acquire new capabilities. Furthermore, they have access to complementary resources that they cannot easily duplicate. An entrepreneurial new venture wisely will consider the development of one or two partnerships consistent with its strategic goals. Going it alone can be a major liability for entrepreneurs. Innovators who get together in alliances can be more successful, especially where uncertainty prevails. Few start-up firms will have all the necessary capabilities and resources, and alliance networks can enable them to move forward effectively. The type of alliance can range from a joint short-term project to a merger, as shown in Figure 5.5.

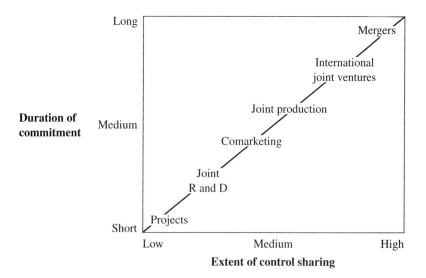

FIGURE 5.5 Range of alliances dependent on commitment and control sharing.

Although a portfolio of alliances can be powerful, too many or too much variety can undermine this power. That's because the ability to manage alliances is valuable and limited. The limit of alliance capacity can be quickly reached if the alliances require significant and time-consuming exchange of knowledge [Hoang, 2003].

5.3 TECHNOLOGY AND INNOVATION STRATEGY

Innovation is commercialized invention. Only about 6 percent of inventions developed by independent inventors actually reach a market [Astebro, 1998]. The success rate for inventions developed in established firms is about four times as high. However, once they are commercialized, inventions from start-ups or established companies experience about the same rate of success. For example, an independent Swiss inventor got the idea for the Velcro fastener based on a hook-and-loop device. He set up a company to license the product, and a Canadian company brought the product to market. Velcro North America bought the rights to the patents in the late 1960s. The patents expired in 1975, and since then, Velcro has become a commonly used device.

It is often a long road from invention to commercialization. Chester Carlson developed the photocopier process—converting an image into a pattern of electrostatic charges that attract a powdered ink—in his kitchen. He patented the process in 1942. After years of little interest from established companies, he obtained help from the Battelle Institute. Then the Haloid Company purchased a license in 1946. Haloid, which became Xerox, successfully demonstrated a working product in 1949. In 1960, Haloid Xerox introduced the first successful office copier, the Xerox 914. The creation of new firms is an important mechanism through which entrepreneurs use technology to bring new products, processes, and ways of organizing into existence. Inventors do not always start firms to exploit technological opportunities. Sometimes they license or sell these opportunities to others who start a firm or a new business unit to exploit them.

Schumpeter asserted that the process by which independent entrepreneurs independently created inventions to produce new goods, services, raw materials, and organizing methods is central to understanding business organization, the process of technical change, and economic growth. An innovation strategy rests on the competencies and knowledge of the new firm. Continual product and process innovation can enable the firm to maintain a strategic advantage.

Three factors influence the decision to exploit an independent invention through firm creation: the interests of the entrepreneurial team, the characteristics of the industry in which the invention would be exploited, and the characteristics of the invention itself. We have discussed the first two of these factors in earlier chapters. Of course, the entrepreneurial team must be interested in the opportunity to be solved by the invention, and it must be satisfied that the industry will welcome and support the commercialized invention. In this section, we will consider the characteristics of the invention itself.

Three dimensions of technological inventions impact the probability that they will be commercialized through a new firm formation: importance, radicalness, and patent scope [Shane, 2001]. *Importance* reflects the magnitude of the economic value of an invention. The *importance* of an invention should increase the likelihood that a new firm will be founded to commercialize it because more important inventions have higher economic value and thus payoff to the entrepreneurs. Many inventions have limited commercial value and thus are not attractive to the entrepreneur.

Radicalness measures the degree to which an invention, regardless of economic value, differs from previous inventions in the field. The *radicalness* of an invention is a reflection of the potential market effect of the commercialized invention. These are often called disruptive innovations. Radical technologies destroy the capabilities of existing firms because they depend on new capabilities and resources. *Patent scope* describes the breadth of intellectual property protection for the invention. These three dimensions of likelihood of commercialization are listed in Table 5.3.

Dean Kamen, the holder of more than 150 patents, is one of the well-known inventors of the past three decades. He invented devices for infant care, insulin delivery to diabetics, and for replacing the wheelchair [Brown, 2002]. Kamen invented the Segway Human Transporter in 2001. It is an electric scooterlike device. Gyroscopes inside the base platform make the scooter highly stable and self-balancing. There is no brake handle, engine, throttle, or gearshift. Users lean forward to go forward and backward to reverse direction. The inventor says the Segway can traverse ice, snow, or even large rocks. The target market is warehouse workers, postal employees, and, eventually, urban dwellers. Only time will tell us the extent of this invention's importance, radicalness, and breadth of patent protection.

We can portray the new business formation process for an invention as shown in Figure 5.6. Many times the inventor will not serve as the entrepreneur. As shown in Figure 5.6, the entrepreneurial activities are distinct from the inventive activities. Using this process to review the potential of the Segway Human Transporter, one can obtain different conclusions for the various proposed uses: postal service, warehouses, or urban dwellers. Perhaps the best application for this device is not yet named.

TABLE 5.3 Factors that influence the entrepreneur to exploit an independent invention.

1. Business interests, capabilities, and experiences of the entrepreneurial team.

2. Characteristics of the industry in which the invention will be exploited.

3. Characteristics of the invention:

    a. Importance of the invention: Economic value and potential payoff.

    b. Radicalness of the invention: Differentiation of the invention from its predecessors.

    c. Breadth of patent protection of the intellectual property.

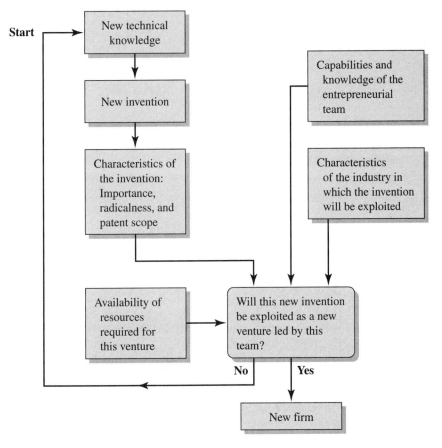

FIGURE 5.6 New business formation process for an invention.

Often a new enterprise is founded on the invention of a new technology. A *technology* is a device, artifact, or tool that can be applied to a need. An example is the microprocessor invented by Ted Hoff of Intel in 1968. The ensuing era of computers and information technology has provided many technologies that enabled new enterprises to provide new solutions to problems. Myriad enterprises have been founded on the basis of a new application of a technology [Carr, 2003].

The difficulty with deciding whether to proceed to commercialize an invention can depend on the radicalness of the invention. Disruptive or radical innovations introduce a set of attributes to a marketplace different than the ones that mainstream customers historically have valued, and the products often initially perform unfavorably along one or two dimensions of performance that are particularly important to customers. As a result, mainstream customers are unwilling or unable to use disruptive products in applications they know and understand. At first, therefore, disruptive innovations tend to be used and valued only in new uncertain markets or applications.

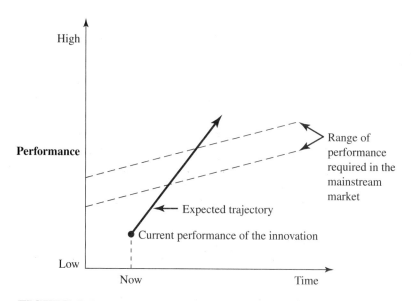

FIGURE 5.7 Expected trajectory of a disruptive innovation.

Often the disruptive technology will not immediately serve a mainstream market, as shown in Figure 5.7. It will initially serve a niche market but will eventually enter the low end of the range of the mainstream market, as shown in Figure 5.7. Some innovations start as high-end solutions and enter the mass market range of performance from the high end. Consider the disruptive innovation of discount stores in the 1960s. The increased mobility of shoppers enabled discount stores such as Kmart to select locations at the edge of town, reducing department stores' competitive advantage of prime city-center locations. The discount store had a new innovative business model: low-cost, high-unit volume and turnover provided at convenient suburban locations. They executed a trajectory from low-cost hard goods to low-cost hard and soft goods, and entered the mass markets in the 1970s and 1980s. Today, Target and Wal-Mart are in the center of the mass market. A new disruptive challenge in the retail industry is Amazon.com, which appeared originally as an online bookstore but migrated on a trajectory toward an online department store. Will Amazon's trajectory take it into the mass market and profitability? Some products are less appropriate for online sales than others. While Internet retailers excel at getting the right product in the right place at the right price, they're at a disadvantage when it comes to delivering physical products at the right time. When shoppers require products immediately, they'll head for their cars, not their computers [Christensen and Tedlow, 2000].

Consider an example of a disruptive innovation: voice recognition software. The current performance of computer software for voice recognition is not always adequate for high-accuracy speaking (dictating) of documents rather than typing them; this might require 95 percent accuracy. Undoubtedly, there are many less-demanding uses for voice recognition software, such as voice-generated e-mails, customer service by telephone, or chat room messages. Thus, this innovation has entered the low end of the range of required performance and is progressing upward toward wider application.

Innovation is based on invention and creativity, and is defined as invention that has produced economic value in the marketplace. Innovation is based on the commercialization of new technology. An innovation can include new products, new processes, new services, and new ways of doing business. Disruptive innovation transforms the relationship between customer and supplier, restructures markets, displaces current products, and often creates new product categories [Leifer et al., 2000]. Competitors create innovative products and offer new value to customers, as shown in Figure 5.8. The struggle is for each competitor to keep up in the innovation cycle.

Many new innovations come from small companies. Consider the ski industry, which stagnated in the late 1980s. The new innovation was snowboards, which were popularized by appealing to the young person who enjoyed skateboards. Jake Burton Carpenter was the early builder of boards, and his firm now has a large share of the market (see www.burton.com). The shift from old-fashioned Schwinn bicycles to mountain bikes is another example of an innovation starting from new, small companies.

Innovative imitation is taking an existing idea and creatively expanding on it. The imitator must understand the customer better than the innovator and creatively act on that knowledge. Portable, battery-driven radios have been used since the 1950s, but Trevor Bayles saw an opportunity to bring information to remote Africa by creating a windup spring- and dynamo-powered radio. Twenty-five seconds of winding gives the user one hour of listening. Bay Gen in Cape

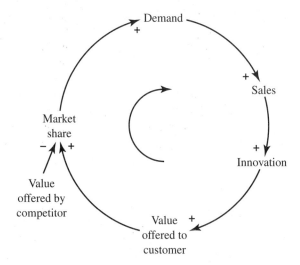

FIGURE 5.8 Innovation to value competition cycle for market share.

Town, South Africa, now manufactures more than 60,000 of these radios a month [Handy, 1999b].

Architectural innovation changes the way in which components of a product are linked together while leaving the core design concepts untouched. Thus, the components remain unchanged, but the architecture of module connection is the innovation. The overall architecture of the product describes how the components will work together. The essence of an architectural innovation is the reconfiguration of an established system to link together existing components in a new way [Henderson and Clark, 1990].

A product design requires knowledge of the components and modules as well as architectural knowledge. Figure 5.9 shows four types of innovation. Architectural innovation requires a design capability to arrange the linkages between modules. Modular innovation is focused on the innovation of new components and modules. Radical innovation uses new modules and new architecture to create new products. The Internet is an example of a network system with new modules and new architecture—a radical or disruptive innovation.

Sources of innovation for new ventures include universities, research laboratories, and independent inventors [Bronscomb and Florida, 1999]. Another possibly important source is the ultimate customer. Unfortunately, most customers have limited skills in predicting new products. They are best at reacting to a potential product and describing the outcomes they desire [Ulwick, 2002]. Customers are best at describing their experiences but limited in describing future needs. Who knew in advance that people wanted the Internet or electric-hybrid cars? Another danger is the common practice of listening to the recommendations of a group of customers called **lead users** who have an advanced understanding of a product and are experts in its use. Lead users can offer product ideas, but since they are not average users, their recommendations may have limited appeal. A good approach is to ask users what results or outcomes they want to see in doing their job using your product. Developing an understanding of what customers value may be a far more useful exercise than merely asking them to

Basic design concepts

		Reinforced	Overturned
Linkages between modules	Unchanged	Incremental innovation	Component or modular innovation
	Changed	Architectural innovation	Radical or disruptive innovation

FIGURE 5.9 Four types of innovation.

submit their own solutions. However, some lead users can be very helpful at identifying valuable solutions. Lead users are numerous. It is the task of the entrepreneur to pick the right one. The process of innovation begins with identifying the outcomes customers want to achieve; it ends in the creation of items they will buy. One method is to ask potential customers to describe their typical day. They may reveal an important gap and a potential need.

Consider the idea of a convenient peanut butter sandwich for children at school and people traveling. You could ask these would-be customers for a solution, or ask them what they are hoping for as an outcome. The answer might be: They want to make a peanut butter sandwich without knives or spoons or having to tote jars. How about "sliced peanut butter" that can just be slapped on a slice of bread. Why not sliced peanut butter? We have sliced bread. It's all convenience. Researchers at Oklahoma State University have perfected a process for creating individually wrapped, fresh peanut butter slices. Then, we might need sliced jelly. [Stires, 2001].

By the mid-nineteenth century, a large natural-ice industry was supplying ice for preserving food, chilling drinks, and making ice cream for the wealthy and restaurants in the northeastern United States. Taken from the lakes of New England, the ice was stored and shipped widely. In the winter of 1879–80 alone, between 8 million and 10 million tons were harvested. The trade in eastern cities remained strong until the 1920s. With the invention of the electric refrigerator, the natural-ice business declined. By 1920, the new, small-motor electric refrigerator was widely available. By 1937, 3 million refrigerators were in U.S. homes, and the natural-ice industry was in full decline [Weightman, 2003].

The probability of success in disruptive innovation is based on the creation of new, important products or services that initially target customers at the low end of a market who don't need all the functionality of existing products or who have no other solution for their problem. With the right resources and capabilities, a new firm can satisfy this initial need in the market. Conversely, if it makes it easier to do something customers weren't trying to do, it will fail [Christensen, 2002]. A "build it and they will come" innovation strategy will most likely fail. Many innovations fail after the innovator assumes that what customers say they want to do is what they actually would do. The successful entrepreneur possesses the capability and resources necessary to effectively shift economic capital out an area of lower productivity and into an area of higher productivity by creating and commercializing an innovation that makes a difference to the customers' outcomes and reduces the required effort. For example, many students will often say they want more detailed and complete information in their textbooks, when they actually want worked examples that will help them pass an exam.

5.4 CREATIVITY, INVENTION, AND INNOVATION

Creativity leads to invention and thus to innovation. **Creativity** is the ability to use the imagination to develop new ideas, new things, or new solutions. Creative ideas flow to invention, and invention flows to innovation. Creative thinking is a core competency of most new ventures, and entrepreneurs strive to have creative people on their team. Creative ideas often arise when creative people look at established solutions, practices, or products and think of something new or different. The successful company generates cost-effective surprises [Schrage, 2001]. This firm is committed to making innovation the underlying focus of its business.

The creative enterprise is based on six resources, as shown in Table 5.4 [Sternberg, O'Hara, and Lubart, 1997]. To create something new, one needs knowledge of the field and of the domain of knowledge required. Domains are areas such as science, engineering or marketing. Fields within a domain might be circuit design or market research. Wise, knowledgeable, creative people avoid being blinded or limited by their knowledge.

The intellectual ability required is the ability to see linkages between things, redefine problems, and envision and analyze possible practical solutions. Creative people use intuitive thinking that reflects in novel ways on a problem. A creative thinker is motivated to make something happen. Creative people are open to taking reasonable risks and acting when advised otherwise. Finally, the creative person understands the context of the problem and is willing to take a risk and advocate change. The person who has most of these skills is often called *intuitive;* that is, he or she has an instinctive ability to perceive or learn relationships, ideas, and solutions.

The intuitive person suspends critical and conventional thinking long enough to consider the possibility of new solutions. One method of creative discovery includes the following steps: 1) slow down to explore different ideas, 2) read about the field, but not too much, 3) look at the available raw data, and 4) cultivate smart friends who have good intellectual skills [Paydarfar and Schwartz, 2001].

Creative thinking involves divergent thinking, which is the ability to see the differences among various data and events. Creativity involves the ability to synthesize, working through information to come up with combinations that are new and useful [Florida, 2002]. Incubation of the issues and time to reflect are im-

TABLE 5.4 Six resources for a creative enterprise.

- Knowledge in the required domain and fields: Knowing what is new.
- Intellectual abilities to recognize connections, redefine problems, and envision and analyze possible practical ideas and solutions.
- Inventive thinking about the problem in novel ways.
- Motivation toward action.
- Opportunity-oriented personality and openness to change.
- Contextual understanding that supports creativity and mitigates risks.

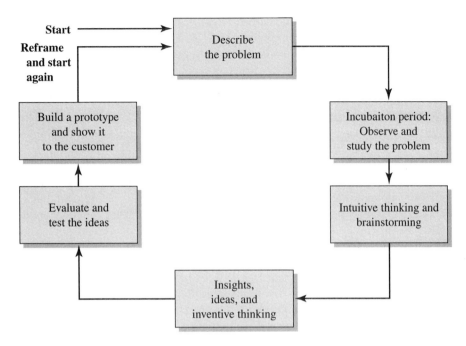

FIGURE 5.10 Creativity process.

portant steps to creativity. One process of creative thinking is shown in Figure 5.10. It starts with a description of a problem and rests for a period of incubation. Then, intuitive brainstorming leads to good insights and ideas that can be evaluated and tested. Finally, a prototype is built and shown to the potential customer. This may lead to a reframing of the question or problem and a second cycle through the process. An iterative process around the loop is followed until the prototype product solves the problem.

Managing for creativity can clash with rational management [R. Sutton, 2001]. Organizing for creativity can be different than organizing for routine work. One way to spur creativity is to find new uses for old materials, products, or concepts. In 1954, Kay Zufall was looking for new things for children to do. She didn't like the modeling clay sold for children because it was too stiff. However, her brother-in-law made a doughy mixture for cleaning wallpaper. Zufall tried it as a modeling medium and discovered it was soft and easy to mold and cut up. She and her brother-in-law reformulated it as a safe and colorful product for children, and they came up with the name: PlayDoh [R. Sutton, 2002].

All firms need a culture that sustains a creative process that enables the team members to engage and interact with ideas and new solutions. Apple was the first to develop the Newton personal digital appliance (PDA) with handwriting recognition software. In practice, few people were willing to wait for the Newton to slowly learn to recognize their handwriting. The Palm used a Graffiti interface and succeeded in capturing the market. Palm recognized that it was much easier to let humans learn to standardize their script than it was to develop software that enabled a computer to recognize all possible script.

A natural conflict exists between creatively generating ideas and inventions and implementing them. Creativity leads to new inventions and ideas. Bringing these inventions to market, however, takes routine processes. The forces of creativity and process can conflict or interact, depending on the firm's culture. Creativity flourishes when companies hire creative people, invest resources in risky projects, and get their workers to critique the ideas. These unconventional practices work because they make companies vary their thinking, see old things in new ways, and break from the past.

Rules and policies stifle creativity, and undisciplined thinking undermines routine manufacturing processes. The conflict is between managing for replication and managing for creativity. A small, emerging firm can accommodate both tendencies within it. As a firm grows, it needs to build a culture that reinforces the best qualities of creativity as well as efficient execution of its business processes [Brown and Duquid, 2001].

Creativity can be seen as the ability to link together two seemingly unrelated ideas or concepts. Many ideas ignite in a free-form environment where people have capabilities and self-confidence. Mixing creative people in unexpected ways to unleash new ideas pays off. Creating new methods, products, or business models requires a powerful vision showing people what the problems are and how to resolve them. Compelling dramatic portrayals or visualizations can help people to see and feel new opportunities [Kotter and Cohen, 2002].

5.5 NEW TECHNOLOGY VENTURES

Often, a new technology becomes available to an entrepreneur, but an economic application of the technology is not obvious. This new technology usually becomes available due to scientific discoveries or a new invention. Entrepreneurs find that this new technology may offer myriad opportunities for new ventures. However, it is unclear what, if any, application will be economically viable. Often, a new technology can be characterized as a solution looking for a problem. Neither the first companies to use the technology nor the companies with the best technology win. Usually the firms that find the right application for the technology succeed [Balachandra 2004].

The elements of an attractive innovation strategy are provided in Table 5.5. Any new venture should have a defined customer, one or two key benefits, a short period to payback, and a proprietary advantage. Finally, the new venture team must possess the necessary core competencies to exploit the new technology.

One way of describing the potential applications is to use the model shown in Table 5.6. The new technology is described briefly, and the key assumptions are listed. Then, the core competencies required for the venture are described. Finally, the possible applications are noted. Table 5.6 shows the summary of two possible new ventures. The Rotary Engine Inc. example illustrates an attractive new technology venture for vehicle engines, marine engines, appliances, and

TABLE 5.5 Elements of an attractive innovation strategy.

■ Well-defined customer.	■ Proprietary advantage that can be maintained or defended.
■ Key customer benefit that is measurable in dollars.	■ Core competencies required to exploit the new technology are present or available to the new venture.
■ Short period until economic payback and positive cash flow.	■ Access to the necessary resources.
■ High benefit-to-price ratio for the customer.	

recreational vehicles. The market challenges are listed, and an attempt is made to fund the best application that will satisfy the required elements of an attractive innovation strategy, as listed in Table 5.5.

A second example of a new technology is provided in Table 5.6. Fuel cell technologies have been of great interest over the past decade. However, an economic application is not yet proven. With the lack of supporting infrastructure, fuel cells have limited automobile applications. On the other hand, fuel cells as energy storage devices serving as battery replacements may be viable soon.

Examples of new technologies that eventually found attractive economic applications include semiconductors, genomics, stents, and wireless telephony.

TABLE 5.6 Two potential new technology ventures.

Potential venture	Rotary Engine Inc.	Fuel Cell Inc.
Technology	Advanced rotary gasoline engine technology	Hydrogen fuel cell technology
Key assumptions and benefits	Improved engine efficiency and reduced pollution	Pollution reduced to near zero
Core competencies required	Engine design and manufacture	Fuel cell design and manufacture
Potential applications	1. Automobiles 2. Marine (ships) 3. Small appliances such as lawn mowers 4. Snowmobiles and off-road vehicles	1. Automobiles 2. Small, local electric generators 3. Battery replacements 4. Marine (ships)
Market challenges	1. Limited acceptance of rotary engines by customer 2. Lack of service knowledge for rotary engines 3. Benefits may be unclear to the customer	1. Limited infrastructure for hydrogen fuel cells 2. Benefits unclear to customer 3. Reliability of fuel cells is unproven

FIGURE 5.11 Four steps to achieve a favorable technology innovation.

All these technologies eventually traversed the four steps necessary for a favorable technology innovation, as shown in Figure 5.11.

Any new attractive technology has to be feasible, manufacturable, and provide valued performance. With a sound business model and strategy, the new technology venture strives to achieve profitability in a reasonably short period.

Future Beef Operations was founded in 1995 by three agriscientists to exploit the latest production technologies in the beef processing industry. They designed and built a fully integrated and coordinated beef production facility. Future Beef used genetic data and production control systems to produce large amounts of top-quality meat [McCuen, 2003]. Future Beef also wanted to make use of all possible value-added products in its fully integrated facility. Its $100 million plant was named *Food Engineering* magazine's plant of the year in 2001. It even had an on-site tannery. However, the plant closed one year after it opened.

This new technology venture failed because it tried to integrate several unproven technologies. The venture lacked most of the elements of an innovation strategy, as listed in Table 5.5. If they had listed the market challenges and technological limitations, as illustrated in Table 5.6, the agriscientists might have tried a more modest, step-by-step path to development of the plant. Too many uncertain technologies overwhelmed their capabilities.

Technology entrepreneurs bring together the technical world and the business world in a profitable way. Entrepreneurship is a fundamental driver of the technological innovation process [Burgelman, Christensen, and Wheelwright, 2004]. In summary, technology entrepreneurship is about the creation of a new business enterprise that generates benefits (wealth, jobs, value, progress) for participating parties by creating unique, new arrangements of resources, including technology, to meet the needs of customers and society.

5.6 AGRAQUEST

Fungi and bacteria are finding their way more and more into California's groves and vineyards. Biofungicides—new products based on naturally occurring microorganisms or other plant derivatives—are bringing growers tough disease-fighting tools while making a very slight environmental impact.

Biofungicides are just one of a larger category of products known as biological pesticides, or "biopesticides." Biopesticides are pesticides derived from natural materials, including animals, plants, and bacteria. Many biofungicides are produced by fermentation, a process in which a microorganism with fungicidal properties is grown, much the same way yeasts grow in the fermentation of beer or wine.

Using a proprietary technology, each week AgraQuest researchers analyze hundreds of potential naturally occurring microbes for a novel ability to destroy or impact various undesired bacteria, fungi, insects, and nematodes, all enemies of crop production. To date, the company has screened more than 20,000 microorganisms and identified 23 that display high levels of activity against insects, nematodes, and plant pathogens. AgraQuest has selected a set of these candidates for further development.

One of the more promising discoveries that AgraQuest has licensed is a stinky fungus from Honduras that may provide farmers with a natural alternative to methyl bromide. A recent federal law requires that use of the ozone-damaging gas be eliminated by 2005, except for very limited purposes. Methyl bromide is used as a soil fumigant by growers of strawberries, tomatoes, and other vegetables that are AgraQuest's prime market.

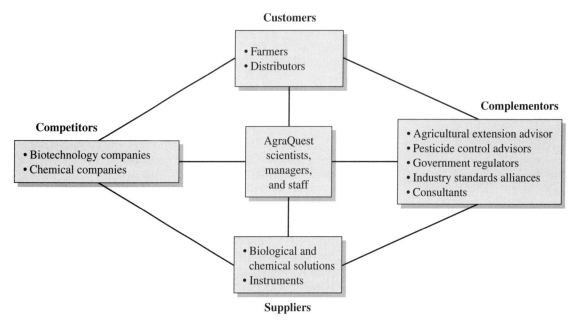

FIGURE 5.12 Value network for AgraQuest.

The registration process for biopesticides and other such bioproducts through the Biopesticides and Pollution Prevention Division of the U.S. EPA's Office of Pesticide Programs tends to be shorter and more efficient than for chemical fungicides. Therefore, AgraQuest has a shorter time-to-market for a new bioproduct.

The biopesticide industry is only emerging and should grow over the next decade. The window of opportunity has opened, and many companies are competing for leadership. AgraQuest experiences significant delays in development and approval of products. It takes about two to three years and $6 million to get one product to market (see Figure 5.1).

AgraQuest's innovation strategy is based on proprietary processes and patents. The firm holds 20 U.S. patents and three U.S. patent applications, along with nine foreign patents and 95 foreign patent applications. The patents cover the microbe and its use as well as novel natural product compounds and mixtures.

AgraQuest's value network is shown in Figure 5.12. AgraQuest is dependent on its complementors helping it succeed. For example, the advice of the extension advisors and pesticide application firms is highly valued by the farmers.

5.7 SUMMARY

Successful innovative firms strive to time their entry into markets. They balance a sense of urgency with a deliberate buildup to action. Working with partners—firms and individuals—most firms can enhance their capabilities and strengths for creativity, invention, and innovation. Almost all firms build an innovation strategy that strives to provide them with a sustainable action plan.

- A first-mover strategy can lead to significant benefits in an emerging market, but is not a guarantee of success.
- An entrepreneur builds a value network of partners that help build a firm's strengths.
- An innovation strategy creates a road map for continual commercialized invention.

Principle 5
An innovation strategy builds on creativity, invention, and technologies, acting within a value network, to effectively commercialize new products and services for its customers.

5.8 EXERCISES

5.1 By 1993, a new way of making movies available for home viewing was introduced as the digital video disk (DVD). As DVD players became more widely available, the rental of DVDs started taking off, and by the late 1990s, the DVD became a challenge to the videotape format. Netflix offers a website (www.netflix.com) where customers set up a password-protected account through which they can order movies to be sent to them by regular mail with a return mailer.

Examine the Netflix website and determine the firm's basic strategy. What are the challenges to its strategy? Consider the timing of the initiation of NetFlix: was it too early or right on time?

5.2 Determine and describe the enabling technology used by Take-Two Interactive to develop its interactive software games (see www.take2games.com).

5.3 In May 1985, a small company, America Online, started in business with a service for connecting to the Internet. On March 19, 1992, AOL sold a portion of its shares to the public. AOL's aggressive pricing and marketing campaigns fostered rapid growth. Why did AOL beat its competitors Prodigy and CompuServe?

5.4 Gentex Corporation designs and manufactures automatic-dimming automotive rearview mirrors. Its safety mirrors use sensors and electronics to detect glare from trailing approaching vehicles at night and darken accordingly (see www.gentex.com) [Palmer, 2003]. Describe the invention and technology that Gentex uses. Draw a value-net diagram for Gentex and name its partners.

5.5 Zebra Technologies Corporation provides bar-code labeling solutions for use in automatic identification and data collection systems (see www.zebra.com). Describe the technology of Zebra in terms of the three dimensions of technological inventions: importance, radicalness, and patent scope.

5.6 Apple Computer introduced iTunes Music Store in 2003 (www.apple.com/itunes). Describe the underlying technology enabling iTunes. Estimate the impact of this innovation on the music industry.

5.7 Netflix rents DVDs by mail through an online service (www.netflix.com). Netflix faces a competitive challenge from Wal-Mart (www.walmart.com). Netflix's advantage comes from having started early, when DVDs were rare, and having developed a big lead in customers, revenue, and brand recognition. However, Wal-Mart can mount a relentless attack. In June 2003, Netflix won a patent covering much of its business model [Thompson, 2003]. Netflix serves consumers who want specialized, foreign, and independent films as well as popular films. Reed Hastings said he founded the company, in part, due to his frustration over a $40 late fee at his local video store. Using the six forces model, determine if Netflix will compete and win against Wal-Mart.

5.8 Daimler Chrysler is offering a car call GEM that runs on a battery and will travel up to 30 miles with a top speed of 25 mph before needing recharging. It sells for $6,000 to $8,000 (www.gemcar.com). What innovation is necessary to make this car a winner? Using that innovation, outline an innovation strategy using Table 5.5.

5.9 Determine the alliance partners for Johnson & Johnson Inc. and describe the benefits that accrue to Johnson & Johnson. Using Figure 5.5, determine the location of these alliances on the diagram.

5.10 An inventor brings you a new design for an electric toothbrush with an oscillating head and a tilted handle that appears to meet the American Dental Association criteria. The inventor has filed a preliminary application for a patent. Also, you have tried the brush and found it easy to use. Using the factors of Table 5.3, provide a brief review of this invention. Would you recommend proceeding with commercialization?

5.11 The E-Stamp Corporation was first to market in 1997 with the ability to sell stamps over the Internet to consumers who print the stamps on their printer (www.estamp.com). By 2001, however, E-Stamps' 31 patents and other intellectual property were purchased by Stamps.com (www.stamps.com). Study this acquisition and determine why being first to market was not a winning strategy for E-Stamp.

6

Risk, Return, and Product Design

Our greatest glory is not in never falling but in rising every time we fall.
Confucius

A new venture that creates a novel solution to a problem will be subject to uncertainty of outcome. An action in an uncertain market is sure to experience a risk of delay or loss. It is the entrepreneur's task to reduce and manage all risks as much as possible.

Attractive new ventures can be designed to grow as demand for their products increases. Furthermore, it is hoped that economies of scale will be experienced so that as demand and sales grow, the cost to produce a unit of product will decline. Additionally, it is desirable to have economies of scope so that costs per unit decline due to the spreading of fixed costs over a wide range of products.

Many industries established on a network format exhibit network economies resulting in a reinforcing characteristic leading to the emergence of an industry standard. Furthermore, product design and development, which is concerned with the concrete details that embody a new product, can add significant value to the product offered to the customer. Often, a new venture can design a product that is a disruptive application that reshapes an industry. ■

6.1 RISK AND UNCERTAINTY

The pursuit of important opportunities and big goals by entrepreneurs requires them to assume more risks than they might take on working for a mature company or the government. Introducing a novel product into a new market has an uncertain outcome. An outcome resulting from an action is said to be **certain** in that it will definitely happen. Something certain is reliable or guaranteed. For example, it is certain that if we drop a rock, it will fall to the ground (and not float upward).

An outcome resulting from an action is said to be **uncertain** in that the outcome is not known or is likely to be variable. **Risk** is the chance or possibility of loss. This loss could be financial, physical, or reputational. When Christopher Columbus embarked on his first voyage to the New World, he risked financial, reputational, and bodily harm. Farmers are a group whose fortunes are vulnerable to unpredictable outcomes due to drought, flood, or other weather conditions.

Most, perhaps almost all, people are risk-averse or risk-avoiders. Logically, an entrepreneur seeks to avoid or reduce the risk of an action. For example, farmers purchase insurance to mitigate the effects of uncertain weather. A simple measure of a person's risk aversion is provided in the following example of a coin toss game. [Bernstein, 1996].

Example: A Coin Toss Game

You have a choice of receiving $50 for certain or an opportunity to play a coin toss game in which you have a 50 percent chance of winning $100. The $50 gift is certain, if you elect it. The game's outcome is uncertain, but the expected outcome over the long run is (if you play it many times) also $50 since, for a true coin, the probability of winning, P, is 50 percent. Surely, you would play the game if the probability, P, of winning were 100 percent. Would you play if the probability were 70 percent? What is the probability, P, that you would require to play? If your P=70 percent, you are risk-averse, and if your P=40 percent, you are a risk-seeker. Of course, your willingness to play this game will also depend on your overall wealth in relation to the $50 outcome of loss and the fun of playing this simple game.

The ability to successfully choose the risks worth bearing is a form of human capital based on experience and good judgment [Davis and Meyer, 2000]. We usually assume that the elevated risk of an entrepreneurial venture may provide a higher return on this human capital. Entreprencurs often have a capability to limit the downside risks of a venture by applying their skills that are useful for mitigating the risks. One mental model may be the circus high-wire walker with a strong net below. In fact, most successful entrepreneurial firms create value by taking calculated risks and possess the core competency to mitigate or manage these risks. All of life is the management of risk, not its elimination.

TABLE 6.1 Four levels of uncertainty.

Uncertainty level	Level of risk	Example
1. A clear, single outcome	Very low	Purchase a Treasury Bill
2. A limited set of possible outcomes	Low	Set up a pushcart for selling sports items at the Olympics
3. A wide range of possible future outcomes	Medium	Launch an improved product in a new market
4. A limitless range of possible outcomes	High	Establish a firm to attempt to design and sell a revolutionary power source based on a fuel cell breakthrough

The entrepreneur is in many ways like an investment manager who chooses to pursue selected opportunities and not others. When we analyze risk, we look forward in time and try to estimate the potential outcome and the variability of that outcome. Risk is a measure of the potential variability of outcomes that will be experienced in the future. Furthermore, risk is the chance or possibility of loss. Table 6.1 lists the four possible levels of uncertainty [Courtney, 2001]. The level of return on investment we might expect should be commensurate will the level of uncertainty.

Entrepreneurs practice in the realm of opportunity in a manner similar to what financial investors do in the stock market [Sternberg, O'Hara, and Lubart, 1997]. They pursue opportunities with expected levels of risk and attempt to "buy low and sell high." Buying low means pursuing ideas that are not widely recognized or are out of favor. Selling high means finding a cash buyer for the successful venture, convincing them that its worth will yield a significant return and it is time to harvest some of the wealth already created. As with any investment, "invest in a business you understand" is a good principle. Thus, pursuing the hope of fusion power should be left to those few who know and understand this big but risky opportunity. The entrepreneur-investor assumes the risk of the venture and should be willing to take on sizable risks with the knowledge that he or she can manage or mitigate them.

The best method for most entrepreneurs is to use a form of experimentation—trial and error. Identify possible new ventures and take a few steps toward a business and then evaluate the early feedback. If it looks good, keep going forward. Entrepreneurs will most likely take the risk of failure if the losses are constrained on the downside and the potential rewards are high on the upside. Henry Ford said: "Failure is the opportunity to begin again, more intelligently."

Entrepreneur-investors should consider the concept of **regret,** which we define as the amount of loss a person can tolerate. People treat regret of loss differently than the potential of possible gain. How much regret people feel varies depending on their circumstances of wealth, age, and psychological well-being. Reconsider the coin toss game on page 124. If you play the game and lose $50 on the first turn, it may be disturbing. Your level of regret may be $200. If you play a sequence of four turns and lose each of them, you will reach your level of regret. Thus, entrepreneurs need to evaluate their regret level of acceptable loss and limit their investment in any entrepreneurial venture. If an entrepreneur can forgo one year of income ($50,000) and willingly invest his or her savings of $60,000 in the venture, then the entrepreneur's level of regret is $110,000.

The amount of risk entrepreneurs will endure varies, but most retain some personal financial reserve so that failure will not equate to homelessness or starvation. It is easier to be a risk taker if one has some reserves to fall back on.

The risk-adjusted value of a venture, V, is

$$V = U - \lambda R \tag{6.1}$$

where U=upside, λ=risk-adjusted constant, usually greater than 1, and R=downside or regret. The larger the value of λ, the more risk-averse is the entrepreneur. We are neutral at λ=1 but risk-averse at λ=2 [Dembo and Freeman, 1998]. If your regret is R=$110,000 and λ=2, to proceed requires V > 0, or

$$U > \lambda R \tag{6.2}$$

Therefore, the required upside, U, is U > $220,000. Entrepreneurs can use scenarios and economic analysis to estimate the potential upside, U, of a venture.

The strategic response to uncertainty is to build a venture in stages, reserving the right to adjust your core competencies and strategies and play again at the next stage. Thus, reconsidering the case of the entrepreneur who is risk-averse with λ=2, he or she can choose to proceed for six months with the venture so that R=$55,000, and the minimum upside required is then only $110,000. After the six months, the entrepreneur adjusts the business strategy to improve the business performance and based on a new calculation of the upside, U, proceeds to a second stage of activity or decides to terminate the venture.

To perform the kinds of analyses appropriate to high levels of uncertainty, most firms will need to enhance their strategic capabilities. Scenario-planning techniques can be useful for determining strategy under conditions of uncertainty.

Risk reflects the degree of uncertainty and the potential loss associated with the outcomes, which may follow from an action or set of actions. Risk consists of two elements: the significance of the potential losses and the uncertainty of those losses. In most new ventures, it is the significance or size of the potential losses, *hazard,* and the *uncertainty* that are estimated by new venture entrepreneurs and their investors. We propose a measure of risk as:

$$\text{Risk} = \text{hazard} \times \text{uncertainty} \tag{6.3}$$

The hazard, H, is the size of the potential losses as perceived by the entrepreneurial team. Hazard is an entrepreneur's income forgone (opportunity cost, OC) plus the financial investment, I, which he or she will need to make. Therefore,

$$\text{Risk} = (I + OC) \times UC \tag{6.4}$$

The uncertainty, UC, is measured by the variability in anticipated outcomes, which may be described by their estimate of the probability of loss (failure). Based on these factors, the entrepreneurial team may make a selection of a new

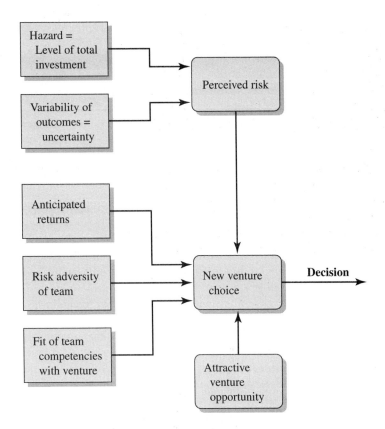

FIGURE 6.1 Risks and new venture choice.

venture (or to proceed or not), as shown in Figure 6.1. High levels of hazard may not deter entrepreneurs from choosing ventures with potentially high levels of returns. The decision to proceed or not depends on the level of risk adversity of the team and their perception of the extent of the uncertainty. Of course, the estimate of potential returns is subject to the assumptions underlying their calculation.

An entrepreneurial venture is launched with a degree of uncertainty due to the novelty of the product to the market, the novelty of the production processes, and the novelty to management [Shepard, Douglas, and Shanley, 2000]. Novelty to the market concerns the lack of uncertainty of the market and customer uncertainty. Novelty of production processes is dependent on the extent of knowledge of the processes by the venture team. Novelty to management concerns the venture team's lack of the necessary competencies. Regulation and legal changes are also a source of uncertainty. Sources of uncertainty are listed in Table 6.2.

The risk of failure or poor performance is significant and should not be understated. According to the U.S. Small Business Administration, about one-half of all small businesses are acquired by another firm or leave the market within four years. Of course, being acquired at a good price may be the success one is

TABLE 6.2. Sources of uncertainty.

1. Market uncertainties
   - Customer
   - Market size and growth
   - Channels
   - Competitors

2. Organization and management uncertainties
   - Capabilities
   - Financial strength
   - Talent
   - Learning skills
   - Strategies

3. Product and processes uncertainties
   - Cost
   - Technology
   - Materials
   - Suppliers
   - Design

4. Regulation and legal uncertainties
   - Government regulation
   - Federal and state laws and local ordinances
   - Standards and industry rules

5. Financial uncertainties
   - Cost and availability of capital
   - Expected return on investment

seeking. It is fair to say that one-fourth of all business start-ups are discontinued within four years, but it is unclear at what cost or return.

Technology entrepreneurs are often less concerned with certainty than with getting into the game quickly and learning how to participate. The uncertainty and associated risks decline as the novelty in the three dimensions of market, production, and management declines. Novelty is synonymous with uncertainty, and thus, we expect uncertainty to decline as knowledge of the market, the processes, and management competencies improve. Thus, the liability of newness declines over time. A process for managing risk and uncertainty is shown in Figure 6.2.

Whenever there is uncertainty, there is usually the possibility of reducing it by the acquisition of information. Indeed, information is essentially the negative

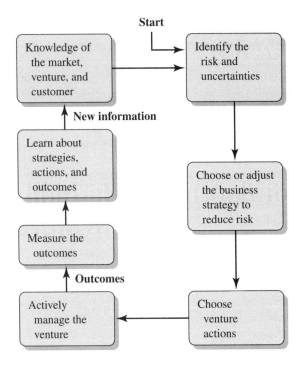

FIGURE 6.2 Managing risk and uncertainty.

of uncertainty. The acquisition of information and knowledge improves an organization's chances of adaptation and performance. The entrepreneurs are continuously making decisions about highly uncertain environments and the new venture's internal structures that in turn modify its performance outcomes. New venture managers may learn from past choices about how to perform better in the future. This learning can facilitate adaptation to changed environmental conditions since the strategic choices of managers help define the outcomes. An appropriate strategy for risk reduction uses new information learned from experience to adjust the business strategy and the actions taken to execute the strategy. Appropriate strategies might include adding new people to the team, creating new alliances, reducing costs, or improving customer relationships, among others.

Arthur Pitney received a patent in 1902 for a hand-operated postage meter that he hoped would be used to replace stamps for mass mailings. Pitney recognized his primary source of risk was regulatory since the U.S. Post Office controlled postage services. By 1918, after continual rejections, Pitney was joined by Walter Bowes to form Pitney Bowes and try again to get Post Office approval. With Bowes's persuasive skills, they finally received a license for the postage meter in 1920. They ultimately overcame the regulatory hazards to build up their firm. Many entrepreneurs underestimate their various risks.

In 1981, Gordon Eubanks, a software pioneer in the personal computer industry, co-founded C&E Software to develop an integrated database management and word processor product. In 1983, John Doerr, who was a venture capitalist with Silicon Valley's Kleiner Perkins Caufield and Byers (KPCB), approached Eubanks regarding one of his investments. Symantec was an artificial intelligence software firm that was struggling to stay afloat despite having an interesting technology known as natural language recognition. Doerr suggested that Eubanks merge with Symantec and incorporate Symantec's technology into C&E's product (see Figure 6.2). Although Eubanks viewed this technology integration as a compromise, he was persuaded by the upside of the deal—a substantial percentage of ownership for C&E in the merged venture, an additional cash investment from KPCB, and a chance to lead the new company. In late 1984, Symantec Corporation was reborn as a firm that developed a natural language database manager with word processing capabilities.

Even with the additional venture capital funds, it became apparent that Symantec would not survive as a one-product company. A few months before the first version of Symantec's database manager shipped in 1985, Eubanks hired Tom Byers to search for new revenue streams and to diversify the product line. Byers realized that a market existed for software utilities that added features to the then-popular Lotus 1-2-3 spreadsheet. Symantec used a strategy similar to book publishing with these products: in return for the rights to package and sell the software, Symantec paid developers a royalty as the product's author. Eubanks had adjusted Symantec's strategy in response to risk and uncertainty.

In 1987, Eubanks further diversified the firm. Symantec purchased Breakthrough Software for minimal cash and a $10 million note payable in cash if and when Symantec went public. Timeline, Breakthrough's project management software, quickly doubled Symantec's incoming revenue, which was important because sales of Lotus utilities were rapidly declining.

Also in the late 1980s, Eubanks quickly acted on two emerging trends in the industry. First was the proliferation of computer viruses as more and more computers were connected in networks. Second was the rapid advancements of graphical user interfaces (GUI) in personal computers such as the Macintosh. To rapidly respond to these changing conditions, Eubanks made several key decisions and acquisitions.

Ted Schlein, who worked in Symantec's publishing division, identified a market for antivirus software and pushed for Symantec to publish Symantec Antivirus for Macintosh (SAM), which became the first successful commercial antivirus product. In following years, Symantec made numerous acquisitions in this category, including Peter Norton Computing;

network security eventually became Symantec's core business. In addition, Eubanks acquired two Macintosh software companies whose developers were very experienced in GUI-based software. These acquisitions helped prepare the company for the subsequent decade in which all PC software was based either on Windows or Macintosh user interfaces.

Finally, marketing and distributing Symantec's growing line of software became vital to its success. Recognizing that the software industry was rapidly maturing with large corporations becoming most important, Eubanks hired John Laing as head of global sales and Bob Dykes as CFO. Laing and Dykes, experienced professionals who had worked in large business operations in the past, created the necessary systems and processes to handle the upcoming rapid growth.

In 1989, Symantec went public with 264 employees, $40 million in sales, net income of $3 million, and 15 products. During the 1990s, Eubanks focused Symantec on network security technology. In 1999, Eubanks stepped down as CEO of Symantec, and John Thompson from IBM took the reigns. In 2004, Symantec had revenues of nearly $2.1 billion, more than 5,000 employees, operations in 36 countries, and a market capitalization of over $14 billion. Thompson continued the tradition of focusing the entire product line around enterprise security.

Business can manage the problem of unpredictable customer behavior by following the ideas of portfolio management. The portfolio of customers should be diversified so as to produce the desired returns at the particular level of uncertainty the firm can tolerate. Customers are risky assets. As with stocks, the cost of acquiring them is supposed to reflect the cash-flow values they are likely to generate. The concept of risk-adjusted lifetime value of a customer has a transforming power [Dhar, 2003].

To the uninitiated, successful new ventures appear to be the right idea at the right time. However, entrepreneurs put the pieces together so that while it looks like a happenstance, it is actually the entrepreneur who makes it happen with good calculations based on sound information. As Charles Kemmon Wilson, the founder of Holiday Inns, said [Jakle 1996]:

> "Opportunity comes often. It knocks as often as you have an ear trained to hear it, an eye trained to see it, a hand trained to grasp it, and a head trained to use it."

One way to calculate an estimate of a new venture's potential risk and reward is to answer the four questions posed in Table 6.3. In general, the entrepreneur seeks a venture where the return is expected to significantly exceed the potential losses.

TABLE 6.3 Estimating risk and reward.

1. Describe the most likely scenario, the expected reward, and its estimated probability.

2. Describe the worst-case scenario, the expected loss, and its estimated probability.

3. Describe the best-case scenario, the expected reward, and the estimated probability of it occurring.

4. Determine how much the entrepreneurial team and their investors can afford to lose. Include their investment and opportunity costs.

6.2 SCALE AND SCOPE

In this section, we consider the strategic impacts of the scale and scope of a firm. The **scale** of a firm is the extent of the activity of a firm as described by its size. The scale of a firm's activity can be described by its revenues, units sold, or some other measure of size. **Economies of scale** are expected based on the concept that larger quantities of units sold will result in reduced per-unit costs. Economies of scale are generally achieved by distributing fixed costs such as rent, general and administrative expenses, and other overhead over a larger quantity, q, of units sold. This effect is portrayed in Figure 6.3. The cost per unit decreases, reaching a minimum at q_m. Often, the cost per unit will increase for $q > q_m$, since the complexity of coordination may increase costs per unit for a high number of units.

When significant economies of scale exist in manufacturing, distribution, service, or other functions of a business, larger firms (up to some point) have a cost advantage over smaller firms. Thus, smaller, new-entrant firms need to dif-

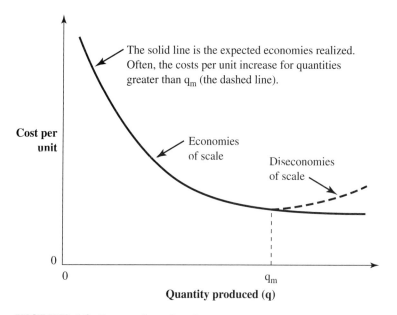

FIGURE 6.3 Economies of scale.

ferentiate their product on qualities other than price. As the smaller, new-entrant firm grows in size, it can also learn to reduce its costs per unit and price competitively with larger firms.

> Google's sustainable competitive advantage is tied to its innovation underlying its search engine. Google offers great results and keeps people coming back. Google may eventually be threatened by a powerful competitor—Microsoft. Revenues for Google were estimated at $100 million in 2001, only three years after starting up. By 2003, its revenues were approaching $1 billion. Its primary revenues are from advertising on its pages. It gets its fees by selling rights to given keywords so an ad shows up first when those words are entered. Search engines for the Web are based on computer science algorithms and need to search without failure [Hardy, 2003]. Google has the potential to benefit from economies of scale.

Another issue related to scale is the concept of scalability. **Scalability** refers to how big a firm can grow in various dimensions to provide more service. There are several measures of scalability. They include volume or quantity sold per year, revenues, and number of customers. These dimensions are not independent, as scaling up the size of a firm in one dimension can affect the other dimensions. Easily scalable ventures are attractive, while ventures that are difficult to grow are less so.

The consequence of growth is the necessity to respond by increasing capacity. **Capacity** is the ability to act or do something. Any firm has processes, assets, inventory, cash, and other factors that must be expanded as the company grows its sales volume. A firm that can easily grow its capacity is said to be readily scalable. For example, as a firm grows, its working capital requirements will grow. **Working capital** is a firm's current assets minus its current liabilities. Sources of working capital for an emerging firm can include long- and short-term borrowing, the sales of fixed assets, new capital infusions, and net income. The ability for a new firm to grow will be influenced by its access to new capital and assets. Managing a firm for scalability is important to its success. To preempt or match competitors, a firm must attempt to foresee increases in demand and then move rapidly to be able to satisfy the predicted demand. This strategy can be risky since it involves investing in resources before the extent of the demand is verified. The total cost, TC, of the production of units is described as:

$$TC = FC + VC$$

where FC is the fixed costs that do not vary with the quantity of production. The variable costs, VC, do vary with the quantity produced where $VC = c \times q$, c being the cost/unit and the quantity being q. This relationship is shown in Figure 6.4. Table 6.4 describes the scalability and economies of scale for four types of businesses.

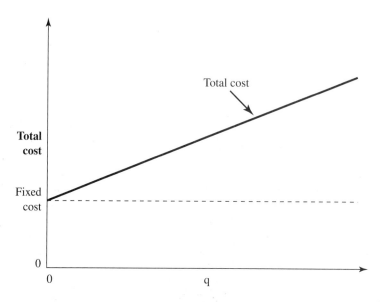

FIGURE 6.4 Total cost as fixed costs plus variable costs.

The advantage of a talent-based business such as a consulting firm is that the start-up funds are low. The firm is scalable as long as new talent can be recruited for expansion, but it has few economies of scale. The advantage of a firm based on a mixture of talent and physical assets is that it can expand as long as it can

TABLE 6.4 Scalability and the effects of fixed and variable costs for four types of businesses.

Type of business	Economies of scale	Scalability	Fixed costs	Variable costs	Primary strategy	Funds required by start-up
1. Based on talent; e.g., consulting	Low	Medium	Low	High	Recruit talent	Low
2. Based on talent and knowledge assets; e.g., plastics, toys	Medium	Medium	Medium	Medium	Secure physical assets and talent	Medium
3. Based on physical assets, knowledge, and materials; e.g., biotechnology, semiconductors	High	Low	High	Low to medium	Secure physical assets	High
4. Based on information with few physical assets; e.g., software, movies	High	High	High	Low	Secure talent to create the software or the movie	High

secure the necessary funds. A physical asset-based business such as a steel company must secure new plants and equipment as it grows, requiring capital infusion. An information-based business must invest funds upfront to create the software or a movie. It has low variable costs and high economies of scale.

The **scope** of a firm is the range of products offered or distribution channels utilized (or both). The sharing of resources such as manufacturing facilities, distribution channels, and other factors by multiple products or business units gives rise to **economies of scope.** For example, the cost per unit of Procter & Gamble's advertising and sales activities are low because they are spread over a wide range of products. Procter & Gamble's disposable diaper and paper towel businesses demonstrate a successful realization of economies of scope. These businesses share the costs of procuring certain raw materials and developing the technology for new products and processes. In addition, a sales force sells both products to supermarket buyers, and both products are shipped by means of the same distribution system. This resource sharing has given both business units a cost advantage compared to their competitors [Hill and Jones, 2001].

Campusfood.com is an information-based business that provides interactive menus from which college students can order meals and snacks for delivery. Launched at the University of Pennsylvania in 1998, the company managed growth and scalability by adding other universities and restaurants to its list. Campusfood.com takes a 5 percent commission on each order. Campusfood.com enables purchasers in their dorm rooms to select a pizza, sandwich, or other food items, pay for it, and have it delivered. Because scalability and capacity were relatively easy to achieve using an online format, the firm now represents 1,000 restaurants at 200 colleges and universities (see www.campusfood.com).

The economies of scale and scope both reduce the cost per unit. For a factory, its **throughput**—the amount processed within a given time—needs to be consistently high. The introduction of the railroad to the United States and Europe reduced the travel time between markets and supply, increasing the flow of raw materials to factories. The revolution in railroad transportation and telegraph communication resulted in great increases in throughput. In 1870, the Union Pacific Railroad joined the Central Pacific at Promontory, Utah, spanning the United States [Beatty, 2001]. The United States became the world's leading industrial producer by 1929. The economies of scale and scope helped the United States to become a low-cost producer and distributor of goods (innovation was another reason for high productivity).

The strategy of a new firm must incorporate plans for economies of scale and scope. A modern example of a company with vast economies of scale and scope is Wal-Mart, the largest firm, by revenues, in the United States. Other examples include General Electric and Exxon-Mobil.

6.3 NETWORK EFFECTS AND INCREASING RETURNS

In recent years, realization has been growing that network economies are an important element of competitive economics for new entrepreneurial firms. **Network economies** arise in industries where a network of complementary products is a determinant of demand (also called network effects). For example, the demand for telephones is dependent on the number of other telephones that can be called with a telephone. As more people get telephones, the value of a telephone increases, thus leading to increased demand for telephones. This process is called a positive feedback loop since as more people use the process, the value to users of the process increases and thus demand goes up, leading to more people using the process.

The positive feedback process for Windows-Intel (Wintel) personal computers is illustrated in Figure 6.5. As the number of Wintel PCs increases, the incentive for the development of software applications increases (a complementary product). With more application software available, the PC is more valuable to a user. As the value of a PC increases, the demand for Wintel PCs increases, leading to an increased number of Wintel PCs.

Networks include telephone networks, railroad networks, airline networks, fax machine networks, computer networks, ATM networks, and the Internet, among others. The overall tendency is toward "bigger is better." As Figure 6.6 indicates, over time, a winner emerges (company A) and the competitors decline. In the PC industry, Wintel has become the winner, with Apple holding only about 10 percent of the market. In general, network effects exhibit reinforcing characteristics, as shown in Figure 6.7.

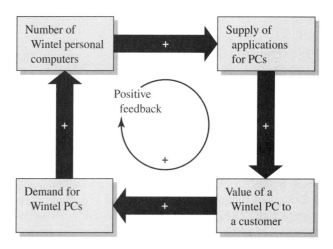

FIGURE 6.5 Increasing demand for Wintel personal computers due to positive feedback.

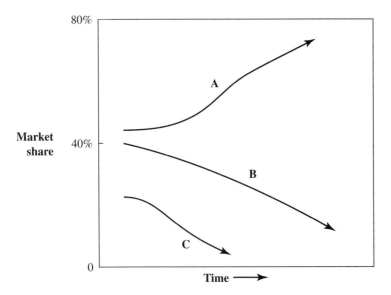

FIGURE 6.6 Emergence of a dominant firm, A.

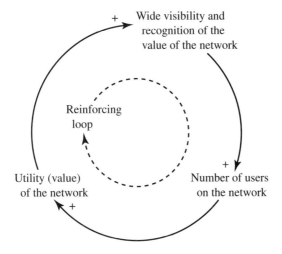

FIGURE 6.7 Reinforcing characteristic of a positive loop exhibiting network effects.

The value of a network, according to Bob Metcalfe, is approximately:

$$\text{value of network} = kn^2$$

where k is a constant and n = number of participants in the network. Based on this model, the value of a network grows rapidly as n grows. This simple model assumes all participants are equally valuable, which is not always true. The incentive to a firm in a network economy is to secure market share, eventually taking off toward dominance. This is the underlying theory behind Amazon.com and other Internet start-ups. Of course, market share can grow rapidly while profitability may be elusive. A balance of both market share and profitability can lead an entrepreneurial firm to eventual success if it has a product that has high value for its customers and strong alliances with its complementors.

Network economies work when revenues grow faster than costs. Webvan tried to become the online grocer of choice in the United States. It had to invest in warehousing, trucks, and logistic systems, which led to increased costs and inventories, which caused its costs to spiral upward faster than its revenues, eventually sinking the firm into bankruptcy.

Increasing returns mean that the marginal benefits of a good or of an activity are growing with the total quantity of the good or the activity consumed or produced [Van den Ende and Wijaberg, 2003]. Increasing returns is the tendency for a company that is ahead, firm A in Figure 6.6, to get farther ahead. The theory is that the firm that has a successful product that is increasingly becoming the standard for the industry will experience increasing returns as increasing quantities are sold. However, there is no guarantee that increasing market share (firm A in Figure 6.6) will experience profitability. Furthermore, it is not possible to predict ahead of time which firm will attain market share dominance. If a product or a company or a technology—one of many competing in a market—gets ahead by quality of offering or clever strategy, increasing returns can magnify this advantage, and the product or company or technology can go on to lock in the market. Microsoft DOS became dominant after a protracted battle with CPM and Apple for PC operating system leadership.

Many new firms may enter a new industry with the potential to eventually dominate, but only one or two survive. Products do not stand alone but depend on complementary products to make their use valuable. Examples of network industries that exhibit increasing returns are airlines and banking. For example, as Southwest Airlines expands the number of the cities it serves, the value to a customer of using the airline increases. Another prominent example is eBay, which has dominant market share in online auctions. Because it was the first to connect individual buyers and sellers over the Internet in an auctionlike format, it grew very quickly. As more rare items came up for auction on eBay, more buyers were attracted to the website to bid on those items. Those extra bidders attracted still more sellers, thus leading to dominance. Note that eBay was profitable from the first year onward.

While Metcalfe's law illustrates the general idea of the value of a network, it is only an approximation to reality. The value of each node (participant) will vary. Furthermore, some links will be strong while others are weak. Customers value the number of nodes in the network but also key links in the network. A network of five nodes and eight links is shown in Figure 6.8. Note that not all nodes are connected by a link to all other nodes in this example. Consider a bank network with 100 branches. Most people do not visit many other branches ex-

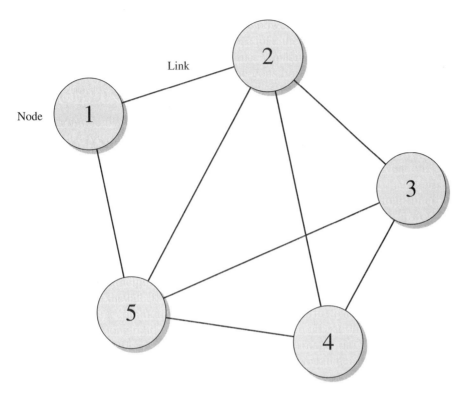

FIGURE 6.8 Network of five nodes and eight links.

cept their local branch and perhaps one near their work. For many customers, the link to their account at their local branch and online or via telephone is what they value. Thus, designers of a network business must analyze their customers' needs and build their business on the best information.

Consider Southwest Airlines. It has no physical branches, nor does it use travel agents; instead it uses telephone and online links to its network. It offers strong incentives to its customers to use the online rather than the telephone links. Physical branches may be necessary for banks but not for airlines. Wells Fargo is putting minibranches in grocery stores on the theory that its customers value physical nodes (branches) as well as Internet links.

In general, knowledge-based products exhibit increasing returns. The up-front development costs are high, but the per-unit production costs thereafter are low. Knowledge-based products also exhibit effects; that is, the more people who use these products, the more valuable they become.

An example of such a start-up is eBay, which eventually became the auction site of choice. eBay experienced a virtuous cycle of network effects. Buyers came to eBay because it was where all the sellers were; sellers came because that was where the buyers were.

Many entrepreneurs have come to the conventional conclusion that you can't sell used cars on the Internet. Many buyers, however, have purchased used autos on eBay [Wingfield and Lundegaard, 2003]. EBay hosted 300,000 used-vehicle sales in 2002. Offering a used car online has big advantages. Classified car listings in newspapers are usually limited to a few lines of tiny text. Used-car websites are jammed with pictures, making it easier to portray all the scrapes and quirks of a vehicle. There is also a better selection online than in the hodgepodge of local newspapers and auto publications in which dealers usually advertise. Most sites just advertise used cars or refer a customer to a dealer. Car buyers on eBay actually commit to buying vehicles on the Internet. eBay sellers now pay $40 to list a vehicle for auction, with unlimited pictures and text, as well as an additional $40 "success fee" if the auction closes with a winning bidder.

A competitor, AutoTrader, has designed its new site to exploit what it sees as the great weakness of eBay: the need to commit to a purchase before seeing the vehicle firsthand. With AutoTrader's auctions, sellers can offer "conditional" listings that let the winning bidder inspect a vehicle before buying it to ensure it was accurately described by the seller (see www.autotraders.com).

EBay, AutoTrader, and AutoNation are competing for dominance in this market and the eventual benefits of network effects. Which of these three companies do you expect to dominate this market and reap the benefits of network effects?

6.4 PRODUCT DESIGN AND DEVELOPMENT

One of the early tasks of a new venture is the design and development of the new product. The entrepreneurial team wants to develop a new product or service that can establish a leadership position. One of the strengths of a new venture is that the leadership of the venture plays a central role in all stages of the development effort. Furthermore, the small new firm is able to integrate the specialized capabilities necessary for the development of a successful product [Burgelman, 2002].

In recent years, product complexity has dramatically increased. As products acquire more functions, the difficulty of forecasting product requirements rises exponentially. Furthermore, the rate of change in most markets is also increasing, thereby reducing the effectiveness of traditional approaches to forecasting future product requirements. As a result, entrepreneurs need to redefine the problem from one of improving forecasting to one of eliminating the need for accurate long-term forecasts. Thus, many product designers attempt to retain flexibility of the product characteristics as the development proceeds. A design and development project can be said to be flexible to the extent that the cost of any change is low. Then, project leaders can make product design choices that allow the product to easily accommodate change [Thompke and Reinertsen, 1998].

Uncertainty is an inevitable aspect of all design and development projects, and most entrepreneurs have difficulty controlling it. The challenge is to find the right balance between planning and learning. Planning provides discipline, and learning provides flexibility and adaptation. Openness to learning is necessary for most new ventures that are finding their way into the market [DeMeyer, Loch, and Pich, 2002].

Design of a product leads to the arrangement of concrete details that embodies a new product idea or concept. The design process is the organization and management of people, concepts, and information utilized in the development of the form and function of a product. The role of design is, in part, to mediate between the novel concept and the established institutional needs. For example, Thomas Edison designed and described the electric light in terms of the established institutions and culture. As a result, he succeeded in developing an electric lighting system that gained rapid acceptance as an alternate to the gas lamp. A new product needs to be advanced, yet it should not deprive the user of the familiar features necessary for understanding and using the product. As new products are designed, the challenge ultimately lies in finding familiar cues that locate and describe new ideas without binding users too closely to the old ways of doing things. Entrepreneurs must find the balance between novelty and familiarity, between impact and acceptance [Hargadon and Douglas, 2001].

A prototype model of a Palm Pilot was built by Donna Dubinsky and Jeff Hawkins and shown to buyers. The key issue was addressed when they showed the docking cradle and explained that their device could connect to a PC with the touch of a button. Nobody had done that before. It seems basic now, but no one had made the logical leap that this organizer was a PC accessory, not a stand-alone PC.

The overall development process is shown in Figure 6.9 [Thompke and von Hippel, 2002]. The overall development process can include design of the product and its architecture, its physical design, and testing. The iPod and the BMW auto are examples of the outcome of a creative, artistic process of design. Part of the user experience is the look and feel of a product. A good product is attractive to see and easy to use and understand. Furthermore, customers want a product that does a few things really well. Fortunately, customers can participate fruitfully in the product design process when the innovations are incremental [Nambisan, 2002]. Many designers think about the qualities of a product but overlook that services consist of soft benefits such as warmth, status, and community.

Design includes aesthetics as well as basic needs. However, design also includes compromises and limits. A beautiful glass must be functional as well as attractive. Even the Maglite flashlight is flawed by the spot in the middle of the beam.

Successful product design and development requires commitment, vision, improvisation, information exchange, and collaboration, as listed in Table 6.5 [Lynn and Reilly, 2002]. These five practices may be easy to achieve in a start-

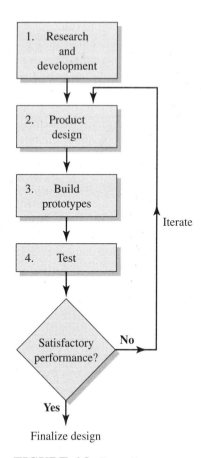

FIGURE 6.9 Overall
development process.

up where collaboration is the order of the day. The product team, which may be
all of the employees of a start-up, needs to clearly understand the vision for the
product and work together effectively.

The product design (step 2 in Figure 6.9) process is shown in Figure 6.10. The
first step is to establish the goals and attributes of the product expressed as the required performance and robustness of the product (step A). The components and

TABLE 6.5 Five practices of good product development.

■ Commitment of senior management to the design process.	■ Improvisation and iteration to develop a prototype.
■ Clear and stable vision and goals for the product.	■ Open sharing of information.
	■ Collaboration of everyone on the team.

Source: Lynn and Reilly, 2002.

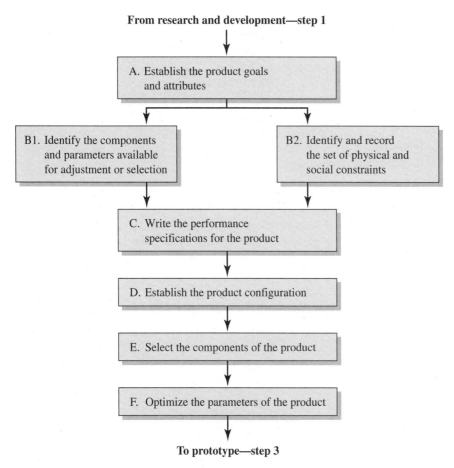

From research and development—step 1

A. Establish the product goals
   and attributes

B1. Identify the components
    and parameters available
    for adjustment or selection

B2. Identify and record
    the set of physical and
    social constraints

C. Write the performance
   specifications for the product

D. Establish the product configuration

E. Select the components of the product

F. Optimize the parameters of the product

To prototype—step 3

FIGURE 6.10 Product design process (step 2 of Figure 6.9).

parameters available for adjustment are identified, and specifications for the product are agreed upon. Specifications are the precise description of what the product has to do. In addition, the set of physical and social constraints should be determined. Next, the product configuration is established, and the components of the product are recorded. Finally, the parameters of the product are optimized to achieve the best performance and robustness at a reasonable cost [Ullman, 2003]. A **robust product** is one that is relatively insensitive to aging, deterioration, component variations, and environmental conditions. Preparing a robust design implies minimizing variation in performance and quality. All designs involve tradeoffs between performance, cost, physical factors, and other constraints [Petroski, 2003]. The success or failure of any design is ultimately determined in the marketplace.

Usability is a measure of the quality of a user's experience when interacting with a product. Usability is a combination of the five factors listed in Table

TABLE 6.6 Five factors of usability.

1. **Ease of learning:**	How long does it take to learn the product's operation?
2. **Efficiency of use:**	Once experienced, how fast can the user complete the necessary steps?
3. **Memorability:**	Can the user remember how to use the product?
4. **Error frequency and severity:**	How often do users make errors, and how serious are these errors?
5. **Satisfaction:**	Does the user like operating the product?

6.6. Examples of a common product with poor usability are most VCR and DVD players. New products should pass the five-minute test, which requires that the product is simple enough to use after quickly reading the instructions and then trying it for a few minutes. Products with excellent usability are the Post-It note and the Zagat Survey Guides.

> A baby diaper is a straightforward product with high usability. It feels like papery underwear wrapped around a pad of cotton. The liner made of polyfilm is the reason the garment stays dry. It has pores that let air flow in but are too small to let water flow out. The chief characteristic of the diaper is its simplicity based on the absorbent polymer. The design of diapers has tended toward smaller dimensions with increased holding capacity. Here is a great example of a technology (polymers) enabling a basic retail firm to excel [Gladwell, 2001]. It is in the perfection of a design using an improved technology that leads to great firms such as Procter & Gamble and products such as Huggies.

Many system designs use a combination of modules within a specified architecture. A **module** is an independent interchangeable unit that can be combined with others to form a larger system. In modular designs, changing one component has little influence on the performance of others or on the system as a whole. An example is the Walkman, which Sony's engineers first developed from a wide range of standard, interchangeable parts and modules. Design methods using independent modules make product design more predictable. Of course, the predictability inherent in modular design increases the chances that competitors can develop similar products.

Realistically, most products consist of modules that possess some dependency between them. For example, an auto is a product that consists of wheels, engine, body, and controls that are relatively interdependent. Products made up of modules with intermediate levels of interdependence are harder for competitors to duplicate and may also provide better performance than a design based on purely independent modules [Fleming and Sorenson, 2001].

Designers strive to create new products different enough to attract interest but close enough to current products to be feasible to make a market. Many new designs flow from changing the components, attributes, or integration scheme to create a new product [Goldenberg et al., 2003]. The designer asks what can be rearranged, removed, or replicated in new ways.

Over time a dominant design in a product class wins the allegiance of the marketplace. A **dominant design** is a single architecture that establishes dominance in a product class. An example is the IBM-compatible personal computer, which is a dominant design because it is viewed as superior in the marketplace. Eventually, a dominant design becomes embedded in linkages to other systems. The VHS system became the dominant design in its competition with the Betamax system.

A **product platform** is a set of modules and interfaces that forms a common architecture from which a stream of derivative products can be efficiently developed and produced. For example, Procter & Gamble's Liquid Tide is the platform for a whole line of Tide brand products. Firms target new platforms to meet the needs of a core group of customers but design them for ready modification into derivative products through the addition, substitution, or removal of features. Well-designed platforms also provide a smooth migration path between generations so neither the customer nor the distribution channel is disrupted. A good example of a platform is Hewlett-Packard's electronics and software used for its wide range of printers.

6.5 FINDING DISRUPTIVE APPLICATIONS

A **disruptive application** is a new product that establishes an entirely new category and dominates that category [Downes and Mui, 1998]. E-mail is an application on the Internet that is a disruptive application (often called a "killer app"). It is e-mail that makes the Internet so widely used. An example of a disruptive application for the personal computer is spreadsheet software.

VisiCalc, created in 1979, was the first electronic spreadsheet that helped make the personal computers widely useful. It is this type of disruptive application that causes an industry to grow exponentially. Entrepreneurs need to discern a possible disruptive application for their start-up firm's product. Great disruptive applications bring significant value to a product. A good example is the invention of the "800" number toll-free telephone call by AT&T, which profited handsomely from this innovation. The Google search engine may turn out to be a disruptive technology. Finally, the online college may be the disruptive application in the education industry of the future.

Akio Marita, the founder of Sony, asked his engineers to design a small portable radio and cassette player that would provide good audio quality and be attached to a person's head. No customer was asking for this product, yet eventually, the Sony Walkman became one of the twentieth century's disruptive applications of miniaturized electronics. In a capitalist economy, success is the ability to anticipate and meet the difficult-to-anticipate needs and wants of customers, and the most successful entrepreneurs are those who do this best.

In the search for disruptive applications, people often look for attributes that will attract users to a new product. Another approach is to recognize that customers "hire" a product or service to get a job done. Home Depot and Lowe's are organized around jobs to be done [Christensen and Raynor, 2003]. Customers become aware of needing to get something done or fix some problem and set out to hire or engage a product or service that can meet their need. Customers will pay a significant premium for products that do a job well. For example, Courtyard by Marriott is designed for business travelers whose need (job) is for a clean, quiet room in which to work into the evening. People don't want a tool; they want a solution or outcome.

> A two-wheel-drive mountain bike provides added traction on hills and in mud. Cyclists now have an all-wheel-drive mountain bike that serves as a disruptive application of a mountain bike for true off-road aficionados (see www.christini.com).

The development of the CT scanner for medical imaging drew upon X-ray technology combined with computer technology to create a new "killer app" for X-ray technology [Adner and Levinthal, 2002].

6.6 RISK AND RETURN

Reaching for higher returns carries higher risk. Assuming the entrepreneurs and their investors are rational beings, they will demand higher potential returns for higher-risk ventures. We illustrate a risk-reward model in Figure 6.11. The expected return varies as

$$ER = R_f + R$$

where ER = expected return, R_f = risk-free rate of return, and R = risk. High-risk and high-return investments will be expected to return in excess of 30 percent annualized over a period of years, T. For a high-risk venture, T may range from three to seven years [Ross, Westerfield, and Jaffe, 2002]. A disruptive application or radical innovation is expected to return in excess of 40 percent annualized over T years.

Alloy Inc. was incorporated in 1996 as a direct catalog and online firm selling clothes and other goods to 60 million females between the ages of 10 and 24 in the United States (see www.alloy.com). The founders, Matt Diamond and Jim Johnson, each contributed $60,000 in cash and a friend, Sam Graders, added another $150,000 when he joined Alloy six months later. By April 1999, Alloy went public with an IPO at $15 per share. Alloy was a high-risk, high-return start-up over the three-year period of 1996 to 1999. When the IPO was completed, the annualized return to the founders exceeded 100 percent.

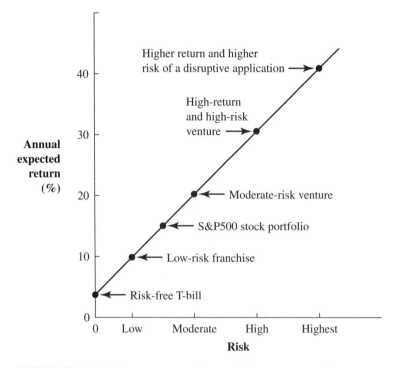

FIGURE 6.11 Return-versus-risk model for a new venture.

6.7 AGRAQUEST

From its inception, AgraQuest has experienced great uncertainty and risk. Risks arise from market uncertainty, organizational uncertainty, and process and regulatory uncertainty. AgraQuest has the potential to provide several environmentally friendly solutions to the pesticide problems of farmers. However, the market size and the willingness of farmers to use a new product are sources of significant risk. In addition, regulatory and legal controls of pesticides are large potential sources of risk.

   AgraQuest stated in its original plan that it expected to receive approval for each product from the Environmental Protection Agency in 12 months. Actually, the EPA required a two-year approval process. On the other hand, from the beginning, AgraQuest has experienced few problems managing its organizational and management uncertainties. AgraQuest underestimated the market and regulatory risks but was very successful in mitigating its organizational and processes risk. Its scientific and technological capabilities and product selection and design led to few risks in the building of its technological success. Unfortunately, the market and regulatory problems resulted in a slower growth curve than planned for.

From its start in 1995 until 2002, AgraQuest has only been able to grow its annual revenue to about $3.4 million. It has not yet been able to achieve economies of scale. Furthermore, only one product, Serenade, was available in 2003. Therefore, its product scope is very limited. The firm is now scalable with proven, solid processes for product development and production; it is ready to grow.

The company's first offering, Serenade, is a foliar biofungicide that is approved for use on crops such as grapes, apples, pears, cherries, peanuts, and tomatoes. A complimentary pesticide, Sonata, is awaiting approval, and AgraQuest is developing an insecticide dubbed Virtuoso. The musical names represent the movements in AgraQuest's integrated pest management symphony.

In addition, to help production buildup, AgraQuest acquired an exclusive license in 2002 for *Muscodor albus* and related strains, which are endophytic fungi (living within a plant) that produce gaseous substances, for use as an all-natural fumigant. Gary Strobel, a professor at Montana State University, discovered *M. albus* during an expedition in the rainforests of Central America. In 2001, Strobel's work on *M. albus* was selected by the editor of *Science* magazine as a key work in the entire field of microbiology. *Muscodor* can be used as a fumigant to control plant pathogenic fungi.

In 2000, AgraQuest purchased a manufacturing plant in Tlaxcala, Mexico, which produces Serenade and, eventually, future products. AgraQuest is seeking to build a product line that will become a disruptive application as chemical pesticides are replaced by natural pesticides. This is an attempt to develop a "killer app" in the agricultural pesticide industry.

6.8 SUMMARY

A new venture is subject to risks due to uncertainty of outcomes in the marketplace. It is the entrepreneur's job to manage and reduce all risks. As ventures grow, they may experience economies of scale and scopes, resulting in lower costs per unit produced. Furthermore, attractive ventures are those that can readily expand their capacity in response to demand and are said to be readily scalable. Many industries operate in a network of participants and exhibit network economies. As some network industries move toward an industry standard, a few firms may experience increasing returns.

Significant value can be added to a product when a creative design leads to the embodiment of critical details. Furthermore, a new venture can strive to design a product that is a disruptive application ("killer app") that reshapes an industry.

Principle 6
The entrepreneur seeks to manage risks, engage in creative design, and attain economies of scale, scope, and networks while achieving scalability of the business.

6.9 EXERCISES

6.1 Amazon.com and Borders are both in the book- and music-selling business. Contrast the scalability of each business.

6.2 Prefabricated housing may be a good option for affordable housing throughout the world. Sears Roebuck started selling prefabricated home kits in 1908. Modular homes look pretty much like regular houses but are made in a factory and delivered in large segments by truck. Homes can readily be constructed of rectangular modules. Describe the design issues and other factors that limit the wide use of modular housing [Akst, 2003]. See www.modern-modular.com for further information about modern modular homes.

6.3 Why hasn't the digital video recorder (DVR) become a hit product? DVRs are similar to videocassette recorders, except that they record onto a hard drive instead of tapes. Users browse a listings grid on the TV and press a record button on the remote to select each show that they want to record. Another press sets the machine to record every episode of that series automatically [Pogue, 2003]. The features on these devices provided by TiVo and others are difficult to understand and use. Examine one of these devices and propose a plan for reducing the features or redesigning how a user controls the features. How would you design a DVR to sell widely and reach the mass market?

6.4 An investor is asked to invest $10,000 in a new venture today. The expected return in three years is $28,000 with a probability of occurrence of 70 percent. Would you recommend this investment? Describe your reasoning.

6.5 The BMW Mini Cooper automobile design is a redesign of the highly regarded Mini of the 1960s. Does this design provide an attractive product for people in their twenties? Will it succeed as a favorite automobile?

6.6 Neopets is a website that offers ownership of virtual pets that inhabit a mythical land called Neopia. The primary user is between 12 and 18 years old (see www.neopets.com). Describe the potential for building a network of users possibly leading to increasing returns.

6.7 Consider the CD jewel box with its easily cracked hinge, cranky release mechanism, and stubborn cellophane wrapping. Suggest a redesign of a storage box for CDs and DVDs.

6.8 Social network websites aim to connect people to new people [Grimes, 2003]. For example, Friendster allows a user to create a personal profile and invite new friends to connect. These sites can be used to find a job, a friend, or a date. Social networking sites exhibit a form of network effects. Visit www.friendster.com and describe the benefits of this activity.

6.9 Vending machines are like unattended stores. New machines that reliably dispense the desired item are needed. Studies show that people want glass-front machines that show the products and accept debt or credit cards [Berk, 2003]. Use the product design process of Figure 6.10 to obtain an initial design for this product.

6.10 A new entrepreneur is relatively risk-averse with a risk-adjusted constant $\lambda = 2$. Her opportunity cost is \$100,000 before earning a regular salary from the venture in its second year. She also invests her savings of \$50,000. Calculate the minimum return after one year that will be acceptable to her in the second or third year.

6.11 Using the product design method of Figure 6.10, complete an outline of the steps necessary to design a small electric car that can compete favorably with the GEM car (www.gemcar.com).

CHAPTER 7

Corporate Technology Ventures

Even if you are on the right track, you'll get run over if you just sit there.
Will Rogers

T here are many types of new ventures, ranging from a small business or consulting firm to a high-growth start-up. Important social organizations are often established as nonprofit ventures. Other organizations start out operating in a niche market and grow into a broader market. Other new ventures emerge within larger existing corporations and are granted autonomy so they can fulfill their promise. Corporate ventures are an important part of the entrepreneurial world and account for many new innovations. Corporations are often constrained by their existing commitments and capabilities and often fail to respond to significant new opportunities. Well-planned corporate new ventures, however, can help refresh and strengthen large corporations. ■

7.1 TYPES OF NEW BUSINESS VENTURES

Many purposes exist for establishing new ventures in different formats. Table 7.1 describes six types of new ventures. Each of these types has a set of characteristics that distinguishes it. We can describe a **small business** as a sole proprietorship, a partnership, or a corporation owned by a few people. Examples include consulting firms, convenience stores, and local bookstores. Typically, a small business has fewer than 30 employees and annual revenues less than $3 million.

A **niche business** seeks to exploit a limited opportunity or market to provide the entrepreneurs with independence and a slow-growth buildup of the business. This business might employ fewer than 100 employees and have annual revenues of less than $10 million. Nevertheless, a niche business can grow over time into a large, important enterprise.

A **high-growth business** aims to build an important new business and requires a significant initial investment to start up. A **radical innovation business** seeks to commercialize an important new innovation and build an important new business.

A **nonprofit organization** is a corporation or a member association initiated to serve a social or charitable purpose. Thousands of new nonprofits are organized every year to serve important social needs throughout the world. Perhaps the most well-known nonprofit is the International Red Cross (see www.ifrc.org).

Another form of new ventures are those started by existing corporations for the purpose of building an important new business unit as a solely owned subsidiary or a spin-off as a new independent company. This activity can be called a **corporate new venture (CNV).**

TABLE 7.1 Six types of new ventures.

Type	Revenue growth	Planned-for most likely size	Description	Objective
1. Small business	Slow	Small	Sole proprietorship, family business	Provide independence and wealth to partners by serving customers
2. Niche	Slow to medium	Small to medium	Slow growth of corporation	Provide steady, lower-risk growth and good income
3. High-growth	Fast	Medium to large	Fast growth; needs large initial investment	Important new business
4. Radical innovation	Fast	Large	Requires R and D and seeks disruptive innovation	Commercialize an important innovation
5. Nonprofit organization	Slow	Small to medium	Serves members or a social need	Serve a social need
6. Corporate new venture	Medium to fast	Large	Independent unit of an existing corporation	Build an important new business unit or separate firm

All of these types of new ventures stand to benefit from considering all the issues of strategy and implementation as well as the preparation of a business plan. Niche markets that start small but grow over time to large markets are less likely to attract large companies. A niche market of $40 million will be too small to enable a large company with $10 billion in revenues to meet its growth goals. Niche market firms must keep focused on their uniqueness and the target market, and resist straying far from their niche. A good example of a niche firm that has grown to be a mid-sized firm is Hot Topic, which specializes in music-influenced apparel, accessories, and gifts for teenage girls (www.hottopic.com). Most of the merchandise is tied to modern rock bands. This concept is unique and very focused.

Anita Roddick, founder of the Body Shop, was 34 when she initiated her first cosmetics store in Brighton, England, in 1976. Her products differed from competitors' because they were made of natural ingredients. The first store was financed by a bank loan using the family-owned hotel as collateral. Upon ready success in her initial niche, she expanded to a second store, raising £4,000 for a half share of the business. She started a franchise system in 1978, and the company went public in 1984. Her business was profitable from the start, and the natural products were a hit within the growing trend of "green" products.

7.2 CORPORATE NEW VENTURES

A new venture started by an existing corporation for the purpose of initiating and building an important new business unit or organization can be called a corporate new venture. Some people refer to this process as intrapreneurship. The building of the new business enterprise depends on an entrepreneurial team leading the effort. Corporate entrepreneurship is focused on the identification and exploitation of previously unexplored opportunities that utilize the resources and competencies of an existing corporation. Corporate venturing is usually involved with the birth of new businesses and the associated revitalization of a corporation. We differentiate a corporate venture from a project by: 1) its newness to the corporation, and 2) its independence from the existing activities, organizational units, and products of the corporation. The characteristics of a corporate new venture are summarized in Table 7.2. Corporate new ventures are distinguished from projects and product development efforts by having a limited relationship to existing business units, autonomy, innovation, and entrepreneurial leadership.

TABLE 7.2 Characteristics of corporate new ventures.

■ Newness and novelty of the product relative to the firm's existing products.	■ High potential for significant innovation.
■ Independence or semiautonomy from existing corporate structure.	■ Unique entrepreneurial team leadership capabilities.

FIGURE 7.1 Corporate new venture model.

The factors for a successful corporate new venture are substantially the same as for the independent new venture formed by an entrepreneurial team: opportunity, vision, commitment, capabilities, resources, innovation, strategy, and execution. Success of corporate new ventures has been shown to be positively associated with growth and profitability of the firm [Antoncic and Hisrich, 2001]. A representation of this relationship and corporate ventures is shown in Figure 7.1. Mature corporations that engage in new business venturing are innovative, continuously renew themselves, and proactive.

Conventional wisdom says that large firms are weak at transforming opportunities into viable new businesses. The perceived reason for this weakness is the effects of bureaucracy and inflexibility that exist in large firms. Although the inertia within large firms does often constrain innovative activity, these firms possess resources and capabilities that would be the envy of individual entrepreneurs trying to strike out on their own. Unlike individuals, firms usually possess many of the resources, capabilities, and knowledge necessary for innovation. Often, the challenge for large firms is to protect the CNV from the pressures and controls of existing units within the firm. Typically, a CNV needs to be established as a relatively autonomous unit that can accommodate the entrepreneurial team and its business plan. The new corporate venture must be seen as a sustained effort to create innovation in new areas of business for the established firm.

Josh Gitlin, head of product development at Whirlpool Corporation's Kitchen Aid unit, wrote a proposal for in-home cooking classes taught by a network of known chefs and received support to launch it in September 2000. Whirlpool then funded a subsidiary named Inspired Chef. Whirlpool now has 35 innovation consultants around the world whose full-time job is to help employees turn ideas into revenue generators. These "I-consultants" are aided by part-time mentors who can also act as sounding boards and get workers around bosses who stand in the way. Once senior managers judge a proposal worth testing, employees and their I-consultant advisers are given just 100 days and $100,000 to get it off the ground by creating prototypes or conducting consumer research [Arndt, 2002]. (See www.whirlpool.com.) Unfortunately, Inspired Chef was closed in late 2002, but it was a try worthy of corporate venturing.

Corporate inertia can flow from successes that reinforce the rigidity of assumptions, processes, relationships, and values. All these factors are difficult to challenge: thus, the need for new independent organizations, such as new venture units, subsidiaries, or start-ups. Existing corporations often are managed on a tightly defined strategy and within highly controlled boundaries. A new corporate venture needs some loosening of these controls and a balance of independence and control [Birkinshaw, 2003]. Important corporate ventures can be developed by combining the entrepreneurial skills embedded in a venture team with the assets of the established corporation.

7.3 THE INNOVATOR'S DILEMMA

Disruptive innovation can revolutionize industry structure and cause existing firms to decline or fail [Christensen and Tedlow, 2000]. Incumbent firms listen to their existing customers, who usually do not express a desire for radical innovation before it is introduced. Eventually, the new innovation improves and starts to challenge the existing methods, but it may be too late for the incumbent firm to catch up. By the time the new innovative firm has improved the innovation and captured the market, the existing firm has lost market share. The disruptive innovation gets its start outside of the mainstream of the market, and then its functionality improves over time. The existing firm ignores new, potential innovations at its own peril.

Another problem for all successful firms is **cannibalization,** which is the act of introducing products that compete with the company's existing product line. When companies decline to try to cannibalize their own products, they operate under the delusion that if they do not develop the new product, no other firm will do so. When new opportunities open up, new entrants into an industry can be more flexible because they face no trade-offs with their existing activities [Burgelman, Christensen, and Wheelwright, 2004].

For many large firms, the pursuit of innovation must take a backseat to the effective exploitation of existing competencies, the satisfaction of existing customers, and the continuous improvement of existing technologies. Innovations eventually become breakthroughs when the web of people, ideas, and technologies that surrounds them grows and evolves, forcing change farther and farther into the established systems they emerged from. Existing organizations will tend to exploit an innovation that is related to their primary forms of work when the innovation enhances their existing competencies. When an innovation is unrelated, new organizations will typically emerge to capitalize on the innovation.

Established companies can offer disruptive new products of their own that capture new customers and produce new revenue growth. Instead of waiting for the threat of new products to appear, the established firm can create its own response to an anticipated threat. It is important for existing companies to invest in the development of disruptive innovations. This is a form of hedging one's bets. A disruptive innovation, however, often requires new competencies, resources, and value net relationships. Thus, the best strategy for an existing firm may be to

TABLE 7.3 **Strengths and weaknesses of a corporate new venture.**

Strengths	Weaknesses
Ready access to capital	May be subjected to a corporate budget process
Access to capabilities of corporate employees	Multiple control and review levels
Suppliers willing to help in the design process	Limited autonomy
Emphasis on the marketing plan	Limited access to strong entrepreneurial talent
Gain from brand equity of the parent firm	Risk-reward may be less attractive than for an independent entrepreneur
Access to processes and technologies of the parent	Limited to parent firm's technologies and processes

establish an autonomous corporate new venture unit or subsidiary with its own mandate, vision, people, and incentives.

The strengths and weaknesses of a corporate new venture are summarized in Table 7.3. Corporate new ventures benefit from ready access to the capital, people, suppliers, technologies, and brand of the parent firm.

On the other hand, CNVs may be limited by the budgetary and control practices of the parent. Furthermore, the parent firm may not have the technologies or people that the new venture requires. Recent studies show that the resource advantages of the existing corporation do not necessarily translate into higher performance for CNVs [Shrader and Simon, 1997]. To enable a CNV to be most successful, it may be necessary to give the new unit more autonomy, separating it from the controls and limitations of its parent.

Capacity for creative innovation within a corporation is a function of the ability of the entrepreneurs to: 1) obtain nonredundant information from their networks and 2) avoid pressures for conformity and sustained trust in developing novel innovations [Ruef, 2002]. CNVs are likely to successfully emerge from entrepreneurial teams drawing on a diverse set of functional roles in a firm. Corporate entrepreneurs can benefit from the knowledge resources of the firm but must avoid excessive conformity to the established means and norms of the firm.

The history of technology-based businesses is marked by large changes in cost effectiveness [Grove, 2003]. Classical competition theory does not account for these transformations, which change the context of the industry as well as the competitive challenges. An example of a large transformation was the introduction of the Boeing 747 in 1966.

Established companies, when confronted by disruptive technologies, should consider creating a completely separate organization and give it a charter to build a new business [Christensen and Raynor, 2003]. When IBM was confronted by minicomputers, it created an autonomous business unit in Florida to develop and sell a personal computer.

The confluence of technologies could transform the health care industry by using genomics and proteomics along with computer technologies to create new health solutions. Perhaps a corporate new venture at Amgen or Johnson & Johnson will create an enterprise that will impact the health care industry.

7.4 INCENTIVES FOR CORPORATE VENTURE SUCCESS

Can once-mighty giants of industry restore their health after they mature and decline in performance? Can these mature companies use corporate new ventures to transform their performance? Many studies point to significant difficulties in transforming large firms [Majumdar, 1999]. Structural factors, such as their intrinsic complexity, formality, and rigidity, are not conducive to either high performance or reorientation. Not only are larger firms and organizations structurally sluggish, but, with the passage of time, their culture becomes rigid and hard to change because of commitments to particular ways of doing things. As Dee Hock, founder of Visa, stated: The problem is never how to get new, innovative thoughts into your mind, but how to get old ones out.

Large firms possess large collections of knowledge and intellectual capital. Furthermore, these firms have many talented staff who have entrepreneurial tendencies and the ability to exploit the intellectual capital of the firm. **Absorptive capacity** is the ability of a firm to exploit external knowledge for the production of innovations. Thus, corporate new ventures can be based on both internal and external knowledge to the extent that the ability to absorb and exploit it is rewarded. A firm's successful use of innovations depends on its ability to exploit its existing base of knowledge while learning about technologies that lie outside its existing competencies [Manla, Keil, and Zahra, 2003].

Majumdar [1999] studied the dynamic performance of large and small companies in the dynamic U.S. telecommunications industry and showed that size was not a material factor in performance. Large firms are able to effect change in a dynamic industry. With a larger variety and pool of resources available, larger firms can undergo transformation as effectively as smaller firms through a process of dynamic learning. Large firms can be transformed and rigid cultures can be made flexible by using the right methods, such as a corporate venture program.

One important incentive for corporate venturing is tying executive compensation to initiation and support of corporate new ventures. Another incentive is to encourage ownership of stock in the firm by its executives. When executives own stock in the companies they manage, they become motivated to increase the long-term value of their firm using corporate ventures [Zahra, 2000].

Mature firms need to exploit opportunities for novel innovations by increasing their commitment to corporate new ventures. Larger, mature firms also need to recognize, however, the barriers to CNV: familiarity, maturity, and propinquity [Ahuja and Lampert, 2001]. Familiarity exhibits itself in a firm's tendency to favor the routine or common knowledge and ways of doing things. Maturity refers to favoring fully developed knowledge rather than novelty. Finally, propinquity refers to favoring a search for solutions similar to existing solutions.

The challenge of identifying applications in the early stage is complicated not only by the limitations in technology performance, but also by the fact that attention focused on search for a market application is attention diverted from immediate development. Many large firms bring together varying groups from time to time to address the questions of what they should do differently and what new products they should develop. People at these sessions are urged to suggest all ideas without hesitation [Drucker, 2002].

Mature firms should strive to build CNVs to experiment with novel, emerging, and pioneering innovation to create new dynamic growth. To pursue opportunities, firms need to identify and encourage corporate entrepreneurs. Corporate entrepreneurs are employees of a firm who take leadership responsibility for driving a venture in the firm. Art Fry, an entrepreneur within 3M, pushed the commercialization of Post-It notes through the firm. 3M has a guideline that its researchers can spend 15 percent of their time working on an idea without approval of management.

Corporations can use old ideas as the raw material for new applications. They can use old ideas and knowledge in new combinations or new ways. They can take an idea commonplace in one area and move it to a context where it is not common at all [Hargadon and Sutton, 2000]. This often can occur by just moving ideas from one division of the firm to another. They can also use a special internal consulting group dedicated to facilitating knowledge brokering within the firm. What distinguishes entrepreneurial firms from others are the actions they take when information is still incomplete. Entrepreneurial ability is not a function of simply gathering information but of having both the ability to make early judgments and the confidence to act on these judgments.

Because leading new CNVs always carry the risk of failure, many potential corporate entrepreneurs avoid joining them. They fear the loss of status that failure can bring. Talented entrepreneurs will only be willing to test new opportunities when the company executives explicitly state that making mistakes is acceptable and refrain from penalizing employees for failures.

Incentives for corporate entrepreneurs can be stock ownership, bonuses, or promotion within the firm if the expected performance is attained. A firm might have a goal that it will launch four new ventures each year and expect that at least one new venture will create an important new business.

A few years ago, an employee of Virgin Atlantic noticed some empty curb space at Heathrow Airport. In a matter of days, he secured the rights to the space and laid out a plan for Virgin to start a curbside check-in kiosk business unit. As a result, Virgin became the first airline at Heathrow to offer its business class passengers the advantage of getting a boarding pass without having to stand in a check-in line. As a result of his effort, the employee received a promotion. [Hamel, 2001].

A key issue is the appropriate reward to a corporate entrepreneur. The corporate entrepreneur who is offered large financial gains is usually resented by his

TABLE 7.4 **Incentives for corporate entrepreneurs.**

■ Support for and recognition of employees who create and champion new ideas and opportunities.	■ Significant degree of autonomy for corporate entrepreneurs.
■ A culture that favors individual or team initiative to create new ideas.	■ Effective rewards such as promotion, stock ownership, or bonuses.
■ Slack time for exploring not-yet-approved projects.	

or her associates because the entrepreneur relies on corporate resources that he or she did not create as an independent entrepreneur would [Sathe, 2003]. A list of possible incentives for corporate entrepreneurs is provided in Table 7.4. Incentives include social incentives such as recognition, support, and a culture that favors individuals or teams to take the initiative to create new ideas and explore new opportunities. Employees can be provided slack time, which is time on the job for exploring as-yet-to-be-approved projects. A significant degree of autonomy and effective financial and promotion awards provide incentives for corporate entrepreneurs. Entrepreneurs are less risk-averse and seek independence of activity [Douglas and Shepard, 2002]. These preferences can be exploited by CNVs.

7.5 BUILDING AND MANAGING CORPORATE VENTURES

An existing company is wise to attempt to exploit a new opportunity through some form of a new business. The types of new business arrangement that an existing corporation can use include a new independent venture, a spin-off of a new corporation, a transfer of the opportunity to the existing company's product development department, or authorization of a small project. Figure 7.2 shows the four types of business arrangements and their relationship to operational relatedness and strategic importance. Operational relatedness refers to how the new business organization couples to the existing operational resources and competencies. Strategic importance refers to the critical nature of the long-term results of this new organization to the success of the parent firm. Internal corporate new ventures are most useful for high operational relatedness and high strategic importance (quadrant 1 of Figure 7.2).

An opportunity with low strategic importance and low operational relatedness (quadrant 3) most likely calls for declining to proceed with this project or proceeding with a modest project until the strategic importance becomes clear.

An opportunity with high strategic importance and low operational relatedness may call for a spin-off to a new company (quadrant 2). A **spin-off unit** is an organization that is established within an existing company and then sent off on its own. The parent provides some resources and capabilities and sets the

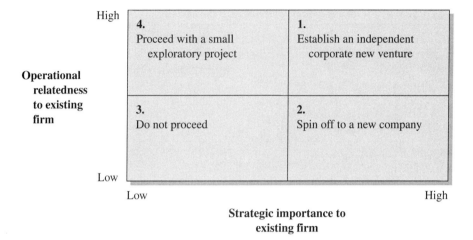

FIGURE 7.2 Four types of new business opportunities and the best business arrangement for each opportunity.

spin-off toward independence. Often the parent retains less than majority ownership of the spin-off. An opportunity with high relatedness and low strategic importance (quadrant 4) is a good candidate for a modest exploratory project.

Cisco Systems established a spin-off Andiamo Systems which makes switching gear. Cisco loaned Andiamo $42 million and committed to up to $142 million more. Cisco held 44 percent ownership of Andiamo. Cisco purchased the remaining 56 percent from the 300 employee-shareholders for $750 million of Cisco stock in 2004. [Thurm, 2002].

For the effective creation of a spin-off or new internal corporate venture, the opportunity needs a champion in the parent company [Greene, Brush, and Hart, 1999]. The **champion** is an executive or leader in the parent company who advocates or provides support and resources as well as protection of the venture when parent company routines are breached. The champion helps, describes, and defends the venture and secures the necessary resources. The champion enables the resource transfer process, as shown in Figure 7.3 [Lord, Mandel, and Wager, 2002].

Corporate ventures are managed differently than traditional in-house corporate research and development. A corporate venture may be riskier and less subject to rigid management of internal costs than conventional corporate product

FIGURE 7.3 Resource transfer process and the champion.

TABLE 7.5 Establishing conditions for corporate new ventures.

■ Increase the sources for innovation: New ideas tend to evolve and expand through conversation. The more people you can get involved, the more high-quality ideas you will generate. ■ Establish a process for collecting and evaluating ideas: Establish a forum for assessing the merits of various	proposals to ensure that the most worthy ideas receive funding. ■ Don't let traditional executives control the budget: Many executives are protecting their own departments and are unwilling to risk small amounts of resources on new and untested ventures.

development. Indeed, protecting venture investments from such controls is one reason that corporate new ventures and spin-offs are often housed outside the corporation's walls.

Furthermore, in corporate venturing, returns are part financial and part strategic, whereas with pure venture capital, investors' expected financial returns are paramount. Corporate ventures should follow the best practices of venture capital, but the dual objectives of financial and strategic returns must be balanced in ways that do not concern the venture capitalists.

Large corporations can make room for radical, low-cost innovations by establishing a process for finding and funding new ideas [Wood and Hamel, 2002]. Table 7.5 describes a three-step process for finding, evaluating, and funding new entrepreneurial entities. First, expand the conversations about new opportunities widely throughout the firm. Then, establish a process for selecting and funding the best ideas. Finally, keep the budgetary control within the new venture and avoid letting traditional managers begin to control the budget of the new venture.

Corporate entrepreneurs gain acceptance for new ideas by influencing organization members' perceptions about the nature of organizational interests. For an initiative to be accepted as part of the official company strategy, belief in the idea must be linked to corporate organizational goals. Most large companies prefer to separate new venture efforts from their core business. This permits the new venture to focus on the new opportunity and readily gather and coordinate the necessary capabilities and resources [Albrinck et al., 2002].

The elements of a business as practiced by the parent firm and the corporate new venture are shown in Table 7.6. In general, the parent firm has developed assets, revenues, reward systems, and management practices that tend to support growth, fairness, and policies that lead to orderly progress. The corporate new venture needs to leverage its assets to create new revenue streams by rewarding entrepreneurial actions and flexibility. Furthermore, the CNV wants to attract the best talent of the parent company to it. Separating the CNV from the parent company enables the CNV to act quickly with flexibility to seize new opportunities.

Many companies use a portfolio strategy for holding ownership in several corporate new ventures as subsidiaries or spin-offs. These companies also use a new venture development process based on the development of a business plan and the analysis of what form of CNV is appropriate. This process is summarized in Table 7.7 [Albrinck et al., 2002]. At each step of the process, the parent

TABLE 7.6 Elements of a business as practiced by the parent firm and the corporate new venture.

Element	Parent firm	Corporate new venture
Assets	Protect and use	Leverage
Revenues and growth	Growth of existing revenue stream	Create new revenue stream
Management	Adhere to policies and procedures	Act decisively with flexibility
Rewards	Maintain fairness and equity	Reward entrepreneurship and performance
Talent—people and knowledge	Retain talent and knowledge	Attract the top talent and transfer the best knowledge from the parent company to the CNV

company must evaluate the best next steps. Step 1 helps the CNV take shape as a venture champion and entrepreneurial team are identified. Step 2 includes the development of an initial concept statement and the outline of the elements of a business plan. The next step is to complete the development of a comprehensive business plan. Step 4 is focused on selecting the best organizational form for the CNV, based on the long-range objectives of the parent [Miles and Covin, 2002]. Finally, in step 5, the corporate new venture is established with the requisite resources, talent, and capabilities transferred from the parent company.

The selection of the appropriate form (step 4) should try to fit the needs and strategy of the parent company. For example, 3M usually incorporates the CNV within an existing or new division. Conversely, Barnes and Noble, when considering the establishment of an online unit, decided to spin off a new company into the stock market.

The Virgin Group, under the leadership of Richard Branson, has created 200 new businesses in many industries such as media, airlines, and music. Business ideas come from anywhere in Virgin Group, and Branson remains accessible to

TABLE 7.7 Five-step process for establishing a corporate new venture.

1. Identify and screen opportunities. Create a vision. Designate a venture champion and an entrepreneurial team.

2. Refine the concept and determine feasibility. Prepare the concept and vision statement. Draft a brief business plan summary or outline for review and to gather support.

3. Prepare a complete business plan. Identify the person to lead the new venture.

4. Determine the best form of the corporate new venture: internal new venture unit, spin-off, subsidiary, or internal project.

5. Establish the corporate new venture with talent, resources, and capabilities transferred from the parent company.

employees who have proposals. Branson also hosts gatherings for employees where they can give him their ideas. One employee proposed a bridal planning service including wedding apparel, catering, air travel, and hotel reservations. She became the CEO of Virgin Bride. Virgin Group works to start independent new ventures (see www.virgin.com).

Landmark Communications launched the Weather Channel as an internal new venture in 1981 [Batten, 2002b]. With Landmark's strong corporate support and commitment, the Weather Channel became a top weather information source. With the help of Landmark's talent, knowledge, resources, and capabilities, the new venture took off through several deals with cable operators. By 1996, the Weather Channel was also available online. The Weather Channel succeeded because of the investment of significant resources of Landmark. The Weather Channel was started in the face of widespread skepticism but prevailed because of the assets and capabilities of Landmark.

Existing firms have the capability to organize a market, turning an idea into something that can be economically produced, marketed, and distributed to the customer. Entrepreneurs are able to explore new technologies quickly and effectively, and make the creative leap from technological possibility to something that meets consumer needs. Effective firms that meet the challenge of change possess people who are capable of both tasks.

Many new innovations are introduced by pioneer firms, and a learning phase is started in the market. Existing corporations can recognize these new innovations and quickly join in the innovation-commercialization phase, exploiting their capability to produce, market, and support new products.

In the early 1990s, the pharmaceutical giant Eli Lilly built a series of internal ventures focused on medical devices. By 1994, Lilly had created four internal corporate venture divisions in the medical devices area concerned with cardiac and vascular issues. By September 1994, Lilly incorporated these units into a new company, Guidant, and consummated an initial public offering of its common stock. By September 1995, Lilly disposed of its ownership, and Guidant was a separate company. Guidant is an excellent example of starting as a group of internal corporate ventures that eventually became a leading company on its own. By 2003, Guidant had $3.4 billion in sales (see www.guidant.com).

Many critics depict incumbent firms as going into decline in the face of radical technological innovation. This tendency is not universal, however, nor should it be inevitable. Corporations can respond effectively to new technological innovations when they are prepared and organized to do so. Firms that have a history of navigating turbulence and creating loosely coupled, stand-alone divisions, and possess a critical complementary asset have a good chance of managing the challenge of a radical innovation [Hill and Rothaermel, 2003].

7.6 AGRAQUEST

Pamela Marrone was the entrepreneurial leader of two corporate new ventures before incorporating an independent start-up, AgraQuest. After she received her doctorate in 1983, she was recruited to start a new corporate venture by Monsanto of St. Louis. Her unit was a separate research and development project within an existing division of the agricultural business organization of Monsanto. She reported to a division head and enjoyed moderate independence for the direction of her new venture unit.

In 1990, Marrone was recruited by Novo Nordisk, a Danish company, to start a wholly owned subsidiary called Entotech. Novo Nordisk and Marrone chose Davis, California, as the location for this new company because of the significant research work being done in entomology at the University of California, Davis. She recruited scientists and staff and built the firm up to 50 people by 1995. Marrone enjoyed day-to-day control of Entotech but was required to travel to Denmark each month to report on plans and progress. By 1995, Novo Nordisk decided that Entotech was outside its core business segment and sold the new venture corporation to Abbott Labs.

Rather than move with the unit to Abbott Labs, Marrone decided to launch her own firm, incorporating it as AgraQuest. Her earlier experiences from 1983 to 1995 served her well in moving on to found a new company. She sought independence as well as the chance to create the contribution that a natural pesticide firm could offer to agriculture around the world. Often, as in Marrone's case, the entrepreneur finds the controls and limits of an existing company are burdens uncompensated by the availability of the resources of the larger parent firm. Many entrepreneurs who seek their own career path find that it requires a difficult but important decision to establish their independence and make their own way in the world of commerce by starting a new independent venture.

7.7 SUMMARY

There are six types of new ventures: small business, niche, high-growth, radical innovation, nonprofit, and corporate. Important contributions to our society have been made by small and niche businesses, especially when they later grow and extend their mission worldwide. High-growth and radical innovation start-ups are very important to creating growth and jobs as well as providing an important service or product that makes a difference. A special form of new venture, a nonprofit, enables an organization to meet an important social purpose. Finally, corporate new ventures make important contributions of novelty and creativity and can provide new vigor for existing large corporations.

- ■ For many firms, the pursuit of innovation and the creation a new venture independent of the existing strictures may renew the vigor of the firm.
- ■ A new corporate venture needs the right amount of slack, independence, and resources to create a novel business venture.

> **Principle 7**
> An important, vigorous new business venture can emerge from a large firm when afforded the appropriate balance of independence, resources, and people to respond to the opportunity.

7.8 EXERCISES

7.1 Identify and describe a corporate new venture at a firm such as 3M, Boeing, or Chevron.

7.2 Why would a firm like Zipcar not likely emerge from a big automotive firm like Ford or Daimler Chrysler? What limits the big firms' ability to start an innovation such as Zipcar? See www.zipcar.com.

7.3 Zimmer Holdings (www.zimmer.com) was incorporated in January 2001 as a wholly owned subsidiary of Bristol-Meyers Squibb Company. Zimmer designs and markets orthopedic products and surgical products. The subsidiary was spun off from its parent on August 6, 2001, with shareholders receiving one share of Zimmer for each 10 of Bristol-Meyers they owned. Study the origins of Zimmer and determine if the spin-off was the right action for Bristol-Meyers.

7.4 Coach Inc. is a designer, producer, and marketer of high-quality handbags, women's and men's accessories, luggage, and other personal items. Coach was acquired by Sara Lee in 1985 and built up internally as a stand-alone division. In October 2000, the shares of Coach were distributed to the owners of Sara Lee. Examine the history of Coach as an internal venture within Sara Lee and the reasons for spinning off Coach as a separate firm (see www.coach.com).

7.5 Walt Disney Company launched a CNV in 2003 to sell a DVD that is priced the same as a rental DVD but does not need to be returned because it stops working after a fixed period of time (see www.ez-d.com). Explore the success of this new product, the EZ-D DVD, and determine the factor that influenced this CNV [Taub, 2003].

7.6 Walt Disney Company tested a CNV product called Moviebeam (see www.moviebeam.com). This product requires hooking a box to a TV set that enables the viewer to connect to movies on demand. The Moviebeam box is refreshed regularly with new movies. The Moviebeam box can be purchased through local electronics stores. Examine this product and determine if and why this is a successful CNV.

CHAPTER 8

Creating New Ventures and the Business Plan

The method of enterprising is to plan with audacity and execute with vigor.
Christian Bovée

Entrepreneurs respond to attractive opportunities by forming new ventures. The appropriate legal/organizational format used to organize the new venture will vary according to several factors such as context, people, legal and tax consequences, and cultural and social norms. In this chapter, we consider the various organizational and legal forms that entrepreneurs employ to achieve their objectives using a five-step process for establishing a new venture. We also describe the tasks and benefits accrued by a socially responsible firm. Then, we consider the benefits of a cluster of interconnected companies operating in a geographic region.

The imitation of an existing business is a common form of entrepreneurship. Building a new business venture is described as a process that can be learned and mastered by talented entrepreneurs. Finally, we describe the benefits and limitations of creating and writing a business plan, a significant and challenging task for most entrepreneurs. ∎

8.1 INDEPENDENT AND CORPORATE VENTURES

An **independent venture** is a new venture not owned or controlled by an established corporation. A **corporate venture** is a new venture started by an existing corporation, as described in Chapter 7. An independent venture is typically unconstrained in its choice of a potential opportunity, yet is usually constrained by limited resources. The corporate venture is usually constrained in choice of opportunities to those consistent with the parent business. The corporate venture, however, usually has access to the significant resources of the parent firm [Shepard and Shanley, 1998]. These corporate and independent ventures are usually established as for-profit entities.

While independent and corporate ventures both face the same external context, their different competencies and resources cause them to develop different strategies. The independent venture has more flexibility and potentially requires fewer resources than the corporate venture. Furthermore, the independent venture has access to a wide range of advisors, while the corporate venture is advised and controlled by the parent company. Thus, the independent venture has the advantage of flexibility, adaptability, and high incentives, while the corporate venture is usually advantaged by its access to valuable capabilities and resources.

Table 8.1 shows the five-step process for starting a new venture. The new venture will follow the steps to prepare a business plan that is suitable for the team as well as for the investors and business partners. This process is broad enough to apply to all types of businesses: independent or corporate; small or large; niche or broad; family or franchise; nonprofit; and one attempting a radical versus incremental innovation.

The corporate venture will also prepare a business plan suitable for review by the parent corporation to secure the necessary resources and assistance from its parent. The five-step process of Table 8.1 can be used for corporate ventures where the investors of step 5 will normally be the parent firm. We defer until Chapter 10 the discussion of the appropriate legal form a venture should adopt.

TABLE 8.1 Five-step process for establishing a new venture.

1. Identify and screen opportunities. Create a vision and concept statement, and build an initial core entrepreneurial team. Describe the initial ideas about the value proposition and the business model.

2. Refine the concept, determine feasibility, and prepare a mission statement. Research the business idea and prepare a set of scenarios. Draft the outline of a business plan and an executive summary.

3. Prepare a complete business plan with a financial plan and the legal organization suitable for the venture.

4. Determine the amount of financial, physical, and human resources required. Prepare a financial model for the business and determine the necessary resources. Prepare a plan for acquiring these resources.

5. Secure the necessary resources and capabilities from investors, as well as new talent and alliances. Launch the organization.

8.2 NONPROFIT AND PUBLIC-SECTOR VENTURES

The purpose of a new venture is to create wealth for society. Often, wealth is seen as financial wealth, but many entrepreneurs seek to provide social wealth for their society. The product of a nonprofit hospital is a healthy patient. Only for the tax collector does it make a difference whether the hospital is nonprofit or for-profit. A **nonprofit organization** is a corporation, member association, or charitable organization that provides a service but does not earn a profit, nor does it distribute dividends or payments to its employees, donors, or volunteers. Today, nonprofit organizations are often called not-for-profit. One out of every two adult Americans is estimated to work as volunteers in the nonprofit sector. See Appendix C for websites about nonprofits.

The product of the Girl Scouts is a mature young woman who has values, skills, and respect for herself. The purpose of the Red Cross is to enable a community hit by natural disaster to regain its capacity to look after itself. In this way, the nonprofit venture forms to respond to a social need.

The decision to start a nonprofit venture will be determined by the nature of the social opportunity and the creation of an innovative response that cannot or should not be performed for profit. Social functions that depend primarily on volunteers or members such as churches, museums, theaters, social clubs, industry associations, credit unions, and farmers' cooperatives are usually formed as nonprofits.

A nonprofit organization is permitted to generate a financial surplus but may not distribute this surplus to officers, investors, or employees. Furthermore, nonprofits have no owners. Any surplus must be used for the approved nonprofit mission of the organization. However, like for-profit firms, these organizations may be technology-based.

An excellent example of an entrepreneurial nonprofit is the Guggenheim Museum, headquartered in New York City. The museum helped build or found a series of museums in Venice, Italy; Bilbao, Spain; Berlin, Germany; and Las Vegas, Nevada. The entrepreneurial leadership was provided by Thomas Krens, who holds a master's degree from the Yale School of Management [Schmetter, 2003]. (See www.guggenheim.org.)

Organizations that satisfy the conditions of section 501(c)(3) of the Internal Revenue Code are called charitable. The purpose of these organizations includes religious, educational, scientific, literary, or charitable aims. Donations to these organizations are tax-deductible for the donor and exempt from estate taxes. Noncharitable nonprofit organizations are primarily set up to serve the purposes of their members and are also tax-exempt, but donations are not normally tax-deductible.

The establishment of a nonprofit organization should follow the five steps of Table 8.1. The vision for the organization is defined in terms of the creation of social value rather than economic value. Many nonprofits can be described as socially conscious service organizations. Thus, the entrepreneurial team must be committed to the social values of the new venture with its attendant risks and uncertainties.

Like any new organization, a value proposition and business model will help to shape the nonprofit organization. Consider Elderhostel, a nonprofit educational organization for mature people. Elderhostel, formed in 1975, became a vehicle for social change by offering educational experiences to older people. Elderhostel's business model is clear and simple. It provides courses in various settings and builds a community of engaged, learning individuals. It charges affordable fees, usually on a good-value basis, and offers a very wide range of instruction and housing at numerous locations.

Nonprofits often find it difficult to agree on who their customer is. For the American Red Cross, is it the hospital, the blood donor, or the financial donor? Who is the ultimate beneficiary of the service?

Once the business plan is written, the required financial and human resources must be determined. How will the necessary funds be acquired and the human talent attracted? The challenge is finding donors whose special interests match those of the new nonprofit venture and who also have the expertise and commitment to provide an independent check on management judgment.

A unique form of a nonprofit corporation is a consumer cooperative, which is a business that belongs to the members who use it. The member/owners establish policy, elect directors, and often receive cash dividends. Examples of cooperatives include credit unions, housing co-ops, food co-ops, and utility co-ops. In 1938, mountain climbers Lloyd and Mary Anderson joined with 23 fellow climbers in the Pacific Northwest to found Recreational Equipment, Inc. (REI). The group formed a consumer cooperative to supply themselves with quality gear, clothing, and footwear selected for performance and durability for outdoor recreation, including hiking, climbing, camping, bicycling, and other sports. After more than six decades, REI has grown into a supplier of specialty outdoor gear currently serving more than 2 million active members through 63 retail stores in the United States and direct sales via the Internet (www.rei.com), telephone, and mail.

Leading a nonprofit requires the competencies needed for most businesses mixed with a commitment to the entity's social cause. Michael Miller leads the Goodwill Industries of the Portland, Oregon, area and has built it to $46 million in revenues by using his entrepreneurial skills [Kellner, 2002].

Nonprofit organizations have traditionally operated in the social sector to solve or mitigate such problems as hunger, homelessness, pollution, drug abuse, and domestic violence. They have also helped provide certain basic social goods such as education, the arts, and health care that society believes the marketplace may not adequately supply. Nonprofits have supplemented government activities, contributed ideas for new programs, and functioned as vehicles for private citizens to pursue their own vision of the good society.

Former U.S. Senator Bill Bradley recently noted that the nonprofit sector has the potential to expand the flow of funds and benefits by increasing their effectiveness and efficiency [Bradley, Jansen, and Silverman, 2003]. The nonprofit sector can spawn very big, high-impact ventures in the same way that the for-profit sector does. For example, the nonprofit sector delivers much of the health care for most nations.

The need for innovation in the public sector (government) can be met in part by entrepreneurial new ventures. Various public-sector reform programs seek to align the public sector with the economic discipline of the private sector [Sadler, 2000]. Public-sector entries are often portrayed as conservative and bureaucratic. However, governments do seek to add innovation to their activities through entrepreneurial ventures. Public-sector entities can seek opportunities that have the potential to generate innovation. Those opportunities may not be commercially advantageous but can be transformed into a new value within the dynamic set of public-enterprise objectives. The application of innovation and new technologies can help to transform nonprofits and public-sector ventures.

Any effort to pursue new ventures in the public sector needs to reckon with the ambiguity of goals, limited leadership autonomy, cautious attitudes, and a short-term orientation. However, public-sector organizations can respond to opportunities that are generated within these barriers. A great example of new ventures in the public sector is in the provision of medical care worldwide.

New ventures in the public sector can be stimulated by clearly understood objectives, effective reward systems, resource availability, autonomy, and participatory decision-making. However, many public-sector managers are risk-averse since they are accountable to elected representatives who themselves are risk-averse, given their need for periodic re-election or reappointment. In addition, public-sector activities may attract significant media scrutiny.

Perhaps the largest public-sector entrepreneurial effort of the twentieth century is the Port Authority of New York and New Jersey, established in 1921. The Port Authority has built or facilitated the construction of ports, railroads, tunnels, bridges, terminals, and the World Trade Center (see www.panynj.gov).

8.3 FAMILY-OWNED BUSINESSES

A **family-owned business** is one that includes two or more members of a family who hold control of the firm. Perhaps 80 percent of the businesses in the United States and Canada are family-owned and -managed. In most of the world, family-owned enterprises dominate the business landscape. Not all of them are small. About 25 percent of the Fortune 1,000 businesses are family-controlled. Mondavi Corporation is still family-led and -controlled, although it is a public corporation. Many of the great twentieth-century companies were originally family enterprises, including IBM, Marriott, Merck, McGraw-Hill, and Wal-Mart. Family businesses have unique advantages when everyone works well together. More trust often exists among the family members than with outside employees, and customers often feel that they are getting good treatment when they can work directly with a family member. Family businesses, however, have a unique set of problems because family issues often carry over into the business operations. As with entrepreneurial couples, businesses operated by the owner and the children range from very small to large. For example, the mutual fund industry includes many family businesses. The Johnson family owns and leads Fidelity Investments, a private company. About 30 percent of family businesses

TABLE 8.2 Some family-owned or -managed businesses.

Anheuser-Busch	Food and beverages
Archer Daniels Midland	Foods
Cargill	Foods
Carlson Travel	Travel
Clear Channel	Radio
Fabbrice D'Armi Beretta	Firearms
Ford Motor	Autos
Gallo	Wine
ILX Lightwave	Laser instruments
NASCAR	Auto races
Pruitt	Health care
S. C. Johnson	Waxes

are transferred to the children. A list of some family-owned or -managed businesses is provided in Table 8.2. See Appendix C for family business websites.

> Fidelity Investments—FMR Corporation (www.fidelity.com)—was founded by the late Edward C. Johnson II in 1946. Ned Johnson, his son, was named president in 1972. Ned's daughter, Abby Johnson (born in 1962), was named president in 2001.

The family business has ownership, family, and business-management activities that overlap, as shown in Figure 8.1 [Gersick and Davis, 1997]. The family member who has ownership (area 4) will have different incentives than the owner-manager in area 5. Another issue is whether to give some ownership to a family member (area 1) when he or she is not active in the business. Should the successor CEO come from area 6 or 7—or elsewhere? These nonalignments can show up during difficult times.

Family businesses have a history of persevering through hard times. They typically have a strong commitment to shared goals and values, and they emphasize continuity and the long-term.

> Cargill was founded as a family business in 1865 with a single grain elevator in Iowa. Now headquartered in a suburb of Minneapolis, Minnesota, Cargill employs more than 90,000 people in 57 countries, all committed to a single, powerful idea: to be the global leader in nourishing people. Cargill is an international marketer, processor, and distributor of agricultural, food, financial, and industrial products and services. It provides solutions in supply-chain management, food applications, health, and nutrition.

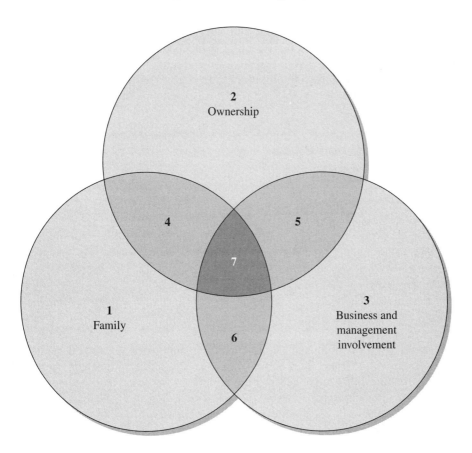

FIGURE 8.1 Model of a family business.

The advantages and disadvantages of family businesses are listed in Table 8.3. Perhaps the most significant disadvantage is family strife over equitable compensation, fair treatment, and succession. The great advantage can be the continuity and commitment of a family to its business partners and employees.

TABLE 8.3 Advantages and disadvantages of family businesses.

Advantages	Disadvantages
■ Long-term orientation	■ Nepotism
■ Independence of action	■ Family strife
■ Resilience and commitment	■ Financial strains
■ Family culture	■ Succession problems
■ Natural succession, if appropriate	■ Limited access to capital markets
	■ Concentrated risk in case of business failure

Entrepreneurs are rediscovering the natural connection between work and family. More people work at home, acting as independent contractors. Work can have a different meaning and be less stressful and more fun when it is conducted at home. Families are the building blocks of a stable, productive society, and a family firm can provide a good base for an entrepreneurial effort.

8.4 FRANCHISES AND DIRECT SELLING ORGANIZATIONS

A **franchise** is a legal arrangement in which the owner of a business format has licensed it to an individual or a local firm, called a **franchisee.** The **franchisor** is the organization that owns and operates a firm that controls the business format and its associated trademarks and logo. The franchisee receives the right through a contract to use the franchisor's business format, brand, and logo in a specific geographic region. For purposes of the legal agreement, the franchisee pays an initial fee and ongoing payments to the franchisor. The franchisor offers the business format and its trade secrets to the franchisee as well as training in the operation of a franchised business. An example of a widely known franchised business is RadioShack electronics stores. See Appendix C for websites about franchising.

A franchisee is obligated by the contract to follow the prescribed methods of operation of the franchise. The franchisee can operate as an independent businessperson and share the benefits of the franchise—its brand and methods. The franchisee is constrained, however, by the format and the rules of operation. Thus, the franchise offers significant benefits to the franchisee but constrains the amount of innovation that the franchisee can introduce into his or her business. The advantages and disadvantages for the franchisee are summarized in Table 8.4. The advantages and disadvantages for the franchisor are summarized in Table 8.5. The franchisee gains a proven method of business and gives up significant independence of operation as well as pays continuing fees. The word *franchise* originally came from the French, meaning to be "free from servitude."

TABLE 8.4 Advantages and disadvantages for the franchisee.

Advantages	Disadvantages
■ Training	■ Franchise fees
■ Continuing guidance	■ Control of the format by the franchisor
■ Proven business format	■ Unfulfilled promises
■ Brand appeal	■ Rigidity of rules
■ Satisfaction and independence	■ High start-up costs
■ National advertising	■ Restrictions on purchasing of supplies
■ Site selection assistance	■ Unfair restrictions of the contract on resale
	■ Financial failure of the franchisor

TABLE 8.5 Advantages and disadvantages for the franchisor.

Advantages	Disadvantages
■ Enhanced ability to expand	■ Potential for uncooperative franchisees
■ Expanded geographic reach	■ Varying regulations from state to state
■ Use of the capital of the franchisee in expanding the business format	■ Unable to innovate quickly
■ Attracts owner-managers for the franchise chain	

Most franchisees are seeking independence through the operation of a reliable, stable business. For some franchisees, this arrangement is a perfect fit with their needs and attitudes. Others may find the controls and fees to be excessive and believe they can do better by themselves.

A **business format franchise** involves the provision of a complete business method, including a license for the trade name and logo, the products and methods, the form of the physical facility, the strategy, and the purchasing system. This type of franchise is typified by Wendy's and Holiday Inn.

A **trade-name franchise** primarily involves a brand name such as Western Auto or ACE Hardware. A **product distribution franchise** is a license to sell specific products under a manufacturer's trademark and brand. Examples of a distribution franchise include auto dealerships and gasoline stations.

The payment to a franchisor may include an initial fee and an ongoing fee based on the revenues of the franchise. The initial fee and percent of sales fees are $45,000 plus 8 percent of sales for McDonald's and $40,000 plus 5.5 percent of sales for Krispy Kreme.

Protections for the franchisee are provided by state laws. Most states review and limit the elements of the franchise contract. Furthermore, federal law mandates a uniform franchise offering circular that requires the disclosure of the full terms and performance of the franchisor.

An entrepreneur may decide to be a franchisee and work toward the development of several locations within his or her franchise territory. Other entrepreneurs may have a proven business that is suitable for forming into a franchise.

Like any other new business venture, the franchise business must possess some unique competitive advantage, such as brand and a quality product. The franchisor, franchisee, and partners strive to find an appropriate balance of operating to suit the needs of each party. The franchisor seeks rapid growth, while the franchisee seeks attention to quality and execution.

At the heart of a franchise agreement is the desire by two parties to earn money while reducing risk. The franchisor wants to expand an existing company without spending its own funds. The franchisee wants to start his or her own business without going it alone and risking everything on a new idea. One provides a brand name, a business plan, expertise, and access to equipment and supplies. The other puts up the money and does the work. In 1898, General Motors lacked the capital to hire salespeople for its new automobiles, so it sold franchises to prospective car dealers, giving them exclusive rights to certain territories. Franchising was an innovative way to grow a new company in a new industry.

Technology companies that have used franchising to expand their business include FASTSIGNS (www.fastsigns.com). Another franchise, The UPS Store (formerly Mail Boxes, Etc.) is heavily dependent on technology applications. An example of a service company based on intellectual property is Dale Carnegie and Associates (www.dalecarnegie.com).

Snap-on Inc. manufactures and distributes mechanic's tools via franchisee dealers who sell at retail from walk-in vans that visit user locations. It also sells power tools, diagnostic equipment, and software (www.snapon.com). The franchise network provides a direct retail channel for the sales of its equipment.

Practitioners often recommend franchising as a method that entrepreneurs can use to assemble resources to create large chains rapidly [Michael, 2003]. Services such as the sales of tools and training typically must be delivered at a particular place. Furthermore, the first product in a market can shape customer preferences. Franchising can be used by entrepreneurs to build large chains effectively.

Direct selling organizations (DSOs) offer family-focused businesses to people who seek to work part-time and build their entrepreneurial skills. A direct selling organization sells directly to the customers through a local representative who typically sponsors a gathering in a home where people try and purchase the product. Examples of DSOs are Tupperware, Mary Kay Cosmetics, and Amway, which all use social networks to serve business ends [Biggart, 1989].

DSOs are often called network marketing organizations (NMOs). These retail selling channels use independent distributors not only to buy and resell at retail but also to recruit new distributors (sellers) into a growing network [Coughlan and Grayson, 1998]. The commission and markup on the products are the means of income for the independent distributors. NMOs are lean organizations with a menu of compensation schemes and rewards for the distributors.

The DSO uses a "party plan," which is a form that exploits the local representative's social network. Each local representative acts as independent contractor (nonemployee) and builds and learns his or her entrepreneurial capabilities by engaging in a regional network of fellow representatives. The network of representatives is often led by a charismatic leader, such as Mary Kay Cosmetics, and ascribe to the moral value of the entrepreneurial network (see www.dsa.org for the Direct Selling Association). The direct selling method offers entrepreneurial selling opportunities to those seeking part-time trial steps into the business world. This can be a valuable training opportunity.

8.5 CLUSTER DYNAMICS

Firms in industrial regions benefit from localization of cost economies derived from specialist suppliers, a specialist labor pool, and the spread of local knowledge [Best, 2001]. Since many new ventures benefit from new knowledge, competent suppliers, and available talent, they should consider locating in a region that offers easy access to all three factors. Entrepreneurial activity differs significantly across regions of a country. The formation of new ventures creates market opportunities for others. For example, a new venture may introduce a new product that creates the possibility of complementary products offered by other local enterprises.

A **cluster** is a geographic concentration of interconnected companies in a particular field. Clusters can include companies, suppliers, trade associations, financial institutions, and universities active in a field or industry. A good example is the Hollywood cluster of firms and infrastructure coming together for the creation of movies. If a new venture wants to enter the movie industry, it is probably wise to consider locating in Los Angeles.

Another excellent example of a cluster is the California wine cluster, which centers in the wine region of Sonoma and Napa counties. A wide-ranging network of industries supporting wine making and grape growing includes suppliers, investors, educational centers, and trade associations. A cluster's boundaries may be defined by the totality of the industry participants and may reach across political boundaries.

The location of a start-up company can be a key determinant of success. It may be wise to locate the new company close to its competitors and its most important customers. A start-up can gain regional advantages by joining a cluster of companies that have complementary or competitive capabilities and resources. A new company within a cluster is more likely to find the employees and infrastructure that it needs. A major success factor is the network of companies with which a new firm does business [Iansiti 2004]

Clusters promote both competition and cooperation. Firms form alliances, recruit each other's talent, and compete, all at the same time. A cluster also provides new firms with a critical mass of talent, knowledge, and suppliers for easy entry into an industry. A cluster of independent and informally linked companies and institutions represents a robust organizational form that offers advantages in efficiency, effectiveness, and flexibility. Furthermore, clusters are conducive to new business formation for a variety of reasons. Individuals working within a cluster can more easily perceive gaps in products or services around which they can build businesses. Beyond that, barriers to entry are lower than elsewhere. Examples of attractive locations for new ventures in the United States are provided in Table 8.6.

Clusters have existed since the beginning of commerce. The merchants of Venice benefited from a natural harbor, a supply of mariners and ships, and ready investors for their ventures. All business relationships and alliances involve their own complex bargaining and governance problems, and can inhibit a company's flexibility. The close, informal relationships possible among companies in a cluster are often a superior arrangement. A multitude of social networks enable a valuable diffusion of information between complementary firms [Ferrary, 2003].

The availability of firms providing complementary products may be critical to a new venture's success in a tourism cluster, such as New York City, for example. The quality of a visitor's experience depends not only on the appeal of

TABLE 8.6 Centers of entrepreneurial activity in the United States.

West	Midwest	Mountain
■ San Francisco Bay Area	■ Minneapolis–St. Paul	■ Denver
■ San Diego	■ Grand Rapids	■ Salt Lake City
■ Seattle	■ Chicago	**East**
Southwest	**South**	■ New York
■ Phoenix	■ Orlando	■ Boston
■ Las Vegas	■ Raleigh–Durham	■ Washington, D.C.
■ Dallas–Fort Worth	■ Atlanta	
■ Austin		

the primary attraction but also on the quality and efficiency of complementary business such as hotels, restaurants, shopping malls, and transportation systems. Because members of the cluster are mutually dependent, good performance by one can improve the success of the others.

Clusters play a big role in improving the development of knowledge and innovation. The Massachusetts Institute of Technology and other Boston-area universities have played a big role in the development of the technology cluster around Boston.

The area south of San Francisco called Silicon Valley is a cluster for electronics, medical devices, and computer companies. Connectedness and mobility of talent and ideas are a way of life in Silicon Valley. The support structure includes entrepreneurs, venture capitalists, attorneys, consultants, board members, universities, and research centers. Technology firms in Silicon Valley prosper in a dynamic environment of novelty and innovation. This network environment is the outcome of collaborations between individual entrepreneurs, firms, and institutions focused on the pursuit of innovation and its commercialization. Silicon Valley has an openness to change and is supportive of the creative and the different [Florida, 2002].

The entrepreneurial attitude prevalent in Silicon Valley is exemplified by the views of T. J. Rogers, founder of Cypress Semiconductor, who said [Malone, 2002]:

> What makes us special and different here in Silicon Valley is that we're truly capitalists. We invest. There is no safety net. You can go out of business. You can crash into the wall. There are companies, you can count them on both fingers every day, that go out of business, and that's life.

Dynamic, growing industrial regions are constantly upgrading their capabilities and resources, and commercializing innovations. Regional clusters can be a virtuous circle leading to better opportunities, more venture capital, increasingly educated talent, and more success. The cluster of independent activities can engender dynamic flows of cost reductions and competitive advantage.

8.6 THE SOCIALLY RESPONSIBLE FIRM

Any strategy adopted by a new venture firm inevitably affects the welfare of its stakeholders: customers, suppliers, stockholders, and the community. While a specific strategy may enhance the welfare of some stakeholders, it may harm others. The leaders of new ventures are challenged to build a strategy that attempts to balance the economic and social needs of stakeholders while protecting the social and environmental needs of its region. An explicit statement of a new firm's strategy for acting responsibly may be an appropriate part of a business plan.

The quality of life on Earth depends on three factors, as illustrated in Figure 8.2. The quality of life in a society depends on equity of liberty, opportunity, and health, and the maintenance of community and households, which can be called **social capital,** or social assets. The growth of the economy and the standard of living are critical needs for all people; we call this **economic capital,** or economic assets. Finally, the environmental quality of a region or the world can be called **natural capital.** The interrelationship between these three factors adds up to the total quality of life. Quality of life includes such basic necessities as clothing, shelter, food, water, and safe sewage disposal. Beyond that, quality of life includes access to opportunity, liberty, and reasonable material and cultural well-being [Dorf, 2001].

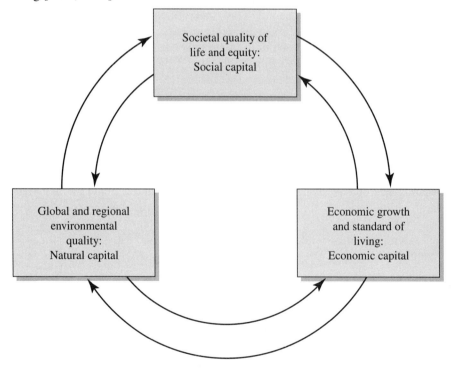

FIGURE 8.2 Three interrelated factors that determine the quality of life on Earth.

Business, government, and environmental leaders need to build up capabilities for measuring and integrating these three factors and using them for decision-making. We define the sum of these factors as the **triple bottom line.**

As they strive to treat nature and society respectfully while enhancing people's quality of life, corporations need to use nature only for what is necessary and in balance with what can be recycled and replenished.

Recognizing the interconnectedness and interdependence of all living things, corporate leaders can seek a balance using the triple bottom line concept. Economics, ecology, and society can be portrayed as a whole that depends on the person, the corporation, cultural values, and the community. Decisions made by corporations or society need to account for all the three factors of the triple bottom line.

For many, there is a presumption that a company exists to enhance the welfare of society at large. For others, the only goal is the maximization of profits. We assert that the public welfare can be in the best interest of the corporation itself. One of the purposes of a firm is to make a profit—but service of society is also an implied expectation. In many ways, socially responsible behavior—remembering its obligations to its employees, its communities and the environment, even as it pursues profits for shareholders—is in a firm's self-interest. A growing number of companies make corporate responsibility part of their value proposition. For example, Henry Ford believed he should pay his workers enough to afford to buy the cars they produced. His decision ultimately benefited Ford Motor Company by making it an attractive employer and stimulating demand for its products.

A **social entrepreneur** is a person or team that acts to form a new venture in response to an opportunity to deliver social benefits while satisfying environmental and economic values. Social entrepreneurs focus on the social welfare of their customers or clients but are cognizant of the economic and environmental costs and benefits.

Social entrepreneurs hold certain advantages over entrepreneurs in other circumstances. With a mission as the guiding vision, social entrepreneurs strive to organize and deploy diverse resources. They can engage volunteers, customers, partners, and investors through a sound business plan that furthers the organization's mission. In the social sector, success in the enterprise equals significance through improved lives and healthy communities. Social entrepreneurs focus on creating social value [Dees, Emerson, and Economy, 2002].

David Green, a social entrepreneur, has started several businesses in developing countries to make inexpensive medical devices such as intraocular lenses and hearing aids. These profitable businesses make low-cost devices that meet the needs of the poor without sacrificing quality [Kirkpatrick, 2003]. See www.aurolab.com.

An excellent example of social entrepreneurship is the nonprofit organization Trees, Water, & People (TWP), which has an environmental and social mission (see www.treeswaterpeople.org). Its mission is to reforest degraded areas and plant fast-growing trees in Central America. It decided to view the problem from the demand side and seek to reduce the demand for fuel wood. TWP teamed up with Aprovecho Research Center in Oregon to introduce a fuel-efficient stove that burns 50 to 60 percent less wood than traditional open fires by using an insulated, elbow-shaped burning chamber. The Justa stove also saves lives by removing toxic smoke through a chimney. TWP gives these wood-conserving stoves to farmers in El Salvador as an incentive to reforest their land. Fuel conservation, health improvement, and reforestation are accomplished together.

Some of the best companies in history have tended to pursue a mixture of objectives, of which making money is just one—and not necessarily the primary one. For Merck, a top priority is patient welfare. For Boeing, it is advancement of aviation technology. Profitability is a necessary condition for existence, but it is not the end in itself for many of the visionary companies. Consider Johnson & Johnson, whose credo, published in the early 1940s, was the basis for its response to the 1982 Tylenol crisis, when a cyanide tampering incident caused the deaths of seven people in the Chicago area. The company quickly removed all Tylenol capsules from the entire U.S. market at a cost of $100 million, though the deaths occurred only in Chicago.

> Ben & Jerry's Ice Cream developed a mission statement that put forward three separate missions: product, economic, and social. The product mission called for a quality, natural product. The economic mission embodied profitability. Finally, the social mission focused on serving the community and the customer (see www.benjerry.com).

The social virtue matrix of Figure 8.3 illustrates the four possible responses to social responsibility challenges. The response of the lower-left quadrant (box 3) is conduct that corporations engage in by choice, in accordance with norms and customs. The lower-right quadrant (box 4) represents compliance—responsible conduct mandated by law or regulation [Martin, 2002]. These two lower quadrants represent the basic commitment of companies to society's values and laws.

The two upper quadrants encompass activities that are not directly beneficial to shareholders. The strategic benefits quadrant (box 1) includes activities that may add to shareholder value by generating positive reactions from customers, employees, or legal authorities. These actions may ultimately benefit the firm by accruing customer goodwill and community support. The upper-right quadrant (box 2) encompasses activities that clearly benefit society or the environment at a cost to the corporation.

An example of a firm active in the upper-left quadrant (box 1) is CitySoft (www.citysoft.com), which develops and maintains websites and intranet

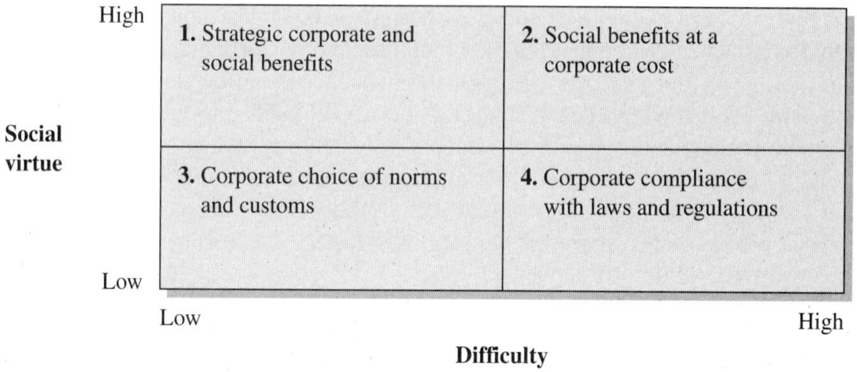

FIGURE 8.3 Social virtue matrix.

applications for businesses and organizations. What made it unique was its fundamental commitment to employ inner-city residents to provide these services. The founders chose to create a for-profit business with a strong social goal. One great opportunity to enter business in the top-left quadrant by offering strategic corporate and social benefits is to stimulate commerce at the bottom of the economic pyramid. While individual incomes may be low, the aggregate buying power of poor communities is actually quite large, representing a substantial market in many countries. In these markets, entrepreneurs need to reconsider their focus on high gross margins and shift toward securing good returns on invested capital while delivering social and environmental benefits [Prahalad and Hammond, 2002].

Actions in the two lower quadrants (boxes 3 and 4) of Figure 8.3 generate little credit since the public expects actions to be in compliance with its laws and norms. Actions in the upper-right quadrant (box 2) may ultimately engender benefits for shareholders. However, actions that provide benefits to society at a cost to a firm are difficult to defend to shareholders. For example, if only one automaker had decided to add air bags, it would lose some profits. When mandated, all automakers can provide added social benefits at a competitive cost.

The public wants information about a company's record on social and environmental responsibility to help decide which companies to buy from, invest in, and work for. As an example, see Starbucks website at www.starbucks.com/aboutus/csr.asp. Good deeds can redound to a company's credit, but they can be overlooked if unheralded. They also can backfire if the company fails to live up to the good-neighbor image it tries to project. Fifteen highly-ranked respected companies are listed in Table 8.7.

The most significant impediment to the growth of corporate virtue is limited vision for actions beyond compliance and allegiance to society's norms. Corporate coalitions acting in the upper right-hand quadrant (box 2 of Figure 8.3) can add social benefits when they agree on the benefits that outweigh the costs. New

TABLE 8.7 Highly respected companies.

■ Coca-Cola	■ Johnson & Johnson	■ Target
■ Dell	■ Lowe's	■ 3M
■ FedEx	■ Medtronic	■ Toyota
■ General Electric	■ Microsoft	■ Wal-Mart
■ IBM	■ Starbucks	■ UPS

Sources: 2001 Harris Interactive/Reputation Institute Survey. Fortune's Most Admired Companies, 2004.

ventures can lead the way in building momentum toward socially responsible leadership enhancing the quality of life for the world.

8.7 IMITATION

Imitation is said to be the greatest form of flattery. Many important new ventures have been based on the replication or modification of an existing business that the entrepreneur encounters through previous employment or by chance [Bhide, 2000]. Entrepreneurs start a business because they believe they can manage the business as well or better than the example they are copying. Sam Walton opened his first Wal-Mart in Rogers, Arkansas, after making numerous trips to study discount retailers in other regions of the United States. Walton once said: "Most everything I've done, I've copied from someone else." Chuck Williams got the idea for Williams-Sonoma during a trip to Paris. He was impressed by the kitchenware such as copper pots and porcelain ovenware sold there. He started Williams-Sonoma and expanded the chain of stores as the rage for French cooking grew. Technologists attend trade shows and conferences and often notice competitors' new products that their firm may readily produce.

Unfortunately, most attempts to replicate excellent businesses fail [Szulanski and Winter, 2002]. The difficulty of imitation springs from the lack of deep understanding of the excellent business example. Furthermore, the transfer of the best business practices from one setting to another can be fraught with unforeseen uncertainty. Imitation by independent entrepreneurs can be difficult because when they look at the existing business, they can't fully understand what makes it work. Thus, the best approach is to copy it in detail but recognize that quick response to customer comments will be necessary.

In 1986, Howard Schultz started his first independent effort as Il Giornale, a store modeled on his experience of Italian espresso bars. He played Italian opera in the Seattle store, and servers wore bowties. Il Giornale was set up as a stand-up bar, as is common in Italy, and it did not offer nonfat milk. Schultz had transferred the Italian coffee bar to Seattle with mixed success. People wanted chairs, and servers did not want ties. Nonfat milk quickly found its way onto the menu [Schultz, 1997].

Close copying may be the best method of imitation. It is important, however, to recognize the value of management and leadership, which is difficult to clone. A talented leader of an excellent business possesses some skills and capabilities that may be difficult to readily understand or copy.

Once the new business is up and running, customer comments can be used to adjust the business procedures to local conditions. Schultz loved the experience of Italian coffee bars but eventually adjusted his coffee café to fit Seattle's desires. Schultz was successful in adapting his Il Giornale store and ultimately created a successful system. He opened a second Seattle store after six months and a third store in Vancouver in 1987. By August 1987, Schultz arranged for his investor group to purchase all the Starbucks stores and its coffee roasting facility. He then merged his Il Giornale and Starbucks under the Starbucks name. By 2004, Starbucks had about 5,000 stores worldwide and revenues of $5 billion.

JetBlue Airways is a good imitation of Southwest Air. Based on a low-cost, all-coach, point-to-point business model, JetBlue started in February 2000 with two aircraft serving New York's JFK airport and Fort Lauderdale, Florida. JetBlue's initial public offering on April 11, 2002, raised $147 million for expansion. JetBlue is a profitable airline and an excellent example of sound imitation.

8.8 BUILDING A NEW BUSINESS

A talented leader of a new venture has a vision and a plan to implement it. He or she can motivate people and manage information and resources so that the business can create a profit. The performance of a new venture is a consequence of factors that encompass the dimensions of a business as portrayed in Figure 8.4.

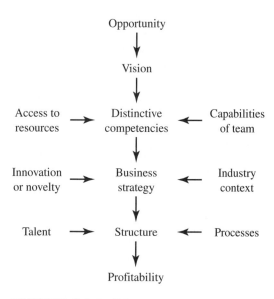

FIGURE 8.4 Building a new venture.

The quality of the opportunity and its fit with the vision lead to the accumulation or creation of the distinctive competencies based on resources and capabilities. Following from a set of competencies, a business strategy is created based on novelty or innovation within an industry context. The attractiveness of the industry with respect to its business opportunities affects the profit potential and the expected return of the venture. Access to resources and the ability to attract the entrepreneurial team will depend on the attractiveness of the new venture. The context of the industry environment will determine the amount of resources available to the venture since capital typically flows to industries in which opportunities are abundant and attractive. Furthermore, the industry's related competencies of the entrepreneurial team, which refers to the level of experience and knowledge with the industry, will lead to the expectation of greater success.

The identification and acquisition of required resources and capabilities are crucial for a firm's success. For a fast-growing firm based on continuing innovation, the intellectual resources are critical to success. The securing of the necessary resources and capabilities may occur in stages, as required.

The creation of a business plan focuses on the stages of building a business, as portrayed in Figure 8.4. The initial formation of a suitable structure for the business follows the business strategy. The ability to remain competitive and innovative while operating through an appropriate structure can lead to enduring profitability.

A business plan may include a life-cycle business matrix, as shown in Figure 8.5 for a color-photo printer [Kim and Mauborgne, 2000]. The matrix shows the business levers active at each stage of the buyer experience. This matrix provides the viewer with a visual portrayal of a business over the six stages of a customer's experience. The photo printer has good qualities, convenience, and assurance that promises will be fulfilled. Delivery is simple and secure. The product qualities and experience are good in the use stage. Add-ons and supplements are easy to get and use, and are environmentally friendly. Maintenance is easy and can be done over the phone and Internet. Finally, disposal of the device is convenient and environmentally friendly.

The greatest risk in the building stage of creating a business is the failure to complete all the steps of Figure 8.4. Some entrepreneurs who possess strong technical skills and capabilities unfortunately skip the formation of a business strategy that will lead to profitability. Another risk is that the business plan may have an inadequate plan for the organizational structure and the management of processes and talent. The road to success includes all the steps of Figure 8.4.

8.9 THE BUSINESS PLAN

Once an entrepreneurial team has selected an opportunity that is attractive and feasible, they need to build a plan for describing the elements of the proposed business. This business plan can be used for many purposes, such as attracting talented individuals and resources to the venture. Of course, there is no one right way to organize and write a business plan. A **business plan** is a document that

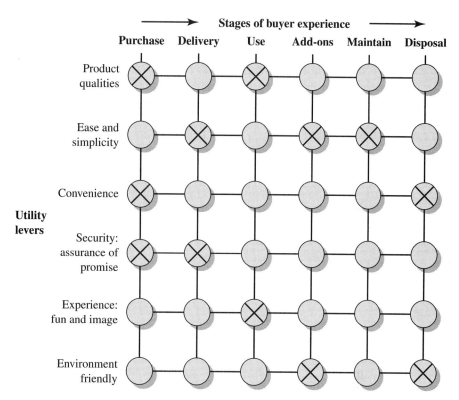

FIGURE 8.5 A life-cycle business matrix for a color-photo printer firm. The utility levers are on the vertical axis, and the horizontal axis displays the buyer experience. The ⊗ indicates this lever is present for the buyer experience stages.

describes the opportunity, product, context, strategy, team, required resources, financial return, and harvest of a business venture. The elements of the business plan are listed in Table 8.8.

As stated by Daniel Hudson Burnham:

> Make no little plans. They have no magic to stir men's blood and probably themselves will not be realized. Make big plans; aim high in hope and work, remembering that a noble, logical diagram once recorded will never die, but long after we are gone will be a living thing, asserting itself with ever-growing insistency. Remember that our sons and grandsons are going to do things that would stagger us.
>
> Let your watchword be order and your beacon beauty.

The business plan is a blueprint for your business. The clarity of this plan will be enhanced by the entrepreneurial team over a period of weeks or months. It enables the team to see clearly the plan of action and understand all the ele-

TABLE 8.8 Elements of a business plan.

■ Executive summary	■ Entrepreneurial team: capabilities, commitment
■ Opportunity: quality, growth potential	■ Financial plan: assumptions, cash flow, profit
■ Vision: mission, objective, core concept	■ Required resources: financial, physical, human
■ Product or service: value proposition, business model	■ Uncertainties and risks
■ Context: industry, timeliness, regulation	■ Financial return: return on investment
■ Strategy: entry, marketing, operations, market analysis	■ Harvest: return of cash to investors and entrepreneurs
■ Organization: structure, culture, talent	

ments. Single entrepreneurs might not need a written plan if they are using their own resources, but they still will benefit from writing the plan. They can show it to advisors who can help them shape a better plan.

If the venture requires outside financing, a business plan will be required by most investors. Listing the assumptions underlying the financial plan will be helpful to all participants. Most useful plans number 10 to 20 pages with backup and supporting material available on request. A business plan is a reflection of its authors and should be the result of a team effort.

A business plan is an important part of the business building process. However, we recognize that many small businesses start with a modestly crafted plan and build it up over the first months after launch. A business plan is a necessity, but it may not be required in a formal and complete form in order to start. For some, the business plan may be conceived in conversations and recorded as notes by the entrepreneur. Informed action is what is needed for entrepreneurship. Eventually, however, the entrepreneur will want a formal business plan to show to investors, bankers, or potential executives. In a dynamic industry such as semiconductors or nanotechnology, flexibility is key to success, and rigid adherence to a plan may be too limiting. The writing of a plan may be a useful exercise for the founder team, but they must recognize that it will need to be updated periodically.

The key function of writing a plan is to record the opportunity and the solution to the need and show that solution can be made to be economically favorable within a reasonable period of time. Scott McNealy (born in 1954) was a cofounder of Sun Microsystems in 1982 with Vinod Khosla. They created a four-page business plan to raise $250,000 in February. They were profitable by May with revenues of $8.9 million in their first year.

After the business plan is outlined or drafted, an executive summary is written. It should be concise and about one to three pages in length. Some investors and potential team members will read the summary to determine if they wish to proceed further. For many investors, the executive summary will be sufficient to initiate discussions. As explained in Section 2.6, often entrepreneurs will write

TABLE 8.9   Ten common mistakes or gaps in business plans.

- Solutions or technologies looking for a problem.

- Unclear or incomplete business model and value proposition.

- Incomplete competitor analysis and marketing plan.

- Inadequate description of the uncertainties and risks.

- Gaps in capabilities required of the team.

- Inadequate description of revenue and profit drivers.

- Limited or no description of the metrics of the business.

- Lack of focus and a sound mission.

- Too many top-down assumptions such as "we will get 1 percent market share."

- Limited confirmation of customer demand or pain.

an initial business summary as a first step and later revise it to become the executive summary.

Creating a business plan teaches the team about the market, customers, and each other. Working through creating a business plan will enable the entrepreneurs to estimate when cash flow will turn positive. A plan will probably turn up at least one or two big gaps, which can be corrected. Table 8.9 lists 10 common gaps that show up in a business plan. The entrepreneurs can respond to these gaps or mistakes and seek to fix them.

An example of a classic entrepreneurial business case is the R&R case. This case describes the efforts of Bob Reiss to respond to an opportunity in the game industry. This case describes the business opportunity and provides a series of letters and documents, that, taken together, represent a business plan. Reiss accomplished this venture analysis and built his team within a few months [Reiss, 2000]. Reiss has a clear business plan recorded in his notes and letters, and he communicated these ideas clearly to his partners in the venture.

Reiss was experienced in the game business and recognized an opportunity for a new game by observing a popular game called Trivial Pursuit in Canada. He conceived a trivia game based on *TV Guide.* Then, he wrote a letter to the publisher of *TV Guide* and followed up with a detailed three-page business plan. With the agreement of the publisher, he proceeded to engage the services, on a royalty basis, of a game designer. He then engaged a long-time friend as a partner, Sam Kaplan. Kaplan handled all the operational issues, including the production of the game as well as the marketing activities. Reiss also contracted for collection of invoices and payments. Reiss designed the marketing plan and launched the product in 1983. The game was well-received and sold 580,000 units in 1984. Kaplan and Reiss each netted $1 million over the two-year life of the venture.

TV Guide's publisher agreed to work with Reiss after calling 13 references and satisfying its concerns. Kaplan was a good partner because he had resources and capabilities complementary to Reiss's assets. The timing of the introduction of the game was excellent, and Reiss constructed a deal for all his suppliers and partners that gave them strong incentives.

Entrepreneurs like Reiss create and build new business ventures. They break old rules and make new ones. They exploit tools and technologies and create new markets. They are particularly adept at matching the opportunities with the competencies of the new venture team. They observe and understand others and how their needs change over time. Their passion emerges as they find opportunity.

Entrepreneurs are advised to follow the five steps to build a business shown in Figure 8.6. Reiss followed these five steps to build the R&R venture. He clearly described the opportunity for a new trivia game based on *TV Guide.* His solution was to build a game using a designer he knew and produce the game and market it widely. The organization was based on a set of well-known, reliable partners who would share in the effort of creating the game and reaping the returns.

The table of contents of a complete, formal written business plan is shown in Table 8.10. Most plans will omit some sections or deemphasize others when they are a relatively less important element.

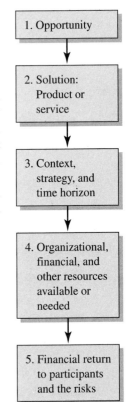

FIGURE 8.6
Roadmap for building a business plan.

TABLE 8.10 Table of contents of a business plan.

	Reference to sections of this book
1. Cover page	
2. Table of contents	
3. Executive summary	Section 2.6
4. Opportunity	
■ Need or problem	Chapter 2
■ Vision and mission statement	Sections 3.1 and 3.2
■ Product and value proposition	Section 3.3
5. Strategy	
■ Core competencies	Section 3.6
■ SWOT analysis	Section 4.3
■ Industry and competitor analysis	Sections 4.2 and 4.3
■ Competitive advantage	Sections 3.7 and 4.5
■ Timing and time horizon	Section 5.1
6. The Concept	
■ Legal form, name, logo	Section 10.1
■ Patents, licenses, copyrights	Chapter 10
■ Research and development	Section 6.4
■ Innovation plan	Section 5.3
■ Prototype	Section 9.4

(continued on next page)

TABLE 8.10 (continued)

	Reference to sections of this book
7. Marketing plan	
■ Objectives and segments	Section 11.2
■ Research	Section 11.4
■ Strategy and brand equity	Section 11.5
■ Marketing mix	Section 11.6
■ Customer relationship management	Section 11.7
■ Sales method	Section 11.10
8. Organization	
■ Leadership team and brief resumes	Sections 12.4 and 12.5
■ Organizational culture	Section 12.7
■ Ownership and stock options	Section 12.10
■ Board of directors	Section 12.11
■ Global strategy	Section 14.3
■ Alliances and partnerships	Section 5.2
9. Operations	
■ Location and facilities	Section 13.3
■ Vertical integration	Section 13.7
■ Outsourcing	Section 13.7
■ Suppliers	Section 15.1
■ Production or process methods	Section 15.2
10. Financial plan	
■ Revenue model	Section 16.1
■ Profit model	Section 16.2
■ Growth rate plan	Section 16.3
■ Profit and loss forecast	Section 17.4
■ Cash flow forecast	Section 17.5
■ Break-even analysis	Section 17.8
■ Harvest plan	Section 16.4
11. Risks and milestones	Chapter 6
■ Critical risks	
■ Potential mitigations of the risks	
■ Performance milestones	

(continued on next page)

TABLE 8.10 (contcluded)

	Reference to sections of this book
12. Investment offering	
■ Amount requested and price of share offering	Section 18.9
■ Valuation of firm and percent of firm offered	Section 18.8
■ Terms of the deal	Section 19.4
■ Uses of funds	Section 18.7

All business plans need a cover sheet if you are going to provide a copy to potential investors or employees. The cover page includes the name of the new firm, its address and phone number, and the lead entrepreneur's name and contact information. Each copy of the plan is numbered to inhibit copying and passing to others.

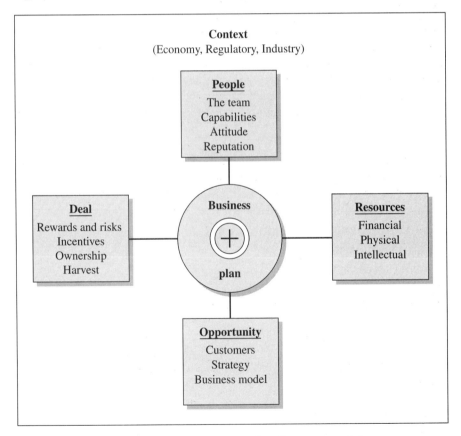

FIGURE 8.7 The business plan serves as the alignment tool for a new business venture.

The business plan can be used to align the interests of all the participants of a new venture, as shown in Figure 8.7. The business plan explains how the people, the resources, and the opportunity can be linked to a deal that will hopefully benefit all stakeholders—employees, investors, suppliers, and allies [Sahlman, 1999].

Two examples of well-prepared and complete business plans are provided in Appendix A.

8.10 AGRAQUEST

Natural pesticides potentially offer a host of benefits, from easier regulation to flexible application timing. But for all growers, the most important element is field performance. You cannot sell a product just on its environmental friendliness. It has to be efficacious, easy to use, and reliable. AgraQuest prepared a plan to develop natural pesticides as an alternative to chemicals.

AgraQuest completed its first business plan dated May 5, 1995, with the purpose of raising $3.6 million from private investors. The 40-page plan had a table of contents, as shown in Table 8.11. The 1995 executive summary is shown in Table 2.16.

TABLE 8.11   Table of contents of AgraQuest business plan (1995).

Table of Contents

- EXECUTIVE SUMMARY
- THE COMPANY—Company Ownership, Products, Facilities
- SIZE OF MARKET AND MARKET TRENDS—World Pesticide Market
- TECHNOLOGY
- DEVELOPMENT STRATEGY AND PROCESS
- MANUFACTURING AND PRODUCT COSTS
- MARKETING STRATEGY
- AGRAQUEST'S COMPETITIVE ADVANTAGE
- BUSINESS STRATEGY AND STRATEGIC ALLIANCES
- MILESTONES/GOALS
- ORGANIZATIONAL STRUCTURE
- PERSONNEL PLAN
- MANAGEMENT TEAM
- SCIENTIFIC ADVISORY BOARD
- BOARD OF DIRECTORS
- KEYS TO SUCCESS
- FINANCIAL PLAN AND PROJECTIONS

TABLE 8.12 Keys to success for AgraQuest from the 1995 business plan.

1. Recruitment of talented, experienced scientists who can function in a team environment.

2. Effective teamwork in the management group.

3. Good board–management team working relationship.

4. Enough money to hire a critical mass of experienced scientists, build sufficient laboratories, and purchase necessary equipment.

5. Development of strategic relationships with large agrochemical companies.

6. Proprietary protection on discoveries.

7. EPA registration as microbials or biochemicals.

8. Aggressive licensing activities to bring in product candidates from outside the company.

AgraQuest's plan described its unique competitive advantage as follows:

Successful natural product discovery is not a simple matter of screening any group of microbes against a range of targets. It requires sophisticated knowledge of the specific microbial groups, locations, types of chemistry, and screening (isolation, fermentation, bioassay) techniques that yield the highest numbers of bio-active natural products. In addition, the type and style of people management can either enhance or hinder natural product discovery and development. For each step of the discovery process, AgraQuest has unique know-how and proprietary techniques that generate more novel, proprietary pesticidal natural products faster than others in the field.

The business plan identified the uncertainties and risks and described the keys to success, as listed in Table 8.12.

8.11 SUMMARY

New ventures are created in various formats that fit the needs of the venture, society, and legal constraints. Suitable forms include:

- Independent for-profit ventures
- Corporate for-profit ventures
- Nonprofit firms
- Family-owned business
- Franchises
- Direct selling organizations

Building a business is described as a process that can be learned and mastered by talented and educated entrepreneurs through accomplishing the following descriptions or tasks:

- Opportunity, vision, value proposition, business model
- Concept, feasibility, executive summary
- Financial plan, legal form, business plan

■ Resource needs
■ Secure resources and launch

Finally, we note that all entrepreneurs have a business plan that helps them to codify and communicate their road map to achievement. Many, if not most, entrepreneurs will benefit from putting this plan in written form to readily and consistently communicate the plan for their business.

Principle 8
Entrepreneurs can learn and master a process for building a new venture and they communicate their intentions by writing a business plan.

8.12 EXERCISES

8.1   Mobile Technologies Inc. was started by Martin Amery in December 1989 and has grown to sales of more than $5 million annually. Amery owns 90 percent of the firm, and his two daughters each own 5 percent. Nicole Amery has held several positions in the firm during the past decade. Amery's other daughter is a pastor in Dallas. The firm has always been cash-flow positive. Amery would like eventually to have his daughter Nicole purchase the firm. How can they establish a fair purchase price and finance the transfer?

8.2   Gary Solomon purchased the rights to an invention and built a business on the new technology for making commercial signs. FASTSIGNS was established in 1985 as a franchisor of sign-making stores (see www.fastsigns.com). After the sign is designed on a computer, a plotter device cuts out the graphic, which is then transferred to a mounting surface. Examine the benefits of the FASTSIGNS franchise to the franchisee and the franchisor. If possible, visit one of the 500 locations of FASTSIGNS and ask the franchisee to describe his or her experience with the franchise format.

8.3   In 2000, three graduate students at Harvard University launched a nonprofit called New Leaders for New Schools (NLNS). This venture recruits, trains, places, and supports principals in U.S. urban school districts (see www.nlns.org). The three founders met while enrolled in a class at Harvard on entrepreneurship in the social sector. Determine the mission of NLNS and describe the accomplishments of the enterprise.

8.4   Determine what types of products are suitable for direct selling organizations. Would it be possible to sell cell phones or home computers through a DSO?

8.5   Select an industry of interest to you and then try to find a good candidate for imitation. Describe the opportunity and tell how you will reap the benefits of imitation.

8.6   Santa Fe, a city of only 65,000 people, sells over $200 million in art every year and is home to over 300 art galleries [Hutchinson, 2003]. Describe the elements that support the art cluster in Santa Fe, New Mexico.

8.7   A popular factory tour is at the Longaberger Company in Dresden, Ohio (www.longaberger.com). This family-owned, privately held firm sells baskets through a direct sales organization. Using the format of Figure 8.7, describe the business alignment of this firm.

8.8   Green Marketplace has positioned its activities in an environmentally friendly niche (www.greenmarketplace.com). Visit this firm and describe its social venture.

8.9   UPS and FedEx are working to use delivery trucks with better mileage and reduced emissions. The goal is lower costs and better environmental results [Haddad, 2003]. They are using "green" hybrid electric-diesel engines to drive their trucks. Visit their websites and compare their achievements. How do their activities fit into the grid of Figure 8.3?

8.10 Dangdang.com is China's biggest online bookseller. Peggy Yu moved from New York to Beijing to co-found the online book store. Dangdang.com is a shameless imitation of Amazon.com, offering 210,000 titles [Chen, 2003]. Visit www.Dangdang.com and determine its business model.

8.11 Movies are edited for showing on airlines by removing explicit scenes or crude language. This approach could be used to sell R-rated movies for family home rental. Using the method of imitation, describe a concept for the editing and rental of R-rated movies.

8.12 A partnership between Stanford University's Social Entrepreneurship Startup Course (ses.Stanford.edu) and the nonprofit Light Up the World Foundation (www.lightuptheworld.org) is working to bring safe, affordable lighting to people in Mexico, China, and India [Snyder, 2003]. Students in engineering and business work to design a lamp appropriate to the needs of villagers. Develop a brief plan for a social entrepreneurship project with an international nonprofit for your university.

Building Knowledge and Learning in a New Enterprise

Knowledge and human power are synonymous, since the ignorance of the cause frustrates the effect.
Francis Bacon

CHAPTER OUTLINE

Knowledge is power. Knowledge assets and intellectual capital are potential sources of wealth. The creation and management of knowledge can lead to new, novel applications and products. Sharing knowledge throughout a firm can enhance the firm's processes and core competences, thus making the firm more innovative and competitive. Most technology ventures are based on knowledge and intellectual property that must be enhanced and managed. A learning organization is skilled at creating and sharing new knowledge and uses this knowledge to do a better job.

Prototypes are models of a product or service and can help a new technology venture to learn about the right form of the product for the customer. Scenarios are used to create a mental model of a possible sequence of future events or outcomes. Knowledge acquired, shared, and used is a powerful tool for the entrepreneur to build a new venture organization. ∎

9.1 THE KNOWLEDGE OF AN ORGANIZATION

Assets are potential sources of future benefit that a firm controls or has access to. Knowledge is an asset that is a potential source of wealth, as described in Chapter 1. The creation and management of knowledge leads to new, novel applications and products that can result in wealth creation. **Knowledge** is the awareness and possession of information, facts, ideas, truths, and principles in an area of expertise. **Intellectual capital** is the sum of the knowledge assets of a firm. These knowledge assets include the knowledge of its people, the effectiveness of its management processes, the efficacy of its customer and supplier relations, and the technical knowledge that is shared among its people. It can be thought of as best practices, new ideas, synergies, insights, and breakthrough processes. Thus, the firm's intellectual capital (IC) is the sum of its human capital (HC), organizational capital (OC), and relationship capital (RC), as described in Chapter 1.

From the generation of new ideas through the launch of a new product, the creation and exploitation of knowledge is a core theme of the new product development process. In fact, the entire new product development process can be viewed as a process of embodying new knowledge in a product [Rothaermel and Deeds, 2004].

Knowledge is one of the few assets of a firm that grows when shared among its people. A new venture is wise to strive to acquire, store, manage, and share its knowledge throughout its organization. Intellectual capital is knowledge that has been formalized, captured, and leveraged to produce an output that has great value. In many ways, we can view a product as embodied knowledge. For example, an ATM machine embodies all the knowledge necessary for completing most banking transactions.

The knowledge creating and sharing activities of a firm can be represented by Figure 9.1 [Leonard-Barton, 1995]. The value of commercial knowledge is in its use, not its possession. The value of knowledge compounds when it is shared. Using current knowledge for cooperative problem-solving is the first of the four knowledge activities of a firm. The second is the implementation of new processes and tools within the firm. The third activity is experimenting and learning in order to build the knowledge base. The fourth activity is acquiring knowledge from outside the firm.

As a result of creating and sharing knowledge, a firm can enhance its peoples' skills and capabilities, as well as the knowledge embedded in its processes and managerial systems. Knowledge finds value in practice and use. The strategic approach of a new venture is linked to a set of intellectual assets and capabilities. Thus, if a firm has an opportunity that requires certain knowledge and it is not yet available, we can state that the firm has a knowledge gap. Acquiring the knowledge to fill the gap will be critical to the future success of this firm.

The knowledge of a firm encompasses: 1) cognitive knowledge, 2) skills, 3) system understanding, 4) creativity, and 5) intuition. The first three forms of knowledge can be codified and stored. The last two forms of knowledge are types of trained intellect that people possess but are difficult to codify.

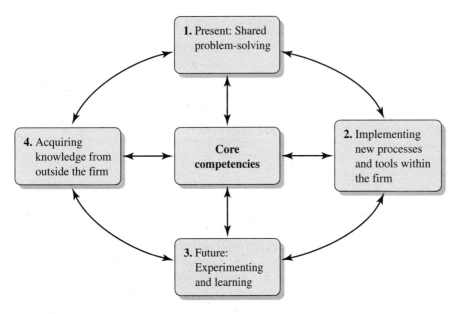

FIGURE 9.1 Knowledge creating and sharing activities of a firm.

Knowledge can be *prepositional,* which comprises beliefs about natural phenomena, such as scientific discoveries and practical insights into the properties of materials, waves, and nature. *Prescriptive* knowledge is all about techniques—the manipulation of processes and formulas, such as how to write a piece of software. The growing interplay between these two forms of knowledge transformed the world economy during the twentieth century [Mokyr, 2003].

Knowledge can be seen as a source of innovation and change leading to action. Also, knowledge provides a firm with the potential for novel action and the creation of new ventures. Knowledge creates real wealth for a new venture through multiple applications. Knowledge applications have breadth across an organization and length in time of use. The knowledge represented by patented inventions, software, marketing programs, and skillful employees comprises 70 to 90 percent of the assets held by corporations like Microsoft, Amgen, and Intel.

9.2 MANAGING KNOWLEDGE ASSETS

The growth of a new venture rests, in part, on the increasing value of a knowledge base within the emerging firm. **Knowledge management** is the practice of collecting, organizing, and disseminating the intellectual knowledge of a firm for the purpose of enhancing its competitive advantages. The four steps for managing knowledge in a new venture are given in Table 9.1. The first step is to identify and evaluate the role of knowledge in the firm. How is knowledge created, stored, and shared? The second step is to identify the expertise, capabilities, and

TABLE 9.1 Managing knowledge in a technology venture.

1. **Role:**	Identify and evaluate the role of knowledge in the firm.
2. **Value:**	Identify the expertise, capabilities, and intellectual capital that creates value in the form of products and services.
3. **Plan:**	Create a plan for investing in the firm's intellectual capital and exploiting its value while protecting it from leakage to competitors.
4. **Improve:**	Improve the knowledge creation and sharing process within the new venture.

intellectual capital that create value for a firm. Then, we examine the uniqueness and value of these intellectual assets.

The third step is the creation of an investment and exploitation plan for maintaining and harvesting the value of the knowledge assets. Finally, the firm improves the process of creating and sharing its knowledge. Though knowledge is one of the few assets that grows when shared, the new venture needs to carefully determine what knowledge to share and what knowledge should be protected or kept secret. This is particularly true for technology ventures for which intellectual property is usually their key asset.

Most professionals are unable to keep up with all they need to know. One method of knowledge access is to embed knowledge into the technologies used by the professionals. For example, when designing a product, the databases required can be linked directly to the design tools [Davenport and Glaser, 2002].

Part of an emerging firm's knowledge base is information about competitors. This knowledge is useful in responding to competitive changes and challenges. **Competitive intelligence** is the process of legally gathering data about competitors. Competitor intelligence may include securing data about competitors' products, services, channels of distribution, pricing policies, and other facts. Legal means of acquiring competitor intelligence include gathering company reports, news releases, and industry reports, and visiting competitor websites and trade show booths.

Knowledge is worthy of attention because it tells firms how to do things and how they might do them better [Davenport and Prusak, 1998]. The key skill for an emerging start-up is the ability to turn knowledge into products and services. Knowledge turns into action as it is embedded in the products, routines, processes and practices of the new technology venture. Knowledge embedded in a company's activities can be a sustainable competitive advantage since imitating it is difficult for competitors.

9.3 LEARNING ORGANIZATIONS

New ventures grow powerful from learning and adapting to new challenges and opportunities. A **learning organization** is skilled at creating, acquiring, and sharing new knowledge and at adapting its activities and behavior to reflect new knowledge and insights. A technology venture creates and acquires knowledge and

shares this knowledge among its people. As a result of this new knowledge, the organization adapts its actions and behavior. Learning organizations are skilled at five activities: systematic problem-solving, experimentation with new approaches, learning from their own experience and past history, learning from the experiences and best practices of others, and transferring knowledge quickly and efficiently throughout the organization. The learning organization is active, imaginative, and participative. It attempts to shape its future rather than react to forces. A learning firm adapts itself to its learning and increases opportunities, initiates change, and instills in employees the desire to be innovative. A learning organization confronts the unknown with new hypotheses, tests them, and creates new knowledge. Thus, the learning organization creates innovation and new knowledge that is used by the technology venture to develop new products and services.

Knowledge is actionable. Information that does not enable an action of some kind is not knowledge. Knowledge comes from the ability to act on information as it is presented. It truly is power, giving an organization the ability to continuously better itself. The power of knowledge depends on the company's ability to provide a supportive environment: a culture that rewards the sharing of knowledge across various barriers. The company that develops the right set of incentives for its employees to work collaboratively and share their knowledge will be successful in its knowledge management effort. Knowledge management has several benefits: it fosters innovation by encouraging a free flow of ideas, enhances employee retention rates, enables companies to have tangible competitive advantages, and helps cut costs.

The decisions of an entrepreneurial firm are the result of the firm's ability to process knowledge and learning [Minniti, 2001]. Knowledge acquired through learning-by-doing takes place when entrepreneurs choose among alternative actions whose payoffs are uncertain. Over time, entrepreneurs repeat only those choices that appear most promising and discard the ones that resulted in failure. Thus, entrepreneurship is based on a process of learning that allows entrepreneurs to learn from successes as well as failures. Jack Welch [2002] described the learning process as: "In the end I believe we created the greatest people factory in the world, a learning enterprise, with a boundary-less culture."

Websense Inc. is an example of a firm that has utilized the methods of Figure 9.1 to grow into an emerging firm (see www.websense.com). Websense was founded in 1994 as a value-added reseller of Internet software solutions such as firewalls. Through that experience, by 1996, Websense had developed and launched its first product. In 2003, the company released a software system for monitoring and managing Internet usage by a customer's employees. As a knowledge-based learning organization, the firm has been profitable since 2001.

Managers and entrepreneurs often have distorted pictures of their businesses and their environments. Busy among the trees, they can lose sight of the forest. They can review the impacts of their actions, however, and then modify their approach accordingly. The greatest asset that entrepreneurs can bring to knowledge and learning is their willingness to seek and make wise use of feedback [Mezias and Starbuck, 2003].

TABLE 9.2 Entrepreneurial learning process.

Step	Question	Outcome or action required
1. Identify the problem or opportunity	What do we want to change?	Desired specific result
2. Analyze the problem or opportunity	What is the key cause of the problem?	Key cause identified
3. Generate potential solutions	How can we make a positive change?	List of possible solutions
4. Select a solution and create a plan	What is the best way to do it?	Establish a criteria, select the best solution, and set a plan to accomplish it
5. Implement the selected plan	How do we implement the plan effectively?	Monitor the implementation
6. Evaluate the outcome and learn from the results	How well did the outcome match our desired result?	Verify that the problem is solved. Why did it work?

The entrepreneur's learning process is based on the six steps outlined in Table 9.2 [Garvin, 1993]. At each stage in the development of the new business, the entrepreneur encounters a set of challenges or problems that require resolution. A firm can use the method shown in Figure 9.2 to resolve issues and learn from its successes and failures.

Organizational learning looks at an organization as a thinking system. Organizations rely on feedback to adjust to a changing world. Thus, organizations engage in complex processes such as anticipating, perceiving, envisioning, and problem-solving in order to learn. This approach is very important for new technology ventures to adopt and improve.

Process improvement projects can produce two types of learning. *Conceptual learning* is the process of acquiring a better understanding of cause-and-effect relationships by using statistics and scientific methods to develop a theory. *Operational learning* is the process of implementing a theory and observing positive results. Conceptual learning yields know-why—the team understands why a problem happens. Operational learning yields know-how—the team has implemented a theory and knows how to apply it and make it work. It is useful to design projects that are more likely to deliver both conceptual and operational learning [Lapre and Van Wassenhove, 2002].

For learning to be widely used in a firm, knowledge must spread widely and quickly throughout the organization. Ideas carry maximum impact when they are shared broadly rather than retained by a few people. A variety of mechanisms enable this process, including written, oral, and visual reports, knowledge bases, personnel rotation programs, education, training programs, and formal and informal networks. The organization needs to foster an environment that is conducive to learning. The new venture must strive to set aside time for reflection and sharing. Furthermore, boundaries that inhibit the flow of knowledge must be

reduced so that learning is shared. A new venture can profit from efforts to eliminate barriers that impede learning and to place learning high on the organizational agenda.

> Neurocrine Biosciences was established in 1992 to develop small-molecule drugs to treat diseases of the immune and central nervous systems. Knowledge of interactions between the nervous, endocrine, and immune systems underpinned the firm's products. Over the years, as a result of new knowledge, the focus has shifted to neural and hormonal disorders. The initial public offering raising $40 million was held in May 1996.

A learning organization, properly managed, can enable a firm to meet the challenge of change by constantly reshaping competitive advantage even as the marketplace rapidly shifts. The learning organization is able to improvise or adapt to balance the structure that is vital to meet budgets and schedules with flexibility that ensures the creation of innovative products and services that meet the needs of changing markets [Brown and Eisenhardt, 1998]. The firm that is most responsive to change and capable of learning is the one that succeeds [Galor and Moar, 2002]. See the Genentech example on page 210.

Knowledge is stored in documents, databases, and people's knowledge. Knowledge created in a learning process as depicted in Figure 9.2 is a social process that leads to increasing knowledge [McElroy, 2003]. Knowledge is shared by people and embedded within the business processes of the firm. Innovation flows through the business processes, products, and services of the firm, as shown in Figure 9.2. As the firm learns and creates new knowledge, new innovation is created.

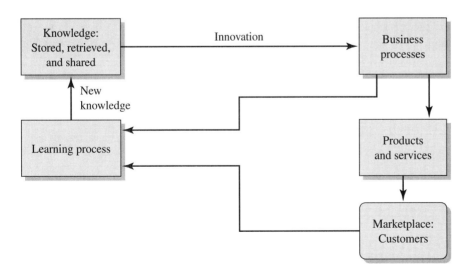

FIGURE 9.2 Knowledge and learning within a technology firm.

9.4 PRODUCT PROTOTYPES

Whenever possible, new business ventures should, early in the design process as described in Chapter 6, create a prototype of their product. A **prototype** is a physical model of the product or service. It is a model that has the essential features of the proposed product but remains open to modification. It can be used to identify and test requirements for the product. Prototypes are incomplete models that can be used by the new venture team to elicit comments from designers, users, and others to learn more about the product. Prototypes can be pictures, sketches, mockups, or diagrams that can be collaboratively studied. New ventures can use prototypes to redefine their business models and strategies. Prototypes can be used to create a dialogue between people that leads to innovation [Schrage, 2000]. Testing a prototype on a small group has been a common approach for many new products.

The computer software industry uses prototypes called beta versions of software to elicit response from lead customers. Microsoft sent out 400,000 beta versions of its Windows 95 operating system to potential users. Prototypes can be physical, digital, pictorial, or some combination of media. Innovative prototypes lead to innovative conversations, which potentially lead to better products.

> Henry Ford planned to build a horseless carriage. However, no one could be persuaded to invest in it. The turning point came when Ford built a prototype car for the Grosse Pointe automobile races. Ford entered the races, drove the car himself, and won decisively. He repeated the feat the following year, in 1902. The victory attracted financiers, and the Ford Motor Company was up and running. On the way, Ford broke the world land speed record for a four-cylinder automobile.

In the creation of a movie or play, many innovators use sketches, storyboards, and videos to describe the product. The designers of a movie or play want to see how it works and engage in a collaborative redesign. The iterative procedure for prototype development is shown in Figure 9.3. Two or three iterations of the process may be sufficient to arrive at a satisfactory prototype.

New technologies such as computer simulations can make the creation of a prototype fast and cheap. **Rapid prototyping** is the fast development of a useful prototype that can be used for collaborative review and modification. An initial prototype can be rough since it enables the team to view the product and improve it. The ability to see and manipulate high-quality computer images helps create innovative designs. BMW uses computers to help engineers visualize automobile design and the results of crash tests [Thompke, 2001].

For the innovator, a prototype is a mechanism for teaching the market about the technology and for learning from the market how valuable that technology is in that application arena. Elisha G. Otis used a prototype of his safety elevator to show his potential customers. He built his elevator out of simple materials, mounting saw-toothed iron bars on the elevator-shaft guide rails and then in-

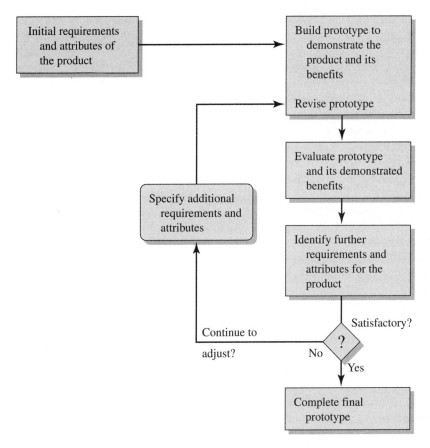

FIGURE 9.3 Prototype development process.

stalling retractable iron teeth on the cab itself. If the cable snapped, a spring attached to the cable would push out the teeth, which would catch on the iron bars and stop the cab in its tracks.

Otis demonstrated the prototype to customers by boarding his own lift filled with heavy boxes. He then ordered the lift rope cut. As the cab began to fall, the safety device kicked in and the elevator stopped. With this evidence, elevators became an item for hotels and office buildings, and Otis Brothers & Co. began to lift off.

It is often best to carry multiple product concepts into the prototyping phase and to select the best of those designs later in the process [Dahan and Srinivasan, 2000]. Keeping multiple product concept options open and freezing the concept late in the development process affords the flexibility to respond to market and technology shifts. It is possible to create static and dynamic virtual prototypes that are displayed at a website for review and testing by suppliers, customers, and designers. Virtual prototypes cost considerably less to build and test than their physical counterparts, so design teams using Internet-based product research can afford to explore a much larger number of concepts. Furthermore, Internet-based prototypes can help to reduce the uncertainty in a new product introduction by allowing more ideas to be tested in parallel.

SaltAire Sinus Relief provides a nose wash to relieve the symptoms of sinusitis and allergies (see www.saltairesinuswash.com). A bottle and pump are used to spray a supersalty solution into the sinus chamber and relieve the symptoms [Ridgway, 2003]. Two New York physicians founded the firm in 1997 and created a series of prototypes for people to try. Based on that knowledge, they showed a revised product to other physicians and launched the product in 2000. The patented dispenser bottle the firm developed won an award for innovative design.

Alfred Butts decided to create a board game that satisfied his urge to challenge his mental skills while having fun. He liked anagrams and crossword puzzles, but he wanted a game more than one person could play at a time. He studied skill-based board games and classified them as three types: 1) number-based, such as backgammon, 2) moving pieces, such as chess, and 3) word games. Butts chose word games and focused on anagrams—letters forming a word. He constructed an initial word game prototype and tried it on his friends and family. He then went on to improve the game. He later chose to expand the game's supply of letters with multiple copies of the letters "e" and "s" based on their frequency in words. His prototype was refined and issued as Lexico. The scoring was based on the length of words. In his next version, he added values to letters and a grid pattern for words. He also moved the starting point to the center square. By 1947, Butts had spent more than 10 years perfecting his game. In 1947, he sold the game to James Brunot, who agreed to pay Butts a royalty on sales. Brunot renamed the game Scrabble. By 2003, Scrabble had sold more than 100 million copies and was the world's popular word game.

Many firms have developed their products by entering potential markets with early versions of the products, learning from the tests, and probing again. These firms ran a series of market experiments, which introduced prototypes into a variety of market segments. The initial product design was not the culmination of the development process but rather the first step, and the first step in the development process was in and of itself less important than the learning and the subsequent, better-informed steps that followed. Software products lend themselves to rapid prototyping and early tests with potential customers.

Mike Marks formed WorkTools Inc. to improve work tools such as screwdrivers and staplers (see www.worktoolsdesign.com). In 1990, Marks and his brother Joel started work on designing an effective, easy-to-use stapling gun that would operate better than the Arrow T-50 of that day. Their first design provided easy, effective use. They immediately built a working prototype and took it with the T-50 to a local lumberyard and set up a table.

All but one of 65 testers preferred the new design. They refined their prototype and retested it at lumberyards and at the National Hardware Show. They showed it to Black and Decker, which signed up to sell it. Today, the product is the Power Shot. The key factors in their success were rapid prototyping and early testing [Lynn and Reilly, 2002].

9.5 SCENARIOS

Any new venture can benefit from creating a set of scenarios to address the complex, uncertain challenges as it develops its strategy. A **scenario** is an imagined sequence of possible events or outcomes, sometimes called a mental model. A few realistic scenarios based on the industrial context and a few associated possible sequences of events help a planner to plan for the future. Each scenario tells a story of how the various elements might interact under a variety of assumptions. It paints vivid narratives of the future. The goal of scenario planning is not to forecast what is going to happen but to encourage an openness of mind, a flexibility of response, and a habit of questioning conventional wisdom. As Stephen Covey [1996] stated: "The best way to predict your future is to create it."

Scenarios lead to learning in a two-step process: constructing a scenario and using the content of the scenario to learn [Fahey, 1998]. The key elements of a scenario are shown in Figure 9.4. A scenario attempts to answer key questions and is

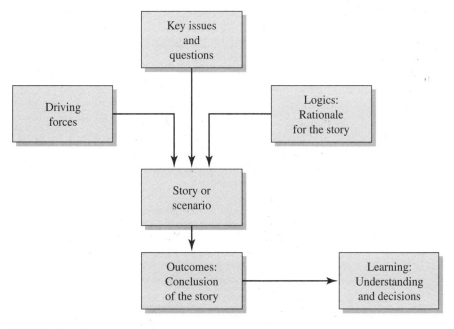

FIGURE 9.4 Elements of a scenario.

based on a statement of the driving forces and the rationale for the story. The outcomes or results of the story lead to understanding and a useful decision. A scenario is an internally consistent picture of what the future might bring. It is not a forecast, but rather one possible outcome. Creating four or five scenarios will help portray the range of potential outcomes to core questions facing any organization.

Entrepreneurs will often weigh whether a new technology will be radical or nonlinear and have a profound impact on the marketplace. A scenario can help define the impact and time frame for a new technology.

An example of the outline of a scenario for the growth of electric auto sales is shown in Figure 9.5. The structure of the story for electric vehicles can be used to build several possible scenarios that can be used to learn about the opportunities in this market.

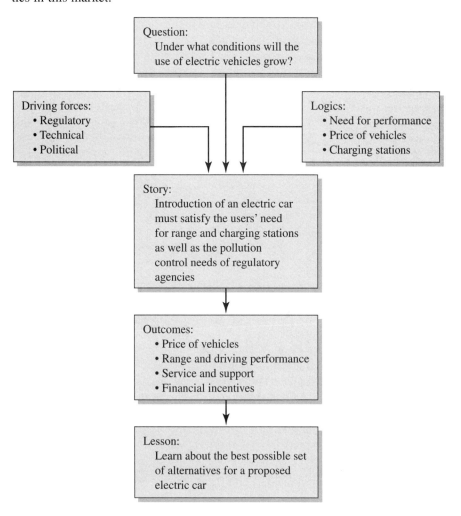

FIGURE 9.5 Elements of scenarios for electric cars.

Scenarios can sometimes become a mirage. By 2001, George Gilder, among others, had created a scenario for the future of telecommunications that was overblown and ill-timed. This rosy, nirvana-like scenario missed the regulatory issues and the concept of excess capacity. Scenarios often can be too rosy [Malik, 2003].

Jake Burton Carpenter created Burton Snowboards in Vermont. With limited market research, he started making many prototypes. By 1978, he was selling snowboards. His first scenario overestimated his early sales potential, however, and he sold just 300 boards in the first winter of production. He started building lower-priced boards and eventually broke into ski resorts. He knew his most likely scenario remained probable, and he persevered. Today, Burton Snowboards holds about one-third of the snowboard market.

9.6 AGRAQUEST

With two other former executives, Duane Ewing and Bruce Holm, and three former Entotech scientists, Pamela Marrone launched AgraQuest in 1995 in a small Davis, California, lab, furnished with $35,000 in used furniture and equipment. By 2003, the company has grown to 50 employees with 19 scientists.

Hunting for new natural products for pharmaceutical purposes has a long history. For example, penicillin and streptomycin are natural-product antibiotics from microorganisms. Little concerted effort has been made to hunt for new natural-product pesticides. Only approximately 200 insecticidal/miticidal, 30 herbicidal, and fewer than 10 nematicidal natural products are known, compared to the tens of thousands of known pharmaceutical natural products. Some natural products have become commercial pesticides. AgraQuest has tapped into the vast diversity of the world's microorganisms to discover and sell novel and environmentally friendly natural products for pest management with the efficacy of chemical pesticides. It is one of the first companies to invest in a continued and sustained effort to find them.

By focusing on unique groups of microorganisms, unique automated in-vivo screening, and technically difficult and unusual types of chemistry, AgraQuest has obtained a greater number of pesticidal natural products with fewer resources than other companies. The company knows how to organize and manage scientists in a work team structure that increases the output of novel molecules.

AgraQuest, like other small companies, can readily share and manage knowledge created by its work. Furthermore, it can quickly add new products to its product line by licensing them. In 2002, AgraQuest licensed *M. albus* from Gary Strobel of Montana State University. This product, when fully developed, will help control many pathogenic plant fungi. AgraQuest uses a computer system for managing the data and results on the screening of 20,000 microbes for 10 diseases. This database is available to all 19 scientists and the executives of the firm.

AgraQuest created its first prototype product, Laginex, for killing mosquito larvae, in 1998. It drew a lot of interest from mosquito control districts, and the product was fully tested in many sites. It was found to have a short shelf-life, however, and alternative production methods for this product remain under study.

By 2002, after three years in development, testing, and government review, Serenade was granted registration by the U.S. Environmental Protection Agency and its California counterpart, and became available for commercial sales a month later. Serenade is an environmentally friendly fungicide that can be used to ward off diseases that attack grapes, vegetables, nuts, and hops.

To satisfy the EPA's regulations and requirements, AgraQuest must accurately maintain test results and reports from toxic laboratories. All this information is provided to the EPA for review.

9.7 SUMMARY

Knowledge is power, and the creation and management of knowledge can lead to novel applications, markets, and products. Sharing and managing knowledge wisely and efficiently with a technology venture can help build competitive and innovative skills. A new entrepreneurial firm seeks to build a sound knowledge management system that supports a learning organization.

Prototypes are models of a product or service and can help a new venture learn the right form and function of a product by showing it to customers and letting them observe it or try it. Furthermore, scenarios can be used to examine possible future outcomes based on specific actions.

Principle 9

Knowledge acquired, shared, and used is a powerful tool for the entrepreneur to build an innovative, learning organization that can compete and grow effectively.

Genentech: Learning from Prior Experience

In the early 1970s, the biotechnology industry was just beginning to emerge when Cetus was formed by Ronald Cape (a PhD biologist with an MBA), Donald Glaser (Nobel laureate in Physics), Peter Farley (a physician with an MBA), Calvin Ward (a scientist), and Moshe Alafi (a venture capitalist). Being one of the first firms in the industry and having the backing of a Nobel laureate, Cetus was able to attract a star-studded advisory board. Unfortunately, neither Cetus's employees nor its advisors knew what a biotechnology firm should look like or what it should do. Therefore, Cetus took money and formed partnerships with whomever was willing, and attempted to be all things to all people. The end result was that Cetus worked on projects that spanned from health sciences to agriculture to finding better processes for making industrial alcohol. In the early 1980s, Cetus recognized that it needed to tighten its focus. It channeled 70 percent of its R and D spending into the health care field, it brought in a professional manager to run the firm, and it was more forthright with analysts and the media. Unfortunately, by then, Cetus had lost many of its supporters and investors.

Frustrated with Cetus's lack of direction and armed with their experience, several of the board members left Cetus to form their own biotechnology companies. They were convinced—and investors appeared to agree—that a more focused strategy was a better way to go. One individual who was familiar with Cetus's business strategy was Robert Swanson, a young venture capitalist with Silicon Valley's Kleiner Perkins. In 1976, just five years after Cetus was founded, Swanson approached Herbert Boyer, then a professor at University of California at San Francisco (UCSF), about starting a new biotechnology company based on work that Boyer had completed with Stanford University professor Stanley Cohen. What Boyer had scheduled as a 20-minute polite conversation turned into a three-hour meeting as Swanson won Boyer over with his enthusiasm. By the time Swanson and Boyer left the bar, they had agreed to form Genentech (for "genetic engineering technology").

Swanson left Kleiner Perkins and delved right into learning the science and becoming a hands-on, deeply involved CEO. Boyer also became deeply involved and took a leave of absence from UCSF. Swanson and Boyer worked hard to build a creative firm that was in many ways the antithesis of traditional pharmaceutical firms. To lure post-doctoral candidates away from academia, they offered their employees stock options and structured the R and D portion of their firm to resemble an academic lab: scientists worked flexible hours, dressed casually, and were allowed to publish their research.

In 1980, Genentech became the first biotechnology firm to go public, pricing its million shares at $35 each. Within half an hour of trading, the stock hit $89 per share and closed the day at $70. Genentech's public offering broke many previous IPO records. Several months later, Cetus, this time learning from Genentech's extremely successful offering, raised $120 million in its public offering.

Sources: Lax 1985; Swanson 1997; Teitelman 1989; Robbins-Roth 2000.

9.8 EXERCISES

9.1 Examples of learning organizations are Microsoft, Hewlett-Packard, Medtronic, Pfizer, and Starbucks. Choose a company and describe its entrepreneurial learning process.

9.2 Flexcar of Seattle and Zipcar of Boston compete for members in their car-sharing enterprises in Washington, D.C. Flexcar uses Hondas, and Zipcar uses Volkswagens, but both firms are operated similarly. Those who join either firm pay a membership fee and a refundable deposit. Hourly rates range from $4 to $8 (see www.flexcar.com and www.zipcar.com). Describe how Flexcar and Zipcar have learned and adjusted their enterprises in the competitive marketplace in Washington, D.C.

9.3 A new format for movie theaters, easyCinema, was tried in a prototype in England in 2003. Stelios Haji-Ioannou, the founder, also owns easyJet and easyCar. All tickets are purchased online and for low prices. easyCinema customers print out a barcoded entry pass from their computer. They then scan the pass at a computer at the low-frills theater. Visit www.easycinema.com and consider the system. How would you adjust this prototype cinema? [*Economist,* 2003c]

9.4 Capstone Turbine is a developer, assembler, and supplier of microturbine technology. Its primary customers are in the on-site power production and hybrid-electric car markets (see www.capstoneturbine.com). Using the format of Figure 9.3, describe a scenario for the growth of Capstone over the next five years.

9.5 During the Iraq war, U.S. soldiers were equipped with hemorrhage-control bandages made by HemCon, an Oregon-based start-up. Fabricated from thin layers of chitosan (a polymer harvested from shrimp shell that has long been known for its clot-inducing properties), the material is treated with acid to increase its positive charge and tighten its bond with negatively charged blood cells. Visit www.hemcon.com and determine the knowledge assets used to develop the product. Also, review the development of a prototype product.

9.6 A new firm plans to design and sell fuel cells for vehicle use. The firm has received a $1 million grant from the U.S. Department of Energy and is free to exploit the intellectual property developed during the research and development grant. Prepare a knowledge management plan that will enable the firm to file for patents.

9.7 Aircraft designers are working on a prototype electric plane that will use batteries and an electric motor for propulsion [Sweetman, 2003]. Using the process of Figure 9.3, develop the initial requirements and sketch an initial prototype (see www.aviationtomorrow.com/highlights.htm).

CHAPTER 10

Legal Formation
and Intellectual Property

When one door closes, another door opens; but we often look so long and so regretfully upon the closed door that we do not see the ones which open for us.
Alexander Graham Bell

When entrepreneurs establish a new business, they must make some critical decisions about the detailed elements of the firm. The first steps for establishing a new firm are illustrated in Figure 10.1. The choice of a legal form, name, logo, and other formal elements are critical to a successful future. The right name and logo can be the key to the path to building a significant brand. Consider Sony, Intel, and IBM as examples of firms that built a big brand.

The legal form of the venture should match the objectives of the entrepreneurs, customers, and investors of the corporation. Furthermore, there should be a plan to build and protect the intellectual property of the new venture. The proper array of trade secrets, patents, trademarks, and copyrights can add up to a set of very valuable proprietary assets. For many new firms built on innovation and technology, intellectual property can provide a temporary monopoly in the marketplace. ■

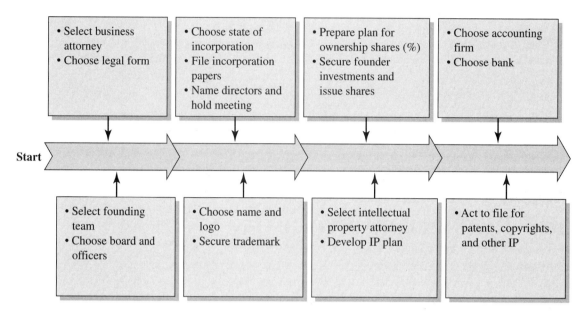

FIGURE 10.1 First steps of establishing a new firm.

10.1 LEGAL FORM OF THE FIRM

When establishing a new technology venture, the entrepreneur needs to choose the legal form of the organization. The entrepreneur should choose a legal form that will facilitate the business, tax, and capital-raising objectives of the new company. The choice of a legal form depends, in part, on how the firm wants to handle federal taxes. For tax purposes, we will address two types of forms: **taxable corporations** and **flow-through entities.** The elements of taxable corporations and flow-through entities are summarized in Table 10.1. A **corporation** is a legal entity separate from its owners. A flow-through entity, sometimes called a pass-

TABLE 10.1 Legal forms of a firm.

Type	Taxation
1. Regular taxable corporation: C-corporation	Taxation of the corporate profits as well as taxation of any corporate distributions to owners
2. Flow-through entities 　■　Partnership 　■　Sole proprietorship 　■　S-corporation 　■　Limited liability company (LLC)	All profits or losses flow through to the owners and are not taxable to the firm

through entity, is one that passes all losses or gains through to the owners of the firm. The profits of a regular corporation are taxable to the corporation, and any distributions are taxable to the owners. This results in double taxation of any distributions such as dividends. Most technology ventures choose the taxable corporation form since they expect to seek venture capital and corporate investors.

There are four main types of flow-through entities to choose from. They are the sole proprietorship with one owner, which is the simplest form; the partnership; the S-corporation, which is taxed much like a partnership and is named for the Internal Revenue Code subchapter that covers it; and the limited liability company (LLC).

Many new flow-through firms are formed as LLCs rather than partnerships or S-corporations. Because they limit liability, it is often wise to use them for sole proprietorships. Thus, the Internal Revenue Service (IRS) treats single-owner LLCs automatically as sole proprietorships and multiple-owner companies as partnerships, unless they elect treatment as corporations.

Most entrepreneurs will, with their attorney, consider the regular corporate form or the LLC. Both of these forms offer limited liability. The personal liability of a regular corporation or an LLC is limited to the amount of capital contributed to the entity by that person.

Many firms that start as a small business should consider a LLC or a subchapter S-corporation form. These forms allow the initial business losses to flow through to the owners, and these losses can be used to offset income from other sources. As the firm grows, it may be wise to consider converting the LLC or S-corporation to a regular corporation since a regular corporation has several potential advantages. It can be sold or merged into another corporation with a tax-free exchange of stock. Other factors that should be considered with an attorney are the number of owners and investors, as well as the need to raise capital and the long-term goals of the business.

A corporation or LLC can be created under state laws and usually require some legal steps, including registering the name of the firm and the owners. Typically, the new firm is registered and incorporated in the state of origin.

A **sole proprietorship** is a business that is owned, and usually operated, by one person. This is a simple form of doing business but exposes the owner to unlimited liability for all debts of the business. A **partnership** is a voluntary association of two or more persons who act as co-owners of a business. Each partner is liable for the acts of the business. Liability for all acts of a business is all-encompassing, and this factor encourages most entrepreneurs to set up a LLC or corporation.

The corporation is seen as an artificial person created by law to act as a business with limited liability. The owners receive shares of the corporation called **stock.** The process of forming a corporation is called incorporation. Usually, a company incorporates in its home state, but some incorporate in another state for reasons of law or ease of doing business. A corporation will issue shares and establish a board of directors and officers as required by law. The limited liability feature of a corporation arises from the fact the corporation is itself a legal

TABLE 10.2 Key elements of the five types of legal form for a new business.

Factors	Sole proprietorship	Partnership	Regular C-corporation	S-corporation	LLC
Owners' personal liability	Unlimited	Unlimited	Limited	Limited	Limited
Taxation	Proprietor's personal tax forms	Partners' personal tax forms	Profits taxed at corporation and owners pay tax on distributions	Profits or losses flow through to owners	Profits or losses flow through to owners
Continuity of business	Terminated by proprietor	Dissolved by partners	Perpetual	Perpetual	Limited to fixed number of years
Cost of formation	Very low	Low	Moderate	Moderate	Moderate
Ability to raise capital	Low	Moderate	High	Moderate	High

"person," separate from its owners. If a corporation fails, creditors have a claim only on the corporation's assets, not on the owners' personal assets.

An **S-corporation** is a corporation that is taxed as a flow-through entity. To qualify, a firm must meet certain requirements regarding its owners and types of stock. The S-status is established by filing with the IRS and may later be converted to a regular C-corporation. Some entrepreneurs prefer this election to the LLC. The key elements of the five types of legal form of a new business are summarized in Table 10.2.

The Hewlett-Packard Company was first established as a partnership in 1937, with William Hewlett and David Packard as equal partners. Their first sale was eight oscillators to Disney Studios in 1939 [Packard, 1995]. In 1947, Hewlett-Packard was incorporated to provide for continuity of life for the firm as well as limited liability for the owners.

In general, most businesses start out as sole proprietorships or partnerships but soon migrate to LLCs or a corporate form. With unlimited liability as a risk of the proprietor or partnership form, it may be unwise to continue in that form beyond the initial period necessary for completing a business plan. Most investors will only be willing to invest in a corporation or LLC since they wish to avoid any liability beyond the amount of their investment.

If the intention of the new business is to raise a significant amount of funds to start the venture and eventually to build it to a significant size, it may be wise to start from the beginning as a regular C-corporation. A **C-corporation** provides limited liability, unlimited life, the ability to accept investments from other corporations, and the ability to merge with other corporations.

The majority of venture-backed companies are incorporated in a handful of states such as California, New York, or Delaware. That is because, relative to other states, California, New York, and Delaware have corporate laws that are well developed, stable, and transparent, all characteristics that ultimately reduce the risk to investors. Moreover, because venture capitalists and their counsel are familiar with conducting financings under such laws, the speed and efficiency with which a transaction moves will likely be improved.

When Howard Schultz started Il Giornale, he immediately received an offer to invest $150,000 from his former business colleague, Jerry Baldwin. Schultz envisioned a chain of stores requiring continuing investment and chose the C-corporation form. Later, in 1987, Il Giornale purchased the name and assets of Starbucks [Schultz, 1997].

The limited liability company offers an ideal form of ownership for small companies. It offers limited liability to the owners along with the tax advantages of a sole proprietorship or partnership. An LLC is particularly attractive to a family business that receives investments of a family's funds since it offers continuity of life, limited liability to participants, and advantages in handling tax issues.

The LLC's articles of organization establish the company's name, its duration, and the names and addresses of organizers. The operating agreement is similar to a set of bylaws in that it outlines the way the LLC will operate. Typically, the LLC will elect limited liability and centralized management. The owners of an LLC are called *members,* and their ownership interests are known as *interests.* These terms are equivalent to *stockholders* and *stock.* Unlike the S-corporation, there is no limitation on the number of members on their status, and the LLC may have foreign investors. Like the S-corporation, the LLC will need to convert to a general corporation to accept any venture capital or issue stock in a public market.

The regular C-corporation form will be often chosen by firms that intend to seek investment from numerous professional investors and other corporations. The C-corporation offers limited liability, a centralized management arrangement, and perpetuity of life. Firms that require large investments from venture capitalists and intend to do business internationally should seriously consider the regular C-corporation form. The C-corporation allows for various classes of stock, such as common stock and preferred stock. In addition, investors in a C-corporation can purchase *convertible preferred stock,* which is the most common form of venture capital investment.

10.2 COMPANY NAME

The name of the new company is important. It should be memorable, related to the product or service, and attractive. It is also helpful if it can be used as the company's website domain address. The right name can evoke a sense of the company's character, bestow distinction, and make a powerful impression. Ideally, the name tells the prospective customer what the product's major benefit is. Many firms are named after their founders; for example, Dell and Wrigley. Others use creative names, such as Kodak or Exxon. Some firms use a locational name, such as Silicon Valley Bank or Allegheny Technology.

The right name can deliver a subtle message about the firm's unique features. Tinker Toys evokes a spirit of play. If possible, a name will serve as a marketing tool and will be easy to remember, spell, and say.

Jeff Bezos liked Cadabra as a name for his online Internet business. When his attorney wisely asked, "Cadaver?", Bezos dropped that name and chose Amazon.com. it conveyed the idea of a huge entity and did not limit him to one product [Leibovich, 2002].

A good example of a name of a firm that immediately conveys its purpose is FreshDirect, which offers online food purchase and delivery (www.freshdirect.com). Kinko's created a memorable name that makes customers assume it is unique. Think of Kinko's versus Speedyprint or Quickprint.

It is best to test the proposed name of your firm on others since it is important to avoid negative connotations. Once a suitable name is chosen, a name search will be required to ensure that no one else has already claimed the name. Then, the name is registered with the appropriate state office. If national and international operation is envisioned, it may be wise to register the name as a trademark with the U.S. Trademark Registry Office.

> Founder Scott Cook of Intuit, the makers of Quicken, tried and rejected the name Instinct for his new company because it sounded like "it stinks" [Taylor and Schroeder, 2003].

In late 1984, Leonard Bozack and Sandra Lerner initially financed their own new firm with funds from their credit cards. They named the company cisco, as in the end of the name of the city of San Francisco. Lerner designed the logo in the form of the Golden Gate Bridge. Eventually, they capitalized the name to become Cisco [Bunnell, 2000].

A new venture should make sure it has the legal rights to use the name it chooses, that the name does not translate into something embarrassing or negative in a foreign language, and that the name carries no other undesirable connotations. Another factor is its pronounceability. Recently, Tricon Global Restaurants changed its name to Yum. Yum is easy to say and remember, and you know what business it is in. A software company supporting Linux chose Red Hat as a memorable name and created a red hat logo. Other examples of good, memorable, robust names include Wal-Mart, Intel, General Electric, and Microsoft.

Once the new venture chooses its name, it should reserve a domain name for its website and e-mail address. New ventures should check into the availability of a domain name early on, because doing so often eliminates certain choices for company names and trademarks. The best situation is when you can use the same legal name and domain name. Good examples of memorable corporate names that are also used as domain names are Google and Yahoo.

10.3 INTELLECTUAL PROPERTY

Within intellectual assets is a subset of ideas, called intellectual property, that can be legally protected [Davis and Harrison, 2001]. Property is defined as something valuable that is owned, such as land or jewelry. Furthermore, we can

TABLE 10.3 Comparison of physical and intellectual property

Factor	Physical property	Intellectual property
Multi-use	Use by one firm precludes simultaneous use by another	Use by one firm does not prevent use by another
Physical depreciation	Depreciates, wears out	Does not wear out
Protection and enforcement from encroachment	Generally can enforce and protect ownership	May be difficult or expensive to enforce and protect ownership

distinguish real property (or physical property) from intellectual property. **Intellectual property (IP)** is valuable intangible property owned by persons or companies. As discussed in Chapter 9, knowledge is the awareness and possession of information, facts, ideas, truths, methods, and principles in an area of expertise. This knowledge is a valuable asset of the firm and is called intellectual property. Intellectual property includes trade secrets, trademarks, copyrights, patents, and other forms. A comparison of the qualities of physical and intellectual property is provided in Table 10.3.

Since knowledge and innovation are keys to competitive success, the management of intellectual property is important to most firms. For many firms, intellectual assets are the wellsprings of wealth and competitive advantage. The protection of intellectual property can lead to the possession of valuable assets; for example, the secret formula for Coca-Cola.

A **trade secret** is a confidential intellectual asset that is maintained as a secret by the owner. A trade secret is limited to knowledge or methods that are not publicly known, derived, or reverse-engineered. The period of life for a trade secret is potentially indefinite. The formula for Coca-Cola has been a trade secret for over a century. Trade secret protection may be lost, however, upon any unauthorized disclosure, such as theft, violation of confidentiality, independent recreation, or reverse engineering.

The protection and enforcement of legal ownership of intellectual property is more difficult than for physical property. How can a firm tell that another firm has used or taken its intellectual property? Copying or illegal use of intellectual property is difficult to discern and prove. The owner of a textbook or a CD purchased at a store has the right to share it with another person but is precluded by law from copying it for sale to another person.

The purpose of intellectual property law is to balance two competing interests: the public and the private. The public interest is served by the creation and distribution of inventions, music, literature, and other forms of intellectual expression. The private interest is served by rewarding people for creating these works through the establishment of a time-limited monopoly granting exclusive use to the creator.

During the course of working for a firm, an employee has an idea for a new product that is outside of the scope of the business of the firm. Who owns this intellectual property: the employer or the employee? Does it matter whether the new idea was conceived on the weekend? Entrepreneurs should avoid any potential complication or dispute of ownership of intellectual property by assiduously following the legal and moral laws of property. They should also reread all signed employment agreements that may state restrictions on ownership of intellectual property.

Clearly, if a firm plans to apply for a patent or use some technical advance of a proprietary nature, the firm and its employees should not reveal the details to prospective investors. If these investors become serious about investing in the new business, then the firm will need to reveal more information about the proprietary asset.

Many technology start-ups take four to seven years to reach the market and become profitable. Typically, these new ventures are founded on a significant array of intellectual property such as patents. When the entrepreneur's personal knowledge is perceived to be a critical portion of the intellectual property, then this person will be expected to remain with the firm for several years. Arrangements such as employment agreements will be required to guarantee this active involvement [Lowe, 2001].

Protecting and managing intellectual property is important. Analysts estimate the intellectual property market to be $100 billion annually. At IBM, patents and licenses represent 15 percent of revenues. A useful reference for an entrepreneur is *Patent, Copyright, and Trademark,* by Stephen Elias and R. Stim (Berkeley: Nolo Press, 2003). Of course, professional legal assistance is also advisable.

10.4 TRADE SECRETS

Secrecy is the state of concealment or being concealed or maintained as a secret. A secret is known only to a few people and intentionally withheld from general knowledge. Only two people are said to have access to the secret formula for Coca-Cola syrup. In any firm, the role of corporate secrets is important. The potential protection offered by secrecy depends on the attributes of the intellectual property and the circumstances of its use. Secrecy is valuable for formulas, algorithms, and know-how that can be implemented by a firm without it being known by other than a few people. If the knowledge must be widely shared throughout the firm, it will be difficult to protect it from those who would copy or imitate it.

Many production processes can be protected behind the walls of the firm. For example, methods for manufacturing integrated circuits are widely available, but the best production process for making them is quite complicated. Several semiconductor firms maintain their competitive advantage by maintaining the secrecy of their methods as well as their processes. Of course, the risk always exists that an employee will learn the secrets of the methods and the process and decide to start a competitor firm.

A firm's efforts to protect its secrets are enabled by the law of trade secrets. The two conditions for protection as a trade secret are that it provides a competitive advantage and that it is managed as a secret by the firm. The firm uses confidentially, nondisclosure, and assignment of inventions agreements as well as physical barriers such as safes and limited access to protect its intellectual assets. The general rule is that a trade secret is lost if it is disclosed to the general public or competitors, or if the person seeking to protect a trade secret does not take reasonable steps under the circumstances to ensure its confidentiality.

A firm will have to balance the need to protect secrets with the necessity to widely share information among employees. Employees must be informed that they are dealing with secrets that are the property of the firm and they are expected to protect these secrets.

For many firms, common knowledge among employees of the methods and procedures is necessary for success. For them, it is the execution of the total business process that provides the competitive advantage. Howard Schultz of Starbucks writes [Schultz, 1997]:

> We had no lock on the world's supply of fine coffee, no patent on the dark roast, no claim to the words *café latté* apart from the fact that we popularized the drink in America. You could start up a neighborhood expresso bar and compete against us tomorrow, if you haven't done so already.

10.5 PATENTS

Abraham Lincoln called the introduction of the U.S. system of patent laws "one of the three most important events in the world's history along with the advent of the printing press and the discovery of America" [Schwartz, 2002]. A **patent** grants inventors the right to exclude others from making, using, or selling their invention for a limited period of time. Patents are granted to new and useful machines, manufactured products, and industrial processes, and to significant improvements of existing ones. Patents are also granted to new chemical compounds, foods, and medicinal products, as well as to the processes for producing them. Patents can also be granted to new plant or animal forms developed through genetic engineering. The U.S. Patent and Trademark Office requires that a patentable invention must be shown to have a specific, substantial, and credible use. See Appendix C for patent web sites.

Utility patents are issued for the protection of new, useful, nonobvious, and adequately specified processes, machines, and manufacturing processes for a period of 17 years. Examples include the patent for the safety razor and the rolling bag that is widely used by air travelers.

Design patents are issued for new original, ornamental, and nonobvious designs for articles of manufacture for a period of 14 years. For example, the new design of a computer case could be submitted for a patent.

Plant patents are issued for a term of 17 years for certain new varieties of plants that have been asexually reproduced.

A **business method patent** is actually a type of a utility patent and involves the creation and ownership of a process or method. Amazon.com has a patent on its one-click method for a purchase. There is a four-part test for patentability of a business method, as there is for any invention. The new way of doing business has to be useful. It has to be new. It cannot be so incremental that it would be obvious to a skilled practitioner. And in the application process, disclosure of the innovation has to be so complete that fellow practitioners can understand it.

In most cases, an invention must be considered novel and useful to receive a patent. It must also represent a relatively significant advance in the state of the art and cannot merely be an obvious change from what is already known. Such requirements are meant to reduce the number of inventions that modify existing products in minimal ways. Patents are often granted for improvements of previously patented articles or processes if the requirements of patentability are otherwise met.

A patent is recognized as a type of property with the attributes of personal property. It may be sold or assigned to others or mortgaged, or it may pass to the heirs of a deceased inventor. Because the patent gives the owner the right to exclude others from making, using, or selling the invention, the owner may authorize others to do any of these things by a license and receive royalties or other compensation for the privilege. If anyone makes use of a patented invention without authorization, their infringement can be brought to court in a suit filed by the patent holder, who may request monetary damages as well as a court injunction to prevent further infringement [Elias and Stim, 2003].

King C. Gillette desired to invent a product that would be used and then discarded so that the customer would keep returning for more. While sharpening a permanent, straight-edge razor, Gillette had the idea of substituting a thin, double-edged steel blade placed between two plates and held in place by a handle. Though the invention was received with skepticism because the blades could not be sharpened, the manufactured product was a success from the beginning. Gillette filed for a patent in 1902 and started sales in 1903. By the end of 1904, Gillette's company had produced 90,000 razors and 12.4 million blades.

The patent registration process requires an application that includes a clear, concise description of the invention and a statement of ownership. It also defines the boundaries of the exclusive rights that the inventor claims. Furthermore, inventors must apply for patents in each country where they wish to manufacture, use, or sell their inventions.

Once a patent has been issued, the owner must start a patent protection program that includes issuing notices and labeling the patent, monitoring uses of the patent, and pursuing known or suspected infringers of the patent. The patent provides the owner the right to exclude others from using the patent without compensation. However, the owner is responsible for enforcing that right by sending official notices to infringers and resorting to litigation, if necessary. Thus, patents grant rights only to sue imitators. Sometimes imitators are able to design new products or methods that circumvent the existing patent.

The patent awarded to Stanley Cohen and Herbert Boyer in 1974 covered gene-splicing techniques, a basic part of biotechnology, which is estimated to have earned over $250 million for its owners, Stanford University and the University of California. Patents have proved to be very effective for inventions in

TABLE 10.4 U.S. patents issued.

Year	1980	1990	2001
Number issued (thousands)	66.2	99.2	166.0

Source: *Statistical Abstract of the United States,* 2001.

the pharmaceutical and medical instruments industries. Revenues in the United States from the licensing of patent rights have grown from $15 billion in 1990 to more than $110 billion in 2000.

In 2001, the U.S. Patent and Trademark Office issued a record 166,000 utility patents. IBM, for example, was granted more than 3,000 patents in 2001. Favorable court decisions have made biotechnology, genes, software, and business methods patentable. For example, Dell Computer now has 77 patents protecting its build-to-order business method. The one-click buying method patent was issued to Amazon.com in 1998. Having a set of strong patents helps to discourage competitors and attract investors.

A laptop computer may include up to 500 patented inventions held by many firms. On the other hand, a pharmaceutical drug will normally be covered by a single patent. In many industries, firms eager to capture gains from their innovations are filing patent applications at an unprecedented rate. The number of patents issued in the United States in recent years is summarized in Table 10.4. Companies are increasingly building their innovation strategy around patents and intellectual property. In the biotechnology field, firms widely use patents to protect their intellectual property as they develop the products for market.

Several legal changes and court decisions in the 1980s provided more protection for patents. The 1985 case in which Polaroid won more than $900 million in damages from Kodak for instant-camera patent infringement provided strengthened precedence for patent infringement litigation.

Patenting an invention can be expensive since a single patent application can cost $20,000. In addition, the cost of infringement litigation can be very high. Proving patent infringement requires documentation and analysis of the infringing product or process.

According to the 2003 Arab Human Development Report, between 1980 and 1999 the nine leading Arab economies registered 370 patents (in the U.S.) for new inventions. Patents are a good measure of a society's education quality, entrepreneurship, rule of law and innovation. During that same 20-year period, South Korea alone registered 16,328 patents for inventions [Friedman 2004].

Thomas A. Edison (1847–1931) was America's quintessential inventor-entrepreneur. His inventions included the stock ticker, phonograph, motion pictures, incandescent lamp, and electric power system. While he spent most of his time inventing, he dedicated years to the struggle for patent ownership. Since inventors are responsible for protecting their patents, the legal struggles can go on for years. Fledgling entrepreneurs cannot expect to have the funds required to litigate all possible infringements of their inventions and need to use a judicious

TABLE 10.5 Thirteen important U.S. patents.

1. Cotton gin—Eli Whitney, 1794
2. Sewing machine—Elias Howe, 1846
3. Barbed wire—Joseph Glidden, 1874
4. Telephone—Alexander Graham Bell, 1876
5. Lightbulb—Thomas Edison, 1880
6. Airplane—Orville and Wilbur Wright, 1906
7. Gyroscope—Elmer A. Sperry, 1916
8. Television—Philo T. Farnsworth, 1927
9. Xerography—Chester Carlson, 1942
10. Transistor—J. Bardeen, W. Brattan, and W. Shockley, 1950
11. Integrated circuit—Jack Kilby and Robert Noyce, 1958
12. Microprocessor—Ted Hoff, S. Mazor, and F. Faggin, 1971
13. Gene-splicing—Stanley Cohen and Herbert Boyer, 1974
14. One-click order process—Amazon.com, 1998

TABLE 10.6 Developing a patent strategy.

1. Identify the goals of a patent portfolio.
2. Identify the intellectual assets and gather supporting documents.
3. Identify those assets most suitable for patent applications.
4. Draft invention disclosures and patent applications.
5. Develop a plan for licensing, enforcing, and enhancing patents.

Source: Fenwick & West, LLP, R. P. Patel.

balance of secrecy and patents to protect their intellectual property. If possible, they need to select the most powerful inventions for patent protection and then protect them aggressively. Table 10.5 lists 14 of the most powerful U.S. patents. Any of these patents were worth every effort required to protect them. A new technology venture should prepare a strategy for developing and maintaining its patent portfolio, as outlined in Table 10.6.

10.6 TRADEMARKS

A **trademark** is any distinctive word, name, symbol, slogan, shape, sound, or logo that identifies the source of a product or service. A registered trademark is awarded for 20 years, renewable indefinitely as long as commercial use is proven. A new venture should consider trademarking its company name, sym-

FIGURE 10.2 Apple Computer logo.

bol, or logo. Commonly known trademarks include Kodak, Apple, Kleenex, the NBC logo, and Yahoo.

Once a trademark has been registered with the U.S. Patent and Trademark Office, the owner has the right to bring legal action against infringement for damages and recovery of profits. Trademark rights are often the most valuable assets of an emerging new venture in today's competitive marketplace. The goodwill and consumer recognition that trademarks represent have great economic value and are therefore usually worth the effort and expense to properly register and protect them.

A good trademark is an integral part of a firm's brand. To possess good value, a trademark should readily be associated with and exclusive to the firm. Excellent examples of a powerful trademark are the Apple logo and the Intel Inside logo. The logo of Apple computer is an apple with a bite removed. The logo (see Figure 10.2) matches the company name.

There were 145,000 trademarks issued in the United States in 2001, and there is an estimated total of 1.6 million registered trademarks in the United States.

A firm may lose the exclusive right to a trademark if it loses its unique character and becomes a generic name. Aspirin, thermos, and cellophane are examples of names that have become generic. Coca-Cola and Xerox have successfully protected their trademarks.

10.7 COPYRIGHTS

A **copyright** is a right of an author to prevent others from printing, copying, or publishing any of his or her original works. The life of the copyright is for the life of the author plus 50 years after the author's death. Because copyright protection automatically attaches upon creation of a work and the process of registering a work with the U.S. Copyright Office requires only the completion of a simple form, the process of obtaining copyright protection demands very few resources.

A copyright extends protection to authors, composers, and artists, and it relates to the form of expression rather than the subject matter. This is important, because a copyright only prevents duplicating or using the original material. This does not prevent use of the subject matter. Therefore, software programs, books, and music are protected from copying, but the ideas in these forms may be used by others.

The protection provided by copyright, however, is limited. In the software field, courts have applied copyright protection narrowly to prevent only the exact copying of code. Copyright is most effective against wholesale copying of software. It has limited protection for functional aspects of software products, such as the underlying algorithms, data structures, and protocols of multimedia technology. Under the 1998 Digital Millennium Copyright Act, no one may create or distribute hardware or software that circumvents a copy-protection scheme.

10.8 LICENSING

Licensing is a contractual method of exploiting intellectual property by transferring rights to other firms without a transfer of ownership. A **license** is a grant to another firm to make use of the rights of the intellectual property. This license is defined in a contract and usually requires the licensee to pay a royalty or fee to the licensor.

Licensing is used to provide cash flows to the owners of intellectual property by enabling others to exploit their patents, copyrights, and trade secrets. A new venture can derive valuable income streams through licensing. The benefits to the licensor include spreading the risk, achieving expanded market penetration, earning license income, and testing new products and markets. Disadvantages of licensing may include risk of infringement and nonperformance of the licensee.

Many firms have a large number of unexploited or underexploited patents that a licensee may be able to exploit. IBM, for example, widely grants licenses, and its royalty income amounted to $2 billion in 2002. IBM holds more patents than any other U.S. company and licenses its software patents widely.

Licensing is widely used to provide software to users. For example, Microsoft derives most of its revenues from license fees for its Office suite and Windows operating system.

Most new firms realize that intellectual property can be among their most valuable and flexible assets. And the licensing market is still in its infancy. Yet most companies remain unaware of the earnings potential of their patent holdings. Those that do begin licensing efforts tend to do so only when pressed.

Instead of developing a technology within a new venture, a firm can save time and resources by licensing another firm's technology to use in its products. The terms of the license grant contained in the contract with the third-party technology owner establish what rights a start-up has to use, distribute, modify, and sublicense the licensed technology. Dolby Laboratories Inc. gets much of its revenue from licensing its products to electronics makers. It succeeds this way partly because of its well-known technology and the reasonable price it charges other firms for its use.

Albany Molecular Research (AMR) was founded in 1991 by chemist Thomas D'Ambra as a pharmaceutical research firm. The firm develops and patents new methods and drug compounds (see www.albmolecular. com). AMR receives royalties on patented technologies that it has licensed. It had revenues of about $150 million in 2004.

New entrepreneurial ventures should consider licensing intellectual property rather than developing it. If the new venture develops new intellectual property, it should consider licensing it to other firms for noncompetitive uses. Licensing is a powerful business tool. License opportunities may be a way to reduce risk, expand a business overseas, or complement an existing product line. The rapid public effects of widely licensing innovations is the positive impact on the growth of GDP [Baumol, 2002].

10.9 AGRAQUEST

Upon its formation, AgraQuest's entrepreneurial team looked for a name that conveyed that they were searching for natural solutions in the agricultural industry. They first proposed the name Agrisearch, but it was already taken. Their next choice was AgraQuest, and it was available. They registered the name as a trademark internationally. Then the AgraQuest team selected a logo depicting a hummingbird searching in a flower for nectar, as shown in Figure 10.3.

FIGURE 10.3 AgraQuest logo of a hummingbird searching for nectar in a flower.

AgraQuest incorporated in Delaware, since it anticipated venture capital investments and an initial public offering. The company possesses many trade secrets, primarily in the manufacturing of its products. It also has a proprietary method for identifying, screening, and developing the microbes for its products.

AgraQuest holds 20 U.S. and nine foreign patents and has filed an additional three U.S. and 95 foreign patent applications. Each patent covers a microbe and its use. In addition, AgraQuest has licensed *M. albus* and a collection of microbes for testing.

The intellectual property management of AgraQuest is managed by Pamela Marrone and the director of research. The firm engages a large law firm for business and intellectual property legal matters.

10.10 SUMMARY

The legal form of the venture should match the needs of the entrepreneurs, customers, and investors. For most high-growth ventures, the regular C-corporation will be most appropriate. However, for many new organizations, other legal forms may be suitable. Ten mistakes with legal matters are provided in Table 10.7.

The plan to acquire, build, and protect the intellectual property of the new venture should be clear to all the participants. The proper array of trade secrets, patents, trademarks, and copyrights can come together as a strong set of very valuable proprietary assets. For high-growth, technology-based companies, intellectual property can be used to build a temporary monopoly in the marketplace.

An important step after deciding to launch a business is to choose a name and logo for the business that conveys the right image and represents the business well. One of the best names and logos is Target, while one less impressive is Cingular. Target is memorable and the logo is a target, in contrast to Cingular, which conveys little meaning.

TABLE 10.7 Ten mistakes with legal matters.

■ Failing to secure legal assistance.	■ Starting a business while employed by a potential competitor.
■ Delaying the handling of legal issues.	
■ Delaying the intellectual property management process.	■ Overpromising and exaggerating claims in the business plan.
■ Issuing founder shares without vesting provisions.	■ Failing to register the name and logo of the firm early in the start-up process.
■ Failing to incorporate early.	■ Failing to develop confidentiality, nondisclosure, and noncompete agreements.
■ Disclosing intellectual property without a nondisclosure agreement.	

Source: Bagley, 2002.

> **Principle 10**
> The name, logo, and intellectual property of a new venture can provide a proprietary advantage leading to a temporary monopoly in the marketplace.

10.11 EXERCISES

10.1   Mayo Clinic has filed an application for a broad method patent that gives it control over a new generation of treatments for chronic sinus inflammation, or sinusitis. The patent, in effect, blocks others from selling an antifungal agent to treat the condition without Mayo's approval. Mayo will soon try to license this patent to a pharmaceutical company. Are patents helpful in the process of developing a cure for diseases?

10.2   Three friends have decided to form a firm to design and manufacture nanotechnology devices for medical technologies. Michael Rogers has worked for Hewlett-Packard for 12 years and on his own has designed and submitted a patent claim for a nanotechnology manufacturing technology. Steve Allegro, a graduate student, has a software design program he has developed for the design of nanotechnology medical devices. Alicia Simmons, CFO of Alletech Software Inc., is a skilled and experienced manager. Shall they incorporate immediately? What is the problem, if any, of using Rogers's patent ideas? Should they incorporate in their home state of Alabama? Simmons has knowledge of several manufacturing trade secrets of Alletech. Can she use these secret methods at her new firm?

10.3   The three founders of the new firm described in Exercise 10.2 are looking for a name for their firm. One idea is Advanced Nanoscience & Technology. Another is Nanoscience Applications. What do you think of these names? Can you suggest a better name?

10.4   Suggest a logo for the firm described in Exercises 10.2 and 10.3.

10.5   Headwaters Inc. develops and licenses technologies for turning coal and other fossil fuels into a higher-value product. Is this an appropriate name for the firm? Revenue is generated through the licensing of the firm's patented chemical processes (see www.hdwtrs.com). Examine the patents of Headwaters and its intellectual property protection. How would you describe the growth potential of this firm?

10.6   Google and Overture (now part of Yahoo) are two powerful search engine firms. Compare the names of the two firms. Do they confirm the offering and convey the value of each firm? Determine if either firm holds any patents on its technological innovations.

10.7 Review the logo of Cisco Systems and describe what it conveys. Is this a suitable logo for a computer network equipment company? Propose a new logo for Cisco Systems.

10.8 The Ananos family spotted an opportunity to offer a local cola drink in Peru in the late 1980s. Establishing Industrias Ananos, they now offer Kola Real throughout Latin America and compete with Coca-Cola and Pepsi. Examine the threat to the intellectual property of Coca-Cola and Pepsi [Luhnow and Terhurie, 2003].

10.9 Vending machines are like unattended stores. New machines that reliably dispense the desired item are needed. Studies show that people want glass-front machines that show the products and work on debit or credit cards [Berk, 2003]. Prepare a design for a reliable vending machine (see Exercise 6.10) and develop a patent development plan for this design (see Table 10.6).

10.10 Patents must be protected to avoid misuse by others. Gillette Company and Schick-Wilkinson compete in the razor market. Gillette sued Schick in August 2003, claiming that Schick's new Quattro four-bladed razor violated a 2001 Gillette patent on a particular arrangement of three blades in a razor cartridge [Forelle, 2003]. The patented technology concerns the blade alignment and exposure. Visit the websites of both companies and examine Schick's potential infringement on Gillette's patent. Describe the possible infringement and the strength of the legal case.

CHAPTER 11

The Marketing and Sales Plan

Successful salesmanship is 90 percent preparation and 10 percent presentation.
Bertrand R. Canfield

A fundamental activity of any firm is to attract, serve, and retain customers for its product offerings. This activity, called marketing, is critical to the success of a new firm since the firm normally starts without any customers. A new business must create a marketing and sales plan, which describes its target customers for its product offering. A sound marketing plan is built on solid information obtained through market research. The new firm creates a product position and a mix of price, product, promotion, and distribution channels that will attract and satisfy the customer. Gaining recognition and acceptance in a target market requires the following steps in sequence:

■ Describe the product offering

■ Describe the target customer

■ State the marketing objectives

■ Gather information through market research

■ Create a marketing plan

■ Create a sales plan

■ Build a marketing and sales staff

In this chapter, we describe each of these activities. ■

11.1 MARKETING

Marketing is a set of activities with the objective of securing, serving, and retaining customers for the firm's product offerings. Marketing is getting the right message to the right customer segment via the appropriate media and methods. It is the task of the marketing function to help create the product and the terms of its offering as well as communicate its value to the customers. The purpose of the **marketing plan** is to describe the steps required to achieve the marketing objective. The marketing plan is a written document serving as a section of a new venture's business plan and contains action steps for the marketing program for the products. Peter Drucker [2002] has said: "Because its purpose is to create a customer, the business has two basic functions: marketing and innovation. Marketing and innovation produce results; all the rest are costs."

In Chapter 3, we have described the creation of a value proposition and business model for the identified customer. In Chapters 4 and 5, we have described the elements of an overall business strategy. Given these business elements, we need to develop a marketing strategy and build a marketing plan. The six elements of a marketing plan are shown in Table 11.1.

The first element of the marketing plan is a clear statement of objectives. The second element is the identification of one or more customer target segments. The goal of target segments is to carefully select the appropriate customers and to focus the marketing activities on those segments. The third element is the description of the product and the terms and conditions of its formal offering. Given our knowledge of the product and its offering, we need to determine what the response of the customer might be and how we can develop a strategy to attract and retain the customer. Next, we describe the marketing mix consisting of price, product, promotion, and place (channels). Finally, we describe plans for relating to our customer in the sales and service activities.

The marketing plan will be implemented through a marketing program. The plan will describe how we will take the product to market and attract, serve, and retain satisfied customers. The marketing and sales activity is portrayed in Figure 11.1. The new venture communicates information about its product and how it sells and services the product for the customer. When the customers purchase the product, they provide some useful information about the purchase and the use of the product for the seller.

TABLE 11.1 Six elements of the marketing plan.

1. Marketing objectives
2. Target customer segments
3. Product offering description
4. Market research and strategy
5. Marketing mix
6. Customer relationship management

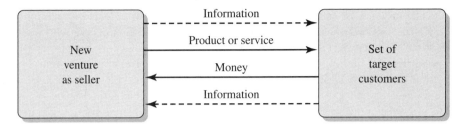

FIGURE 11.1 Marketing and sales activity of the new venture.

The marketing and sales plan will flow from the opportunity and the business model, as shown in Figure 11.2.

11.2 MARKETING OBJECTIVES AND CUSTOMER TARGET SEGMENTS

The **marketing objectives statement** is a clear description of the key objectives of the marketing program. Objectives may include sales goals, market share, profitability, regional plans, and customer acquisition goals. Objectives should be quantified and given for a time period, such as "the firm will sell 1,000 units in the initial sales phase in Texas and Oklahoma in the first year of activity."

Market penetration for selected markets is often one of the objectives of the marketing plan. Selected markets are often called customer target segments. A **market segment** consists of a group with similar needs or wants and may include geographical location, purchasing power, and buying attitudes. **Market segmentation** divides markets into segments that have different buying needs, wants, and habits. Different segments will require different marketing strategies. Often a new venture identifies one target segment for its initial marketing effort, carefully describing the customer in that segment in terms of geographic, demographic, psychographic, and other variables [Winer, 2000]. Geographic variables include city, region, and type, such as "urban." Demographic variables include age, gender, income, education, religion, and social class. Psychographic variables include lifestyle and personality variables that influence a customer's wants and needs.

InVision Technologies designs and manufactures electronic baggage screening systems (see www.invision-tech.com). InVision Technologies was founded in 1990 to provide airport security devices by using computed tomography to detect explosives. Its target segment, U.S. airport baggage screening, was slow growing before the terrorist attacks on September 11, 2001. It sold 250 machines in the preceding decade but 750 machines in the two years following the attacks. InVision also plans to enter the international market.

FIGURE 11.2 Building a marketing plan and a sales plan.

The target segment for Starbucks can be described as customers seeking a very attractive coffee café in a suburban or city setting serving people who want affordable luxury and casual social interaction in a clean, comfortable place with a good range of high-quality coffee and bakery products. The target customer for Starbucks can be described as an upper-middle-income working person between the ages of 25 and 55 who enjoys coffee and a comfortable gathering place for a break. With a clear description of the target customers, a new venture can devise a plan to attract and retain them. Marketing to a segment enables the new firm to narrow the marketing strategy and put all its effort into acquiring new customers in the target market. Often, new firms try to reach too many market segments in their early efforts, thus dissipating their resources before they can build up a customer base.

Car sharing is aimed at people who want to use a car for a short time but do not need to own it. This scheme is particularly attractive in urban centers. In general, people are moving toward access rather than ownership of autos in dense urban areas with high auto costs. Neil Peterson started Flexcar in 1999 in Seattle and has expanded to Los Angeles, San Francisco, Washington, D.C., and San Diego (see www.flexcar.com).

The average cost of owning or leasing a new car totals $625 a month. The average member in a car-sharing program spends less than $100 a month on car expenses. Flexcar identified its target segment as individuals. It soon discovered, however, that the biggest growth came not from individuals but from small and midsize companies that did not want to maintain their own fleets of vehicles [Stringer, 2003]. Just when you think you have identified your target segment, the market shows you a better one.

An interesting target segment is adult women. Women are the majority of purchase decision-makers today, not only in the traditional areas of fashion, food, and cosmetics, but also in automobiles, financial services, computer electronics, and travel. Women have different buying characteristics than men and are a good target segment for many products [Quinlan, 2003].

11.3 PRODUCT AND OFFERING DESCRIPTION

The product is described in an early section of the business plan. In this section, the features of the product and the product offering are described. The product should be described in terms of its primary attributes. If possible, a product positioning map should be developed. All products can be differentiated to some extent by communicating the most highly valuable benefit to the buyer. **Positioning** is the act of designing the product offering and image to occupy a distinctive place in the target customer's mind [Ries and Trout, 2001].

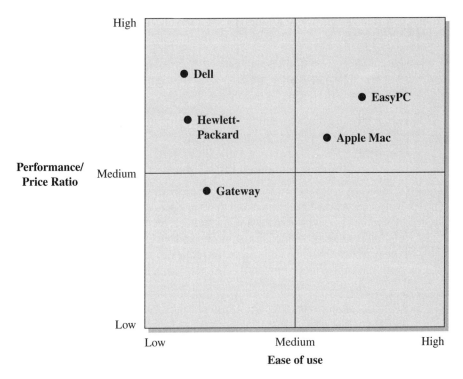

FIGURE 11.3 Positioning map for personal computers showing the position of a new product, EasyPC.

Positioning of a product enables the firm to differentiate it in the mind of the prospect. Volvo connotes safety, and FedEx owns "overnight." A product positioning map shows the product characteristics in relation to its competitors. Figure 11.3 shows a new personal computer called EasyPC that is positioned as having high ease of use and a high performance-to-price ratio. It would then be the task of the marketing effort to clearly communicate that position to the target customer.

Positioning a product focuses on a few key attributes of the value proposition. A positioning statement, as shown in Figure 11.4, helps to define the positioning of the product. Once we have a product position, we seek to build a powerful product offering [Moore, 2002]. A **product offering** communicates the key values of the product and describes the benefits to the customer. The **unique selling proposition** is a statement of the key customer benefit of a product that differentiates it from its competition. The unique selling proposition for the EasyPC is:

EasyPC delivers high performance at a reasonable price and is easy for anyone to use.

- **Positioning statement**
  - **For** (target customer)
  - **Who** (statement of need or opportunity)
  - **Product name is a** (product category)
  - **That** (statement of benefit)   **(a)**

- **Differentiation**
  - **Unlike** (primary competitive alternative)
  - **Our product** (statement of primary differentiation)

- **Positioning statement for Microsoft .NET**
  - For companies
  - Whose employees and partners need up-to-date information when they need it,
  - Microsoft .NET is a new generation of software
  - That enables an unprecedented level of software integration through the use of XML Web services.   **(b)**

- **Differentiation**
  - Unlike Java,
  - Our product is infused into the Microsoft platform, providing the ability to quickly and reliably build, host, deploy, and utilize connected solutions with the protection of industry-standard security technologies.

FIGURE 11.4 (a) Positioning statement format, and **(b)** for Microsoft .NET.

The unique selling proposition of FedEx is:

We deliver your packages overnight—guaranteed.

It is wise to figure out who will be your best customer and then pursue that segment. A **best customer** is one who values your brand, buys it regularly whether your product is on sale or not, tells his or her friends about your product, and will not readily switch to a competitor. Entrepreneurs identify their customer segment and position their product to serve them very well [Ettenberg, 2002].

> The tools and utensils industry has two obvious segments: home and workplace. People thought ergonomic tools were only for the workplace. Oxo showed that people at home wanted easy-to-use tools for the kitchen by focusing on utensils that felt comfortable to handle (see www.oxo.com).

11.4 MARKET RESEARCH

Market research is the process of gathering the information that serves as the basis for a sound marketing plan. Once a target market is selected, the entrepre-

TABLE 11.2 Market research process.

1. Define the product and its unique selling proposition. Identify the customer segment. Develop a set of questions that will provide the necessary data on customer preferences and behavior.

2. Collect the data using surveys, published sources, focus groups, interviews, and other means.

3. Analyze and interpret the data to determine if the product meets the needs or wants of the customers and determine whether they will pay the selected price.

4. Draw conclusions about the customers and their needs, preferences, and behavior.

neur needs some information about customers' preferences and behavior as well as competitors' products. The objective of market research is to learn how to attract and retain customers for a product. Market research can provide critical information to the new venture team. Without complete information, a new venture may launch a product and ultimately determine that the customer does not value the product. A key question is: Does the target segment want the perceived value that our positioning is trying to deliver more than other segments? If so, how can we reach this segment efficiently?

The market research effort can consist of four steps, as shown in Table 11.2. Using these steps, entrepreneurs can develop an understanding of their customers, including their preferences and behavior. The first step is to determine the needed information and research objectives. Then, the new venture team develops a plan for gathering information from the targeted customer segment. It is helpful to use printed sources such as trade data, magazines, and trade journals. Corporate reports and news about the industry will be available. A search on the Internet will lead to many valuable sources. See Appendix D for marketing research sources.

Primary data are collected for your specific research objective. Secondary data sources that were collected for another research purpose are already available. Primary data sources are very valuable, and entrepreneurs should avoid relying solely on secondary sources. Talking to the actual customer and other channel participants is very important. A popular form of research uses the **focus group,** which is a small group of people from the target market. These people are brought together in a room to have a discussion about the product. This discussion can be led by someone from the venture team or a professional moderator. Other methods of collecting data include surveys and observation of customers. The movie business uses free previews to test viewers' reactions. The studio then uses that information to revise or take out scenes or characters or change the ending. Focus groups have limits since no potential customer ever asked for ATMs, traveler's checks, or personal computers. Many technology companies use the "a day in your life" format that takes customers through their day before and after the new product launch to expose the benefits of the new product.

> Consider the efforts of a fast-food chain that wanted to sell more milk shakes. It first used focus groups to talk about what people wanted, but after making the suggested changes, it noticed no improvement in sales. Then it tried to see what job customers were hiring it to do.
>
> The researchers spent a day in a restaurant. What they found was surprising: Nearly half of all milk shakes were bought in the early morning, the shake was the only item purchased, and it was rarely consumed in the restaurant. The customers were hiring the milk shake for their commute and needed a food that could be managed with one hand and lasted for the whole drive [Christensen and Raynor, 2003]. Market research is most useful when it finds out what people are *really* hiring a product for.

An important use of market research is to estimate the **market potential** for maximum sales under expected conditions. Then the new venture team can estimate a realistic **sales forecast.** Market potential for a new product is often overestimated. For example, we might estimate that the potential market for the EasyPC is 10 million units per year, based on the sales of Dell and HP (see Figure 11.3). Then, an optimistic estimate of actual sales in the first year might be 1 million units. Clearly, that forecast is subject to many unstated assumptions—which may be in error. A sales forecast should be a realistic estimate of the amount of sales to be achieved under a set of assumed conditions within a specified period of time. Many sales forecasts are unrealistic. A sales forecast for a new venture needs to be conservatively developed within a statement of assumed conditions.

Know your customer. Being first, being best, and being correct may not matter as much as providing what the customer really wants right now. Finding out what the customer really wants is a very important and difficult task. Often people will not or cannot verbalize their true motivations and attitudes. In creating a marketing plan, the attitudes and preferences may not be clearly reported. One widely used method is **conjoint analysis,** which provides a quantitative measure of the relative importance of one attribute as opposed to another. In conjoint analysis, the respondent is asked to make trade-off adjustments and decisions. This method requires an investment of time and money in the research process but may be worth it to avoid misreading the customers' preferences [Aaker, Kumar, and Day, 2001].

11.5 BRAND EQUITY

The new venture should have a competitive advantage such as low cost, quality, customer relationship, or performance advantage. Many new technology ventures differentiate themselves from competitors by doing a better job of convincing their customers that they have a better product characteristic, such as performance, reliability, or quality. A brand is a combination of name, sign, or

TABLE 11.3 Four dimensions of brand equity.

- Brand awareness or familiarity.
- Perceived quality and vitality of the product.
- Brand associations: connects the customer to the brand.
- Brand loyalty: a bond or tie to the product.

symbol that identifies the goods sold by a firm. A brand accurately identifies the seller to the buyer. A **brand** is something that resides in the minds of consumers. Well-known brands include Intel, HP, and Dell. Brand equity is the brand assets linked to a brand's name and symbol that add value to a product. **Brand equity** is the perceived worthiness of the brand in mind of the customer and may be portrayed as the sum of four dimensions, as shown in Table 11.3 [Aaker and Joachimsthaler, 2000]. Brand awareness or familiarity is the first step in building a brand. The perceived quality of the product and respect for the product will help build brand equity. The quality of the product and its perceived vitality will build an image of the brand. A brand association is how the customer relates to the brand through personal and emotional associations. This dimension is present in the emotional relationship that Harley-Davidson owners have with the motorcycle brand. In other words, brand loyalty responds to promises kept by the seller.

A brand's promise of value is the core element of differentiation. This promise of value is tied to the customer, and loyalty will follow from good customer experiences. Many customers are willing to pay more for some badge of identification—Apple's rainbow-colored logo, for example—that makes them feel they are part of a community. A strong corporate brand lets customers know what they can expect of the whole range of products that a company produces. The most successful corporate brands are universal and facilitate differences of interpretation that appeal to different groups. This is particularly true of corporate brands whose symbolism is strong enough to allow people across cultures to share symbols even when they don't share the same meaning. Two examples of a trusted brand that is a promise fulfilled are Dell Computer and IBM.

Some brands, such as Nike, Harley-Davidson and Volkswagen, become an icon [Holt, 2003]. A brand becomes an icon when it offers a compelling story that can help people resolve tensions in their lives. One of the most potent stories is the depiction of a group of rebels. For example, Nike appeals to rebel youth who want to stand out as different from the crowd.

One approach to building a brand is to identify the differentiating benefit that is important to the target customers and describe the attributes that imply this benefit. Intel identifies superior quality as its benefit. Successive marketing campaigns have informed consumers that Intel integrated circuits have reliable high performance and are leading-edge products promising superior quality and performance.

11.6 MARKETING MIX

The four elements of the marketing mix are shown in Table 11.4. The **product** is the item or service that serves the needs of the customer. The marketing plan describes the key methods of differentiating the product. Coca-Cola, for example, differentiates regular Coke by using a distinctive trademarked bottle with ribbing. Some auto companies use their warranty to distinguish their product. Nordstrom distinguishes its products by quality and its liberal return policy. Kodak's EasyShare digital camera is distinguished by its ease of use. Intel's Pentium chip is distinguished by its high-speed performance.

Pricing policies can be used to distinguish a firm's offering. Warren Buffet said it clearly, "Price is what you pay, value is what you get." For example, Amazon.com offers 30 percent off most books' list prices and free shipping for orders over $25. Price is a flexible element, and various discounts, coupons, and payment periods can be tested in test markets. The price can be initially set by estimating demand, costs, competitors' prices, and a pricing method to select the price. The pricing method or strategy can seek market share, premium pricing, or maximum profit. The cost to make the product is a floor under the price, and an estimate of the total value to the customer sets a ceiling on the price (see Figure 3.5). After studying competitors' prices, the new venture can test a price on a set of test customers.

Consider the setting of a price for a textbook where the total market demand is 10,000 books per year. Competitors have established a retail price in the range $60 to $80, and the demand per year for a new textbook may be described by

$$D = 10,000 - kP \tag{11.1}$$

where D = demand in units, k = estimated sensitivity constant, and P = price in dollars. The fixed cost to produce the new book is $30,000, and the variable cost is $10 per unit. The book is differentiated from its competition by quality and clarity. What price, P, would you select within the established range of $60 to

TABLE 11.4 Four elements of the marketing mix.

Product	Price	Promotion	Place
Product variety	List price	Public relations	Channels
Quality	Discounts	Advertising	Locations
Design	Credit terms	Sales force	Inventory
Features	Payment period	Direct messages	Fulfillment
Brand name			
Packaging			
Warranties			
Returns policy			

TABLE 11.5 Gross profit for selected values of the sensitivity constant, k, and the price of the new book (gross profit in thousands).

		Price		
		$60	**$70**	**$80**
Sensitivity constant, k	**90**	$276	$259	$166
	80	$230	$234	$222

$80? To maximize market penetration, one would select the lowest price, P = $60, since this will result in the largest demand. If market research shows that the market is price-sensitive and k = 90, then when the price is set at $60, the demand is for 4,600 units. Then the gross profit = (revenues – cost of goods) is

$$GP = R - (VC \times D) = (D \times P) - (VC \times D) = (P - VC)D \qquad (11.2)$$

where R = revenues and VC = variable cost. When P = $60 and k = 90, the gross profit is $276,000. As shown in Table 11.5, if you raise the price to $70, the gross profit declines. The calculation of the best price to obtain the maximum gross profit depends on the estimated sensitivity constant. If we change our assumptions so that k = 80, then we obtain the gross profit for the book as shown in Table 11.5. Note that $70 would be the price to maximize profit when k = 80. Note that k is an estimate obtained through experience and research and will change over time.

In many industries, customers demand low prices, and the competitors have little pricing power. Pricing power accrues to companies without wide competition, such as universities that raise tuition, hospitals that increase fees, or virtual monopolies such as cable-television operators. Most mature companies operate in a world of flat or falling prices due to an excess number of providers. A new venture can pick its pricing strategy from the three shown in Figure 11.5. Many new ventures use value pricing since demand will be sensitive to price and the new firms possess little brand equity. Demand-oriented approaches look at the demand for the product at various price levels and try to estimate a price that will provide a good market share and profitability for the long term. Many technology ventures with a new breakthrough product will use a premium pricing strategy.

New technology companies usually offer new, value-oriented products. A new product or service is, by its unknown nature, difficult to price. Many new products are characterized by quality and performance uncertainty. To attract customers to a new product, it may be useful to offer a warranty—a contract or guarantee of a specified performance. Another possibility is to offer quality-contingent pricing that specifies a price rebate for poor performance [Bhargava, 2003]. Three pricing methods are offered in Figure 11.5.

Using a traditional model for growth, firms can take advantage of the demand for new goods and services by creating and marketing products that

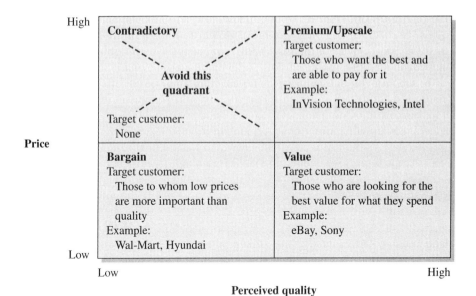

FIGURE 11.5 Three pricing methods.

satisfy a demonstrated need in the marketplace. As their customer bases grow and the products become more and more popular, profits begin to emerge. The profits are then reinvested in projects that will provide new sources of revenue and income. A portion of the profits is retained to build brand value, which can be created through a variety of techniques, not the least of which is aggressive pricing, savvy promotion, and advertising.

Promotion includes public relations, advertising, and sales methods. Selecting the message for advertising and the media for transmitting the message is a complex activity. *Advertising* is the art of delivering a sales proposition and positioning the product uniquely in the customer's mind. [Roman 2003] The initial product message is used to attract customers to the new venture. Advertising can use print, radio, or television media. Charles Revson, co-founder of Revlon, once said: "In our factory, we make lipstick. In our advertising, we sell hope." Many products sell hope. All the purveyors of weight-control products sell hope. Matchmakers and dating services also sell hope. By contrast, Microsoft and Intel sell reliable performance.

A list of marketing media is given in Table 11.6. Sending direct messages via mail or telemarketing can be a useful method. Public relations normally takes the form of an article in the print media or an interview on radio or television that delivers the product message. Many firms find the use of a sales force necessary to carry their message to the customer. Word-of-mouth (buzz) promotion is particularly important for movies, toys, recreational activities, and restaurants. The buzz around the Harry Potter movies was large. Other products such as pharmaceuticals can generate a lot of buzz [Dye, 2000]. Products that merit a

TABLE 11.6
Marketing media

Radio

Newspapers

Magazines

Television

e-mail

Telemarketing

Catalogs

Infommercials

Websites

Presentations

buzz campaign have some unique, attractive attribute, such as the Chrysler PT Cruiser or a new anticancer drug. Furthermore, they should be highly visible. The latest fashion in clothes or accessories often runs on buzz with teenage girls.

Word-of-mouth marketing is often called **viral marketing.** The concept is based on an age-old phenomenon: People will tell others about things that interest them. The Internet is an important avenue for finding passionate tastemakers who will carry a message forward. They have e-mail lists. They have their own networks. JetBlue Airways, Starbucks, and Linux are good examples of brands that became successful because of positive word-of-mouth.

The buzz promotion campaign follows a process of seeding the message and rationing the supply of the product. The PT Cruiser was visible, talked about, and not readily available—perfect for a buzz campaign. Another tactic is to use a celebrity to start the campaign. To profit from buzz, however, you need to act first and fast. As wireless phones and messaging builds, the forum for the latest buzz grows.

Place means selecting the channels for distribution of your product and, when appropriate, the physical location of your stores. Channels of distribution are necessary to bring your product to the end user. A publisher sells a book through multiple channels such as bookstores, direct to the end user, and via Internet bookstores, such as BarnesandNoble.com. Each industry has a transitional distribution system. Differential advantages can accrue to sellers who creatively use different channels.

In the personal computer industry, Dell Computer sells direct to the end user via phone or the Internet, while Hewlett-Packard sells primarily via retail stores and value-added resellers. When several parallel channels are used, channel conflict can occur due to the divergence of goals between channels and domain (territory or customer) disagreements.

Many technology ventures will sell their product to other manufacturers who will incorporate the product as a module or component within the final product. For example, Intel provides microprocessors that Dell incorporates within its PC.

The use of the Internet as a distribution channel will cause a shift in the relationships between consumers, retailers, distributors, manufacturers, and service providers. It presents many companies with the option of reducing or eliminating the role of intermediaries and lets those providers transact directly with their customers. Before launching an e-commerce effort and bypassing its traditional distribution channels, however, a business should analyze which products are appropriate for electronic distribution. Those most appropriate are digital, such as information products.

In the late 1980s, Intel decided to redirect some of its advertising efforts away from computer manufacturers to actual computer buyers. The consumer's choice of a personal computer was based almost exclusively on the brand image of the manufacturer, such as Compaq, Dell, and IBM. Con-

(continued on next page)

(continued from page 243)

sumers did not think about the components inside the computer. By shifting its advertising focus to the consumers, Intel created brand awareness for itself and its products, as well as built brand preference for the microprocessor inside the PC.

The first step was to create a new advertisement using the "Intel: The Computer Inside" slogan. Second, Intel chose a logo to place on a computer—a swirl with "Intel Inside." Then it chose a name for its new microprocessor: Pentium. As a result, Intel became a leader in the PC boom of the 1990s. Intel was successful at branding a component.

Many firms are using the Internet for selling and experimentation. Procter & Gamble (see www.pg.com) is using its corporate website to invite online customers to sample and give feedback on new prototype products [Gaffney, 2001]. This approach permits P&G to test new products and their marketing mix. P&G conducts at least 40 percent of its tests online. In August 2000, when P&G was ready to launch Crest Whitestrips, a home tooth-bleaching kit, it tested its proposed price of $44. P&G ran TV and magazine ads to attract people to the test. It also sent e-mails to people who had signed up to sample new products. In the first eight months, it sold 144,000 whitening kits online.

An emerging firm has to decide how and where to spend its marketing dollars. It may have several product categories and numerous regions on which to spend its limited marketing budget. An emerging firm should collect information on each regional or international market and allocate its resources based on the regions and products that offer the best opportunity for profit [Corstjens and Merrihue, 2003].

11.7 CUSTOMER RELATIONSHIP MANAGEMENT

The quality of the relationship that a firm has with its customers directly influences the intrinsic value of the firm. **Customer relationship management** (CRM) is a set of conversations with the customer. These conversations consist of 1) economic exchanges, 2) the product offering that is the subject of the exchange, 3) the space in which the exchange takes place, and 4) the context of the exchange [McKenzie, 2001]. For the customer relationship to be fruitful, the attraction of the customer, the conversion or sale of the customer, and the customer retention process must all be managed well. These relationships are managed through conversations in real time—that is, without delays. The firm and the customer usually engage in a series of brief conversations that help build a relationship. The conversations take place between the customer and the firm in a relationship space, as shown in Figure 11.6. The first part of a conversation is the economic exchange based on a product offering that is communicated to the customer. The space in which the exchange takes place could be physical, such as grocery store or furniture showroom, or a website displaying goods (e.g.,

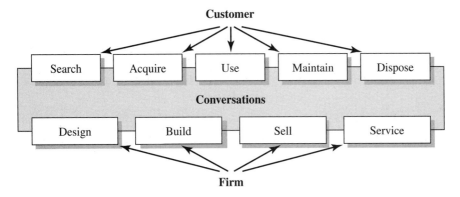

FIGURE 11.6 Customer-firm relationship as a conversation.

Amazon.com). The context of the exchange includes all that is known about the customer and the situation with the customer.

A necessary step to a CRM system is the construction of a customer database. This is relatively easy for banks and retail firms since they have a high frequency of direct customer interaction. It is more difficult for firms that do not interact directly with the end customer, such as semiconductor and auto manufacturers [Winer, 2001]. The customer database can be used for CRM activities such as customer service, loyalty programs, rewards programs, community building, and customization.

CRM, when properly constructed, allows firms to gather customer data quickly, identify valuable customers, and increase loyalty by providing excellent service. Through the CRM process, customers become a new source of competencies engaged in the building the firm's products and services as they provide ideas via the conversational process [Prahalad and Ramaswamy, 2000]. Unfortunately, too many companies distance themselves from customers by using phone loops that trap and frustrate customers seeking aid.

Progressive Corporation uses CRM to relate to its customers 24 hours a day. It sells auto insurance both directly and through agents. Progressive's information systems allow customers to manage their accounts online, including paying their bills electronically and adding a vehicle or driver. It has a highly functional website, telephone call centers, and claims service available 24 hours a day, seven days a week. Progressive's claims agents travel quickly to the scene of an accident. The agents are equipped with notebook computers that communicate wirelessly to the corporate network, which lets them key in information on site.

The CRM and the total marketing effort are depicted in Figure 11.7. CRM helps improve marketing research, customer retention, and the marketing mix.

Customers who say they are satisfied are not necessarily repeat customers because satisfaction is a measure of what people say, whereas loyalty is a measure of what they actually do. Customer surveys measure opinions but are unreliable predictors of future behavior. **Loyalty** is not a matter of opinion [Klein and

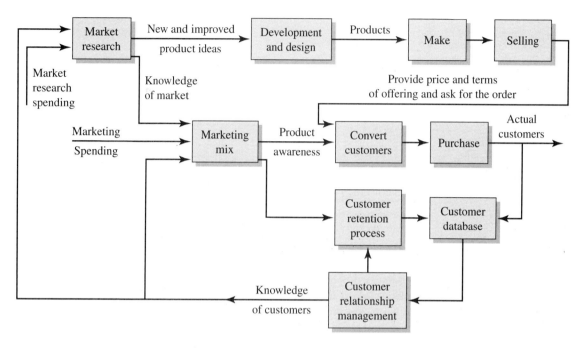

FIGURE 11.7 CRM and the total marketing effort.

Einstein, 2003]. Loyalty is a measure of a customer's commitment to a product or a company's product line. Loyalty measurements are more difficult to obtain than satisfaction measures. Good satisfaction measurement can help identify what's broken in your business today. Good loyalty measurement is a forward-looking tool that firms can use to devise strategies to hold on to customers they want to keep.

> CRM tools can be used to collect and organize the activities of a firm's customers. For example, FrontRange (www.frontrange.com) enables a firm to track current or potential customers and provide service, sales, and support management.

Customization, sometimes called one-to-one marketing, is a process that enables a product to be customized (changed) to a single customer's specifications. A firm uses a CRM system to elicit the information from each customer specific to his or her needs and preferences. Customization allows the company and the customer to learn together about the customer's needs. Dell Computer popularized the concept with its build-to-order website. Other companies such as Levi Strauss and Nike have developed processes and systems for creating customized products according to customers' tastes. For a good example of customization, see Dell Computer at www.dell.com. Customization is easy to do with digital goods such as music files, but other manufacturers can also tailor products to provide customization [Winer, 2001].

Procter & Gamble spun out Reflect.com in 1999 to enable people to customize their own beauty products and to use customer feedback to adjust its products (see www.reflect.com).

11.8 DIFFUSION OF TECHNOLOGY AND INNOVATIONS

Most entrepreneurial ventures have some novelty or innovation embedded in their product. Customers will adopt one innovation earlier than others based on their perception of its advantages and its risks. The **diffusion of innovations** describes the process of how innovations spread through a population of potential adopters. An innovation can be a product, a process, or an idea that is perceived as new by those who might adopt it. Innovations present the potential adopters with a new alternative for solving their problem, but they also present more uncertainty about whether that alternative is better or worse than the old way of doing things. The primary objectives of diffusion theory are to understand and predict the rate of diffusion of an innovation and the pattern of that diffusion. Innovations do not always spread quickly. The best ideas are not always quickly adopted. The British Navy first learned in 1601 that scurvy, a disease that killed more sailors than warfare, accidents, and all other causes of death, could be avoided. The solution was simple (incorporating citrus fruits in a sailor's diet), and the benefits were clear (scurvy onboard was eradicated), yet the British Navy did not adopt this innovation until 1795, almost 200 years later [Rogers, 1995].

The diffusion of innovations depends on a potential adopter's perception of five characteristics of an innovation, as listed in Table 11.7. The adopter's perceptions of these characteristics strongly influence his or her decision to adopt

TABLE 11.7 Five characteristics of an innovation.

- **Relative advantage:** the perceived superiority of an innovation over the current product or solution it would replace. This advantage can take the form of economic benefits to the adopter or better performance.

- **Compatibility:** the perceived fit of an innovation with a potential adopter's existing values, know-how, experiences, and practices.

- **Complexity:** the extent to which an innovation is perceived to be difficult to understand or use. The higher the degree of perceived complexity, the slower the rate of adoption.

- **Trialability:** the extent to which a potential adopter can experience or experiment with the innovation before adopting it. The greater the trialability, the higher the rate of adoption.

- **Observability:** the extent to which the adoption and benefits of an innovation are visible to others within the population of potential adopters. The greater the visibility, the higher the rate of adoption by those who follow.

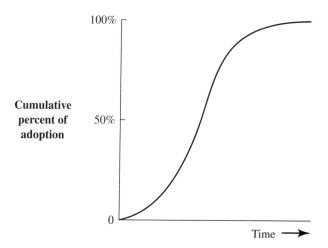

FIGURE 11.8 S curve of adoption of an innovation.

or not. Consider the introduction of black-and-white television in 1947. By 1950, 10 percent of all households had adopted this innovation, and by 1960, 90 percent of households had a TV. This rapid adoption of TV was due to its relative advantage compared to radio, its high compatibility within the home, its relatively low complexity, its easy trialability, and its ready observability in TV store windows and friends' homes. On the other hand, consider the slow adoption of the personal computer in the home. PCs were introduced into the home market by 1982, but by 2002, only 50 percent of households had a PC. The high complexity of a PC discourages many consumers from adopting it in the home. Also, the perceived advantage is not clear to many would-be users.

The adoption of an innovation usually follows an S curve, as shown in Figure 11.8. When the adoption follows the S curve, then the distribution curve of adopters follows a normal distribution, as shown in Figure 11.9, where Sd =

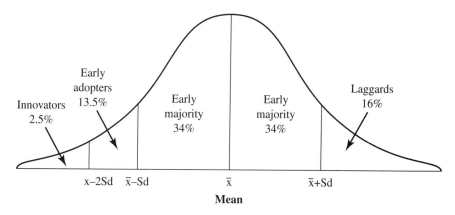

FIGURE 11.9 Innovation adoption categories when Sd = standard deviation.

TABLE 11.8 Five categories of adopters of an innovation.

- **Innovators** want to be on the leading edge of business and are eager to try new innovations. They have an ability to work with complex and often underdeveloped ideas as well as substantial financial resources to help them absorb the uncertainties and potential losses from innovations.

- **Early adopters** are more integrated with potential adopters than innovators and often have the greatest degree of opinion leadership, providing other potential adopters with information and advice about an innovation. They are visionaries.

- The **early majority** adopts just ahead of the average of the population. They typically undertake deliberate and, at times, lengthy decision-making. Because of their size and connectedness with the rest of the potential adopters, they link the early adopters with the bulk of the population, and their adoption signals the phase of rapid diffusion through the population. They are pragmatic.

- The **late majority** is described as adopting innovations because of economic necessity and pressure from peers. While they make up as large a portion of the overall population as the early majority, they tend to have fewer resources and be more conservative, requiring more evidence of the value of an innovation before adopting it.

- **Laggards** are the last to adopt a new innovation. They tend to be relatively isolated from the rest of the adopters and focus on past experiences and traditions. They are the most skeptical when it comes to risking their limited resources on an innovation.

standard deviation. The five categories of adopters are shown in Figure 11.9 and described in Table 11.8 [Rogers, 1995].

11.9 CROSSING THE CHASM

The transition from the early adopters to the early majority is difficult since it requires attracting pragmatists, as shown in Figure 11.10. This large gap between visionaries and pragmatists is called a **chasm** [Moore, 2000]. The early adopters or visionaries are independent, motivated by opportunities, and quickly appreciate the nature of benefits of an innovation. However, the early majority or pragmatists are analytical, conformist, and demand proven results from an

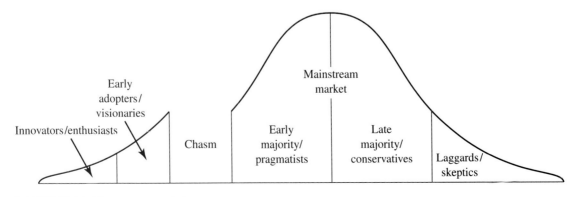

FIGURE 11.10 Chasm model.

An epidemic is spread by a message that is meaningful and emotional to the receiver and motivates the buyer who finds the message sticks in his or her mind and passes it on to others. The persuasive message is communicated by a trusted agent, and it moves the buyer to act. A tipping point can occur as a result of viral marketing. The message of *Sesame Street* made a rapid entry into the children's television market. Who can forget Big Bird? With the right message and agent delivering it, the product may quickly cross the chasm. A contagious message that is memorable, motivating, and delivered by a trusted agent can take a product across the chasm by helping a product reach its tipping point.

> TiVo is a service and a proprietary architecture delivered in a box directly to consumers and licensed to others such as DirectTV, Sony, and Pioneer. It personalizes TV viewing by allowing users to record, play programs, and skip commercials and other unwanted content. The experience is novel and more than the sum of its parts, consequently being hard-to-explain to non-users. Adopted by only 1 percent of US TV users in 2004, it has a way to go to wide acceptance. However, TiVo has the potential to reach a tipping point because its installed based of satisfied users are avid fans who actively persuade others to try it (see www.tivo.com).

The entrepreneur is the agent who creates the vision for the new product and builds a marketing plan that will help potential adopters to understand and value the innovation and respond to the message communicated by the firm. This message must be persuasive, believable, and understandable. New ventures create the resources and necessary strategy to overcome the potential adopters' lack of knowledge and understanding of the product. They are more likely to receive funds from investors and succeed in bringing their product across the chasm and to the remaining adoption categories.

Marketing can be described as taking the actions necessary to create, grow, and maintain your firm's place in your chosen market. To cross the chasm, the marketing strategy has to attract and retain pragmatists. They care about quality, service, ease of use, reliability, and the infrastructure of complementary products (often called the whole product).

To get across the chasm, determine the characteristics of the pragmatist customers and build a marketing plan for them. For a small start-up, this may mean acting locally and then expanding as sales grow. Crossing a chasm means assembling the whole product, with partners, to satisfy the pragmatic buyer in a specific target market [Moore, 2002].

11.10 PERSONAL SELLING AND THE SALES FORCE

All businesses involve **selling,** which is the transfer of products from one person or entity to another through an exchange mechanism. It includes identifying customer needs and matching the product or solution to those needs. For many technology ventures, this process is called business development. Most firms employ a sales force to make the contacts with the purchaser. New ventures should develop a selling strategy and a plan of action. Then they locate target customers and recruit, train, motivate, compensate, and organize a field sales force. They also have to manage the interactions between customers and salespeople. This dialogue is influenced by the buyer's needs and the salesperson's skills. The results of successful salesperson-customer interactions are orders, profits, and repeat customers. The salesperson implements the marketing strategy. In a small start-up, the salespeople may have other roles, such as product development or market planning.

Especially in industrial markets where the customer are other businesses, the buyers may actually be multiple decision makers. The ultimate user of the technology product or service is certainly one of them. However, others could include those who make the recommendation of which solution to buy such as the information technology staff and those who actually negotiate the contract such as the purchasing agent. This can complicate and delay the sale. This length of time from the first contact until a sales transaction is completed is called the **sales cycle.** It can be as short as one day (e.g., purchasing ad listings on Google or auction listings on eBay) or many months (e.g., evaluating and choosing sophisticated MRI equipment for a hospital's lab). Technology ventures must estimate the length of their sales cycle as they develop their business model and financial plan.

IBM's success from 1955 to 1990 was due to a very knowledgeable, well-trained, and highly motivated sales force. IBM's salespeople had real experience with computers as well as understanding of their clients' needs. Thomas Watson, Sr., former CEO of IBM, noted that great technological innovation combined with a powerful sales force was unbeatable.

For a new technology venture with an innovative product, the salesperson must fully understand the product and the idea of creating a solution for the customer. It is the responsibility of everyone in a new venture to 1) identify and create a purchaser, 2) offer a creative solution, and 3) make a profitable sale. In many start-ups, the staff is comfortable with steps 1 and 2 but shies away from step 3. Without actually making the sale, the start-up is destined to fail [Bosworth, 1995]. The goal is to determine the purchaser's needs, or latent pain, create a solution to meet that need, and then sell it to the purchaser.

The solution sales process is shown in Figure 11.11. The salesperson identifies the target market and makes contact with a potential purchaser. Then the salesperson determines the customer's problem and needs. Based on these needs, the solution to the customer's problem is created and presented. The benefits of the solution must be clearly communicated. Then the salesperson asks for the order and, with a positive response from the customer, confirms the order. You may have experienced this process when shopping for new clothes. The salesperson makes a contact and determines your needs. Then the salesperson shows you one or more solutions (options), and you try them on for size and appearance. The salesperson aligns his or her comments on each solution in a

FIGURE 11.11
Solution selling process.

TABLE 11.10 **Selling via company salespeople versus independent representatives.**

	Advantages	Disadvantages
Company salespeople	■ Know your product well ■ Relatively easy to manage ■ Provide feedback from customers ■ Paid salary plus commission	■ High fixed cost ■ Low geographical dispersion ■ Time and costs to hire and train ■ Travel costs
Independent representatives	■ Paid on commissions ■ Lower hiring and training costs ■ Geographical dispersion ■ Have established relationships with customers ■ Low fixed costs	■ Sell for several firms, making it difficult to get their attention ■ Difficult to manage ■ Low feedback from customers ■ May have limited understanding of your complex product

discussion with you. When you both see a solution, the salesperson asks for the order. If you agree, the salesperson writes the order at the cash register. This process is the same, although more complex, for a purchaser seeking a new computer system for a government agency or an electronics firm. Salespeople sell themselves, show they care, and provide proof of product, consistency of message, authority, and scarcity. The sales process rests, in part, on the skills of persuasion, as later described in Section 13.2.

New ventures often use their own people to manage the sales process but engage others, called sales representatives, under contract, to actually sell the product. The advantages and disadvantages of using company salespeople versus independent representatives are shown in Table 11.10. The choice of the right balance of company salespeople and independent representatives is a critical issue for a new business.

The emerging technology business may initially use a focused, direct sales force to create demand and penetrate to the primary target segment. Then, as growth accelerates, a transition to other segments and sales channels may be appropriate. It is important to clearly identify the primary target segment and key customers. [Waaser et al. 2004]

New businesses encounter sales resistance due to competition and lack of knowledge of their product and its quality. One method to overcome this is to utilize trial periods, warranties, and service contracts. Many new ventures do an excellent job of building a good product and developing a solid marketing plan but then fail to make the forecasted sales.

We will cover the issues of international marketing and sales in Chapter 14.

11.11 AGRAQUEST

AgraQuest prepared a sketchy marketing plan in 1998 as it looked forward to launching its first product. Based on a review of this plan, prepared by Pamela Marrone and the vice president for product development, it moved to recruit a vice president of marketing and sales. The marketing plan described the market segment (customers) as farmers in the United States and Chile. AgraQuest planned to use distributors as the channel to distribute the natural biopesticide serenade. The positioning map for Serenade is shown in Figure 11.12. The effectiveness of Serenade and chemicals was the same for most fungi; Serenade, however, had higher resilience (retention of effectiveness over time).

AgraQuest did some market research and realized that many farmers distrusted new bioproducts because of excessive or unfulfilled claims by other natural-product firms. It used field trials to demonstrate the product's effectiveness and overcome the distrust. It also carried out a pricing study and decided to price its product at the same level as those of its chemical company competitors.

In 1998, AgraQuest hired its first vice president for marketing, who launched a large, traditional advertising campaign that consumed $500,000 over two years. The positioning statement was:

The Best Biopesticide on Earth!

The advertising and sales campaign was viewed by farmers as arrogant, considering the product was unproven, and failed to attract buyers. The first marketing VP left, and a second was engaged in 2000. He then proceeded to redo an advertising campaign to build image, brand, and acceptance, which also failed. Both VPs had worked for large chemical pesticide companies that used advertising campaigns to launch new products. AgraQuest, however, was a small

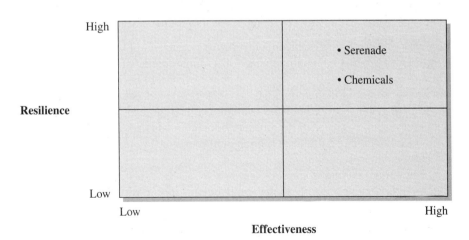

FIGURE 11.12 Positioning map for the biopesticide Serenade versus chemical pesticides and fungicides.

start-up that needed a marketing plan that would help it attract farmers to try its product.

AgraQuest's natural pesticides need to cross the chasm to the pragmatic farmer who is slow to adopt new tools and products. This slow, deliberate adoption process cannot be accelerated by advertising. It needs a series of deliberate trials and proof of its key advantage: resilience of the effectiveness of the product. AgraQuest installed a third vice president of marketing in 2003.

AgraQuest has a huge chasm to cross to convince mainstream customers that they should use Serenade. Success has come slowly by doing extra field trials and demonstrating the efficiency and effectiveness to farmers in their own fields. Farmers will not use something if they haven't seen it work. Breaking into the pesticide market with a natural product is very difficult. The industry is dominated by chemical pesticides that work very well. Farmers like them, and they have a 50-year history of use. A lot of companies have overpromised natural products that have not lived up to the pitch. So farmers are very, very skeptical about natural products.

AgraQuest uses a sales force of seven on a salary and bonus compensation scheme. Its markets are, in order of priority: grapes (United States and Chile), tomatoes, lettuce, bananas (Costa Rica), and apples.

AgraQuest had sales of $2.5 million in 2002 and $3.5 million in 2003. Serenade has a unique advantage because only natural products can be used in the weeks just prior to harvest. If it rains during that time, Serenade, unlike chemicals, can be used to prevent fungi. Harvest seasons in 2000 to 2003 were dry, however, so this competitive advantage is not constant over time or likely to apply very often.

11.12 SUMMARY

Any new firm needs to build a marketing plan that describes how it will attract, serve, and retain the customers targeted for its products. Since a new firm normally starts without established customers, it must carefully identify the target market that will value its product. Market research can provide the information about the customers, appropriate distribution channels, and communication methods for attracting the customer.

The new firm creates a product positioning statement and selects a mix of price, product, promotion, and channels to attract and satisfy the customer. Most new firms are challenged to cross a chasm in the diffusion process that enables their product to attract the pragmatic and skeptical potential customer. The marketing process consists of describing or implementing the following sequential elements:

- Product offering
- Target customer
- Marketing objectives
- Market research
- Marketing plan

> **Principle 11**
> A sound marketing plan enables a new firm to identify the target customer, set its marketing objectives, and implement the steps necessary to sell the product and build solid customer relationships.

- Sales plan
- Marketing and sales staff

11.13 EXERCISES

11.1   Only about 5 percent of Americans have a high-definition television (HDTV), which receives a significantly improved picture. People will not buy the HDTV unless there are enough high-definition shows to watch. In 2003, ESPN launched a HD channel, and this may be the "killer app" needed. Visit an electronics store and determine when and if you would purchase a HDTV. Use the categories of Table 11.8 to describe your response to HDTV.

11.2   Synopsys is a supplier of electronic design automation software to the global electronics industry (www.synopsys.com). Its customers are the designers of integrated circuits. Its primary competitor is Cadence Design (www.cadence.com). Prepare a positioning map for these two firms.

11.3   Harrah's Entertainment (www.harrahs.com) is the most profitable gaming company in the United States. It has the most devoted clientele in the casino industry, a business notorious for fickle customers. Its success is based on mining customer data, running experiments using customer information, and using the findings to develop and implement marketing plans that keep customers coming back for more [Loveman, 2003]. Describe another example of a firm with fierce customer loyalty and how it builds that loyalty.

11.4   Zipcar gives its members short-term, on-demand use of a fleet of cars (www.zipcar.com). Before launching this service, Zipcar proposed a pricing scheme of a $300 refundable security deposit, a $300 annual subscription fee, and $1.50 per hour plus 40 cents per mile. Create another pricing scheme that will be more profitable.

11.5   Powerful brands are built on innovativeness and advertising [Kotler 2001]. Examine the brand value for DuPont and Merck and describe the reasons for their brand power.

11.6   What is the best way to reward salespeople: salary, commission, or a mix of the two? What is the best method for a new emerging life-sciences technology business, such as Enzo Biochem (www.enzo.com)?

11.7   Robotics companies such as iRobot naturally attract a technology-enthusiast customer. Describe a strategy to bring iRobot or a similar company across the chasm.

11.8   One way to create buzz is to make the experience of your company fun. Airlines such as JetBlue and Southwest Airlines use this to create buzz. Find another example of a firm that uses this approach and describe its success.

11.9   Buzz for a new firm can be started with an advertisement. Honda U.K. aired a TV ad called "Cog" in 2003. This ad shows a dance of parts coming together as a Honda car. Visit www.honda.co.uk/multimedia and describe the captivating elements of the ad.

11.10 A new venture intends to prepare and sell a lightly sweetened, bottled cinnamon tea to compete with Snapple and other bottled teas. Prepare a positioning map for the product and its competitors.

11.11 A new e-mail program, Bloomba, was first offered in 2003 for $50. Bloomba has several favorable advantages over Microsoft Outlook Express. Consider the pricing of Bloomba versus Outlook Express, and determine if it is appropriately priced (see www.bloomba.com).

11.12 Advertising can be useful for building brand recognition. Describe the ad campaign for Aflac that uses a duck [Thaler and Koval, 2003]. Why does this ad work?

CHAPTER 12

The New Enterprise Organization

Two people working as a team will produce more than three working as individuals.
Charles P. McCormick

After recognizing an opportunity and deciding it is attractive, usually one or two leaders assemble a new venture team to build a plan and an organization to execute it. This initial team creates or designs an organizational arrangement to respond to the opportunity. The leaders of the venture are identified early in the organization's development. These leaders are able to inspire and motivate others to join the new venture and work on tasks of the venture. They build a team that is collaborative and possesses diverse competencies. As an organization grows, managers will be needed to carry out the tasks that keep the organization running well. A leader is a team's emotional guide and exhibits solid emotional intelligence.

As the organization grows, the firm works to build an organizational culture and trust among team members. Leaders and teams strive to build social relationships and networks to foster collaboration. One of the methods of creating an ownership culture is to facilitate ownership for all people in the firm through stock options or restricted stock. A new firm also builds a set of advisors and a board of directors to help it move forward. ■

12.1 THE NEW VENTURE TEAM

The first step toward forming a new venture is often taken by one or two individuals who recognize a good opportunity and then develop a business concept and vision to exploit it. After a short period, it becomes clear that a team is required in order to have all the necessary capabilities in the leadership group. The **new venture team** is a small group of individuals who possess expertise, management, and leadership skills in the requisite areas. Thus, the team will incorporate people with skills and knowledge in finance, marketing, product development, production, and human resource management. Typically, a team of two to six people is required to develop a business plan, secure the financing, and launch the firm into the marketplace. We define a **team** as a few people with complementary capabilities who are committed to a common objective, goals, and approach for which they hold themselves mutually accountable. New team-founded ventures tend to achieve better performance than individually founded ventures [Chandler et al., 2002].

The capabilities of the one or two lead entrepreneurs are critical to the new venture since others are willing to join the team based on the integrity, experience, and commitment of these lead entrepreneurs. Often we call the lead entrepreneurs the **founders.** In other cases, all the members of the initial leadership team are called the founders. The founders display all the characteristics of capable entrepreneurs: passion, commitment, and vision. Entrepreneurs understand the long-term implications of the information-based, knowledge-driven, and service-intensive economy. They know what new ventures require: speed, flexibility, and continuous self-renewal. They recognize that skilled and motivated people are central to the operations of any company that wishes to flourish in the new age.

Furthermore, these new ventures must exhibit adaptability and readiness to change as the context of the business evolves. The organizational arrangement of the firm must evolve as the market and the customer change. Newly formed firms are challenged by this necessity to change since they usually have limited resource and capability bases. Strong teams ensure that the firm is able to constantly reorganize in terms of strategies, structures, systems, and resources. The new venture team must have the skills to balance the needs for change, efficiencies, alignment of effort, and timeliness. The team must include one or more persons who can gain access to external sources of funds. One of the advantages of a new organization is that people can use active thinking rather than precedent as a basis for action [Pfeffer and Sutton, 2000].

Most of the members of the new venture team are known to the lead entrepreneurs and usually include family, friends, and business associates. The members of a team should be selected for their demonstrated skills and abilities. They also can be selected for their ability to gain access to information, knowledge, financial capital, and other resources required by the new venture.

The leadership team at Apple Computer in 1978 consisted of three diverse persons, all with specific skills and personalities. Steve Jobs was the charismatic leader who could motivate the employees and talk directly to computer lovers.

Mike Markkula was the business and marketing leader. Stephen Wozniak was the engineering leader and creator of the company's computers. This balanced and powerful team created Apple Computer's place in business history.

12.2 ORGANIZATIONAL DESIGN

Organizational design is the design of an organization in terms of its leadership and management arrangements; selection, training, and compensation of its talent (people); shared values and culture; and structure and style. The nine elements of an organization are listed in Table 12.1. Note that the last four elements in the list can be considered as the elements of an organizational design.

The **talent** are the people, often called employees, of an organization. The leadership team and the firm's managers are responsible for communicating and leading the firm in the appropriate direction. The shared values and corporate culture are the guiding concepts and meanings that the members of an organization share. The structure of a firm is its formal arrangement of functions and activities. Style is the manner of working together—for example, collegially or team-oriented.

A good organizational design leads toward the reduction of bureaucratic costs so that a low-cost advantage can be achieved. Furthermore, a good design can maximize a firm's value creation capabilities, leading to differentiation advantages and good profitability.

What is the best way to organize a group of people so as to maximize productivity and innovation? Most successful innovative organizations include many small units having free communication with each other, significant independence in pursuing their own opportunities, and freedom from central micromanagement. Innovation grows most rapidly under conditions of an intermediate degree of fragmentation. Excessive unity and excessive fragmentation are both ultimately harmful. The best organization design is one of teams or units that compete and generate different ideas but maintain relatively free, open communication with each other [Diamond, 2000].

TABLE 12.1 Nine elements of an organization.

1. Mission and vision

2. Goals and objectives

3. Strategy

4. Capabilities and resources

5. Processes and procedures

6. Talent

7. Leadership team and management

8. Shared values and culture

9. Structure and style

TABLE 12.2 Objective, tools, and resources of organizational design.

Objective	Sustainable competitive advantage
	Continuous renewal of the talent
Tools and methods	Vision
	Values
	Flexibility
	Innovation
	Entrepreneurship
Strategic resources	Human capital
	Organizational capital
	Intellectual capital

Competency-based strategies depend on talented people operating in a loose-tight structure. Thus, hierarchical structures need to be replaced by networks, bureaucratic systems changed into flexible processes, and control-based management roles evolved into relationships [Bartlett and Ghoshal, 2002]. Flexible organizations that effectively adapt to change are often called **organic organizations.** As summarized in Table 12.2, the strategic resources necessary for sustainable competitive advantage are human and organizational capital.

New ventures serious about obtaining profits through people will expend the effort needed to ensure that they recruit the right people in the first place. The organization needs to be clear about what are the most critical skills and attributes needed in its applicant pool. The notion of trying to find "good employees" is not very helpful. New ventures need to be as specific as possible about the precise attributes they are seeking. Technology start-ups tend to seek people who have strong technology expertise but also are flexible, and willing and able to assume a number of key roles in a new venture. The talent in a new venture needs to know what to do and be capable of doing it. The idea is to hire a "great athlete who has already run the race before." The qualities of a good new member of a new venture include flexibility, experience, technical knowledge, and self-motivated creativity. Robert Noyce and Gordon E. Moore spawned Intel, William H. Gates and Paul Allen built Microsoft, and Herb Kelleher and Colleen Barrett built Southwest Airlines. All these leaders possessed the requisite characteristics.

Organizational performance is the result of individual actions and behavior. Successful firms have people who take the right actions in concert with others. Thus, the form of the new enterprise is often a network characterized by relationships within the firm. In general, the firm starts off as a single team, and as it grows, it evolves into a series of cross-functional teams.

Southwest Airlines has been productive due to its effective use of its major assets—its aircraft and people. Southwest uses **relational coordination** (RC),

FIGURE 12.1 Model of an innovative organization.

which describes how its people act as well as how they see themselves in relationship to one another [Gittell, 2003]. RC requires frequent, timely problem-solving carried out through shared goals, shared knowledge, and mutual respect. Three conditions that increase the need for RC are reciprocal interdependence, uncertainty, and time constraints—all common to new business ventures.

One model of an organizational design is shown in Figure 12.1. The three activities of an organization are operations, innovation, and customer relationship management (CRM). These activities all support the key objective of the organization to create and maintain a sustainable competitive advantage. The integration of innovation, operations, and CRM can lead to a strong competitive advantage. The Internet and an intranet can help provide low interaction costs between the three activities. Most new ventures have an advantage because their newness permits them to easily integrate the three activities shown in Figure 12.1, thus rapidly gaining a competitive edge.

Most, if not all, new enterprises design a flat organization that facilitates speed of action. They avoid layers and bureaucratic structures, and keep communication flowing and a bias toward action [Joyce, Nohria, and Roberson, 2003].

A new venture normally starts out with a team or a **collaborative structure** that primarily consists of teams with few underlying functional departments. In a collaborative structure, the operating unit is the team, which may consist of five to 10 members. The best collaborative structures are self-organizing and adaptive. A **self-organizing organization** consists of teams of individuals that benefit from the diversity of the individuals and the robustness of their network of interactions. This collaborative effort coupled with the self-organizing behavior of the network can lead to benefits that exceed the sum of the parts—often called synergy.

> Perlegen Sciences is a biotechnology firm that was spun off from Affymetrix in 2000. Venture capitalists invested $100 million in Perlegen. Perlegen's mission is to apply biochip technology to scan human genomes and illuminate DNA variations that render people genetically different [Stipp, 2003] (www.perlegen.com). Perlegen's leading advantages are its founders and leadership team, which includes Stephen Fodor, who coinvented the Affymetrix biochip process, and Alejandro Zaffaroni, a founder of biotechs, including Alza.

12.3 LEADERSHIP

Leadership is the process of influencing and motivating people to work together to achieve a common goal by helping them secure the knowledge, power, tools, and processes to do so. Leadership is critical to an entrepreneurial venture and is normally provided by one or two leaders of the new venture. The leader of a new venture can be thought of as a leader of a jazz band that is known for its ability to play familiar and new music while creating and improving new variations and collaborative music. Mobilizing an organization to adapt its behaviors in order to thrive in changing business environments is critical. Responses to challenges reside in the collective intelligence of employees at all levels who need to use one another as resources, often across boundaries, and together learn their way to those solutions.

The leader of the new venture responds to routine work and challenging work issues in different ways, as shown in Table 12.3 [Heifetz and Laurie, 2001]. The most important capability of the entrepreneur-leader is the ability to cultivate and make use of the competencies of the talented team members [Davidsson, 2002]. Responding to challenges and adapting the effort of the talent is the role of a leader. Challenging problems confronting an organization require the members of the organization to take responsibility for solving the problem. Thus, the leader helps the team members confront the challenges and learn new ways to solve the problems.

TABLE 12.3 Leadership of routine and adaptive work.

A leader's role in:	Routine issues	Challenging and adaptive work
Direction	Define problems and possible solutions	Define the challenge and the issues
Team and individual responsibilities	Clarify and define roles and responsibilities	Define and discuss the necessity to adapt roles and responsibilities to changing needs
Conflict	Restore order and reduce conflict	Accept useful conflict and use it to define new approaches and strategies
Norms and values	Reinforce norms and values	Reshape norms and values
Teaching and coaching	Provide training and skill learning for existing employees	Teach and coach new people

Leaders build companies through a blend of personal humility and professional will [Collins, 2001]. They are ambitious, but primarily for the organization, not themselves. Leaders have the drive to build great companies through the efforts of the team members. They seek sustained results and facilitate new approaches to challenging situations while maintaining clear goals and methods for routine work. Leadership is the ability to acquire new organizational methods and capabilities as the situation changes. The leader stimulates discussion so that people contribute and understand the issues, ultimately leading to a shared strategy for sustained advantage.

There are four styles of leadership, as shown in Figure 12.2 [Northouse, 2001]. A leader's behaviors are both directive (task) and supportive (relationship). *Directive behaviors* assist group members in goal accomplishment through giving directions, establishing goals and methods of evaluation, setting time lines, defining roles, and showing how the goals are to be achieved. Supportive behaviors involve two-way communication and responses that show social and emotional support to others.

The *supporting* style is used when a leader does not focus exclusively on goals but uses supportive behaviors that bring out the employees' skills around the task to be accomplished. The *directing* leader gives instructions about what and how goals are to be achieved by the subordinates and then supervises them carefully. The *coaching* style calls for a leader to focus on goal achievement and give encouragement to subordinates. The *delegating* style occurs when the leader is less directive and facilitates employee confidence. Most effective leaders adopt all four styles of leadership depending on the needs of the situation and the team members while operating within the central target area. The leader of a new technology venture will most likely use a directing-supporting style in the

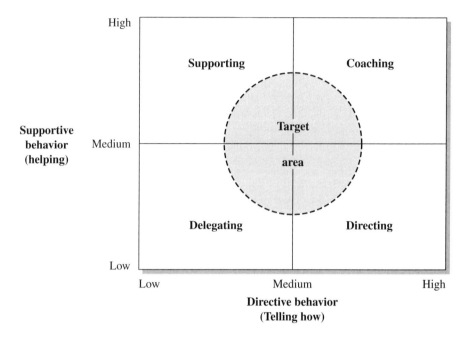

FIGURE 12.2 Four leadership styles.

early period of the venture. Later, in the growth period of the firm, the leader will probably use a mix of all four leadership styles.

   Leaders exhibit seven traits, as recorded in Table 12.4. Leaders are seen as authentic, decisive, focused, caring, coaching, communicative, and improvement-centered [Collins and Lazier, 1992]. Leaders articulate a clear, compelling vision for the venture and stimulate the team to achieve high performance. They avoid letting talk substitute for action. They also strive to develop a sustainable competitive advantage through building new competencies and products in a timely way. Leaders have a bias for simple concepts that can be clearly understood and

TABLE 12.4 Seven traits of leaders.

■ Authenticity: consistent actions and words.	■ People skills: offer helpful feedback and good coaching to all team members.
■ Decisiveness: willing to act on limited, imperfect information.	■ Communication: stimulate conversation and communicate vision.
■ Focus: create a priority list and stick to it.	■ Continuous improvement: keep learning and energy flowing in the firm, retain optimism.
■ Care: build relationships and social capital.	

Source: Adapted from Collins, 1992.

acted on. Informed action is their goal. The entrepreneurial leader uses a collaborative style while setting high standards and driving toward achievement.

Teaching is at the heart of leading. In fact, it is through teaching that leaders lead others. Teaching is how ideas and values are transmitted in an organization. Leading is helping others to see a situation as it really is and to understand what responses need to be taken so that they will act in ways that will move the organization toward where it needs to go. Organizational performance is the result of individual actions and behaviors. In successful companies, people do the "right things." Those companies have effective leaders who create conditions under which their people have the information, authority, and incentives to make the right decisions. When leadership is effective, behavior at all levels of the organization is both aligned and adaptable; thus, the organization performs to its potential. Leaders have a guiding vision and passion that allow them to communicate a sense of hope to followers. The key to this communication is integrity and credibility. Leaders with both have a heightened sense of self-awareness and a strong understanding of what they believe in and value. The five functions of leadership are challenging the status quo, inspiring a shared vision, enabling people to act, modeling through personal example, and motivating people to act.

Leaders display an inner strength and a constant set of values that everyone knows and can rely on. They avoid self-aggrandizement, inspire others, and exhibit a combination of modesty and extraordinary competence.

What makes a great leader? One study reveals a key attribute: the capability to handle adversity and to learn from such experiences [Bennis and Thomas, 2002a]. Bennis and Thomas call these shaping experiences *crucibles*. These experiences make leaders stronger and more confident. Leaders are formed by a combination of their individual personalities and the events of the era in which they spend their formative years, which are then transformed in a crucible of experience and challenge. Leaders organize the meaning of these experiences into capacities to respond to future challenges. These great leaders also evidence a capability to adapt as well as resilience. Good leaders are not necessarily charismatic [Khurana, 2002a]. Good leaders are likely to be capable people with solid experiences that helped shape and build their leadership skills.

12.4 TEAMS

A team is a small number of people with complementary skills who are committed to a common set of goals and tasks for which they hold themselves mutually accountable. Teams are appropriate for firms that seek innovative solutions to dynamic opportunities and challenges. Innovation is a function of both creativity and teamwork. Team members are given the power, freedom, and responsibility to control their work. Teams develop direction, momentum, and commitment by working to shape a meaningful purpose, such as designing a new product.

Teams offer several advantages. First, teams substitute peer-based control for hierarchical control of work. Instead of management devoting time and energy

TABLE 12.5 Characteristics of an effective team.

■ All members share leadership and ownership of the team's tasks.	■ Feedback on performance is frequent.
■ Communication is continuous among members in an informal atmosphere.	■ The division of tasks and work effort is clear.
■ Tasks and purposes are well understood.	■ A collaborative effort is the norm.
■ People listen to each other and are comfortable with disagreements within the team.	■ Members set their own, shared interim deadlines for project stages.
■ Most decisions are reached by consensus.	■ Team members rely on each other and hold each other accountable.
	■ Teams learn and share their learning.

to controlling the workforce directly, workers control themselves. Second, teams permit employees to pool their ideas to come up with better and more creative solutions to problems. Third, by substituting peer for hierarchical control, teams permit removal of layers of hierarchy. A strong team performance rests on good communication, shared purpose and values, and group pride. The characteristics of an effective team are summarized in Table 12.5. Teams act effectively and learn to improve when each member feels comfortable making suggestions, trying things that might not work, pointing out potential problems, and admitting mistakes.

Much of the success of new ventures can be explained by the social interaction (collaboration) within entrepreneurial teams [Lechler, 2001]. In knowledge-intensive, dynamic industries, entrepreneurial teams outperform single entrepreneurs, since the new venture requires more capabilities than one individual is likely to have. It is the combination of complementary capabilities that leads to success. The advantage of a team is the benefits obtained from a combination of people with diverse characteristics, skills, knowledge, and capabilities. The quality of collaboration within an entrepreneurial team is positively correlated with a new venture's success. The collaboration or social interaction within a team can be described by the six factors listed in Table 12.6. Teams that attain a high level of quality of social interaction tend to achieve new venture success.

Team members can support entrepreneurship when they have the necessary competencies and experience and believe that they collectively have the ability to perform the necessary tasks. Furthermore, the team members support the team effort and engage in "buy-in" without respect to the entrepreneurial task [Shepard and Krueger, 2002].

Teams can excel at collaborative problem-solving. Using an open process designed to generate multiple alternatives, teams test the alternatives and select a well-tested solution. With shared values and goals, teams can provide excellent results. To be successful, teams need to excel at creative tasks. Creative teams have members of diverse backgrounds and change membership periodically.

TABLE 12.6 Six factors of the social interaction of a team.

1. Communication: frequency, formalization, structure, and openness.

2. Cohesion: degree to which members desire to remain in the team—interpersonal attraction, commitment, team pride.

3. Work norms: shared work expectations.

4. Mutual support: cooperation and assistance.

5. Coordination of tasks: timeliness and harmonization of tasks.

6. Balance of member contributions and conflicts: resolution and fairness.

They also use trained facilitators when needed to help them with important creative tasks [Leigh Thompson, 2003].

Geese heading south for the winter fly in a **V** formation. As each bird flaps its wings, it creates uplift for the bird immediately following it. By flying in a **V** formation, the whole flock can fly at least 71 percent farther than if each bird flew on its own. People who share a common direction can get where they are going more quickly and easily if they cooperate.

Whenever a goose falls out of formation, it feels the resistance of trying to go it alone and quickly gets back into formation to take advantage of flying with the flock. Teams will work with others who are going the same way. When the lead goose gets tired, it rotates back in the formation, and another goose flies on the point. It pays to take turns doing hard jobs for the team. Perhaps the geese honking from behind are even the cheering squad to encourage those up front to keep up their speed.

Source: Association for Quality and Participation, Louisville Kentucky Chapter, September 1992. Original source unknown.

12.5 MANAGEMENT

Management is a set of processes such as planning, budgeting, organizing, staffing, and controlling that keep an organization running well. Managers are concerned with the allocation of resources and may be particularly focused on routine tasks. The management of a new venture firm will work to accomplish all the tasks required to keep the company operating. Management of a new firm is important and should not be undervalued compared to entrepreneurial leadership—both are valuable. The goal of the manager is make a business carry out activities efficiently and on time. Managers work hard on focused goals in order to implement the strategy of the firm. They make deliberate choices about resources. One reason that purposeful managers are so effective is that they are adept at husbanding resources. Aware of the value of time, they manage it carefully.

Managers use their personal contacts when they need information or help. These informal networks create part of the social capital of an organization. People who link and connect people through a business network can be valuable managers who cross boundaries, help build subnetworks, and make the organization work [Cross and Prusack, 2002].

Managers are good at pattern recognition—making generalizations out of inadequate facts. Good managers are also prepared, frugal, and honest. Getting employees to stick to important strategic initiatives—and to give those initiatives their undivided attention over time—is crucial to competing successfully today. Great organizations have outstanding strategies and capabilities as well as the mechanisms necessary for executing the firm's strategies. The follow-through with operations and processes is part of a firm's competitive advantage. Managers focus on performance, feedback, and decision-making. Managers are also synthesizers who bring resources together in a timely manner. Given the rate of change of the competitive environment, they can make a great contribution by managing resources on a rolling time scale of 12 to 18 months.

Managers set expectations by describing the desired outcome, not the path. Talented people will determine the best path once they know the desired result. Most good managers motivate their people by offering them the opportunity for achievement, recognition for achievement, the work itself, responsibility, and growth or advancement [Harzberg, 2003].

Good managers engage people in decisions that directly affect them, explain why decisions are made the way they are, and clarify what will be expected of them after changes are made. Managers also provide feedback to team members regarding their performance. Organizations profit when employees ask for feedback and deal well with critical criticism. Openness to feedback is critical for all staff in a new venture facing many transitions. As people begin to ask how they are doing relative to management's priorities, their work becomes better aligned with organizational goals [Jackman and Strober, 2003].

All successful managers excel in the making, honoring, and remaking of commitments. A commitment is any action taken in the present that binds an organization to a future course of action. Managerial commitments take many forms, from capital investments to personnel decisions to public statements, but each exerts enduring influence on a company. A commitment may impede a response to changing conditions. Managers can learn to recognize when commitments have become roadblocks to needed changes. The managers can then replace those roadblocks with new, rejuvenating commitments. The ability to make and rewrite commitments is an important managerial skill [Sull, 2003].

Management of a new venture is complex. Managers balance five perspectives: reflection, analysis, contextual dynamics, relationships, and change [Gosling and Mintzberg, 2003]. Reflection can lead to understanding. Analysis can result in better organizational arrangements and performance. Understanding of context (or environment) can lead to regional and global activities. Collaboration with other firms and sound industry relationships are important. Also, a bias toward action and beneficial change is another valuable perspective for a manager.

The effective manager weaves together all five perspectives into a fabric for growth and performance.

12.6 EMOTIONAL INTELLIGENCE

The emotional task of the leader is primal; that is, it is both the original and the most important act of the leader [Goleman, Boyatzis, and McKee, 2002]. The leader is a venture's emotional guide. **Emotional intelligence** (EI) is a bundle of four psychological capabilities that leaders exhibit: self-awareness, self-management, social awareness, and relationship management, as described in Table 12.7. Self-awareness refers to the ability to understand one's moods, emotions, and motivations, as well as their effect on others. Self-management is the ability to control emotions, as well as exhibit optimism and adaptability. Social awareness is the empathetic sensing of other people's emotions and awareness of the social currents within an organization. Relationship management includes inspiration, influence, conflict management, and teamwork.

According to Goleman and colleagues, leaders and managers who possess these capabilities have a high degree of EI and tend to be more effective. Their self-awareness and self-management help to elicit the trust and confidence of their colleagues. Strong social awareness and relationship management skills can help to earn the loyalty of their colleagues. Empathetic and socially adept persons tend to be skilled at managing disputes between people, better able to find common ground and purpose among diverse constituencies, and more likely to move them in a desired direction than leaders who lack these qualities.

People with high emotional intelligence tend to: 1) behave authentically, 2) think optimistically, 3) express emotions effectively, and 4) respond flexibly in their relationship styles.

Leaders and managers who are in touch with their colleagues are said to be in resonance, which is the reinforcement of emotion. People are in resonance when they are "in synch" or on "the same wavelength." Resonant leaders use their EI skills to spread their enthusiasm and resolve conflicts [Goleman, Boyatzis, and McKee, 2001]. Good teams work to establish group resonance by building emotional and social awareness and management.

TABLE 12.7 Four elements of emotional intelligence.

- Self-awareness of one's emotions, emotional strengths and weaknesses, and self-confidence. The ability to read one's own emotions.

- Self-management of honesty, flexibility, initiative, optimism, and emotional self-control. The ability to control one's emotions and act with honesty and integrity.

- Social awareness of empathy and organizational currents, and recognition of the needs of followers and clients.

- Relationship management achieved by communicating through inspiration, influence, catalyst actions, conflict management, and collaboration.

12.7 ORGANIZATIONAL CULTURE

Organizational culture is the bundle of values, norms, and rituals that are shared by people in an organization and govern the way they interact with each other and with other stakeholders. An organization's culture can have a powerful influence on how people in an organization think and act. **Organizational values** are beliefs and ideas about what goals should be pursued and what behavior standards should be used to achieve these goals. Values include entrepreneurship, creativity, honesty, and openness.

Organizational norms are guidelines and expectations that impose appropriate kinds of behavior for members of the organization. Norms (informal rules) include how employees treat each other, flexibility of work hours, dress codes, and use of various means of communication such as e-mail. **Organizational rituals** are rites, ceremonies, and observances that serve to bind together members of the organization. Examples of rites are weekly gatherings, picnics, awards dinners, and promotion recognition.

In innovative firms, the values and beliefs favor collaboration, creativity, and risk-taking [Jassawalla and Sashittal, 2002]. These firms employ stories and rituals to reinforce these values and beliefs.

Intel, which has become the semiconductor standard for personal computers over the last 25 years, uses three principles to keep its entrepreneurial spirit alive. First, leaders must be willing to solve complex problems. There is no substitute for deep domain knowledge in technology industries, and at Intel, new products are continually being released. Second, the most effective managers are not afraid to change the rules. The "Intel Inside" branding campaign was a highly unusual marketing idea for a company with little previous direct contact with consumers. Lastly, Intel rarely tries to convince anyone to take a job or assignment. Managers must display a raw enthusiasm to try something new and a clear passion of their own [Barrett, 2003].

Culture is expressed in community. Communities are built on shared values, interests, and patterns of social interaction. The fit of the company culture in the business environment is critical to the company's competitive advantage. Most entrepreneurial start-ups have the founders and employees bound together by strong values and norms that include long hours working closely together. The sense of solidarity of a new venture is very high. The goals are uniformly shared—survival and early success being foremost. Start-ups are often founded by friends or former colleagues and exhibit high sociability within the organization. Employees possess a high sense of organizational identity and membership. In the early days, employees of Apple thought of themselves as "Apple people." Entrepreneurial firms often sponsor social events that take on ritual significance.

Hewlett-Packard Company fostered a culture known as the "HP Way" [Packard, 1995]. This culture was captured in a statement of objectives, values, and norms regarding fairness and justice. Employees were promised opportunity for security, job satisfaction, and sharing in profit. For example, during the 1981 recession, rather than lay people off, Hewlett-Packard introduced a 10 percent cut in pay and hours across every rank.

T. J. Rodgers, founder of Cypress Semiconductor, explained his view of a favorable workplace of a technology venture [Malone 2002]:

> The goal of starting the company was that I wanted to make a comfortable living. But the primary goal was to control my own environment. I wasn't happy working for other people. I wasn't happy working on projects I didn't find interesting, working with people I didn't enjoy spending time with. So it was an environmental control system I was trying to create, where I'd enjoy going to work everyday, and enjoy the people I was working with, and enjoy the projects I was working on, and make a decent living.

Entrepreneurial firms usually demonstrate high sociability and solidarity in their early years, built around the leadership of the founders—such as Hewlett and Packard. As the founders leave and other challenges come to bear, it may be difficult to retain the communal culture of a start-up. Yahoo built a fast-growing firm based on a set of founders who worked furiously at their jobs. Their communal culture helped them build the firm, but their insularity may have made them see the marketplace through a Yahoo lens in 2000 [Mangalindan and Hwang, 2001]. Starbucks has succeeded because it offers a consistently good customer experience. The culture of Starbucks is maintained at all its sites and keeps the consistency that customers value.

Perhaps the strongest element of an organizational culture is trust [Handy, 1999b]. **Trust** is a firm belief in the reliability or truth of a person or an organization. It is critical that we can trust those with whom we work. Handy provides seven principles of trust, as listed in Table 12.8. Teams are formed with many people we already trust, but new members will need to earn our trust. By *trust,* organizations mean confidence in a person's work, competence, and commitment to the organization's goals and tasks. Every team member must be capable of self-renewal, learning, and adaptability. When trust is broken, the person needs to be reassigned, leave the organization, or have his or her boundaries constrained. Teams need to build bonds among their members to enable trust. People need to meet in person to restore the group bonds. Finally, organizations, like people, need to continually demonstrate that they are trustworthy.

Good communication is part of any sound and trusting community. Knowing when to speak up and when to keep silent is an important skill for any worker. A new enterprise needs openness and creativity to grow and become fruitful. Carefully choosing the right issues to raise and avoiding ill-chosen conflict can be important. When emotions are high and the new enterprise is challenged with obstacles, it may be wise to keep silent. Nevertheless, managers of any enterprise must welcome ideas and comments. Organizations of all sizes and

TABLE 12.8 Seven principles of trust.

1. Trust is not blind. It is unwise to trust people whom you do not know well, whom you have not observed in action over time, and who are not committed to the same goals.

2. Trust needs boundaries. It is wise to trust people in some areas of life but not necessarily in all.

3. Trust requires constant learning. Every individual of a team must be capable of self-renewal and learning.

4. Trust is tough. When trust proves to be misplaced because they do not live up to expectations or cannot be relied on to do what is needed, then those people must go, be reassigned, or have their boundaries severely curtailed.

5. Trust needs bonding. Teams of people need to build their own bonds.

6. Trust needs touch. Personal contact is necessary, and teams need to meet in person to renew their trust and bonds.

7. Trust has to be earned. Organizations that expect their people to trust them must continually demonstrate that they are trustworthy.

Source: Adapted from Charles Handy, *The Hungry Spirit* (New York: Broadway Books, 1999).

arrangements must strive to keep the ideas flowing. Breaking the silence can bring an outpouring of fresh ideas from all levels of an organization—ideas that might just raise the organization's performance to a whole new level [Perlow and Williams, 2003].

In a study of successful companies, the building of a performance-based culture was central [Joyce and Roberson, 2003]. The study showed that winning companies built a culture on the four principles described in Table 12.9. Examples of successful firms that have a performance-based culture include Intel, Cisco Systems, and General Electric. Just about any high-potential firm needs to focus on performance, define it, and build a culture that reinforces it.

Cisco Community Statement

Cisco's culture requires that all employees, at every level of the organization, are committed to responsible business practices. Additionally, our business strategy incorporates our dedication to corporate citizenship, which includes our commitment to improving the global community in which we operate, empowering our workforce, and building trust in our company as a whole.

Cisco was founded on, and still thrives today in, a culture based on the principles of open communication, empowerment, trust, and integrity. These values remain at the forefront of our business decisions. We express these values through ethical workplace practices; philanthropic, community, and social initiatives; and the quality of our people.

Source: *Cisco Systems Annual Report*, 2003.

TABLE 12.9 Four principles of a performance-based culture.

1. Inspire everyone to do their best.

2. Reward achievement with praise and pay-for-performance, and keep raising the performance goals.

3. Create a work environment that is challenging, rewarding, and fun.

4. Establish, communicate, and stick to clear values.

Source: Joyce and Roberson, 2003.

12.8 SOCIAL CAPITAL

Social capital consists of the accumulation of active connections among people in a network [Cohen and Prusak, 2001]. Social capital was considered in Chapters 1 and 6. Social capital refers to the resources available in and through personal and organizational networks [Baker, 2000]. These resources include information, concepts, trust, financial capital, collaboration, social structure, and emotional support. These resources reside in networks of relationships. Social capital depends on whom you know. Social capital, like financial capital, can lead to increased productivity, if used wisely. A firm can build up or deplete its social capital by its actions. Relationship networks are often as important an asset of a venture as technology, land, or capital.

Social capital may be depleted by declining trust among a firm's people and the effects of people working off-site or on their own [Prusak and Cohen, 2001]. Social capital tends to be self-reinforcing and cumulative. Successful collaboration in one endeavor builds connections and trust—social assets that facilitate future collaboration in other, unrelated tasks. Firms that build up social capital demonstrate a commitment to retaining people and promoting from within. They also enable far-flung teams to meet in person periodically. They give people a common sense of purpose and keep their promises to people. Employees need to hear the same messages that an organization sends out to vendors and customers.

Social capital can be described as consisting of three dimensions: 1) structural, 2) relational, and 3) cognitive. The structural dimension concerns the overall pattern of relationships found in organizations. The relational dimension of social capital concerns the nature of the connections between individuals in an organization. The cognitive dimension concerns the extent to which employees within a social network share a common perspective or understanding [Bolino, Turnley, and Bloodgood, 2002]. Social capital is valuable because it facilitates coordination, reduces transaction costs, and enables the flow of information between and among individuals. In other words, it improves the coordinated effort and organization.

Better knowledge sharing can lead to increased trust and better decisions. Teamwork can lead to inventiveness, creative collaboration, and a good spirit. Trust is the fuel of a social capital engine that in turn engenders more trust. When people in an organization say their firm is "political," they often mean that trust is low throughout the organization. An organization with strong capital is a community with shared values and good trust. Conversations help to bind communities and build social capital.

Stories have the power to build and support social capital since stories convey meanings, norms, and values that define social groups. Storytelling is an important leadership skill since it helps to build organizations and galvanize and engage people.

Capitalism is based on a residue of promises fulfilled. Trust is important to a new firm, and it is built on promises kept. A healthy dose of doubt is prudent as trust is continually tested on the job. Doubt can help keep organizations out of harm's way. Some healthy doubt and internal questioning might have averted the tragic fall of Enron [Kramer, 2002].

Louis Gerstner joined IBM in January 1993 to bring the firm back to its roots and success. In his first months, Gerstner created a set of principles for the firm that included [Gerstner, 2002]:

- At our core, we are a technology company with an overriding commitment to quality.
- We operate as an entrepreneurial organization with a minimum of bureaucracy.
- We think and act with a sense of urgency.
- Outstanding, dedicated people make it all happen, particularly when they work together as a team.

Gerstner also described the IBM culture as:

In the end, an organization is nothing more than the collective capacity of its people to create value.

By 2002, Gerstner left IBM a powerful technology company focused on its entrepreneurial principles.

12.9 ATTRACTING AND RETAINING TALENT

All firms know that attracting and retaining the best people is key to their future success. However, open competition for other companies' people is now an accepted fact. Leaders know that in entrepreneurial markets, fast-moving firms are competing for the best people. New ventures pursuing important opportunities can attract talented people. Typically, these people are found in the social networks of the founding entrepreneurial team members. New employees can expect to have a direct stake in the new enterprise and participate in an open, trustworthy team that will build the new business.

Attracting and retaining key employees depends on compensation, work design, training, and networks. Compensation systems include wages, incentives, and ownership options. Successful new enterprises look for people who can thrive in environments in which people trust each other and are willing to debate assumptions, share information, and express feelings. Although winning is important to them, the goal is to win as an enterprise, rather than as individuals. The goal is to get the talent to act as owners of the firm by engaging them in the setting of goals and objectives and then organizing to achieve them. Talented people in new ventures work hard and fast, trying to get to market with the hope of big payoffs. They are motivated by significant responsibility, participation, and the possibility of big financial gains. Their internal commitment is, in fact, aligned with the realities of their incentives, and the potential results can be motivating.

Keeping faithful, loyal employees and partners in a new venture can be based on six principles, as shown in Table 12.10 [Reichheld, 2001]. A simple reliance on financial rewards is insufficient. Good communication, trust, and treatment are essential to retaining talent and partners. High standards of decency, consideration, and integrity are necessary for all members of the new venture. Through loyalty to ideals, the firm becomes worthy of loyalty from its people and partners.

Many, if not all, high-technology start-ups use an organizational design and culture that involve challenging work, peer group control, and selection based on specific task abilities. Few imperatives are more vital to the success of young technology companies than retaining key technical personnel, whose knowledge often represents the firm's most valuable asset [Baron and Hannan, 2002]. Therefore, the leaders of the enterprise need to hold to their commitments to their employees and retain their trust.

Perhaps the most important qualities of a new hire are proficiency in the skills required and demonstrated capability to acquire the attributes needed for future situations. Focusing on the raw ability to learn may be most critical for technology ventures [McCall, 1998].

To the extent possible, as a company grows, finding the best people should be accomplished by recruiting through existing team members and the firm's supply chain. To be sure about the hire, new ventures often use a trial period in a consulting role or temporary relationship on a project basis. The right team members are critical to a new firm's early success. Venture capitalist John Doerr

TABLE 12.10 Six principles for retaining loyal people and partners.

1. Preach what you practice. Communicate the vision, goals, and values of the organization. Practice what you preach is also required.

2. Partners must win also. Enable your vendors and partners to participate in a win-win venture.

3. Be selective in hiring. Select people with values consistent with the firm's. Membership on the team is selective.

4. Use teams of talented people. Use small teams for most tasks and give them the power to decide. Provide simple rules for decision-making so teams can act.

5. Provide high rewards for the right results. Reward long-term values and profitability. Provide solid compensation, benefits, and ownership.

6. Listen hard, talk straight. Use honest, two-way communication and build trust. Tell people how they are doing and where they stand.

Source: Adapted from Reichheld, 2001.

of Kleiner Perkins says that when he looks at the business plan, "I always turn to the biographies of the team first. For me, it's team, team, team. Others might say, people, people, people—but I'm interested in the team as a whole." [Fast Company 1997]

The growth and survival of an emerging business depends on star players and good supporting talent [DeLong and Vijayaraghavan, 2003]. Star employees can make important contributions to performance. Yet a firm's future also depends on the capable, steady performers. These steady performers bring stability and depth to an organization's resilience. Furthermore, they are more likely to be loyal to the organization.

12.10 OWNERSHIP AND STOCK OPTIONS

New ventures are able to offer stock ownership to their people. They also need to provide reasonable compensation and benefits. People in a start-up may have to sacrifice financial compensation for an initial period while building the firm toward financial break-even. Thus, it is important to offer reasonable benefits and ownership opportunities. Offering health benefits may be necessary to attract people to the new venture. Most technology ventures offer health benefits to their employees before reaching break-even.

Ownership interest in the new firm will be of great interest to most new employees. **Stock options** are offered in a plan under which employees can purchase, at a later date, shares of the company at a fixed price (strike price). Stock options take on value once the market price of a company's stock exceeds the exercise (strike) price. Stock options should vest with the recipient over a period of years. For example, a new hire may receive an option for 10,000 shares exercisable at $10 per share vesting over four years. This is equivalent to 2,500 shares each year. Stock options give employees the right to buy the company's stock in the future at a preset price, thus motivating them to work to increase productivity and innovation—and eventually, the market value of the firm.

The purpose of employee stock options is to create a noncash substitute for part of the wage compensation the firm must provide to attract and retain employees. A new, entrepreneurial firm may not be able to provide the cash compensation needed to attract outstanding workers. Instead, it can offer stock options. From the beginning, Starbucks decided to grant stock options to every employee in proportion to their level of base pay: it called these options "bean stock." Microsoft created thousands of millionaires over its first 20 years due to the issuance of wide stock options.

Stock Options Basics

■ Options give employees the right to buy a certain number of their company's shares at a fixed price (strike price) for a certain period of time.

■ The strike price is usually the market price of the stock on the date the options are granted.

- Options usually begin vesting after one year and fully vest after four years. If an employee leaves the company before his or her options fully vest, the remaining options are canceled.

- Once an option is vested, the employee can then exercise it, that is, purchase from the company the allotted number of shares at the strike price, and then either hold the stock or sell it if the company is public.

- The difference between the strike price and the market price of the shares at the time the option is exercised is the employee's gain in the value of the shares.

- When an employee exercises an option, the company must issue new shares of stock.

An alternative to stock options is **restricted stock,** which is stock issued in an employee's name and reserved for his or her purchase at a specified price after a period of time—say, one, two, or three years. Some call this type of stock "reserved stock." New firms need to increase the level of share ownership through such means as offering restricted shares and requiring that employees hold their shares for certain periods of time. For example, an employee could have 10,000 shares reserved to be purchased at $2 per share as they vest in two years. If the price of the stock appreciates to $10, the gain in the stock's value is $80,000. Restricted stock still has value if the stock price falls, while options expire as worthless if the stock does not appreciate. In 2003, Microsoft switched from offering stock options to restricted stock.

12.11 BOARD OF DIRECTORS

An incorporated firm or LLC will have a board of directors. A **board of directors** is a group composed of key officers of a corporation and outside members responsible for the general oversight of the affairs of the entity. This board will normally consist of the founders of the firm and one or more investor-owners. A new start-up might find a board of three owners is adequate. As other investors are added, one or more additional owners may be added to the board. The board is the overseer of the corporation with responsibility to select and approve the appointment of the chief executive officer (CEO) and other officers of the corporation. Directors should possess significant knowledge and competencies in the industry of the company. The board of directors is a legally constituted group whose responsibility is to represent the stockholders. It is usually made up in part by the founders, but it should include one or more outside shareholders. A board of five might consist of two insider executives, two representatives of investors, and one independent director. This board must approve the bylaws, officers, and annual report to the shareholders, as well as any financial offerings to investors or banking activities. The members of this board have fiduciary responsibility, meaning that they are under a legal duty to act in the best interests

TABLE 12.11 Five goals for an effective board process.

1. Engage in constructive conflict—especially with the CEO.

2. Avoid destructive conflict.

3. Work together as a team.

4. Work at the appropriate level of strategic involvement—avoid micromanagement.

5. Address decisions comprehensively.

Source: Finkelstein and Mooney, 2003.

of the corporation and its stockholders. Directors need to be knowledgeable, interested, and shareholder-oriented.

The **board of advisors,** if any, is constituted to provide the firm with advice and contacts. The members have extensive skills and knowledge and provide good advice. The board of advisors is nonfiduciary and does not engage in the legal or official actions of the corporation. Thus, the advisors are free from liability as long as they refrain from any legal or official role.

The board of directors should consist of people who are available for the necessary official meetings and have the skills to make the appropriate decisions. The board of advisors should consist of persons of good reputation who can give help and advice from time to time. Good boards are those that function and work well. They are distinguished by a climate of trust, respect, and candor. Members feel free to challenge each other's assumptions. They should feel a responsibility to contribute significantly to the board's performance. In addition, good boards assess their own performance, both collectively and individually [Sonnenfeld, 2002]. Table 12.11 lists five goals for an effective board process [Finkelstein and Mooney, 2003].

In selecting directors, a premium should be placed on a wide range of expertise and backgrounds but, above all, on people who will seek to expose the downsides as well as the upsides of every major decision. Savvy start-ups look for directors who are fluent in one or more of the following: audit and finance, strategy, marketing, and sales. Compensation of directors will normally be in the form of stock. The members of the board should, after a few years, have a reasonably substantial stake in the firm. This can be achieved with the use of stock options or restricted stock that vests over a period of time.

Conservation International (CI) is a nonprofit corporation whose mission is to conserve the Earth's living natural heritage and biodiversity. CI applies innovations in science and economics to protect the world's natural capital. CI's board of directors is well known (www.conservation.org). The chair of the board is Gordon Moore, co-founder of Intel. Others on the board include actor Harrison Ford, Queen Noor of Jordan, and Orin Smith, CEO of Starbucks.

12.12 AGRAQUEST

The initial leader of AgraQuest was Pamela Marrone, who left Novo Nordisk's Entotech to found AgraQuest as a biotechnology start-up in 1995. She recruited three other experienced biotech leaders to join her team, which then drafted a business plan. They spent from January 1995 to March 1997 rewriting the plan and attempting to raise several million dollars in venture capital. During that 18-month period, all four members of the lead team deferred compensation and agreed to accept stock options as an alternative. One of the original team members could not financially accommodate the 18-month deferment and left for another job. Eventually, they were funded by a venture capital firm in mid-1997. At that time, three scientists from Entotech joined AgraQuest and remain there today.

AgraQuest is a firm with a strong science and technology bias and culture. Most of the staff were attracted by the science and the potential for biopesticides. The organization structure is nonhierarchical, and most employees enjoy an opportunity to provide their own views and input on decisions. By 2001, the firm grew to 70 people, with most acting in a scientific or technical support role.

Marrone uses primarily a directing leadership style, with a secondary style of coaching (see Figure 12.2). From the beginning, stock options have been awarded to all employees based on their salary level and when, in terms of risk, they joined the firm.

Since 1997, AgraQuest's board of directors has consisted of Marrone and representatives of the investors (venture capitalists). No independent directors have been appointed. A scientific advisory board, consisting of professors at prestigious universities, has served AgraQuest since 1996.

Marrone and the director of research and new business development have built a strong network of scientists and technologists that gives AgraQuest access to help and information in the agricultural biotechnology industry.

AgraQuest has good leaders who have created a scientific organization with a collaborative culture and a compensation scheme based on options (whose value has declined). The firm has been unable to achieve profitability, despite a decade of hard work.

12.13 SUMMARY

Early in the development of a firm, a leadership team is created to build a business plan and organizational plan. The organizational plan is structured to align the culture of the firm with its goals and values. The firm's leaders strive to motivate and inspire team members and foster a collaborative, innovative culture. As the firm grows, managers join to build the structure and carry out the detailed tasks of the new firm. As the firm develops, the leaders strive to exhibit emotional intelligence. The firm also puts together a compensation scheme that emphasizes "buy-in" or ownership—normally achieved by awarding stock options or restricted stock. Finally, a firm creates a board of directors and a board of advisors to help monitor and enhance the growing firm.

> **Principle 12**
> Effective leaders coupled with a good organizational plan, a collaborative performance-based culture, and a sound compensation scheme can help align every participant with the goals and objectives of the new firm.

12.14 EXERCISES

12.1   Genentech has a unique culture known for rigorous science, guarding of industry secrets, and rigid rules. Its key principle is: good scientists make for good science make for good products make for a good company. Describe Genentech's culture in terms of norms and rituals. See www.gene.com.

12.2   Ben & Jerry's was dedicated to social values and, from the beginning, had a key policy that the ratio of the highest-paid individual to the lowest-paid should not exceed 5 to 1. By 1990, it had sales of $77 million and needed to recruit managers. What would you recommend to its board of directors regarding the firm's salary policy? See www.benjerry.com.

12.3   Your firm is seeking a new CEO and is interested in a person who just left a firm that is now under investigation by securities regulators. Is it wise to continue considering this person?

12.4   Matthew Smith knows his talented people are the reason for the success of ElectroMag. Smith has noticed that the costs of medical and dental benefits are escalating, and he needs to control them. With 800 employees, the firm's profitability is threatened. He has three options: 1) eliminate health benefits, 2) try to find a cheaper plan that covers fewer medical and dental procedures, or 3) withhold a fixed amount of each person's salary to use to fund the benefits [Gossage, 2003]. Which would you recommend he choose?

12.5   REI is a cooperative of members offering recreational clothing and equipment (see www.rei.com). REI attracts outdoor-oriented employees who fit into its culture. Describe how the REI culture is affected by the co-op form of organization.

12.6   Take-Two Interactive develops entertainment software games. Revenues grew from $19 million in 1997 to $970 million in 2003. Describe the organizational design for Take-Two, which has about 750 employees. Use Figure 12.1 to help determine the organizational model. See www.take2games.com.

12.7   A new stock incentive plan at Microsoft was introduced in late 2003 [Langley, 2003]. The stock award plan is based on performance. Study the new plan and describe its benefits for the employees who participate.

12.8   Getty Images, founded in 1995, offers still and moving images distributed via its website (www.gettyimages.com). Examine its 2003 annual report and determine how the firm motivates its employees. Describe its employee stock plan and its organizational design using Table 12.2.

12.9   Red Hat is the leading distributor of Linux software and services. The firm has about 600 employees and revenues of $145 million. Using the concepts of Section 12.7, describe the firm's organizational culture.

12.10   A new firm, 24/7 Customer, provides services to U.S. firms that wish to outsource call centers and CRM activities to India [Vogelstein, 2003b]. Using the information provided at the firm's website, describe the founding team of this firm. Also, describe its organizational design (www.247customer.com).

CHAPTER **13**

Acquiring, Organizing, and Managing Resources

To get profit without risk, experience without danger, and reward without work,
is as impossible as it is to live without being born.
A. P. Gouthev

Entrepreneurs need to acquire capital, people, intellectual property, and physical assets to successfully launch and operate their venture. To tap the required resources, they need to build credibility and legitimacy in the marketplace of resources and talent. Influence and persuasion can help entrepreneurs build their case for securing scarce resources for their venture.

Choosing a physical location and operating as a virtual organization are viable options for a firm today. Using the Internet and teleconferencing, new organizations can build a powerful venture. All firms need to create a plan for outsourcing functions while maintaining critical functions within the firm. As firms strive to be innovative and competitive, they seek to control costs by outsourcing functions to those who can do them better and cheaper. However, these firms are challenged to retain the cohesion and coordination required to effectively manage these supplier partners. ∎

13.1 ACQUIRING RESOURCES AND CAPABILITIES

Another definition of entrepreneurship is the pursuit of opportunity without regard to resources currently controlled [Stevenson, et al., 1999]. This view stresses the idea that the entrepreneur can locate and access resources when they are needed. For example, when an entrepreneurial team needs legal counsel, they can engage a lawyer. When they need a circuit designer, they can hire one. In fact, resources are usually scarce, and the attraction of talented employees or financial investors is not easy or guaranteed. A firm's competitive advantage flows from the combination of resources and capabilities executing a unique strategy. If these resources and capabilities are scarce, then the new venture needs to compete to secure them. This task is one of the most critical for any new venture.

The founders of a new venture attempt to acquire resources and capabilities by contacting key organizations and people and asking them to support their venture. For example, they ask their bank, suppliers, and sources of financial capital to take some risk and support the new venture. This resource-seeking activity can be represented by the cycle shown in Figure 13.1 [Birley, 2002]. The founders are asking all the participants in the credibility cycle to believe in their opportunity, vision, and story, and invest in their venture. To move forward, entrepreneurs need to persuade someone in the cycle to believe in them. If the entrepreneurs get some talented people to commit to the venture, this will help convince the suppliers. If the entrepreneurs get some customers to tentatively commit to purchasing the product, the sources of financial capital (bankers and investors) will become more interested. The entrepreneurs travel around the cycle, slowly building their credibility. In other words, the entrepreneurs demonstrate the legitimacy or truthfulness of the new venture to the members of the credibility cycle [Zimmerman and Zeitz, 2002]. Legitimacy or credibility is evidence of a social judgment of desirability and enables a new venture to access resources. The holders of scarce resources provide resources to new ventures

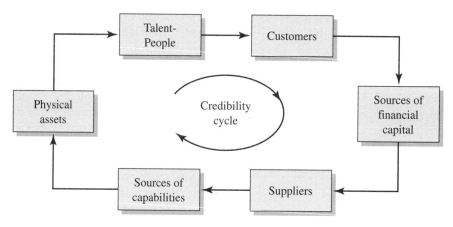

FIGURE 13.1 Credibility cycle.

TABLE 13.1 Sources of legitimacy.

■ Regulatory: legal actions, accreditation, credentials.	■ Talent: known, respected people.
■ Social: fair treatment, endorsements, networks, image.	■ Location: within an industry cluster, favorable location, visible.
■ Industry: attractive, respected industry, known and understood business model.	■ Intellectual property: trade secrets, patents, copyrights.

only if they believe that the ventures are efficient, worthy, and needed, and the teams are competent. The greater the level of a new venture's legitimacy, the more resources it can access.

A new venture can build its legitimacy by tapping the sources of legitimacy listed in Table 13.1. A new venture can join industry associations, secure endorsements, and get commitments from talented, respected individuals. Patents, copyrights, and trade secrets also help to build legitimacy. New ventures should focus on actions that have the greatest payoff for legitimacy. A certain amount of legitimacy is required to make the credibility cycle build up the investment of resources. A successful new venture needs to acquire, build, and use legitimacy to secure the necessary resources to commence operations and grow successfully. Creating, building, and retaining a firm's credibility or legitimacy is a critical task for the leadership team of any business. They determine the key influencers in your industry and reach out to them about the opportunity. The leadership team of a new venture recognizes and creates new economic or social opportunities and makes decisions on the location, form, and use of resources [Wennekers, Uhlaner, and Thurik, 2002].

Entrepreneurs who exhibit higher social competencies and emotional intelligence greatly improve their ability to access resources. Entrepreneurs possessing a high level of social capital (e.g., a favorable reputation and an extensive social network) gain access to persons important for their success. Once such access is attained, their social competence influences the outcomes they experience [Baron and Markman, 2003].

Jeff Hawkins founded Palm Computing in 1992 to build a handheld computing device. Hawkins was an experienced technologist, but he possessed little business experience. He had a patent on a handwriting algorithm, however, and an outstanding reputation in the field of software design, human intelligence, and technology development. With a limited business plan and no prototype, Hawkins was able to raise nearly $2 million from two venture capitalists. He then recruited Donna Dubinsky, an experienced executive, as president and CEO, while he served as chief technology officer and chairman [Brush, Greene, and Hart, 2001].

13.2 INFLUENCE AND PERSUASION

Influence and persuasion play a role in the entrepreneur's acquisition of resources. They are part of the process of selling, acquiring resources, and structuring deals regarding acquisitions and investments. Every entrepreneurial team ideally needs a person who is a master of persuasion [Cialdini, 1993]. Persuasion skills exert greater influence over other's behavior than formal power relationships do.

Persuasion is governed by basic principles that can be taught, learned, and applied [Cialdini, 2001]. The six principles of persuasion are provided in Table 13.2. The principle of *liking* states that people like to please or work with those who sincerely like them also. One can uncover shared interests and bonds and offer sincere praise and compliments.

The second principle of *reciprocity* states that one can elicit the desired behavior from others by displaying it first. Offering help or information to others first can encourage them to reciprocate. The principle of *social proof* states that people look for and respond to a display of endorsements from those they trust.

The fourth principle of *consistency* states that people stick to their verified commitments—those they make as voluntary, public statements. People who

TABLE 13.2 Principles of persuasion.

1. Liking

- People like those who like them.
- Uncover shared bonds and offer sincere praise and compliments.

2. Reciprocity

- People respond in kind to others.
- Give to others what you want to receive.

3. Social proof

- People respond to a display of endorsements by people they trust.
- Use testimonials and endorsements from trusted leaders.

4. Consistency

- People adhere to their verified commitments.
- Ask for voluntary, public commitments.

5. Authority

- People highly regard experts.
- Show and state your expertise.

6. Scarcity

- People want scarce products.
- Describe unique benefits.

Source: Cialdini, 2001.

make verbal public or written commitments are likely to stay with them. The principle of *authority* states that people highly regard experts. Therefore, it is useful to show and display your firm's expertise and competencies.

Finally, the principle of *scarcity* states that people want scarce or unique products. Therefore, it is necessary to explain and demonstrate the unique benefits and the chance to gain exclusive advantages.

The person who is attempting to persuade others needs to combine all the principles into a powerful series of conversations and interactions. William Bratton, police chief of Los Angeles and formerly of New York City, has displayed a strong set of persuasive skills. Bratton has instituted change by using the six principles to mobilize the commitment of the key people to support his efforts [Kim and Mauborgne, 2003].

13.3 LOCATION

Choosing a location for a business can have long-lasting effects on a new venture. As discussed in section 8.5, location can provide significant benefits to a technology startup. Entrepreneurs need to choose their location with their customers and future employees in mind. The quality of infrastructure, transportation, and lifestyle can be important factors. Location is critical to start-ups in the retail and restaurant business. Furthermore, a knowledge-based business may need to locate in a cluster where employees and complementors are readily available (see Section 8.5). Some new firms may choose to locate where the founders reside. Criteria for location selection are listed in Table 13.3. The cost of doing business will be an important factor in location selection. Also, housing that is affordable to future employees is important. A company's home should be a place that current and future employees will like. Thus, it should have good schools, a feeling of safety, and good transportation links. It is important for companies to be physically near some key constituents. These include competitors and other companies in related fields, as well as the venture capital firms that back them.

New ventures that rely on innovation need to identify a location that will enable them to compete effectively. Innovation takes place in clusters, which are geographic concentrations of interconnected companies and an existing infrastructure system. The characteristics of a cluster that support innovation are provided in Table 13.4 [Porter and Stern, 2001].

TABLE 13.3 Criteria for location selection.

- Availability of potential employees and consultants.
- Availability of complementary firms.
- Road and airplane transportation.
- Quality of life—education, culture, recreation.
- Costs of doing business.
- Availability of suitable facilities.
- Proximity to markets.
- Availability of support services.
- Affordable housing.

TABLE 13.4 Characteristics for innovation in a cluster setting.

■ High-quality human resources.	■ Suppliers and complementors.
■ Research in local universities.	■ Competitors.
■ Availability of investment capital.	■ Consultants, attorneys, and accountants.
■ Representative customers.	

Locational advantages are based on the flow of knowledge, relationships, and access to institutions such as universities. Commonly known ideas and technologies that can be accessed from anywhere will be widely available to all competitors. Thus, they cannot serve as a competitive advantage. Local advantages such as access to knowledge and research at a university therefore can become a unique advantage. Local companies that supply required components and technologies could also be very advantageous.

Companies should attempt to proactively link into strong local institutions such as trade associations, universities, and professional societies. Entrepreneurs should take advantage of opportunities in their local region. Long-term competitive advantage relies on being able to avoid imitation by competitors. Location-based advantages in innovation may prove more sustainable than implementing corporate best practices [Porter and Stern, 2001].

For most new ventures, leasing space in an existing facility is the most economic choice. For some more established firms, owning rather than renting can pay off. However, most start-ups need to use their financial resources for innovation and marketing.

Boeing, a mature, innovative firm, moved its headquarters from Seattle to Chicago in 2001. It sought a place that would minimize travel time throughout the country and internationally [Forethought, 2001]. It looked carefully at airline travel, the business climate, ground transportation, personal lifestyle factors, and available facilities. A new venture start-up will be well served by building a location and facility plan into its business plan. This will help to access all the potential advantages of a particular region.

13.4 FACILITY PLANNING

Once the entrepreneur chooses the city or region of location, the next task is to find a suitable facility. A building must fit the needs of the organization and allow for expansion as the firm grows. The next challenge is choosing a proper layout within the facility. A **layout** is the arrangement of the facility to provide a productive workplace. This can be accomplished by aligning the form of the space with its use or function.

Today's facilities have replaced the private office and laboratory with public spaces and open-plan areas without walls. Since innovation is, in part, a social or collaborative activity, the work spaces are laid out to host team activity. New ventures are best started in an open facility with few walls and doors in order to promote collaboration. Studies have shown that communication between people declines inversely proportional to the distance between them. We are four times more likely to communicate with someone who sits six feet away as we are with someone who sits 60 feet away. Thus, a firm needs to avoid separate facilities for as long as possible. The center of the facility might be a public area with a coffee bar and meeting tables. The goal is to design the facility for flexibility and collaboration.

13.5 TELECOMMUTING AND TELECONFERENCING

The idea that community no longer means location has become accepted by most people. By 2002, 55 million Americans maintained an office in their home. **Telework** refers to all kinds of remote work from home, satellite offices, and on the road. Technology has made it practical for people to work at places other than a central office. It has also made it possible to work asynchronously—not at the same time. Asynchronous access through e-mail, voice mail, and groupware enables groups to communicate with one another at any time.

It is important to reinforce the fact that social capital is based on face-to-face communication. Trust requires personal, face-to-face experiences. Videoconferencing and audioconferencing are useful technologies for virtual meetings. Virtual meetings work best, however, for the discussion of focused topics such as budgets, schedules, and facts. Presence is necessary for soft functions such as negotiations, planning, and restructuring of plans. People's social lives take place in a physical world, and the virtual world, may be best reserved for information and content transmission.

13.6 THE INTERNET

The **Internet** is a worldwide network of computers linking businesses, organizations, and individuals. **E-commerce** involves digitally enabled commercial transactions between and among organizations and individuals. The Internet enables worldwide communication. Computer networks link people, organizations, and knowledge, and primarily support social networks [Wellman, 2001]. Networks of workers communicate, in part, using technological means such as the telephone, fax, and Internet. Many workers participate in many networks where teams collaborate on a given task.

The World Wide Web is a popular service that runs on the Internet and provides access to pages and documents via a system of addresses, standards, and protocols. In many ways, browsers for the Web are the "killer apps" of the Internet, serving as a valuable communication and information channel. We use *Web* and *Internet* interchangeably to describe the Web application on the Internet. The Web is a communications infrastructure and information storage system. Because transaction costs can be much lower on the Internet than in traditional channels, companies are shifting some or all of their business and supplier functions onto the Web.

E-commerce was built using Internet technology starting in 1995. Between 1998 and 2000, venture capitalists invested $120 billion in about 12,500 Internet start-ups, often called "dot.coms." Some of the most visible dot.coms include Amazon.com, eBay, and Yahoo. Jeff Bezos founded Amazon in 1995 as an online bookstore, raising several million dollars for the venture. Amazon offers convenience, low prices, and a wide selection of books and other items. In May 1997, Amazon raised $50 million by selling its shares to the public for the first time. By 1999, Bezos's goal was to build an important channel of communication with Amazon's customers, while recognizing the value of other traditional channels. Another important use of the Internet is the distribution of information in the form of a magazine or a newsletter. The *Wall Street Journal* has successfully offered an online version in addition to its print edition.

The Internet also facilitates three important functions: personalization, customization, and versioning. **Personalization** is the provision of content specific to a user's preferences and interests. It uses software programs to find patterns in customer choices and to extrapolate from them. For example, Amazon provides personalized book and music recommendations. **Customization** is providing a product customized to a user's preferences. Dell Computer offers products customized to the preferences a user indicates online. **Versioning** is the creation of multiple versions of a products and selling these modified versions to different market segments at different prices. The *New York Times* offers a free online version of today's newspaper but charges per article for archived material.

While personalization and customization are good ideas, there are some concerns. Many users report that they do not notice any differences in the site due to personalization. Often attempts to customize are cumbersome and imperfect. Dell Computer, Orbitz, and Amazon appear to be good examples of successful personalization and customization. Customization is a more powerful approach if executed well since the customer is actively involved in the selection process. Customization is a powerful tool when customers want their preferences converted directly into a specific form of product.

A central activity of the Internet is the search process that enables users to find information on a seemingly infinite set of items, ideas, terms, and issues. For a price, a buyer of airline tickets can use Expedia, Travelocity, and Orbitz to search for a bargain. Travel services appear to be an ideal service/product for the Internet since travel is an information-intensive product requiring significant consumer research. Table 13.5 lists 15 exemplary Web sites. Wells Fargo is an online standout. It offers all of its banking transaction services online and has fully integrated its online, in-bank, and other services.

Not all customers want to do business online; most prefer having a choice of various ways. The **hybrid** model, sometimes called "bricks and clicks" or "clicks and mortar," utilizes the best of the Internet as well as other channels. A hybrid model can extend a company's reach to new market segments as well as its global reach. The hybrid model is exemplified by the combination of Sears and Land's End [Pottrack and Pearce, 2000].

TABLE 13.5 Exemplary websites.

Website	Address (www.)	Primary offering
Amazon	amazon.com	Books, videos, CDs
AOL	aol.com	Information and e-mail
Dell Computer	dell.com	Computers
Ebay	ebay.com	Auctions
Expedia	expedia.com	Travel
Google	google.com	Search
Land's End	landsend.com	Clothing
L. L. Bean	LLBean.com	Clothing
New York Times	nytimes.com	News
1-800-Flowers	800flowers.com	Flowers and gifts
Red Envelope	redenvelope.com	Gifts
Wall Street Journal	wsj.com	Business news
Wells Fargo	wellsfargo.com	Banking
Williams-Sonoma	Williams-Sonoma.com	Home furniture and housewares
Yahoo	yahoo.com	Search and information

Alliances are set up to combine the functions served by each company, such as the Toys-R-Us and Amazon alliance to make "the Earth's biggest store" offering toys, books, DVDs, music CDs, videos, and electronics. Another example is the alliance of Drugstore.com and Rite Aid.

The advantages of e-commerce are low transaction costs, ubiquity, wide reach, and massive information. Since product and price information are readily available on the Web, the pricing power of many industries has diminished. Furthermore, many early e-commerce ventures underpriced their products in order to secure customers but never showed a profit and eventually failed. A firm must have a competitive advantage to sustain itself, and very low prices may not permit profitability. Because of the wide reach of the Internet, many competitors can imitate successful offerings, thus eroding the competitive advantage of any one firm.

Eventually, all Web-based companies are forced to show profitability. In many cases, the Internet complements, rather than cannibalizes, a company's traditional activities [Porter, 2001]. Consider Land's End, now part of Sears, which has a print catalog, an online catalog, a phone inquiry system, and some of its products in Sears stores. The Land's End website provides information and the ability to order online or connect via telephone to a salesperson while viewing the catalog online.

Perhaps the best approach for a new venture is to consider the Internet as one channel in which its core competence can be applied. For example, Toys-R-Us selects and prices its merchandise while Amazon manages the website and the online sales function. This alliance allows each company to leverage each other's strengths.

All start-up companies should create a website. An early Internet presence can help build credibility and provide the look of an established firm. Most customers will visit a website to learn about the firm and its products. The website should clearly describe what the business does, explain the products and services, and provide complete contact information. Many new firms will also use their computer network to link with their suppliers and customers.

A **portal** is a place or gate of access to a company's offerings. Firms such as Yahoo and AOL offer their users search tools and access to a wide range of information. Other companies such as Hewlett-Packard (www.hp.com) offer a gateway to their range of product offerings. Most firms offer a portal rather than a set of separate websites. A new venture should consider building a portal.

The use of the Internet is widespread. It is estimated that about 58 percent of the U.S. population used the Internet in 2003. Similar growth of the use of Internet is occurring worldwide. The Internet is ubiquitous, cheap, and standardized, and it accommodates data, voice, video, and e-mail. It readily allows individuals to search, collaborate, coordinate, and transact. Thus, a new business venture can use the Internet to synchronize its activities through different channels, different stages of the value experience, and across different offerings [Sawhney and Zabin, 2001]. Perhaps the most revolutionary aspect of the Internet is that it gives virtually everybody access to the same information. It is this transparency that has caused the shift in power from sellers to buyers.

Concept

Technology development

Product design

Logistics and manufacturing

Marketing

Sales and distribution

Service

Customer

FIGURE 13.2
Value chain from concept to customer.

13.7 VERTICAL INTEGRATION AND OUTSOURCING

New ventures normally have limited financial resources and are unable to internally provide all the functions required for operation of all activities. One way to identify resources and activities that have the potential for creating competitive advantages for a firm is to consider the value chain. The **value chain** of a firm is a sequence of business activities for transforming inputs into outputs that customers value, as depicted in Figure 13.2. The issue for the new venture is to decide which of the activities on the value chain will be accomplished by the firm and which activities will be provided by other firms (outsourced). A new or emerging firm will necessarily focus on a few of the activities on the value chain and outsource the others. **Vertical integration** is the extent to which a firm owns or controls all the value chain activities of a business.

In this chapter, we discuss the value chain and the decision to outsource or retain an activity within in the firm. In Chapter 15, we return to the value chain and discuss how to manage and operate it to attain a competitive advantage.

TABLE 13.6 Questions for selecting value chain activities that will be carried out by the new venture firm.

1. **Value**	Is the activity a primary source of product value for the firm?
2. **Rarity**	Does the activity include a resource or capability controlled by the firm and rarely available to competing firms?
3. **Ability to be imitated**	Do competing firms have a cost disadvantage when imitating the scarce resource or capability held by the new venture firm?
4. **Organizational mission**	Is the activity critical to the mission of the firm, and is the firm organized to exploit this valuable, rare, and costly-to-imitate resource or capability?

The decision by a new venture to choose to carry out a value chain activity can be based on four questions, shown in Table 13.6 [Barney, 2002]. The decision to focus on a few value chain activities can be aided by an analysis of the four issues: 1) value, 2) rarity, 3) ability to be imitated, and 4) mission and organization. In many cases, it may be necessary to extend the activities of a firm to control a specific activity on the chain.

Many companies are operating in a hypercompetitive global market where there is overcapacity in most industries. In such an environment, they are being called upon to achieve profitability by relentless cost-cutting. This often entails heavy outsourcing to lower-cost labor and moving business abroad.

If a firm decides to outsource an activity, it plans on gaining a cost advantage or access to the suppliers' superior competency or economy of scale. Access to a supplier's superior competency and cost advantage may be favorable. For example, a new venture cannot offer its employees a superior, low-cost, internally operated cafeteria and food service. When it is cheaper and easier to conduct an activity internally, then new ventures may consider taking it on. The best reason, however, to carry out an activity on the value chain is that it is strategically critical to a firm's success. Typically, product design and marketing cannot be outsourced since they are critical to the success of most new technology ventures.

A new venture in the personal computer business may choose to control product design and marketing while outsourcing the other functions after answering the four questions of Table 13.6. On the other hand, a packaged food business would probably attempt to control all product design, manufacturing, and marketing functions while relying on other firms to provide the technology development, distribution, and service activities [Aaker, 2001].

As companies outsource more functions, the scope for competitive differentiation narrows. Almost all routine activities are of low value, not rare, easily imitated, and not central to the mission of a firm. Thus, most new ventures outsource routine services such as payroll, accounting, and other administrative services. The transaction costs of managing outsourced services have recently declined, thanks to cheaper communication and the standardization of Web-based tools. Managing a relationship with a strategically important outsourcing agent is far more complex than coping with an ordinary supplier. In many cases, a partnership arrangement with a critical outsourcing agent is required. Reasons for failure of outsourcing include poor contracts, poor control of the function, and not planning for a termination strategy [Barthelemy, 2003]. Failures can be avoided or managed if accountability for managing the outsourced functions is clearly assigned to one or two persons in the new venture.

Transaction costs with suppliers and customers can be the most important types of costs. Transaction activities are time-consuming and prone to errors. Thus, companies use technology to automate their purchasing and contracting transactions with their suppliers and providers. Discount broker Charles Schwab & Company entered the securities brokerage market by offering a transaction cost advantage for both its customers and suppliers. Another firm that offers lower transaction costs to its customers and suppliers is FedEx [Spulker, 2004].

While a new venture should consider outsourcing many of its activities, out-sourcing can lead to problems. If an activity is outsourced and the supplier fails to deliver the required result (or activity) on time and with the required per-formance, the new venture can experience great difficulties. The conditions that favor the internalization of the activity within the new venture are summarized in Table 13.7. If the demand forecast for an activity or component is highly un-certain, it may be better to do the task internally. If there are only a few, power-ful suppliers of a service or component, the danger is that they might use their power and not meet the required needs at the agreed-upon cost and time. If a firm's technology is valuable and proprietary, it may decide to keep it internal for reasons of secrecy.

Sometimes firms are led to outsource those value-added functions in which most of their profits will be made in the future. In the struggle for a sustainable competitive advantage, it is important to retain these value-added functions that will be critical to the firm's advantage in the future [Christensen, 2002].

TABLE 13.7 Conditions that favor an activity operating internally within the new venture.

Factor	Internal activity favored due to:
Costs	Total cost to produce internally is lower
Demand forecast	High uncertainty
Number of suppliers	Few powerful suppliers
Proprietary, nonpatented technology	Need to keep secret
Value-added function	Source of the firm's future sustainable competitive advantage

Industries with modular products and standards established for interfaces and interconnections support the use of modular architecture. As a result, vertical integration is less necessary. The personal computer industry is an industry where outsourcing of many activities takes place. A **virtual organization** manages a set of partners and suppliers linked by the Internet, fax, and telephone to provide a source or product. In this case, the value provided by the company is primarily the networking of the participating partners and outsourcing agents. This value, however, is often not rare and can be readily imitated.

Henry Ford built a vertically integrated facility at River Rouge near Detroit in 1917. The self-reliant plant made steel for auto bodies as well as all parts from engines to windshields. From Ford's own forests came wood for the paneling. For Ford, integration meant control of all the activities. A new firm cannot afford this type of vertical integration. It is too expensive and too risky. It is risky because with vertical integration comes control and commitment to a large investment in one way of doing things. Loss of flexibility is risky in a continuously changing economy. Today, Ford Motor Company is responsible only for the design, assembly, and marketing of its vehicles. All the modules and parts are provided by an array of suppliers and partners. In the near future, automobile companies may do only the core tasks of designing, engineering, and marketing vehicles. Everything else, including final assembly, may be done by the parts suppliers.

Salesforce.com proposes to offer a service called "utility computing." The firm offers software for rent that stays on their computer and is delivered to customers online through a Web browser [Clark, 2003b]. Salesforce.com rents software online that companies use to manage their salespeople and plans to let other software developers rent out their software using Saleforce's computers. This is a big idea. One obstacle to this service is that most corporations do not want sensitive information stored on any computer other than their own. Should a start-up firm use this outsourcing service?

It seemed like one of those great business ideas: Do your grocery shopping online. Webvan was founded in 1999 to offer an extensive line of grocery and nongrocery items for selection online with delivery to the customer's door. Following several rounds of venture capital investments and an initial public offering, same-day delivery was offered in several cities. To provide this service, Webvan found it necessary to build distribution centers. With margins as low as 2 percent and expensive packaging and delivery functions, it struggled to make a profit. Webvan hoped to minimize costs by setting up a string of futuristic, $35 million warehouses with motorized carousels and robotic product-pulling machines. This was to help offset the enormous cost of its delivery fleets. After burning its way through more than $1.2 billion, Webvan closed its doors in July 2001. Webvan failed because it had a flawed business model that required a large investment in expensive distribution and service activities, as shown on the value chain of Figure 13.2.

Tesco of Britain offers an online grocery service that is profitable. Tesco used a new channel, the Internet, to reach its existing customers as well as new ones. Tesco provided online ordering combined with customers picking up their selected and boxed groceries at a Tesco store or paying a delivery charge. Tesco used existing stores, while Webvan built new warehouses and extensive delivery services. This illustrates why efficient operations along the value chain are critical to profitability.

13.8 INNOVATION AND VIRTUAL ORGANIZATIONS

A virtual organization manages a set of partners and suppliers in order to provide a product. Innovation and creativity can flourish with teams forming to build new ventures. These firms, form, merge, and change form as required by the demands of a fast-changing industry and exhibit a high diversity of organizational forms and high rates of turnover. Ideas flow with people and recombine for new opportunities. The Hollywood movie industry demonstrates these flexible characteristics. Hollywood firms form and disband nearly continuously, causing a continual reshuffling of the human participants. The vitality of the organizational community appears to reflect its flexibility.

Movies are basically project-based enterprises that rent all their resources. Filmmakers develop their core competency in the identification and recruitment of talented people and the management of a complex project, a movie, through the steps of a value chain. Temporary organizations, such as film projects, capitalize on the specialized skills of their members while controlling the costs of coordination. Coordination of a film project is based on continuous, public, redundant communication. Additionally, as crew members carry out their roles on a project, they are strongly socialized via a culture of direct feedback, excessive gratitude, and role-directed humor, which further reinforces and makes clear role expectation [Bechky, 2002]. An excellent example of a theater company operating as a coordinated project is shown in the film *Shakespeare in Love*.

Amazon.com now operates the online stores and fulfillment activities of the online operations of Target, Toys-R-Us, Circuit City, and Borders. Amazon has become an outsourcing agent of Web services, logistics, and customer service for "brick and mortar" giants. Global Sports (www.globalsports.com) is an outsourcing company that operates e-commerce businesses for 25 sports retailers, including Sports Authority and the Athlete's Foot. Global Sports owns the merchandise and manages inventory, fulfillment, and customer service. Each retailer, in turn, receives a single-digit percentage cut of any revenues that come from its site. To consumers, it just looks like they're buying from their local retailer. Global Sports relieves the retailers of the burden of building and maintaining costly e-commerce infrastructure. Moreover, because Global Sports bears all the costs, it allows a client's e-commerce operations to be profitable from day one. Global Sports, in turn, makes money because of the scale of its business. Moreover, Global Sports spends nothing on advertising and marketing. The retailers take care of that activity.

Companies using outsourcing and networks can pull together resources to address specific projects and objectives without having to build permanent organizations. Virtual organizations use computers and networks to build an integrated system. Software applications can be "rented" from an application service provider (ASP).

> A new U.S. start-up, 24/7 Customer received $22 million in a round of investment in 2003. The firm caters to U.S. corporations that want to outsource backoffice operations to India. It provides customer service and technical support via the Internet from India. 24/7 Customer is already profitable and growing. It attracts solid, middle-class college graduates to its operations in India [Vogelstein, 2003b]. (See www.247customer.com.)

New innovative firms can access new ideas from an "idea" marketplace through licensing, alliances, and renting (subscriptions). Flexible, dynamic firms know that the best ideas are not always within their own boundaries. Importing new ideas is a good way to multiply the building blocks of innovation [Rigby and Zook, 2002]. Furthermore, more companies are willing to outsource their ideas and technologies to serve this market.

The creation of virtual organizations brings several challenges with it. Virtual firms encounter difficulty in building trust, coordination, and cohesion among the partner firms and outsource suppliers. New ventures need to invest time and resources in the task of keeping the coordination, trust, and synergy active in a virtual firm [Kirkman et al., 2002].

13.9 ACQUIRING TECHNOLOGY AND KNOWLEDGE

For many firms, the effective use and management of the outsourcing function can be a competitive advantage. An open-architecture value chain can be a powerful business model [Moore, 2000]. The prudent use of other firms that provide significant contributions to a given need can be productive. For example, an electronic system might be built with an Intel microprocessor, Micron memory, and EMC storage, assembled by Solectron, and distributed by Ingram.

The new venture needs to identify which tasks are *core* and which are *context*. A task is core when its outcome directly affects the firm's competitive advantage. Everything else is context [Moore, 2000]. The core/context ratio is a direct measure of effectiveness at generating shareholder value.

The asset base that a firm seeks to leverage through entrepreneurship has shifted over the past decade. The key assets are no longer plants and physical assets but instead are technology, science, and knowledge [Hill 2002]. Entrepreneurs strive to see where new products have become feasible due to the availability of

TABLE 13.8 Seven key characteristics of a new basic technology.

- Functional performance: an evaluation of the performance of the basic function.

- Acquisition cost: initial total cost.

- Ease of use: use factors.

- Operating cost: cost per unit of service provided.

- Reliability: service needs and useful lifetime.

- Serviceability: time and cost to restore a failed device to service.

- Compatibility: fit with other devices within the system.

new technologies. With the advent of the new technologies such as genomics, entrepreneurs see the opportunity for important new medical drugs, for example.

New enterprises also need to import leverage and recombine knowledge bases. Imported knowledge bases can include licensed technologies, purchased technologies, and knowledgeable employees.

A new technology venture will have developed or acquired a basic new technology. This new technology will, it hopes, be the basis for developing and contributing to countless products for many industries. For example, a new venture may possess powerful competencies and knowledge in the science and technology of superconductors. This new technology venture would look for applications in the electric power industry and the electronics industry.

The success of the embodiment of any new basic technology depends on seven characteristics, as shown in Table 13.8 [Burgelman 2002]. These categories or characteristics are useful because they apply in all industries. For example, a powerful new superconducting technology might provide superconductivity of metal at a low initial cost to semiconductor manufacturers and be easy to use in integrated circuits with very low operating costs, highly reliable operation and serviceability, and high compatibility with normal circuits. In this case, we would have a very powerful new technology for the electronics industry.

One of the most active purchasers of technology is Cisco Systems. Between 1993 and 2003, Cisco acquired more than 80 high-tech companies. More than half of those acquisitions were made in 1999 and 2000; during this period, entrepreneurs would talk of founding companies in hopes that they would be acquired by Cisco. Although the number of acquisitions that Cisco made dropped dramatically after that two-year period, purchasing nascent technology remained an important part of its growth strategy.

To facilitate the often arduous task of integrating the acquired firm, Cisco developed and used a documented template that focused on integrating both the people and the technology. Cisco thought carefully about who it would acquire, often taking months to make the decision. The firm did not believe in hostile takeovers and usually acquired geographically proximate companies with "market congruent" visions. Cisco preferred companies that were old enough to have a first product but still young enough that they were not entrenched in their ways or enmeshed with a broad base of customers. Second, Cisco had a no-layoff policy and made a point of keeping the acquired firm's staff. In the late 1990s, Cisco's turnover rate for acquired employees was well below 5 percent, and senior executives were often folded into Cisco's senior ranks.

On the technology side, the R & D and product organizations were integrated with Cisco's other products and immediately labeled with the Cisco brand. In addition, any nonstandard technology was eliminated from the acquired firm, and its employees were given immediate access to Cisco's own infrastructure and core applications. The result was that most of Cisco acquisitions were fully integrated within 60 to 100 days.

13.10 AGRAQUEST

AgraQuest was formed in January 1995, but it did not attract venture capital investors until March 1996. AgraQuest raised $50,000 from its three founders and $420,000 from friends and family. It used those funds to work to build credibility and access to formal venture capital.

AgraQuest choose to locate in Davis since it was the home of University of California, Davis, a world leader in agricultural biotechnology. The three founders and the three scientists that came over to AgraQuest from Entotech moved into a small (750 square feet) laboratory that had been left by a biotech firm that required more lab space. AgraQuest purchased used laboratory equipment for $30,000. Eventually, AgraQuest occupied the adjoining offices as they became vacant. After three years, it moved to a new facility of 13,000 square feet built for it.

Within nine months of its founding, AgraQuest developed its first product, Serenade, and tested it at a northern California vineyard of the E. & J. Gallo Winery. In this test, it proved to control bunch rot and powdery mildew. With the prototype product and proof of concept, venture investors paid more attention to the new venture, which had successfully traversed the credibility cycle of Figure 13.1.

AgraQuest, once funded, had little trouble recruiting talented scientists. In the first four years, it outsourced the design of the production process as well as the manufacturing itself. This proved to be expensive and complex, so AgraQuest hired two production technologists in 1999. Then, in 2000, it purchased a production plant in Mexico and staffed it with a manager and workers. It was convinced that control of the design of the production process and the operation of production were critical to its success. By late 2000, only payroll, accounting, and legal services were outsourced.

13.11 SUMMARY

Successful entrepreneurs are good at locating and acquiring the resources they need to start and build their firm. They need capital, people, and intellectual and physical assets to launch and grow their business. They do this by building credibility and legitimacy with the sources of these scarce resources. Typically, they are good at telling persuasive stories about their vision and its potential. They use their skills of persuasion to acquire the required resources in a timely way.

Entrepreneurs also create a plan for outsourcing some functions while retaining critical functions, such as product design and marketing. They use the Internet to help communicate and manage their relationships with their partners and suppliers in a virtually integrated firm.

Principle 13

Effective new ventures use their persuasion skills and credibility to secure the required resources for their firm in order to build a well coordinated mix of outsourced and internal functions.

13.12 EXERCISES

13.1 Hewlett-Packard offers a service called HP Blade Servers that can reduce a corporation's cost of owning and managing personal computers by shifting processing power to a central bank of shared computers. Users sit at workstations—each with a screen, a keyboard, and some local memory—sign in with passwords, and gain access to both processing power and individual work files. This service could provide new ventures with computer power. Visit www.hp.com and examine the potential of this service for a new venture.

13.2 A new firm, Chix with Stix Golf, offers golf lessons for women. Golf pros at golf courses, however, are hesitant to recommend the firm's lessons [Thomas, 2003c]. (See www.chixwithstixgolf.com.) Provide a recommendation for this firm that will enable it to reach out to female would-be golfers and attract them to its service.

13.3 Blockbuster offers movies for sale and rent online and competes with Netflix.com and Walmart.com. One advantage of the Blockbuster.com service is the ability to return a rented DVD to one of its stores. Visit the three web sites and rate the three offerings in terms of price, on-time delivery, and range of selection.

13.4 A Massachusetts firm called InterLand offers a hosting service for websites for small and new firms (www.interland.net). For a monthly charge, a firm secures a domain name, a website, and 30 e-mail, accounts. InterLand offers templates for about 200 website designs, the ability to add and edit pictures, and publish and update text without having to program [Lohr, 2003a]. Examine this service and provide an analysis of its advantages and disadvantages. Is this a valuable outsourcing service?

13.5 Four of the most popular U.S. locations for technology firms are Boston, San Francisco Bay Area, Austin, and Seattle. Using Table 13.3, determine the most attractive location for an orthopedic medical devices start-up.

13.6 A new start-up in the photovoltaic industry has a new technology developed by a professor at MIT. His former student will join the start-up as marketing manager but must stay in Chicago for family reasons. The business manager for the firm resides in Los Angeles but has an excellent relationship with the marketing manager. The business manager will move to the new firm's location as soon as funding is available. Where should the headquarters of this firm be located? Will this firm be able to readily build its credibility and acquire the necessary resources?

13.7 Determine who is or was one of the most persuasive people you have known. Using Table 13.2, describe how this person exercised his or her sources of legitimacy.

13.8 Research In Motion develops and manufactures wireless handheld devices and provides the Blackberry service. The firm is located in Waterloo, Ontario, Canada. Study and describe the locational advantages and disadvantages of the firm. Is it located in a cluster setting?

13.9 Identify a firm that operates on a hybrid model of using the Internet as well as physical facilities and stores. Describe the advantages and disadvantages of the hybrid model for the firm.

CHAPTER 14

Acquisitions, Mergers, and Global Business

Opportunity is rare, and a wise person will never let it go by him.
Bayard Taylor

Entrepreneurs can create a new business by acquiring an existing firm and then improving it. The acquirers attempt to create growth and new value for the firm. Another strategy is for entrepreneurs to start and build their own firm and then expand the company by acquiring other firms. A series of successful acquisitions can help build a firm into a powerful leader in an industry. The integration of the newly acquired firm within the existing firm is a large challenge, however, especially when the cultures of the two firms differ significantly.

Most new businesses develop, at the appropriate time, a plan for building an international strategy for growth. The forces for globalization are powerful, and new business ventures need to plan for them. ∎

14.1 ACQUISITIONS AND MERGERS AND THE QUEST FOR SYNERGY

An entrepreneur can enter a new business by acquiring an existing firm. An **acquisition** is when one firm purchases another. Usually the acquired company gives up its independence, and the surviving firm assumes all assets and liabilities. Purchasing an established business has the advantages of ongoing businesses: location, employees, equipment, and products. Obvious potential disadvantages include poor location, depleted assets, obsolete inventory, depreciated brand, and lack of profitability. Buying a business can be less risky than starting new because the business's operating history provides entrepreneurs with valuable data on the chances of its success. Finding and evaluating an acquisition candidate, however, can be time-consuming. After finding a good acquisition, the acquirer needs to arrange financing and negotiate the terms of the deal.

Often acquisitions end up eroding the value of the acquired company due to difficulties with the transition to new ownership and overestimation of the value of the acquired firm. When a transition to the acquirer is attempted, difficulties may occur in working with or changing the established culture of the acquired firm. The three main steps for acquiring a company are: 1) target identification and screening, 2) bidding strategy, and 3) integration or transition to the acquirer.

Most, if not all, acquisitions are justified on the basis of an expected **synergy**, which is the increased effectiveness and achievement produced as a result of the combined action of the united firms. Suppose that you identify a firm that you determine is worth V and that a bid for its acquisition is accepted at a price of V. Then you estimate that after adding the value created by you and the entrepreneurial team, the value of the newly revitalized firm will be V_N. We then expect a synergy (Syn) as defined as:

$$\text{Syn} = V_N - V$$

The synergy is the expected value added by the acquiring party. The source of the synergy may be revenue enhancements and cost reductions due to capabilities and resources introduced into the firm by the new entrepreneur team. When new entrepreneurs acquire an existing business, the synergy is the added value of the entrepreneurial team that often replaces the management team of the acquired firm. The new team strives to add value to the acquired team's product that will be rarely available to competitors and is difficult to imitate.

Acquirers try to find firms with valuable and scarce product innovations that can be enhanced by the capabilities of the acquirer's management team. Nevertheless, it is a good rule to avoid bidding contests and to close the deal in a timely way.

American Online (AOL) and Time Warner announced their merger in January 2000, promising multiple synergies. The press release stated:

The merger will combine Time Warner's vast array of world-class media, entertainment and news brands, and its technologically advanced broadband delivery systems with America Online's extensive Internet franchises, technology and infrastructure, including the world's premier consumer online brands, the largest community in cyberspace, and unmatched e-commerce capabilities. AOL Time Warner's unparalleled resources of creative and journalistic talent, technology assets and expertise, and management experience will enable the new company to dramatically enhance consumers' access to the broadest selection of high-quality content and interactive services.

Source: AOL Time Warner press release, January 10, 2000.

We will consider three common methods of valuation of a firm used by acquirers: 1) book value, 2) price-to-sales ratio, and 3) price-to-earnings ratio. The **book value** is the net worth (equity) of the firm, which is the total assets minus intangible assets (patents, goodwill) and liabilities. The price-to-sales and price-to-earnings ratios are obtained for comparative firms in a specific industry.

Consider a firm that designs and makes orthopedic devices for injured and disabled people. An accounting consultant determines that the net worth of the firm is $800,000. Annual revenues have remained at $1.2 million over the past two years. The firm has several patented products, but it has not fully exploited its marketing opportunities. Therefore, the net worth or book value sets a base value of $800,000 for the firm. With no growth in revenues, the accountant suggests a purchase value of one-half of sales, or $600,000. Earnings have held steady at $100,000 per year for the past several years. Assuming a comparable price-to-earnings ratio of nine for a zero-growth firm, the valuation could be nine times earnings, or $900,000. assuming these three valuation methods ($800,000, $600,000, and $900,000), the buyer chooses a target price—say, $700,000—and tries to determine a suitable deal structure. One starting arrangement could be $200,000 in cash with the remaining $500,000 as a loan from the seller of the firm set at the prime rate for four years. Ultimately, the valuation and the final deal are a result of negotiation between buyer and seller.

Conrad Hilton grew up in New Mexico helping his father turn the family's large home into an inn for travelers. After his father's death in 1918, Hilton sought to continue to expand the family business. In Cisco, Texas, where he had gone to negotiate the purchase of a bank, he bought the Mobley Hotel. Quickly finding the hotel business lucrative, he bought other hotels in Dallas, Fort Worth, Waco, and elsewhere in Texas. By 1946, the Hilton Hotels Corporation was incorporated. Starting with the acquisition of one hotel, Hilton had learned to improve each hotel and created a profitable chain of hotels. Hilton perfected management systems that enabled him to realize the full value of each hotel he purchased. After World War II, Hilton purchased several premier hotels, including the Palmer House in Chicago and the Waldorf Astoria in New York.

Quite often, a would-be entrepreneur is offered the opportunity to purchase a small business and then build it up to a significant business. John Yokoyma was an employee at a small fish stand in Seattle. His employer offered to sell

Yokoyma (who was 25 years old) his Pike Place Fish Market. Using a powerful marketing approach, Yokoyma built this fish stand over the years into a nationally known business (www.pikeplacefish.com).

14.2 ACQUISITIONS AND MERGERS AS A GROWTH STRATEGY

Acquisitions and mergers can serve as a growth strategy in fragmented industries. A **merger** refers to the fusing together of two companies. An **acquisition** is when one company buys another. The difference between a merger and an acquisition is the degree of control by one of the two firms; a merger may result in 50-50 control. Mergers involve a much higher degree of cooperation and integration between the partners than do acquisitions. Most of the time, mergers occur between relatively equal-sized organizations, while one organization tends to be larger and more established in acquisitions. Many mergers suffer from insufficient integration of the functions and activities of the two firms. An example of poor integration of two merged firms is AOL Time Warner.

In fragmented industries, numerous small companies are differentiated as specialists and compete for market share. Powerful forces are driving industries to consolidate into oligopolies. An **oligopoly** is an industry characterized by just a few sellers. The incentives to consolidate are significant in the technology, media, and telecommunication industries, where fixed costs are large and the cost of serving each additional customer is small.

An oligopoly, a market in which a few sellers offer similar products, is not always undesirable. It can produce efficiencies that allow firms to offer consumers better products at lower prices and lead to industrywide standards that create stability for consumers. But an oligopoly can allow some businesses to make big profits at the expense of consumers and economic progress. It can destroy the vital competition that prevents firms from pushing prices well above costs. Many industries also face large fixed costs. A typical semiconductor fabrication plant now costs between $2 billion and $3 billion, compared with $1 billion five years ago. A maker of basic memory chips must sell far more integrated circuits (chips) to justify an investment of that size. This is why makers of memory chips are eager to merge.

Industries tend to become more efficient as they undergo consolidation [Sheth and Sisodia, 2002]. In a fragmented market, the consolidator firm within the industry can realize the synergy of the economics of scale. A new venture in a mature industry ripe for consolidation can offer good opportunity. The new entrant can concentrate the resources and use them effectively in one niche and then acquire small competitor firms as resources to do so become available. An example of a fragmented industry is the Internet service provider (ISP) sector. Every town has many independent ISPs as well as large competitors, such as Earthlink and SBC Global. Earthlink is currently consolidating its strengths, while the small competitors are fading.

Most entrepreneurs would not be attracted to junkyards of wrecked cars. Willis J. Johnson, however, recognized an opportunity to consolidate many small businesses into one large business by purchasing dozens of auto junkyards and building them into a nationwide chain. Starting in 1993, Johnson built his Benica, California-based Copart into a leading share (30 percent) of the market for wrecked vehicles [Hof, 2002]. By 2003, Copart had over 100 sites in 40 states. Sales in the 12 months through December 2003 were $350 million. A big reason for Copart's ascent has been its early adoption of the Internet. Copart became a public company in March 1994 (See www.copart.com).

A merger can stimulate growth if the new conjoined firm has a sound business plan for the near-term future that includes a few key measures of profitability. The merged firms should work to redeploy unproductive assets and focus on optimizing their joint activities. Acquisitions are one form of corporate entrepreneurship that can be particularly useful as an established company tries to innovate and infuse the organization with more entrepreneurial behavior or new product lines.

Most studies show that about two-thirds of mergers do not pay off with any synergy or gains. To realize the full value of a merger, the merged organizations must be appropriately integrated. A **horizontal merger** is a merger between firms that make and sell similar products in a similar market. The merger between Exxon and Mobil is an example of a horizontal merger. A **vertical merger** is the merger of two firms at different places on the value chain. An example is the merger of Lucent and Chromatics Networks. The union of AOL and Time Warner is an example of a merger that has features of both horizontal and vertical mergers. By merging, AOL has enhanced its potential delivery of Internet content, since it can now offer customers some of Time Warner's variety of entertaining and informative products. In the same way, Time Warner has found a partner that can deliver its content to a large existing audience. The idea is that AOL Time Warner would control both the content—music and movies—and the distribution of that content through cable television as well as the Internet. It proved difficult, however, to convince AOL users to buy Time Warner content.

In 1976, Jim McCann got an urge to own his own business. He bought a flower shop in Manhattan for $10,000 and brought in a day-to-day manager. He then opened 12 new stores over the next decade. He was doing well and left his regular job to build up his floral business. He purchased the troubled 1-800-Flowers in 1987 for $2 million and assumed $7 million in debt. He then moved the firm to New York and merged it with his flower store chain. McCann then made a good set of acquisitions, enabling the expansion of 1-800-Flowers. It moved beyond flowers to include gifts in the 1990s. Flowers in 1999 acquired Great Foods, a specialty food unit. In 2001, Flowers acquired Children's Group, a maker of toys and dolls. Flowers has expanded its candy business by offering

TABLE 14.1 Five different types of mergers and acquisitions.

Type:	Overcapacity reduction	Geographic extension (roll-up)	Product or market extension	Technology acquisition	Industry convergence
Objective	To reduce excess capacity and increase efficiencies	To extend a company's reach geographically and build economies of scale and scope	To extend the product line or reach a new market segment	To quickly add new technologies and capabilities	To establish a position during the convergence of an industry or sector
Examples	Daimler-Benz and Chrysler	Bank of America and Nations Bank	Tyco and Raychem	Cisco Systems acquired 80 companies between 1993 and 2003	AOL and Time Warner
	Hewlett-Packard and Compaq	Waste Management and numerous local firms	J. M. Smucker and Jif peanut butter	Medtronic acquired seven companies between 1998 and 2000	Disney and ABC-TV and radio
Issues	What to eliminate in the merged company and how to get it done in a timely way	How to merge two firms with different cultures	Merging two cultures and distribution channels	Overvaluation of the technology acquisition and loss of leaders from the acquired firm	The convergence may not materialize or be of low value

more gift-box products, higher-price candy brands, and express delivery of items. Nonfloral items such as baked goods, sweets, and jewelry are offered on its website (www.800Flowers.com).

Five different types of mergers and acquisitions are listed in Table 14.1 [Bower, 2001]. The first type aims to reduce overcapacity in a relatively mature industry. The acquirer tries to close inefficient plants and reduce costs while retaining the acquired firm's technologies and customers so that economies of scale can be realized. The Hewlett-Packard and Compaq merger in 2002 is, in part, an example of this type.

Many deals are based on acquiring the customers of the acquired company and reducing overcapacity. The goal is to identify the best customers and retain them while trying to attract new customers from the acquired firm's customer list [Selden and Colvin, 2003a].

A geographic extension or roll-up occurs when a successful company rolls up (buys up) local or regional firms into a nationwide powerhouse. Roll-ups are designed to achieve geographic reach and economies of scale and scope. The third category, merger extension, is designed to extend a company's product line or its entry into unserved markets. To extend its product line, J. M. Smucker, maker of fruit spreads, purchased the Jif peanut butter unit from Procter & Gamble.

A technology acquisition aims to quickly acquire new technologies and capabilities by purchasing a small firm. As discussed in section 13.9, Cisco Sys-

TABLE 14.2 Rules for the acquiring firm and the acquired firm.

Rules for the acquiring firm
■ Use your highly valued stock as payment.
■ Identify the key people of the acquired firm and get their agreement to stay.
■ Decide who to keep and build relationships fast.
■ Contain the tendency to act with hubris.
■ Integrate the culture and the operations of the two firms.
■ Appoint an integration manager or team to lead the acquisition process.

Rules for the acquired firm
■ Demand cash, not stock, as payment from the acquiring firm.
■ Key people will agree to stay for a short period.
■ Avoid signing a noncompete agreement or keep its duration short.
■ Explain the benefits that accrue to the employees and managers of the acquired firm.
■ Tell people who will and will not have a job.
■ Restructure with respect for people.

tems built its performance during the 1990s on a long string of acquisitions of small firms. The final type of merger is based on a perception of a future convergence of industries. Disney purchased ABC-TV and Radio, envisioning a convergence of content and media channels.

The integration of two firms is an important task of any merger or acquisition. For example, the merger of Land's End and Sears put the shirts and sweaters of Land's End into Sears stores. Will the quality of service in Sears stores be consistent with the Land's End online and phone service? Will the Sears customer buy the Land's End product line? The merger could result in a large synergy due to Sears gaining customers and a better online presence while Land's End extends its market segment to the Sears customer.

The Hewlett-Packard and Compaq merger was based on both overcapacity reduction and technology acquisition. In this case, the surviving firm, Hewlett-Packard, acquired Compaq's technologies. The goal of the merger was to improve HP's competitiveness since the firm had largely missed out on the personal computer and Internet transitions [Anders, 2003].

Rules for integrating an acquired firm into the acquiring firm are provided in Table 14.2. The key step is to appoint an integration manager who works full-time for a period on integrating the two firms. The integration effort starts with a strategy and an integration plan. Then the goal should be to achieve a majority of the integration by a short period after the close of the deal—about six weeks. An important step is to build an integration team working with the integration manager. The team helps build the social connections for the merger and get early results. The role of the integration manager is to inject speed into the

TABLE 14.3 Four roles of the
integration manager.

1. Inject speed into the process.
   - Push for decisions and progress.
   - Set the pace.
2. Create a new structure.
   - Create joint teams.
   - Lead an integration team.
   - Provide the new structure framework.
3. Make social connections.
   - Interpret the cultures of both companies.
   - Actively be present in both companies.
   - Bring people together.
4. Build success.
   - Identify and communicate synergies.
   - Show short-term benefits.
   - Demonstrate corporate efficiency gains.

process, create a new structure, make social connections, and build success, as summarized in Table 14.3 [Aiello and Watkins, 2000]. The number of mergers and acquisitions in any year in the United States varies depending on market conditions, as shown in Table 14.4.

The leaders of entrepreneurial firms often continue to play a vital role after an acquisition by a larger firm closes. The leaders of the buying firm are often too busy with their own business to provide effective direction to the acquired employees. Moreover, they initially may not understand the acquired business well enough to make good decisions. This creates a need for the acquired managers to continue to lead their companies, even after the deal closes. Acquired leaders can add value by focusing their employees on specific goals and timelines, and by helping to resolve problems that arise as employees are assigned to their new positions and supervisors [Graebner, 2004].

Acquisitions and mergers are as likely to destroy value as to create it. They only add value if they make strategic sense, if their fair value is paid based on

TABLE 14.4 Number of mergers and acquisitions in the United States*

	1985	1990	1995	2000	2002
Number of mergers and acquisitions	3,000	1,600	4,000	10,500	1,500

*Firms with market values exceeding $100 million.

realistic expectations, and if management stays focused on executing the plan. AOL and Time Warner failed to properly mesh the organizations and destroyed value [Klein, 2003].

14.3 GLOBAL BUSINESS

The current globalization phenomenon dates from the fall of the Berlin Wall in 1989, which ended the principle division that ruled the world since 1945, when the Iron Curtain descended. If the metaphor for the Cold War period was the Berlin Wall, the metaphor for globalization is the Web. Globalization is the triumph of free-market capitalism worldwide. The previous era of globalization was built around falling transportation costs. Today's era of globalization is built around falling telecommunications costs [Freedman, 1999]. **Globalization** involves the integration of markets, nation-states, and technologies, enabling people and companies to reach around the world to offer and sell their products in any country in the world. With the Internet and globalization, a company can sell anytime, anywhere. Globalization is characterized by speed, modernization, movement, and the removal of distance.

A new venture should consider a globalization strategy, even if it is to only decide that its initial strategy is to remain local. A new design-automation firm may choose to only serve the United States in its initial years but later consider expanding internationally. We use a classification system for the strategies for globalization, as shown in Figure 14.1.

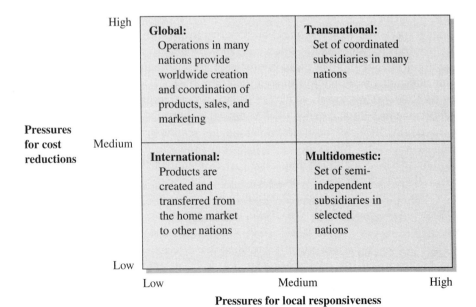

FIGURE 14.1 Strategies for globalization.

A **local** or **regional** strategy focuses all a firm's efforts locally since that is its pathway to a competitive advantage. Another reason to remain local initially is limited availability of resources. Restaurant, retail, and any local opportunity is best started and built locally. Starbucks started in Seattle and moved to other regions of the United States only after perfecting its operational capabilities locally. A new technology firm may start locally and fashion its marketing and sales methods. Then it can go global.

The **multidomestic** strategy calls for a presence in more than one nation as resources permit. In this case, the firm offers a separate product and marketing strategy suitable for each nation. This strategy is not cost-efficient but does enable a firm to have independent subsidiaries in many nations. Examples of companies using a multidomestic strategy are Nokia and Ericsson.

To exploit cost economies while creating differentiated products, a **transnational** strategy can be used. This strategy rests on a flow of product offerings created in any one of the countries of operation and transferred between countries. Examples of companies using a transnational strategy are ABB and Caterpillar.

Trend Micro Inc. has spread its employees around the world and operates with a transnational strategy for its antivirus software [Hamm, 2003]. Its main center is in the Philippines, with other centers located in Munich, Beijing, and Mexico, and its financial headquarters in Tokyo. Its sales office is in Cupertino, California, and its chairman, Steve Chang, is Taiwanese. The management challenge is formidable since the firm's people are separated by oceans, languages, and cultures. (See www.trendmicro.com.)

An **international** strategy tries to create value by transferring products and capabilities from the home market to other nations using export or licensing arrangements. One benefit of international activities can be the exposure to new business environments where the firm can learn about different methods, products, and innovations. Examples are Microsoft and IBM. Microsoft has regularly tried to bring the same business model to other countries. Headquartered in Silicon Valley, Intel's international business has grown to represent 70 percent of its total revenues. Craig Barrett, Intel's CEO, anticipates the largest growth areas in the coming years will be India, China, and Russia. These markets represent almost half the world's population and have just recently become available to U.S. technology companies [Barrett, 2003].

Fargo Electronics is a leading maker of instant ID-card systems that is competing for sales throughout the world (see www.fargo.com). As many as 800 million identity cards with a microchip (smart cards) may be used in China alone. The smart card carries all sorts of data about the individual.

The IDs essentially are microcomputers containing a 32-bit processor and 32 kilobytes of memory that are encrypted to provide access to buildings, computers, e-mail, and other functions. They also can store information about the card holders, such as their medical history. Fargo, based in Eden Prairie, Minnesota, markets its product in 80 countries using an international strategy, as described in Figure 14.1.

A **global** strategy emphasizes worldwide creation of new products, sales, and marketing. The company uses facilities and organizations in several nations to create products for worldwide sales. Examples are General Motors, Intel, and Hewlett-Packard. The advantages and disadvantages of each of the global strategies are listed in Table 14.5 [Hill and Jones, 2001].

In general, a new or emerging firm should choose one of the strategies for entering the global marketplace and then determine which foreign markets to enter and when. The determinants of entry, timing, and costs will lead to the selection of a sound strategy. Some industries are local in nature, and others are international or global. A new manufacturer of integrated circuits is immediately required to quickly build an international strategy since the competitive marketplace is global. For many firms, expansion from local to regional and then to national markets will follow a natural progression. The first step is to expand to selected nations, establishing appropriate distribution channels and supply chains.

International opportunities exist in many industries. Microsoft and Intel sell their products worldwide. London's hit plays, such as *Phantom of the Opera*, move to New York and eventually on worldwide tours. Global opportunities to reduce costs, improve capabilities, and match local needs call for a transnational or global strategy. The resources necessary to mount these strategies can be large, however.

TABLE 14.5 Advantages and disadvantages of the four global strategies.

	Advantages	Disadvantages
■ Multidomestic	Ability to customize products for local markets	Failure to reduce costs and to appropriate learning from other nations
■ Transnational	Ability to reduce costs and learn from other nations Ability to be locally responsive	Difficult to implement due to many independent organizations
■ International	Transfer of unique products and competencies to other nations	Low-local responsiveness Failure to reduce costs
■ Global	Ability to reduce costs and gain worldwide learning	Lack of local responsiveness Difficult to coordinate

TABLE 14.6 Reasons to develop a global strategy for a new venture.

■ Access low-cost labor or materials.	■ Access attractive markets for the firm's products.
■ Work around trade barriers.	■ The industry is global and all competitors are worldwide.
■ Access unique capabilities and learning from others.	■ Possession of a strong brand that is known worldwide.
■ Obtain economies of scale.	

Cisco Systems uses an international strategy that enables it to derive about 50 percent of its revenues from foreign sales in 150 countries. Many new ventures or emerging firms need to consider the development of a global strategy to access unique capabilities or advantages. Reasons for considering entering the global marketplace are listed in Table 14.6.

It is conventional for a start-up to fully develop and test its product in a local or national marketplace before taking it global. Another view is that the best, most profitable markets for a new product may be in a nation other than the home of the new venture firm [Aaker 2001]. However, small, emerging firms often have limited export efforts because they have limited ability to acquire information and knowledge about foreign markets and to manage foreign activities [Julien and Pamangalahy, 2003].

A successful global start-up usually has an international vision from inception, a strong worldwide network, and a unique product in demand worldwide. Consumers throughout the world increasingly demand the same selection of consumer goods, particularly automobiles, clothing, many food and beverage products, and consumer durables such as appliances and electronics. For many businesses, this global consumerism has focused sharp attention on the development of global brands, which are rapidly creating brand equity positions for their companies. Howard Schultz held the view that Starbucks would be a global brand. By 1998, Starbucks acquired a London-based chain of 60 cafes in Britain, gaining its first step into Europe. By 2003, Starbucks had 330 outlets in Britain. Starbucks entered Japan in 1996 and had 467 cafes by 2003.

The mode of entry into another national or regional market offers five possibilities, as listed in Table 14.7. Exporting is an easy method to start up elsewhere. However, high transportation costs can be a disadvantage. Licensing to another party for a fee can be an inexpensive method and can provide some control of the marketing and manufacturing carried out by the licensee. Franchising is a form of licensing with an agreement to follow rules and procedures of operation. This method may result in loss of control over the quality of the product.

A joint venture with a foreign company affords access to the partner's capabilities but also results in diminishing control. A wholly owned subsidiary enables the parent company to exercise full control but can be a costly approach.

McDonald's and Hilton Hotels are usually set up as franchises in other nations. Intel and Hewlett-Packard act through wholly owned subsidiaries. Fuji-Xerox is a joint venture between Xerox-USA and Fuji Photo Film of Japan.

TABLE 14.7 Five forms of entry mode into international markets.

Mode	Description	Advantages	Disadvantages
1. **Exporting**	Send goods abroad for sale in another nation	Ability to sell elsewhere	Transportation costs Difficulties with the agent
2. **Licensing**	Legal permit to use knowledge or patent to make product	Low-cost entry	Weak control of the licensee
3. **Franchising**	Rights granted to a business to sell product using brand, name, and methods of operation	Low-cost entry	Lack of control over quality
4. **Joint venture**	Corporation held jointly with a local entity	Access to partner's capabilities Shared costs	Lack of control
5. **Wholly owned subsidiary**	Company incorporated in another nation	Ability to act directly	High costs

Harley-Davidson exports about 25 percent of its motorcycles directly to dealers and distributors in other countries. Solectron is a manufacturing firm operating globally while headquartered in the United States.

Eastman Kodak is the world's largest producer of photographic products. It has major plants in the United States and eight other countries. About one-half of its employees and one-half of its sales are outside the United States. To retain its worldwide market share, Kodak is moving aggressively into the worldwide filmless digital photography field.

Globalization is spreading as markets open and deregulate. As shown in Figure 14.2, the forces of globalization are powerful [Barkema, Baum, and Mannix, 2002]. Many new firms will need to consider the global factors in their industry and create a plan for responding.

Yves Doz of INSEAD defines a **metanational** company as possessing three core capabilities: 1) being the first to identify and capture new knowledge

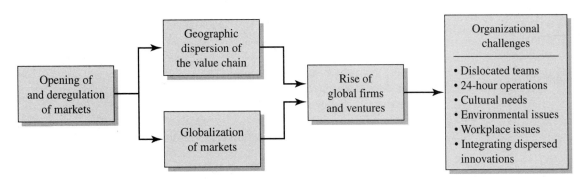

FIGURE 14.2 Forces and consequences of globalization.

emerging all over the world; 2) mobilizing this globally scattered knowledge to out-innovate competitors; and 3) turning this innovation into value by producing, marketing, and delivering efficiently on a global scale [Fisher, 2002]. Doz identifies one example of such a firm as Nestlé. Other examples are IBM and Hewlett-Packard. Perhaps the most challenging task is to access knowledge and innovations everywhere and then integrate them into the firm's competencies and products.

Competition is similar to a three-dimensional game of global chess. The moves an organization makes in one market may achieve goals in another market in ways that are not immediately apparent to its rivals. Where this strategic interdependence between markets exists, the complexity of the competitive situation can quickly overwhelm ordinary analysis. A firm needs an analytic process for mapping the competitive landscape and anticipating how its moves in one market can influence its interactions in others.

Leading firms in technology-intensive industries are based on core capabilities that can be leveraged globally. Innovative companies strive to draw knowledge and coordinate capabilities worldwide. They build and discover new assets and capabilities, and work to achieve economies of scale and scope. As the firms grow, they move toward a global network of capabilities and assets [Tallman and Fladmoe-Lindquist, 2002].

The promise of fish farming (aquaculture) is the potential for helping feed the world's population without depleting natural fisheries. Using new technologies, aquaculture can supply a large source of food while maintaining environmental protection [*Economist,* 2003d]. Salmon and scrimp are examples of fish farmed and marketed widely in the United States and Europe. Aquaculture has the potential to be a global growth industry.

14.4 AGRAQUEST

AgraQuest is currently looking for its first acquisition candidate so that it can grow revenues more quickly. Since it takes several years to develop a new biopesticide and obtain approval for its use, it is very attractive to acquire companies that have one or two approved products. AgraQuest is hoping to acquire one of these firms by issuing its common stock to the owners of the firm. Most likely, the owners of the acquired company would prefer to receive cash, since AgraQuest stock is not yet publicly held.

AgraQuest has a strategy for selling its products internationally through distributors in other countries. Serenade is sold via distributors in Japan, Mexico, Chile, and New Zealand. A distributor in each of those countries has an exclusive sales arrangement. Serenade is sold directly to the banana producers in Costa Rica. AgraQuest has filed for approval of Serenade in Europe and hopes to distribute the product there soon.

14.5 SUMMARY

A new venture may start as a purchase of an existing company by a team of entrepreneurs. Alternatively, entrepreneurs can start their new venture and acquire other small firms to grow their own company. Most, if not all, acquisitions are justified on the basis of an expected synergy, which is the increased effectiveness of the combined firms. Acquisitions can be used for efficiency improvement, geographic expansion, product or market extension, and technology acquisition.

Most new firms start with a regional or national strategy and later develop a plan to export their product internationally. As these firms grow, they may shift from exporting to establishing a wholly owned subsidiary in other nations.

Principle 14
All new technology business ventures should formulate a clear acquisition and global strategy.

14.6 EXERCISES

14.1 An acquirer looks for a company with a good profit margin, a proven history, and a fair price. Choose an industry of interest and list five criteria for selecting candidates for acquisition.

14.2 Hewlett-Packard purchased Compaq Computer Corporation in May 2002 for $19 billion in stock. Prepare a list of six key concerns you would use to lead a team to integrate Compaq into Hewlett-Packard. These concerns or issues should be chosen so that if all six are resolved favorably, the integration effort will be effective. Use Tables 14.2 and 14.3 as a start on your list.

14.3 Wal-Mart is expanding internationally. Using Table 14.7, describe the entry mode to international markets that Wal-Mart has selected.

14.4 Palm Inc. agreed to acquire Handspring on June 4, 2003, for Palm stock. Palm also agreed to lend Handspring $10 million while the deal was pending. The stock offered to Handspring was worth less than the market value of Handspring. Examine the outcome of this merger. Did it create synergy?

14.5 A software firm is available for purchase, but it has experienced no growth for several years. The firm provides a cash return annually before taxes and owner's salary of $100,000. It has annual sales of $1 million. As a purchaser of this firm, you select a discount rate of 14 percent. Calculate the price you would offer for the firm. Assume that you can increase and maintain a growth rate of sales and cash flow of 2 percent annually.

14.6 Starbucks has a goal of 10,000 stores in 60 countries by the end of 2005. Starbucks has almost 500 stores in Japan. It is finding stiff competition and a resistance to the Starbucks experience in Europe [Holmes, 2003]. Study Starbucks in Europe and Asia, and estimate the global strategy of Starbucks using the format of Figure 14.1.

14.7   Harley-Davidson built three 3 motorcycles in 1903 and 263,000 in 2002. Over those 99 years, the firm has grown worldwide. European sales are managed from Oxford, England, and Asian sales from Japan. Harley-Davidson is the leading heavy-weight motorcycle in Japan, the United States, and worldwide. The firm has a network of 1,300 independent dealers worldwide, with 700 in the United States and Canada, 400 in Europe, and 210 in Asia. The firm enjoys a profit margin of about 14 percent. Determine the reason for Harley-Davidson's worldwide success. Describe the strategy for globalization (Figure 14.1) used by the firm.

14.8   Neopets (www.neopets.com) offers its virtual world of virtual pets globally. The firm offers its service in two languages, English and Chinese. Study the Neopets site and determine if it should be offered in many languages. What are the advantages of establishing a Neopets site for each language?

14.9   Logitech International has dual headquarters in Switzerland and California (see www.logitech.com). Describe the global strategy of Logitech using the format of Figure 14.1.

14.10   Earthlink is one of the world's largest Internet service providers, with over 150,000 customers. Its strongest competitor is AOL. Compare the two companies' global strategy using the format of Table 14.5. What form of entry into international markets did each firm use (see Table 14.7)?

CHAPTER 15

The Management of Operations

Real intelligence is a creative use of knowledge, not merely an accumulation of facts.
D. Kenneth Winebrenner

Every business is built on a set of operational processes that serve to create, make, and provide the product to the customer. Most businesses build a chain of activities that add value at each section of the chain. Each element of the value chain has a capability that provides value added to the product. A new venture manages its value chain to provide the ultimate product to the customer. The firm also moves, stores, and tracks parts and materials to its value-adding partners and strives to ensure timely, efficient production of the service or product. Information flow along the value chain enables the coordination of the distributed tasks. An effective enterprise manages for operational excellence by trying to develop and communicate measurements of efficiency and timeliness.

Another way to describe a set of interrelated tasks is as a network of activities. A value web (or network) can use an Internet-based system to communicate with all the network participants. With a common schedule and associated tasks, the venture can manage the value web to achieve on-time production. ∎

15.1 THE VALUE CHAIN

The purpose of a firm is to provide products that customers value. A value chain is a series of activities for transforming inputs into outputs that customers value. Each value chain activity adds value, as shown in Figure 15.1. Information flows back from the customer and the sales and service activities so that the value chain can maximize value for the customer. More than merely things with features, products are increasingly viewed as things with features bundled with services. Products and services are grounded in activities and relationships in a value-creating system. Furthermore, each element of the value chain has certain capabilities that can be improved over time. Capability development along the chain and the design of the chain can lead to a powerful core competency for a new venture. Furthermore, customers participate in this value-creating process by communicating their preferences and priorities. Understanding the customer enables the producer to better match the customer's needs, as described in Table 15.1.

A highly integrated company provides most, if not all, of the functions along a value chain. This approach is most suitable when proprietary interdependent activities occur at each stage of the chain. As many industries mature, the functions along the chain become independent so that modular subproducts are available at each stage. At that point in time, the value chain breaks up, and a number of independent firms participate in the activity chain. For example, today's automakers are adopting modular architectures for their mainstream models. Rather than putting together individual components from diverse suppliers, they are procuring subsystems from fewer suppliers. The architecture within each subsystem (braking, steering, chassis) is becoming progressively more interdependent as these suppliers strive to meet the auto assembler's performance and cost demands [Christensen, Raynor, and Verlinden, 2001].

Every industry has its own rate of evolution that erodes its competitive advantage. In a fast-changing industry, a firm must have the ability to readily redesign its value chain to find new sources of competitive advantage [Fine, et al., 2002]. In designing or redesigning a value chain, each stage of the value chain can be assigned an economic value-added measure (EVA), which accounts for knowledge assets and strategic assets. Strategic assets are those in which the firm has relative competitive advantage. Strategic assets may include logistics, manufacturing, and distribution assets. Knowledge assets will exist primarily in the research design, marketing, and service functions. A new firm should retain the

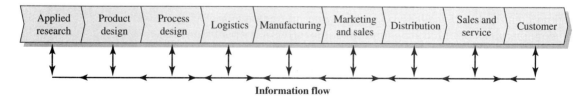

FIGURE 15.1 Value chain and information flow.

TABLE 15.1 Understanding the customer.

■ Preferences	■ Buyer behavior
■ Purchase criteria	■ Functional needs
■ Decision-making process	

functions with high EVA and outsource functions with low EVA. If the industry is changing rapidly, a firm may decide to retain a key function to strategically respond to the change internally. Often a sound approach is to retain the high EVA activities and key strategic assets while outsourcing the low EVA activities.

Vertical integration along the value chain provides firms with an opportunity to choose the value-added stages in which it will compete. Intel is a manufacturer of both integrated circuits and circuit boards, but it also assembles personal computers under an original equipment manufacturer (OEM) agreement with PC companies.

Zara, a European clothing retail firm with 519 stores, has retained its manufacturing capability rather than outsourcing it so it can respond quickly to changing fashion demands. Other firms may be able to make the clothes more cheaply, but the strategic asset retained by Zara is the ability to quickly deliver new fashions to its stores. Information flows (see Figure 15.1) from the store floors back to the designers, who redesign the products to fit the customers' changing ideas and tastes. In fashion, nothing is as important as time to market [Helft, 2002]. Zara has new designs arriving in its stores every two weeks. Many new designs arrive at the stores within a few days.

Logistics is the organization of moving, storing, and tracking parts, materials, and equipment. Logistics can be the basis of competitive advantage since a firm with fast, accurate logistics will be first to respond to customers. Logistics systems usually are based on electronic networks, such as a supply chain intranet. Companies look for unique ways to service customers quickly through improved tracking, transporting, handling, and delivery in an effort to create unique competencies. Nokia, for example, uses a logistics management system to obtain all the parts for its mobile phones, build them, and get the phones out to the market where and when they are wanted

Logistics might sound like a simple business of moving things around, but it is growing more complex as customers demand timely, customized services. New technology and greater use of the Internet opened up new ways of sharing information. Companies are also trying to build only after receiving orders from customers (known as built-to-order, or BTO), rather than estimating what will be in demand and supplying it from accumulated stocks. The BTO concept tries to avoid producing any item without a firm purchase order. Dell Computer is a leading example of a company based on BTO.

The information flow along the value chain can be facilitated by the Internet [Hammer, 2001]. Using this information and working closely with partners along the chain to design and manage the activities, efficiencies can be improved. Who captures the profits from these efficiencies? Often companies such as Wal-Mart and Home Depot have assumed the distribution, sales, and service functions, and captured significant value created by their suppliers. Furthermore, by using the Internet, companies like Amazon and eBay are becoming electronic sales and distribution channels. At the same time, companies like Nokia are producing a seamless manufacturing, distribution, sales, and service offering. Managing the value chain is a challenging task for all new business ventures, which can ill afford to assume many of the value chain activities.

Wal-Mart has a profit margin less than 4 percent, and many supermarket chains have a profit margin less than 2 percent. Clearly, every penny saved is a penny earned. The bar code became a mainstream success after Wal-Mart adopted it in 1980. Now Wal-Mart will require that its suppliers use radio-frequency-identification (RFID) on their supply pallets. RFID relies on a computer chip to hold and convey information. Wal-Mart manages its supply chain as if it were an orchestra.

Intermediaries in a value chain make sense only for exchanges in which the parties to the transaction can save more money by hiring the intermediary than it costs. It can cost 20 percent of revenues to "hire" a retailer to sell a product. Is it cheaper to sell direct? Can the Internet reach the same customers effectively?

Value chain speed is important to any new venture. With a long lead time from design to the customer, a new firm may wind up with inventory it cannot sell. Since many products have short life cycles, the chain must be able to move fast. Zara can design and manufacture a new clothing fashion in one week, if necessary. On the other hand, a movie studio can take a year to produce a new film and introduce it into the market. One of the best examples of value chain management is the Dell Computer system, which enables customers to specify their desired product and pay for it before Dell starts assembling the computer.

While goods and services flow largely down the chain, information flows in both directions. For example, information about what is demanded at successive stages passes up the chain, while information about supply conditions such as availability, pricing, time-to-manufacture, and so on passes down. Because this is an information-intensive process, the Internet holds the potential to significantly increase the amount of value created.

> A critical part of the success of the Krispy Kreme doughnut chain is its value chain technology. Its system uses radio units in the warehouse to tell pickers when and where to pick and ship items. No paper is generated; it is all electronic information transfer (see Figure 15.1).

A new venture's lasting core competency is its ability to continuously assess industrial and technological dynamics, build value chains that exploit current opportunities, and select the high value-added activities for operation by the firm itself.

15.2 PROCESSES AND OPERATIONS MANAGEMENT

An **operation** is a series of actions, and **operations management** is the supervising, monitoring, and coordinating of the activities of a firm carried out along the value chain. Operations management deals with processes that produce goods and services. A **process** is any activity or set of activities that takes one or more inputs, transforms and adds value to them, and provides one or more outputs [Krajewski and Ritzman, 2002]. Processes are the series of operations, methods, actions, tasks, or functions leading to the creation of an end product or service. Processes are used to transfer value to customers in the form of products. A product could involve delivering a tangible good to customers, it could involve performing a service for customers, or it could be, and usually is, a combination of the two [Melnyk and Swink, 2002]. At a factory, a process transforms materials into products. At an insurance company, a process transforms client information into an insurance agreement. A network of processes helps create the value provided at each stage of the value chain.

The business processes of an organization should be aligned with its strategies and employee competencies, as shown in Figure 15.2. Alignment requires a continual rebalancing of strategy, business processes, and the competencies of the people to satisfy and retain the customers in order to keep the business theory as the clear driving force of the business, as shown in Figure 15.2. The business theory should be expressed as a clear statement of vision and purpose. Recall the vision statement of eBay: "We help people trade practically anything on Earth through an online system."

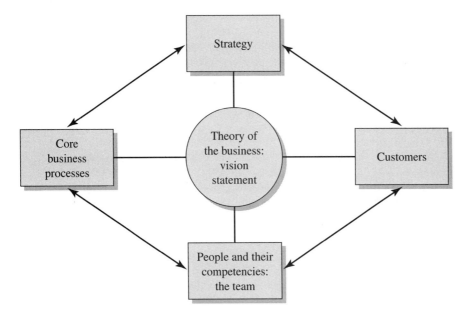

FIGURE 15.2 Business alignment.

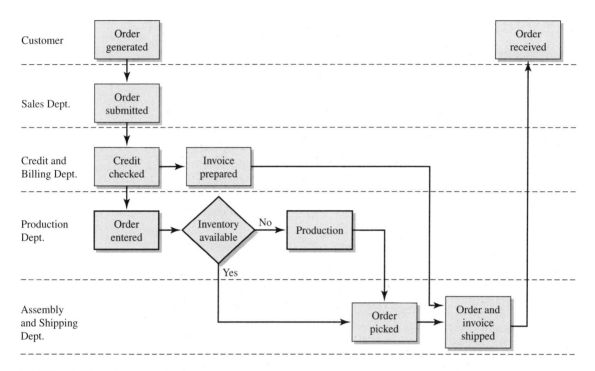

FIGURE 15.3 Common business process.

Ergonomics is making a physical task easier and less stressful to accomplish. Products should be designed ergonomically, and tools used in factories need to be ergonomic. For example, Maytag makes a washing machine with an angled tub and larger door so it easier to reach inside and retrieve items.

Processes bring value to customers and stakeholders. Business success comes, in large part, from a company's process performance. Therefore, a firm should strive for superior process design. An example of a simple business process is shown in Figure 15.3. Part of this process can be automated.

Gordon Moore and Robert Noyce founded Intel in 1968. Their first act was to recruit a director of operations. They offered the job to Andrew S. Grove, who became Intel's third employee. Though Grove had no manufacturing experience, they recognized his innate intelligence and drive. Grove was responsible for getting products designed on schedule and built within budget. The scope of this position extended into nearly every functional area at Intel, from marketing to sales to engineering. His influence, presence, and attitudes pervaded the company, and within three years of Intel's formation, it became clear that the majority of daily decision-making was passing to Grove.

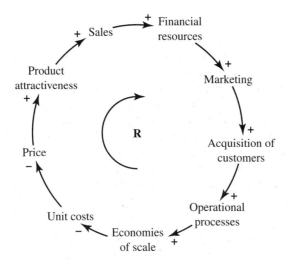

FIGURE 15.4 Self-reinforcing growth through acquisition of customers. A self-reinforcing loop, R, is assumed to work as long as the economies of scale are actually realized through effective operational processes.

Functions in operations management include design of processes, quality control, capacity, and operations infrastructure. The new venture needs to plan for its operations or production function which will be led by one of the entrepreneurial team. Both service and product firms need to design and control operational processes to achieve efficiencies, throughput, capacity availability, inventories, capital expenditures, and productivity. Unique operational management competencies can be part of the competitive advantage of a firm [Vonderembse and White, 2004].

The best return on investment is often found in companies that combine operational excellence—consistently outstanding performance for customers that is brought to the bottom line—with sustained rapid growth. Operational excellence is a necessity today. Investors mercilessly punish companies that fail to meet these expectations [Lucier and Dyer, 2003]. Operational excellence can lead to lower costs, as shown in Figure 15.4. With economies of scale, unit costs will drop, and thus, prices can decline. As prices drop, the product is more attractive and sales increase. As financial resources increase, investments in marketing and operational processes will lead to better economies of scale. This can be a very powerful self-reinforcing loop.

A firm can implement four competitive capabilities: low cost, high quality, speed, and flexibility. **Quality** is a measure of a product that usually includes performance and reliability. Performance is the degree to which a product meets or exceeds certain operating characteristics. **Reliability** is a measure of how long a product performs before it fails. **Speed** is a measure of lead time, on-time delivery, and product development speed. **Flexibility** is a measure of a firm's ability to react to a customer's needs quickly.

Many firms have adopted a six-sigma quality goal, with the aim of getting rid of defects in a process. Six sigma is a statistical term that measures how much a process deviates from the ideal. Six-sigma quality equals just 3.4 flaws per million. The six-sigma method attempts to build low-defect products with low costs [Harry and Schroeder, 2000]. Six-sigma quality is the result of a well-defined and structured process that is highly repeatable—a process with well-defined tasks and milestones.

Consumers often react favorably or unfavorably to the experience they have with a product or service, such as the packaging, clarity of an operating manual, or ease of use. Many customers view consumption as an experience rather than a singular purchase event [LaSalle and Britton, 2003]. Most products or services include *objective value,* such as performance, and *subjective value,* such as experience. A candle purchased for light might cost only $1, while a second candle with shape, form, and scent may provide a richer experience (and sell for $5 to $10). Thus, for many products, the value of the product involves the experience of it and the associated purchase and fulfillment process. The designer needs to identify, design, and fulfill a customer experience that will register with the customer as positive.

Satisfaction with service is difficult to obtain and retain. A survey showed that the companies achieving the highest satisfaction are FedEx, Southwest Airlines, BellSouth, DirecTV, Southern Company, and Hyatt [Barta and Binkley, 2003].

Supply-chain management is focused on the synchronization of a firm's processes and those of its suppliers to match the flow of materials, sources, and information that meets customer demand. Today, the goal is to minimize the stock of goods or inventory required to support variability of customer demand. As products become more susceptible to changing demands, the risk grows that a given product line will have disappointing sales. But if a manufacturer decides to go lean on inventories, it risks running out of stock, lost sales, and endangered relationships with its customers.

Operations systems that are designed to create efficient processes by using a total systems perspective are called **lean systems.** Flexible or lean systems aim to reduce setup times and increase the utilization of key processes. Flexible systems can quickly respond to changes in demand, supply, or processes with little cost or time penalty. They often use a **just-in-time** (JIT) approach that focuses on reducing unnecessary inventory and removing non-value-added activities. This system uses a pull method in which the customer activates production. For example, an order for an auto chassis at a carmaker activates the manufacturing processes.

The Taguchi method is used to design and improve production systems. The Taguchi method is a technique for designing experiments that converge on a near-optimal solution for a robust system. The method uses the term *noise* to describe uncontrolled variations and states that a quality product should be robust to noise factors. The design of a firm's production system is critical to its overall success [Ulrich and Eppinger, 2004].

Companies strive for quality service at low cost. Google, the Web search firm, handles 170 million page views a day with a hardware plant of 12,000 servers (computers) that cost about $2,000 apiece. When a server fails, Google pulls it and replaces it immediately. Google has no fix-it department; it just pulls and inserts. With this approach, Google saves funds and keeps its system up 99.9 percent of the time [Karlgaard, 2003].

Entrepreneurial firms seek to develop unique capabilities by fostering an interactive dynamic between their capabilities and market opportunities. The market provides them with signals of opportunities, and they respond with new products. Adjustment of business models, production capabilities, and skill formation enable the firm to respond to opportunity. In value networks, firms focus on their core competencies and use others for complementary capabilities.

The goal of operations **throughput efficiency** (TE) may be measured by the formula

$$TE = \frac{VA}{VA + NVA} \qquad (15.1)$$

where VA = value adding time and NVA = non-value-adding time. Examples of NVA are waiting in a queue or system downtime. The goal is to reduce NVA.

Gentex, which makes rearview mirrors and other auto equipment, uses technology to control costs and increase throughput (see www.gentex.com). Gentex continually adds automation and monitoring systems in its factories and increased its throughput 30 percent from 2001 to 2003 [Green, 2003].

Jefferson Pilot Financial, an insurance company, formerly used the batch process for processing its applications and claims. To increase throughput, it used lean manufacturing principles to improve service and reduce costs. It built a "model cell"—a fully functioning model of the entire process. This approach allowed managers to experiment and work toward an optimal design. The team applied lean manufacturing practices, including placing linked processes near one another, balancing employees' workloads, posting performance results, and measuring performance and productivity from the customer's perspective [Swank, 2003].

15.3 THE VALUE WEB

A series of business activities can be thought of as a business process, as shown in Figure 15.5. Another way to describe a process is as a set of interrelated tasks accomplished in a network of activities. Instead of a value chain—a linear series of processes—the value-creating process can be organized as a **value web.** Webs are grids with no center but allow open communication and movement of items

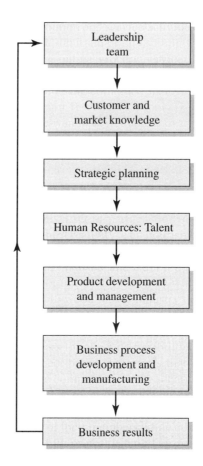

FIGURE 15.5 A business process is a series of activities.

and ideas. In a web, each participant focuses on a limited set of core competencies [Tapscott, Ticoll, and Lowy, 2000]. A value web is usually based on an Internet infrastructure to manage operations dispersed in many firms. The value web consists of the extended enterprise within a network of interrelated stakeholders that create, sustain, and enhance its value-creating capacity. The long-term success of a firm is determined by its ability to establish and maintain relationships within its entire network of stakeholders. It is relationships rather than transactions that are the ultimate sources of organizational wealth [Post, Preston, and Sachs, 2002]. The value web of a typical firm is shown in Figure 15.6.

The value web organized and operated by Amazon includes participants such as Ingram, Target, and Toys-R-Us. Amazon takes responsibility for choosing and offering the product selections, setting prices, and ensuring fulfillment. Cisco Systems leads a value web that provides its routers and computers to its customers. Cisco designs and markets the product, while others do most of the

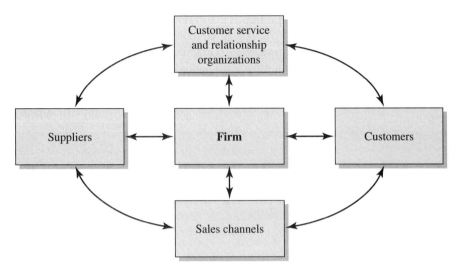

FIGURE 15.6 Value web for a firm.

manufacturing, fulfillment, and on-site customer service. The Cisco Systems value web is shown in Figure 15.7. Recall from earlier chapters that CRM means customer relationship management. Cisco defines the goals and coordinates the integration of the value web providers. Many new ventures will use the Internet to coordinate their value web effectively.

Consider the operations management strategy of IKEA, the Swedish furniture company that has 175 stores in 32 countries. Its business strategy is to make and sell inexpensive, solid, well-designed furniture through large stores. Its business process starts with the identification of a needed product and the specification of a low target price for this item. Next, IKEA determines what materials will be used and what manufacturer will do the assembly work. IKEA buys from about 1,800 suppliers in 55 countries. The next step is to design the item and select the parts. After manufacture, the item is shipped disassembled in a flat cardboard box to one of IKEA's 18 distribution centers and ultimately to one of its stores. IKEA sells its unassembled furniture without salespeople. The customer selects the item, gets the correct box from its rack, and brings it to the checkout counter. IKEA implements a low-cost, quality strategy through a far-flung value network that provides design, parts, and manufacturing in a coordinated manner.

In the past as companies grew, they added assets. As companies grow today, they tend to add relationships and enhance their value web. Orchestrating a value web is a powerful process for growth. Of course, it is possible to develop too complex a web of partners and lose control. Effective management of the value web enterprise requires a new conception of the firm as a network, rather than a hierarchy. The key to effective implementation is recognition of value web management as a core competence.

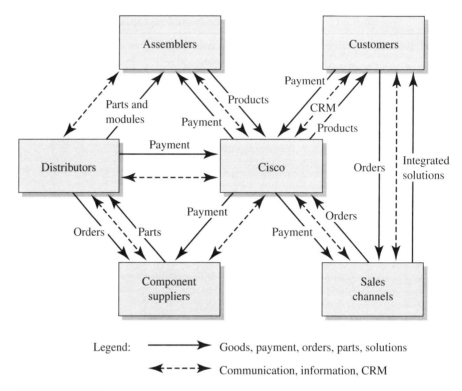

FIGURE 15.7 Value web of Cisco Systems.
Source: Adapted from Slywotzky, 2000.

In a fast-paced, competitive world, competitive advantage can result from the effective concurrent design of products, processes, and capabilities. Designing the product, how it is produced, and a supply chain that works harmoniously is critical to a firm's success. The coordinated product, process, and supply chain system is depicted in Figure 15.8 [Fine, 2002].

15.4 STRATEGIC CONTROL AND THE BALANCED SCORECARD

Strategic control is the process used by firms to monitor their activities, evaluate the efficiency and performance of these activities, and take corrective action to improve performance, if necessary. The goal is to keep the firm's operations on track with the performance goals of efficiency, quality, and responsiveness to customers.

To evaluate the effectiveness of their strategies, some companies are developing **balanced scorecards,** a set of measurements unique to a company that includes both financial and operational metrics. This gives managers a quick yet comprehensive picture of the company's total performance. The balanced

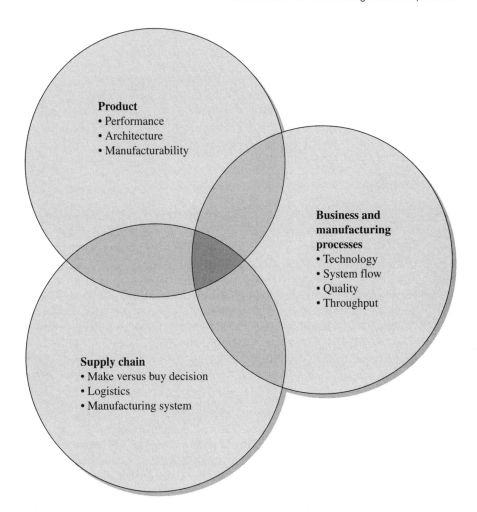

FIGURE 15.8 Coordinated design system of product, processes, and supply chain.

scorecard is a strategy formulation device as well as a report of performance. A successful balanced scorecard measures the tangible objectives that are consistent with meeting an organization's goals. The business operations area indicates how the operations and processes should work to add value to customers. The customer area indicates how the company's customer-oriented strategy and operations add financial value. The financial area measures the company's success in adding value to shareholders. The learning and growth area indicates how the infrastructure for innovation and long-term growth should contribute to strategic goals. A balanced scorecard is shown in Figure 15.9 [Kaplan and Norton, 2000].

   To build an effective scorecard, a firm needs to determine the fundamental drivers of performance and measure them. Finding the right measures such as reliability, quality, or customer satisfaction is challenging.

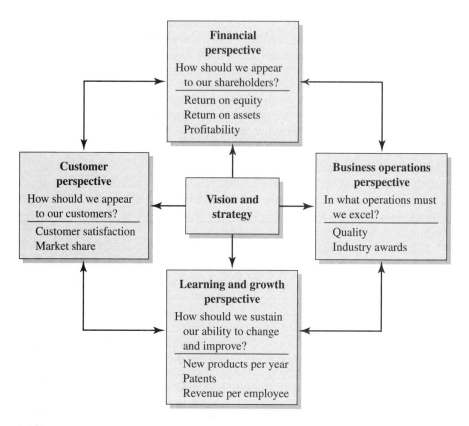

FIGURE 15.9 Balanced scorecard. Each perspective has a question and a set of measures.

Source: Adapted from Kaplan and Norton, 2000.

General Electric Vice Chairman Gary Rogers created the idea for a digital dashboard—the continuously updated online display of a company's vital statistics. GE's new "digital cockpits" now give 300 managers instant access to the company's essential data on desktop PCs and Blackberry PDAs [Tedeschi, 2003].

Jack Welch of General Electric created the idea of a boundaryless company, which eliminated the walls between suppliers, customers, and units of GE. GE's slogan was "Finding a Better Way Every Day," which addressed improvement of business processes. Then it added the idea of measuring performance of the processes. As a result, operating margins went from 1.2 percent in 1994 to 13.8 percent in 2000 [Welch, 2002]. Welch believed that the system of operations was the key to understanding, learning, and improving results.

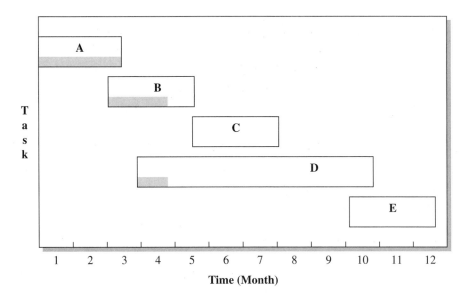

FIGURE 15.10 Gantt chart for five tasks. The actual progress is indicated by the shaded bars.

15.5 SCHEDULING AND OPERATIONS

New firms should develop diagrams and flowcharts that show how their operations work. Diagrams help communicate the process system to all concerned.

An operational plan outlines a number of actions that will be taken in the future. To consolidate the timing of events, the firm should prepare a schedule, in chart form, of all of the important milestones that the firm expects to reach in the near and intermediate term. A *Gantt chart* is a way to depict the sequence of tasks and the time required for each. Gantt charts, by using shaded bars on a grid, compare what was done with what was planned over time, as shown in Figure 15.10. Timelines are a visual means of comparing the actual and planned progress of a project or activity. Timelines allow participants to envision the ending of an otherwise open-ended plan [Yakura, 2002].

Effective management of operational processes requires schedules and coordination. Since Gantt charts are a means of contrasting the actual and planned progress of a plan over time, they depict both scheduling and coordination of separate tasks.

Enterprises of any size can profit from using Gantt charts to depict schedules and milestones. For example, completion of task B could represent the completion of a prototype, and task C could represent the testing of the prototype (see Figure 15.10). It is important to set milestones and then strive to reach them on time.

Another, perhaps more graphic, form of representing the activities, outcomes, and schedule is to use a milestone picture, as shown in Figure 15.11. This depicts a road map to achieve a scheduled outcome.

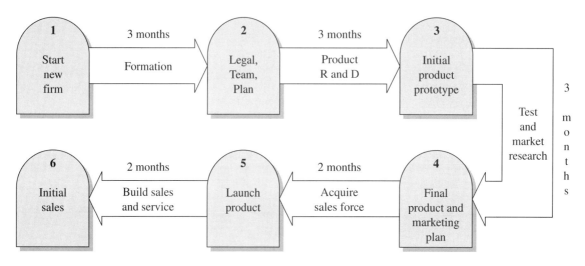

FIGURE 15.11 Example of a roadmap for a new technology firm.

15.6 AGRAQUEST

AgraQuest has two key operations: 1) product development and 2) product manufacturing. The product development process is shown in Figure 15.12. The two steps to develop a new product are discovery of the new natural microbe and

Developing effective natural pest management products

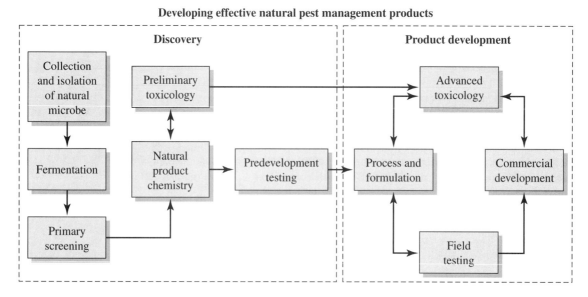

FIGURE 15.12 Product development process for AgraQuest.

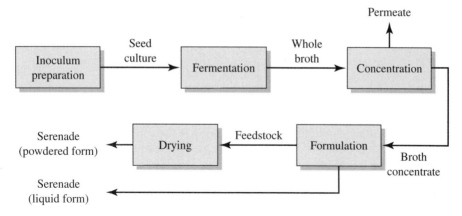

FIGURE 15.13 Manufacturing process of Serenade products.

development of a product based on that microbe. This process leads to microbial products to serve as pesticides or fungicides.

The design of the manufacturing processes takes place at the Davis, California, facility and is led by John Lin, director of process development. AgraQuest operates its own plant in Mexico, where it produces and ships its products. AgraQuest purchased the plant in December 2000 for $7 million and had 16 employees at the plant in 2003. The 208,000-square-foot plant sits on 35 acres near Tlaxcala, Mexico. The manufacturing process is shown in Figure 15.13.

AgraQuest has a system of business relationships based on the value web shown in Figure 15.14.

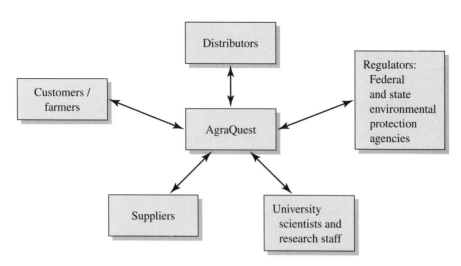

FIGURE 15.14 Value web for AgraQuest.

15.7 SUMMARY

A new venture needs to design a set of operational processes that will enable it to build, store, and ship the products provided to the customer. New businesses build a supply chain of partners that add value at each stage of the assembly or manufacture of the product. Service companies use business processes to put together their service outputs. The new venture manages its value chain to effectively provide the final product or service to its customer. The firm also needs to effectively manage the logistics of parts and materials. It strives to achieve the best possible coordination of its partners as well as its internal processes.

Many firms establish a set of interrelated activities as a network facilitated by an Internet-based value web. With a common schedule, associated tasks, and synchronization, the venture can manage the value web to maintain an efficient, on-time business process.

Principle 15
The design and management of an efficient, real-time set of production, logistical, and business processes can become a sustainable competitive advantage for a new enterprise.

15.8 EXERCISES

15.1   JetBlue Airways provides a laptop computer for its pilots with updated online flight manuals (www.jetblue.com) [Green, 2003]. What other functions shown in Figure 15.1 should be connected to a pilot's laptop computer?

15.2   Consider the challenge of managing the value chain and logistics of a home-delivery food and grocery service such as FreshDirect (www.freshdirect.com), Safeway (www.safeway.com), and PeaPod (www.peapod.com). The customer pays for high-quality groceries delivered reliably [Kirkpatrick, 2002]. FreshDirect and PeaPod use plants and logistic systems to package and ship orders accurately. Try a home delivery service, and then visit the facility that supplies your order. What suggestions do you have for the supplier of your groceries?

15.3   Southwest Airlines is profitable while other airlines are struggling. Southwest works to keep costs low and is prepared to respond to negative surprises. Low costs have kept its margins up. Talk to a Southwest employee about the airline's efforts to maintain operational excellence. Prepare a report of your findings.

15.4   Yellow Corporation has used technology to automate most of its trucking operations and schedules [Green, 2003]. Visit Yellow at www.yellowcorp.com and describe some of its scheduling and monitoring systems that enhance its abilities to attract and retain customers.

15.5 Consider a process to assemble a table or desk offered by IKEA or a local store. Obtain the instruction sheet for the table or desk, and draw a flowchart for the assembly process. Then create a Gantt chart for the assembly process.

15.6 Target Corporation uses technology to control and monitor its supplier chain and value web. Target offers receiptless returns and online shopping. It uses electronic signatures at checkout and electronic coupons for holders of a Target Visa card. It has a value-web replenishment system. Visit a local Target and report on its operational performance (see www.target.com).

15.7 The use of a new technology can bring new life to a mature industry such as the plastics industry. Logistics, supply-chain, and scheduling software enable large productivity increases in several mature manufacturing industries. Examples of such software firms are Moldflow, Quad, and Keane. Select one of these firms and describe an actual application for operations productivity improvement.

15.8 About 5 percent to 10 percent of pharmaceuticals produced do not meet specifications and have to be reworked or discarded [Abboud Hensley, 2003]. Quality testing is done by hand, and the batch process method is widely used. A new venture has been launched to design new processes for drug makers. What new methods and approaches should it develop to sell to drug makers?

15.9 Running a new venture is somewhat like piloting an experimental aircraft. Pilots need an instrument panel, and new companies need a business scorecard or company instrument panel. NetSuite Inc. was founded in 1999 to offer such a product (www.netsuite.com). Another firm that offers a business dashboard is Best Software (www.bestsoftware.com). Visit the websites of both companies and study the two dashboard products. Describe and contrast the price, attributes, and features of each product.

15.10 Using the format of Figure 15.6, prepare a value web diagram for Microsoft Corporation.

15.11 Prepare a road map diagram for the development and launch of a new model of a hybrid automobile to compete with the Toyota Prius.

15.12 Newman's Own is a for-profit firm that gives all its profits to charities. The firm was founded by Paul Newman and A. E. Hotchner on the basis of its first product, salad dressing [Hotchner and Newman, 2003]. The private company makes a line of foods including lemonade, popcorn, salad dressings, and spaghetti sauce. The firm had a profit of $18 million in 2003 and has donated $150 million over its 21-year history. The firm outsources most of its operations. Study Newman's Own (www.newmansown.com) and determine what operations it does not outsource. Why is this firm successful?

CHAPTER 16

The Profit and Harvest Plan

Profit is the product of labor plus capital multiplied by management. You can
hire the first two. The last must be inspired.

Fost

A new firm creates a sales model describing how it will generate revenues from its customers. Then it determines how it will generate profits from its revenues. The revenue and profit engines show how the firm will create powerful value for its customers and how customers will enable the new firm to profit. Many new ventures assume that profit will flow naturally from sales but discover that profits are not guaranteed. It is difficult to operate in a market that is chronically unprofitable.

A new firm seeks positive cash flow as soon as is feasible and acts to move to profitability early in its life. Managing revenue growth is important since uncontrolled growth can lead to negative cash flow and the need to constantly raise new funds from outside investors. Furthermore, a firm needs a plan to harvest the benefits of its growing venture for all owners. Entrepreneurs must also be realistic and accept that termination of the new venture is a possibility. ■

16.1 THE REVENUE MODEL

A firm's **revenues** are its sales after deducting all returns, rebates, and discounts. A firm's **revenue model** describes how the firm will generate revenue; five models are listed in Table 16.1. Most firms generate revenues by selling a product in units to a customer using a **product sales model.** For example, Dell sells its personal computers to one customer at a time, and Intel sells its chips to electronics companies.

In the **subscription revenue model,** a business offers content or a membership to its customers and charges a fee permitting access for a certain period of time. This model is used by magazines, information and data sources, and content websites. *Consumer Reports* offers its information to magazine subscribers (members) as well as to subscribers to its online service for a fee. This model is also used by clubs, cooperatives, or other member-based organizations.

The **advertising revenue model** is used by media companies such as magazines, newspapers, and television broadcasters that provide space or time for advertisements and collect revenues for each use. The media entities that are able to attract viewers or listeners to their ads will be able to collect the highest fees. MSN, Google, and Yahoo collect most of their revenues through the sale of advertising space.

Some firms receive a fee for enabling or executing a transaction. The **transaction fee revenue model** is based on providing a transaction source or activity for a fee. Examples of firms based on transaction fees are Charles Schwab, Visa, and eBay.

The **affiliate revenue model** is based on steering business to an affiliate firm and receiving a referral fee or percentage of revenues. For example, this revenue model is used in the real estate business and by companies that steer business to Amazon.

Magazines and newspapers such as the *New York Times* use both the subscription and the advertising models to generate revenues. Most new ventures use a mix of the five revenue models. Amazon.com, for example, uses the sales product, transaction fee, and affiliate revenue models.

Most start-up firms use a sales metric such as sales per employee to track their performance. Their goal is to exceed $100,000 per employee as soon as possible. Most mature growth firms exceed $200,000 per employee, and technology companies often exceed $400,000 per employee.

Google is a good example of a firm with multiple sources of revenue. Google generates revenues through advertising and licensing its technology to

TABLE 16.1 Five revenue models.

■ Product sales model.	■ Transaction fee revenue model.
■ Subscription (member) fee model.	■ Affiliate revenue model.
■ Advertising revenue model.	

others for a subscription fee. Other competitors collect fees from higher place-ment of a Web address on its list, but Google avoids that model.

Campusfood.com was challenged from inception to determine how to grow to be a billion-dollar company. Michael Saunders founded the firm in 1998. The firm receives an order online from a student on a campus and relays the orders to its participating restaurants. The firm's initial revenue model was advertising on its website. After struggling with that model, it switched to the affiliate revenue model. Each restaurant pays Campus-food.com a percentage of every order (www.campusfood.com).

Any new business needs to determine its revenue model and test it on po-tential customers. Who is the customer? Will the target customer pay for the of-fering? Consider the case of the Weather Channel, which was initiated in 1982 when cable television was building across the country, but lost money for sev-eral years. The Weather Channel assumed the TV viewer at home was its cus-tomer and tried to develop a revenue model. Perhaps, it thought, advertisers would pay to advertise on the channel. Having committed more than $30 million for satellites and the transmission system, the Weather Channel in a few years was on the brink of bankruptcy as a result of few advertisers signing on. At that point, the cable operators agreed to pay a subscriber fee to help keep the chan-nel operating. It turned out that the customer was the cable operator, not the viewer, and while the Weather Channel is not glamorous television, it is prof-itable today [Batten, 2002b].

16.2 THE PROFIT MODEL

Profit is the net return after subtracting the costs from the revenues. The **profit model** is the mechanism a firm uses to reap profits from its revenues. Block-buster's profit model is based on the receipt of rental fees and overdue charges on video and DVD rentals. Sears makes most of its profits from finance charges. General Electric generates one-half of its profits from its financial arm, GE Cap-ital. Newspapers make most of their profits from classified ads. Hewlett-Packard and Xerox make most of their profits from replacement toner cartridges.

Figure 16.1 shows the value of a product and the distribution of the value to the customers and the profit captured by the firm. To remain profitable, a firm strives to reduce its costs while maintaining or increasing the value of its prod-uct to the customer. To generate profits, a firm needs to examine all its activities on the value chain and determine if its cost versus value generated is in line.

Profit accrues to a company that maintains a competitive advantage as con-ditions change. When the PC industry took off in the 1980s, IBM ceded the op-erating system's rights to Microsoft when it incorrectly determined that profit

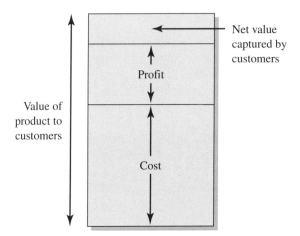

FIGURE 16.1 Value of a product and profit.

FIGURE 16.2
Revenue and profit
flows from the
firm's operations.

would flow to the branded integrator of hardware and software components. Key to profit capture is ownership of the unique, value-added element of the value chain or the product makeup. Examples would be ownership of an essential pipeline, control of the customer interface, or ownership of unique locations for a retail operation. It pays to hold the largest "value-added" step in a value chain or the unique innovation that no one else can match.

During the early years of a firm, the firm may be patient for growth but should be impatient for profitability. As a firm works to gain profitability, it is testing its assumptions that customers will pay for a profitable product [Christensen and Raynor, 2003].

The revenue and profit engines are driven by the firm's business model, strategy, resources, capabilities, operations, and processes, as shown in Figure 16.2.

The best conditions for profit occur when the perceived value of a product to a customer is high and the cost to produce the product is low. Figure 16.3 shows a value grid that enables a firm to determine its potential to reap large profits. Low cost to produce a product leads to low price per unit of perceived value to the customer. The upper-right quadrant is a high profit location that many firms seek to occupy [Chatterjee, 1998].

Most start-ups initially invest time and energy in learning about their customers. Then they use that knowledge to create improved solutions for them. They lose money initially but make money after a period we will call T, as shown in Figure 16.4. Of course, it is best to keep T relatively short and the peak negative profit (NP) small [Slywotzky, 2002]. The profit curve shown in Figure 16.4 is often called a "hockey stick" expectation.

It is useful to try to estimate the attractiveness (the potential profitability) of a market segment. It is of little value to win market leadership in a market seg-

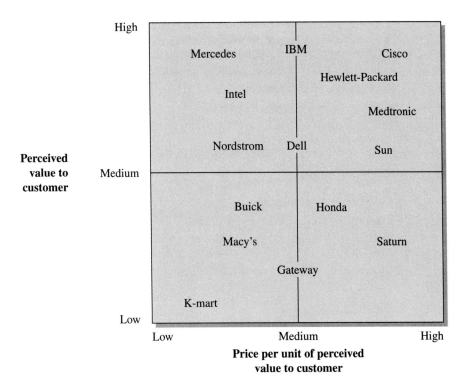

FIGURE 16.3 Value grid.

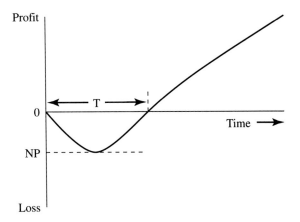

FIGURE 16.4 Early losses of a successful start-up turn profitable after a time, T. The peak negative profit is NP.

TABLE 16.2 Metric for profitability performance for selected firms.

x	Metric	Example firm
Customer	Profit / Customer	Gillette
Employee	Profit / Employee	Abbott Labs
Customer visit	Profit / Customer visit	Walgreens
Tons of output	Profit / Tons	Nucor
Revenue	Profit / Revenue	Safeway

ment that is chronically unprofitable [Ryans et al., 2000].

A firm can create a metric for its profitability as

$$\text{metric} = \frac{\text{profit}}{\text{x}}$$

when x is chosen to fit the firm's goals and business model. As shown in Table 16.2, firms chose the variable x so as to illuminate their profitability performance. A commonly used profit metric is **profit margin,** which is the ratio of profit divided by revenues.

During the telecommunications boom of 2000, companies often used poor indirect metrics of their growth such as the number of building leases. They reasoned, wrongly, that access to office buildings would translate to customers [Malik 2003].

A business model designed for high customer relevance that delivers high value will have the best chance of capturing profit. The best business model helps the customer in the difficult or time-consuming areas of their purchasing process. One of the most powerful profit models is the **installed base profit model.** The supplier builds a large installed base of users who then buy the supplier's consumable products. This is the model used by Gillette, which sells a razor at a modest price while building a large base of users who purchase consumable razor blades. Nine types of profit models are listed in Table 16.3 [Slywotzky et al., 1999]. A new venture will wisely select its needed profit model and work hard to build its strength and resiliency in the competitive marketplace.

Managers have always known that some customers are more profitable than others. For some emerging firms, 20 percent of the firms' customers may provide most of the firms' profits. Furthermore, the firm's worst customers may be costing more than they pay for products or services. Securing profitable customers while getting rid of unprofitable customers can help to double a firm's profits. It pays to be attentive to the best customers and ignore the worst [Selden and Colvin, 2002].

Recall that profit is

$$\text{Profit} = (P–VC)Q – FC \tag{16.1}$$

TABLE 16.3 Nine types of profit models.

Name	Description	Examples
1. **Installed base**	Build a large installed base of customers and sell consumables or upgrades	Gillette Hewlett-Packard printers
2. **Protected innovation**	Create a unique, innovative product and protect it using patents and copyrights	Merck Microsoft
3. **New business model**	Find unmet customer needs and build a new business model	Starbucks Google
4. **Value chain specialization**	Specialize in one or two functions on a value chain	Nucor Intel
5. **Brand**	Create a valued brand for your product	Intel Coca-Cola
6. **Blockbuster**	Focus on creating a series of big winners	NBC Schering Plough
7. **Profit multiplier**	Build a system that reuses a product in many forms	Disney: movies, and theme parks Virgin Group
8. **Solution**	Shift from product to unique total solutions	General Electric Microsoft
9. **Low cost**	Create a low-cost product to offer low price per unit of value	Southwest Airlines Dell

where P = price, VC = variable costs, Q = total number of units sold, and FC = fixed costs. Managing profitability can be achieved by lowering fixed or variable costs, raising units sold, or raising price. One may be able to find a customer segment that is willing to pay a higher price or purchase more units. Otherwise, costs need to be lowered.

Another important measure of performance is **cash flow,** which is the sum of retained earnings minus the depreciation provision made by the firm. Without positive cash flow, a firm may eventually use up all its cash and close its doors. A profit and cash flow model focuses attention on the nature of the driving forces of the revenue and profit models.

Amazon.com has used aggressive price discounting and free shipping to boost its revenues. However, its low operating margin (P–VC) made profitability elusive.

Many mass-market retailers such as Wal-Mart continually lower prices by squeezing inefficiencies from their operations and sacrificing profit margins on products in favor of selling in high volumes. Hewlett-Packard tries to avoid lowering prices to keep its profit margin robust.

All entrepreneurs need to find a suitable profit model for their firm. If profitability appears to be highly elusive and at best in the distant future, it may be wise to not proceed with the venture. We discuss the matter of terminating a venture in Section 16.5.

Candice Carpenter and Nancy Evans met at a business meeting in 1994 and got together in 1995 to consider starting an online start-up. Both had worked in the TV-magazine industry and wanted to build a media-based business using the Internet. They founded iVillage (www.ivillage.com), an online community for women. The site addresses issues such as parenting, health, fitness, food, money, and careers. The company issued an initial public offering in 1999 at $24 per share. The company used sponsorships by companies while avoiding using ads for revenues. Its profit engine is based on subscriptions, which are subject to the volatility of economic conditions.

16.3 MANAGING REVENUE GROWTH

New businesses normally strive to build up revenues and profits so that they can meet their goals. Most entrepreneur teams are naturally inclined to grow their business rapidly. Other entrepreneurs limit the growth of their firm for personal or lifestyle reasons. The degree of commitment of the entrepreneurial team to the growth of the firm can be called **entrepreneurial intensity** [Gundry and Welsch, 2001].

Commitment to growth leads to sacrifices an entrepreneur is willing to make. High growth can require significant financial resources, leading the entrepreneurs to seek outside capital and often give up majority ownership of the firm. Low growth would include firms growing revenues at a rate less than 10 percent per year, and high rates of growth would exceed 25 percent per year. Many high-growth firms grow at 50 percent or higher each year for several years after founding. Technology entrepreneurs who seek a high-growth strategy will usually select a team-based organizational structure and exhibit high entrepreneurial intensity. Furthermore, they are willing to endure the burdens associated with the demands of high growth. The characteristics of a high-growth entrepreneurial team are listed in Table 16.4. High-growth entrepreneurs are willing to put

TABLE 16.4 Characteristics exhibited by entrepreneurial teams that seek high growth rates.

■ Strong entrepreneurial intensity.	■ Emphasis on a team-based organizational structure.
■ Willingness to incur the costs of growth.	
■ Willingness to use a wide range of financing sources.	■ Focus on innovation.

aside some of their personal or family goals and make sacrifices because they are committed to the growth of their ventures.

Leasing of a firm's products to customers can delay revenues into the future and provide long-term cash flow. Because IBM leased computers and office machines, customer service representatives remained close to customers and keep the relationship strong.

A growing business requires cash for working capital, assets, and operating expenses. If a company grows too fast, it will need to continually raise additional cash from investors. The cash required will depend on the operations model of the firm, which depends on its accounts payable cycle, as well as assets and working capital required [Churchill and Mullins, 2001]. In general, most growing firms are unable to sustain a growth rate of sales exceeding 15 percent using their internally generated cash. Some service businesses are less asset-intense and may be able to self-finance a sales growth rate of 20 percent to 30 percent. Very few, if any, firms can self-finance if they plan to grow at 50 percent or more per year. For high growth, a financing plan for outside cash will be required.

Since service businesses are often less asset-intensive and more labor-intensive than production firms, growth typically adds to costs and may not produce the economies of scale. Some companies with a heavy emphasis on service such as IBM, Dell Computer, and Southwest Airlines have managed to successfully combine growth and profitability while other companies have not. Successful service-oriented companies are able to design and implement the right strategies to keep costs low, strengthen customer loyalty, and gain competitive advantage.

The profitability of a firm may be a function of the growth rate of revenues, as shown in Figure 16.5. At a low growth rate such as G_1, the firm is unable to meet demand and loses sales to its competitors. At a very high growth rate such as G_2, the firm is unable to efficiently manage its operating systems, and profitability P_2 is achieved. Growth rate G_m maximizes a firm's profitability P_m. An emerging or new firm should try to estimate its growth rate G_m that would maximize profitability. For many new firms, G_m ranges between 20 percent and 40 percent.

The profitability of a firm may be represented by its return on capital or return on equity (ROE). Thus, one quick estimate of a firm's ability to grow is to state that a firm may grow organically at a rate less than its return on equity without turning to outside financial sources. We define **organic growth** as growth enabled by internally generated funds.

A more complete equation for a sustainable change in sales-to-sales ratio, ΔS/S, for a firm is [Ross, Westerfield, and Jaffe, 2002]:

$$\frac{\Delta S}{S} = \frac{PM\,(1 + L)}{T - [PM(1 + L)]} \tag{16.3}$$

where PM = profit-to-sales ratio, L = debt-to-equity ratio, and T = the ratio of total assets to sales. If a start-up firm has no debt (L = 0), we have

$$\frac{\Delta S}{S} = \frac{PM}{T - PM}$$

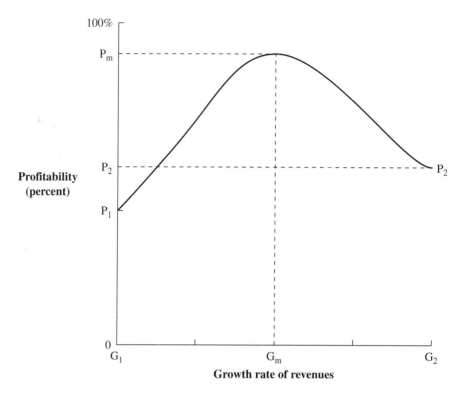

FIGURE 16.5 Profitability of a firm as a function of the growth rate of revenues.

For example, if PM = 0.10 and T = 0.5, then

$$\frac{\Delta S}{S} = \frac{0.10}{0.5 - 0.10} = 0.25$$

or the sustainable sales growth rate is 25 percent. Consider an asset-intensive business with T = 1.0 and determine the sustainable growth rate when PM = 0.10. Then, we have

$$\frac{\Delta S}{S} = \frac{0.1}{1 - 0.1} = 0.11$$

and the sustainable growth rate is 11 percent. If this asset-intensive firm uses debt so that L = 0.8, then

$$\frac{\Delta S}{S} = \frac{0.1(1 + .8)}{1 - [0.1(1 + 0.8)]} = \frac{0.18}{1 - 0.18} = 0.22$$

and the sustainable growth rate is 22 percent.

A start-up will need to examine its expected growth rate and its financing needs carefully. As an example, consider the growth of Extended Stay America

(ESA), which was formed in 1995 and raised $68 million via an initial public offering later that year. Its sales were $130 million in 1997 and $518 million in 2000, growing at an annualized rate of 44 percent. During the same period, the firm's long-term debt rose from $135 million to $947 million, and its debt-equity ratio rose from 0.16 in 1997 to 0.96 in 2000. ESA was able to grow sales at a rate of 44 percent by increasing its debt (and associated risk) significantly.

Service firms require less money to start and expand than asset-intensive industries. Building and growing a service business requires adding employees. A service business has low capital and asset intensity and requires little debt. Consider Robert Half International (www.rhi.com), which grew revenues from $220 million in 1992 to $2.45 billion in 2001 at an annual rate of 29 percent. The firm has negligible debt, thus L=0. The profit-to-sales ratio averaged 0.06 during 1992 to 2001. The ratio of total assets to sales was approximately 0.15. Then, the sustainable growth in sales was

$$\frac{\Delta S}{S} = \frac{PM}{T - PM} = \frac{0.06}{0.15 - 0.06} = 0.67$$

or the sustainable growth rate was 66.7 percent.

To grow steadily and avoid stagnation, a company should learn how to scale up and extend its business, lengthen its expansion phase, and accumulate and apply new knowledge to new products and markets faster than competitors. Entrepreneurs should choose a plan that fits with the knowledge, learning skills, and assets that the organization possesses or is developing.

Rapid growth and good profitability can often cover up some underlying problems of an organization. They can provide a cushion for wasteful decisions regarding the allocation of financial, human, and other resources. The excitement of growth can also veil inadequacies in leadership or management skills. Growth can mask a lack of planning or an inadequate orientation toward long-term issues. Success can disguise a variety of shortcomings while breeding a dangerous form of arrogance.

The incentives for growth by a new venture are several, as summarized in Table 16.5. One incentive to grow is to attract capital investment to expand markets and product lines. Also, growth creates a sense of pride among the employees and provides opportunity for expanding financial reward.

A firm's ability to use and coordinate new assets and activities depends on its organization and managerial capabilities. Rapid growth can challenge those capabilities severely. It is important for a publicly held company to have consistent, predictable financial growth. Consistency requires controlled growth of

TABLE 16.5 Incentives for growth by new ventures.

■ Attracting capital for market expansion.	■ Development of a reputation and brand.
■ Attracting capable team members.	■ Growing profitability and financial rewards for owners and employees.
■ Achieving economies of scale.	

assets and personnel additions. Growing a staff at a rate greater than 15 percent a year will challenge any organization. Paychex Inc. is a $1.3 billion payroll-processing firm that has increased revenues at an average of 18 percent for many years. With economies of scale, it has increased profits by 20 percent each year.

Most companies employ a mix of organic growth using both inside resources and external sources of resources. A balanced approach to growth attempts to break down barriers to growth and improve the company's core competencies. Some firms, such as Linear Technology and Dell, have been successful in building revenues at 30 percent or more per year while maintaining a balanced mix of internally and externally financed growth.

Most highly innovative firms become high-growth firms, compared to low-innovation firms [Kirchhoff, 1994]. Microsoft, for example, grew its revenues at 29 percent per year for the decade 1991 to 2001 and grew its profits at 35 percent per year for the same period.

Sam Walton bought a store in Bentonville, Arkansas, in 1950. By 1952, sales reached $95,000. Walton wanted to grow a chain of stores, so he opened a second store in Fayetteville in 1952. By offering them percentage of the profits, he lured store managers away from competitors. Walton, with his brother, went on to open another store in 1954. The success of his first three stores impelled him to set his sights on becoming a discount store magnate. Walton started opening stores in rapid succession. These new stores were financed by cash flow from the existing stores and investments from friends, store managers, and his family. Walton personally borrowed much of the cash needed as he launched the first Wal-Mart in 1962. As Walton built the chain of Wal-Marts in the 1960s, he used internal and personal cash.

Walton had plans to open at least a dozen stores per year. He knew that he would need a more permanent source of capital. In late 1969, Walton arranged with Stephens Inc. of Little Rock, an off-Wall Street investment banking house, to underwrite an initial public offering of Wal-Mart common stock. The IPO was delayed, so Walton turned to Mass Mutual Insurance Company, which agreed to lend $2.5 million. In October 1970, the IPO was completed, raising $4.6 million to fund growth. Walton never faltered in his growth plans, and his successful performance enabled him to receive the funds for further expansion. Volume buying and a low-cost delivery system enabled Wal-Mart stores to offer name-brand goods at discount prices in locations where there was little competition from other retail chains. As a result, the Wal-Mart chain experienced tremendous and sustained growth, growing to 190 stores by 1977, 800 stores by 1985, and 3,500 stores in 2004.

The sources of revenue growth for a firm can include increasing brand recognition and international expansion, as shown in Table 16.6. Very few firms are able to increase prices in the face of tough competition. New offerings of valuable products are a good source of revenue growth. Perhaps the most successful companies active in new product offerings are Microsoft and Intel.

The market value of a firm can be described as a result of three drivers, as shown in Figure 16.6 [Rappaport, 2001]. Changes in volume, price, and sales

TABLE 16.6 Sources of revenue growth.

■ Increasing brand recognition	■ Acquisition of other firms
■ Intellectual property licensing	■ Price increases
■ International expansion	■ New product offerings

mix lead to changes in sales growth rate. Operating profit margin (profit before taxes divided by revenues) is driven by four factors, as shown in Figure 16.6. Incremental investment rate is the result of investment efficiencies. Operating leverage is the ratio of profit margin increases to upfront preproduction expenses for new product development and capacity expansion. Accounting for all these factors, a firm seeks to increase its efficient use of investments and reduce its operating costs. Furthermore, increased volume, operating leverage, economies of scale, and an improved price and sales mix can lead to an improved sales growth rate and operating profit margin.

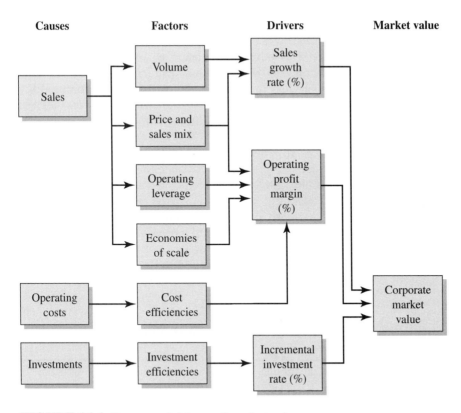

FIGURE 16.6 Causes and drivers of market value.

Note: Operating leverage = $\dfrac{\Delta \text{Profit margin}}{\Delta \text{Costs for product development and manufacturing}}$

Wireless devices such as cell phones and laptop computers use batteries for power. The replacement for a battery may be a small hydrogen fuel cell. Stephen Tang and his team at Millennium Cell are developing a miniature fuel cell that can run a laptop for up to eight hours. Their fuel cell can be discarded or recycled and may cost about $2 [Wolley and Hardy, 2003]. Using the factors of Figure 16.6, Tang must increase his investment and cost efficiencies and improve his product development capabilities before Millennium is able to launch its product (See www. millenniumcell.com).

16.4 THE HARVEST PLAN

Assuming a successful venture, any investor will want to know the plan for providing a cash return to all investors in a timely way. Assuming the venture has a favorable outcome in a few years, how will the investors reap or harvest their fair share of the wealth created by the venture? A **harvest plan** defines how and when the owners and investors will realize or attain an actual cash return on their investment. It delineates how and when they will extract some of the economic value from the investment. Professional investors will expect a return on their investments within five to seven years. Thus, investors will expect a plan for cash liquidity for themselves. Note that "harvest" does not mean the challenges and responsibility of the business are over.

For high-growth firms, the value created by an innovative venture can lead to significant returns over a five-year period. Both the founders and the investors will desire to access the financial return accrued by the new growing firm at the end of that period. This will mean some action will be necessary to yield a cash flow from the firm to the investors and owners. Table 16.7 lists five methods of harvesting a firm. The sale of the entrepreneurial firm to an acquiring firm is an attractive route for the founders and investors. Proceeds from the sale of a private company usually consist of cash, shares of the acquiring company, or a combination of shares and cash.

TABLE 16.7   **Five methods of harvesting the wealth created by a new firm.**

■ Sale of the firm to an acquiring firm.	■ Sale of the firm to the managers and employees.
■ Sale of the firm's stock on a public market through an initial public offering.	■ Transfer of the firm through gifts and sales to family successors.
■ Issuance of cash dividends to the owners and investors.	

Fast-growing companies with annual revenues greater than $20 million may find a solution in the public stock market by using an initial public offering (IPO). If the investors are patient, the issuance of cash dividends to individuals can serve to provide cash to the investors. Of course, it is often possible to arrange a sale of the firm to the managers and employees of the firm. Finally, many owners of relatively small firms will consider passing the firm on to family successors.

The selection of a harvest strategy will depend on the interests of the founders and investors. Professional investors such as venture capitalists expect large annualized returns and normally seek the issuance of an initial public offering by year five or six. Alternately, the venture may be acquired by another, typically larger, firm that provides the liquidity sought by the professional investors.

The entrepreneurial team may describe a plan to harvest their venture after a specified period. This plan will be part of the negotiation with the investors.

A planned sale to employees and managers may be outlined in the business plan. This transfer can use an employee stock ownership plan (ESOP). The firm first establishes an ESOP and guarantees any debt borrowed by the ESOP for the purpose of buying the company's stock. Then the ESOP borrows money from a bank, and the cash is used to buy the owner's stock. The shares of the firm are held by a trust, and the company makes annual tax-deductible contributions to the trust so it can pay off the loan. As the loan is paid off, shares are released and allocated to the employees. While an ESOP benefits the owner by providing a market for selling stock, it also carries with it some tax advantages that make the approach attractive to owner and employees alike.

SafeRent Inc. was a small but growing company that performed credit and background checks on prospective tenants for apartment building owners in the Denver area. SafeRent was founded in 1998 by Linda Bush. As the firm grew, its primary competitor was First American Corporation of Santa Ana, California. Bush and her investors reluctantly sold their firm to their biggest competitor in late 2002. Sometimes, while tough to do, it may be best to harvest the wealth from a firm by selling to a competitor [Bailey, 2002].

Few events in the life of the entrepreneur or the firm are more significant than the harvest. Without the opportunity to harvest, a firm's owners and investors will be denied a significant amount of the value that has been created over the firm's life. The founders may need a harvest strategy due to a desire to retire or diversify their portfolio of assets. Investors may need to realize their returns to invest them elsewhere or benefit in other ways. The timing of the harvest may be uncertain, but a harvest strategy does help the entrepreneur team plan together for the future.

A good time to sell a company is when it is very successful. When that time arrives, it may be best to review or exercise the harvest plan. At that time, a firm needs to determine a realistic valuation for the firm and obtain advice from its board of directors. Entrepreneurs often choose to sell for personal, nonfinancial reasons, such as burnout from the long hours and high stress of running their own businesses. Entrepreneurs who have raised money from family and friends may be especially eager to sell, since they feel a heightened pressure to return their investors' money. Moreover, entrepreneurs typically have much of their personal wealth tied up in a single company, making them eager to sell so they can diversify their holdings.

When entrepreneurs decide to sell, their choice of buyers is about more than price. Company leaders often choose buyers based on "soft" criteria such as strategic and organizational fit. They care about the fate of their employees as well as whether their strategic vision will be carried on. It is rare for entrepreneurs to hold a formal auction for their company; instead, they hold informal discussions with a small number of potential buyers with whom they see a good fit [Graebner, 2004].

Home Depot was formed in 1976 by Bernie Marcus and Arthur Blank. They raised $2 million from an investor group. By 1980, Home Depot was profitable, and it issued an IPO in September 1981, raising $3 million. Subsequently, the two founders sold some of their shares, with Marcus grossing $8.7 million and Blank grossing $6.5 million. This was their way to harvest some of their gains. They held on to the majority of their shares but always had the advantage of liquidating shares over time [Marcus and Blank, 1999].

In 1994, David Edwards, a postgraduate researcher in Robert Langer's lab at MIT, began working on a novel drug delivery system that used large, porous particles to deliver drugs directly to the lungs. Although the technology was entering a seemingly crowded market, Edwards's idea had the potential to perform significantly better than other forms of inhalation delivery systems. From the technology's initial stages, Langer, who had founded several successfully biotechnology firms, recognized the commercial potential of Edwards's idea, but an attempt in 1995 to license the technology to a public drug delivery firm was unsuccessful.

Reluctant to start his own company when the technology was still preliminary, Edwards left MIT in early 1995 for a faculty position at Pennsylvania State University. While at Penn State, he continued to refine the delivery system and to visit Langer at MIT about once a month. By early 1997, Edwards and Langer were sufficiently pleased with their progress to approach Terry McGuire, a Harvard Business School graduate and recent founder of Polaris Ventures. Langer knew McGuire because he had invested in several of Langer's ideas. McGuire was initially hesitant to invest in the novel technology, which was entering a crowded market. His faith in

Langer, however, as both a stellar scientist and entrepreneur, along with an influential *Science* article and external affirmation of the importance of Edwards's technology, convinced McGuire to put aside his uncertainty. In the summer of 1997, McGuire invested $250,000 in return for 11 percent of Advanced Inhalation Research (AIR) and an option to purchase an additional 9 percent. McGuire took on the role of temporary CEO, and Edwards took a leave of absence from Penn State to return to Boston to work on the idea full-time. In January 1998, the company opened its headquarters in Cambridge, Massachusetts. The first three employees were all previously affiliated with the chemical engineering department at MIT.

Once established in their new offices, Langer and Edwards quickly went to work on their first human clinical trial. As their science progressed, so too did interest from outside parties. The founders decided on a two-tier strategy. For drugs that had gone off patent, they would manufacture generic versions of the drugs themselves. For newly developed drugs, AIR would partner with leading pharmaceutical firms to exclusively manufacture the drugs for the delivery system. McGuire focused on successfully closing the deals with minimal dilution of Polaris's stake in the firm.

For a nascent firm, AIR was extremely successful. It had raised capital for an early-stage technology, was making progress in its clinical trials, and was forming favorable partnerships. Nevertheless, when suitors began knocking at the door, AIR needed to consider their offers seriously. To continue to grow, AIR would need to scale up its operations significantly and move from a small R and D firm to a larger manufacturing operation. This task was extremely difficult and one that only a small handful of biotechnology firms had achieved. In late 1998, just a year and a half after AIR was founded, Alkermes offered to acquire the firm. In February 1999, AIR agreed to Alkermes's offer of an all-stock deal valued around $125 million. The deal was considered a success on all sides. Polaris received a healthy return on its investment, AIR's technology was a strategic fit for Alkermes's existing drug delivery systems, and Alkermes's savvy management team and established reputation in the industry allowed it to find the most appropriate partner for AIR.

All three founders remain active in the biotechnology industry. Edwards, now a professor at Harvard University, co-founded Pulmatrix in the spring of 2003 and has been involved in the formation of a number of nonprofit organizations. McGuire remains a general partner with Polaris Ventures and has invested in other early-stage life science firms. Finally, Langer continues to be wildly prolific. He holds more than 380 patents, has licensed his technology to more than 80 firms, and was named one of the 100 most important people by CNN and *Time* magazine.

Sources: Roberts and Gardner, 2000 and www.alkermes.com.

16.5 EXIT AND FAILURE

A large percentage of new ventures shut down within a few years of initiation. Some terminate their efforts when they fail to achieve the original goals; others terminate when they simply run out of cash. Most entrepreneurs and investors assert that new ventures fail because of inadequate management skills, a poor strategy, and inadequate capitalization as well as poor market conditions [Zacharakis, Meyer, and DeCastro, 1999]. For most entrepreneurs, an inadequate team with inadequate past experience leads to failure.

Many people learn from business failure by revising their knowledge and assumptions about their skills [Shepard, 2003]. Learning can be measured in terms of an increased understanding of why the business failed, in an effort to prevent repeating the same mistakes. Often entrepreneurs are overconfident and unrealistically overrate their knowledge and skills. They discount the risks inherent in a new venture. Furthermore, they may exaggerate their ability to control events and people. Examples of highly visible, big failures are Planet Hollywood and Webvan.

Knowing when to stop or terminate a venture may be as important as knowing when to start. The concept of **sunk costs** is that a cost that has already incurred cannot be affected by any present or future decisions. In other words, funds and time invested on a new venture are already gone, regardless of any action you take today or later. If a new venture has not worked out as planned, one should look at proceeding with the venture as a new decision, as shown in Figure 16.7. The decision to terminate or continue should, if possible, be looked at afresh with the information at hand. If a venture has run out of cash and the market has not responded to the new venture, it may be wise to exit the venture.

Pandesic was a joint venture of SAP and Intel designed to develop information architectures for Internet companies. Founded in 1997, Pandesic intended to create a unique e-business solution that would automate the entire business process for companies doing business on the Internet. By March 1997, Pandesic

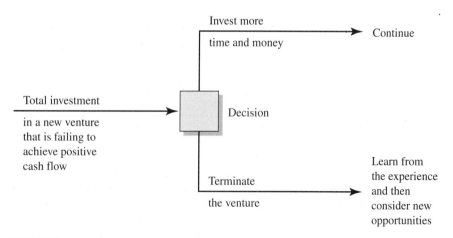

FIGURE 16.7 Decision tree for the sunk cost dilemma.

had grown to 100 employees. By the end of 1997, sales were slowly appearing, and by 1998, it was still struggling without many sales. By April 1999, the sales force was reorganized. As a result, sales started to grow, but the firm was still unprofitable. Pandesic had 400 employees and 100 customers by mid-2000, but it was still experiencing negative cash flow (said to be $80 million per year) and closed its doors in July 2000. The decision to terminate was based on large cash losses and recognition that profitability was not within reach [Girard, 2000].

If the decision is to continue, it may be wise to consider the next phase of the company as a turnaround and devise deliberate interventions that increase the level of communication, collaboration, and respect among all the participants [Kanter, 2003].

If the decision is to terminate the new venture, it is best to try to learn from it. Ideally, it is the venture that failed, not the people in it. Picking oneself up, learning from the venture, and then moving on is the best process. Every exit is an entry somewhere else.

The alternative of investing more time and money should be based on rational recognition of the sunk costs and the potential for recovery and success. An entrepreneur can examine the situation afresh and decide if the opportunity still looks good enough to invest more time and money.

16.6 AGRAQUEST

AgraQuest is a premier social responsibility company. It won the Green Chemistry Award from the U.S. Environmental Protection Agency in 2003 and the best presentation award at the CleanTech Venture Conference in 2003. It has a code of ethics and a sustainable business statement that all employees are required to read and sign.

Regrettably, a solid reputation is only one component of a revenue engine. AgraQuest has a sales revenue model based on selling units (pounds or gallons) of biopesticides. Its profit model is based on a protected innovation via patents and its emerging brand (reputation) in the biopesticide industry. To become profitable, AgraQuest needs to increase the quantity of units sold (Q) and reduce its variable costs (VC), as described in Equation 16.1. It has high fixed costs (FC), and is underutilizing its manufacturing plant.

AgraQuest's growth plan calls for doubling sales revenues each year for the next several years. With unused capacity at its plant, this should be achievable but will require increased working capital. Therefore, several million dollars of working capital will be sought through a line of credit from its bank.

Getting accepted in the pesticide marketplace has been AgraQuest's toughest challenge. Conventional, nonorganic growers are wary of "greener" solutions because of a long history of untruthful claims made by companies peddling unreliable products.

The harvest plan for AgraQuest is either an initial public offering of its shares of common stock or acquisition by a large agricultural technology firm such as DuPont or Monsanto. This may be achievable once AgraQuest reaches sales exceeding $10 million and is profitable.

16.7 SUMMARY

A new firm formulates its revenue model to clearly describe how it will generate and grow its revenues. Revenues are important, but positive cash flow and profitability are critical to ultimate success. Thus, a profit model that can be readily implemented must be created early in a firm's planning. Profit does not occur naturally but rather flows from value shared with the customer. Thus, enough customers must experience high value and find that the firm is the best, if not the only, provider of this value.

Managing revenue growth to match the growth of cash flow is important to achieve organic, internally funded growth. Otherwise, the new firm must constantly seek new financial resources from investors and lenders. Many firms find it necessary to terminate their activities when they are unable to access new sources of funds. With positive results and reasonable growth, a firm should consider a plan to harvest the rewards so that all owners receive a financial return on their investment.

> **Principle 16**
> A new firm with a powerful revenue and profit engine and a reputation for ethical dealings can achieve strong but manageable growth leading to a favorable harvest of the wealth for the owners.

16.8 EXERCISES

16.1   Ballard Power Systems has been developing fuel cells for car engines since 1983 in Burnaby, British Columbia, Canada (www.ballard.com). Fuel cells are seen by many as the replacement for the internal combustion engine. Nevertheless, profitability may still be years in the future [Carlisle, 2002]. For all their clean-air attributes, fuel-cell cars won't appeal to consumers until their price comes into line with that of other automobiles. What revenue model and profit model is Ballard using? What measures should Ballard adopt to improve its revenue and profit models?

16.2   Honest Tea was founded in 1998 to offer bottled teas and tea bags (see www.honesttea.com). Its vision was a bottled tea low in calories, healthy, and organic with good taste. By 2000, it was also selling tea in bags. Try to find one of its products at your local grocery store. Honest Tea's harvest strategy is to build the company to a significant sales figure and sell it to a large firm like Pepsi. What level of revenues will it need to attract Pepsi or Coca-Cola?

16.3   Whole Foods Markets Inc. owns and operates the largest chain of natural food supermarkets in the United States. Started by John Mackey in 1980, it operates 130 stores in 23 states. Sales grew at an annual rate of 18 percent over 10 years, reaching $3.2 billion in 2003. Earnings grew at 19 percent over the same period. Determine the debt-to-equity ratio of the firm and its profit margins, and calculate the firm's sustainable growth rate. Discuss the methods used by Whole Foods to grow sales at 18 percent over the past decade. (See www.wholefoods.com.)

16.4 E. & J. Gallo Winery was founded in 1933 in Modesto, California, by Ernest and Julio Gallo and remains a privately held, family business. Gallo is the largest wine producer in the world, with $1.5 billion in sales. Currently, the second generation of the founder's family is running the winery while the third generation is rising in leadership of the firm. Describe the challenge of succession in this important firm. What strategies does the firm use to enable its owners to participate in the harvest? What methods for harvest would you suggest?

16.5 Iceberg Industries of Newfoundland, Canada, is focused on harvesting icebergs and processing them for pure water. It believes people will pay for pure water for drinking and for use in vodka production [Curtis, 2002]. Is the extra cost of pure iceberg water worth it to customers? (See www.icebergindustries.com.) What is the revenue and profit engine for this firm?

16.6 Two magazines, *Time* and *Consumer Reports,* are financially successful but have different revenue models. Using Table 16.1, describe the revenue model for each magazine.

16.7 The Web business iVillage has had a difficult time building a revenue model and a profit model that works. Determine its revenue and profit models and describe improvements for iVillage. (See www.ivillage.com.)

16.8 Salesforce.com sells software as a service delivered online. Corporations pay about $60 per month for each user. Salesforce was founded by Marc Benioff in 1999 (www.salesforce.com). Describe its revenue model and profit model.

16.9 Google, the search engine firm, uses a complex revenue model and a related profit engine. It has a large base of users and advertisers, and works to link its users to retailers. It also offers Web logs, or "blogs," where users can post messages or comments. Describe its revenue and profit model.

16.10 Amgen, the world's largest biotechnology company, is also one of the world's most profitable. Annual revenue and profits have grown faster than 18 percent from 1998 to 2003. Amgen's debt-to-equity ratio (L) is about 0.19. Using Equation 16.3, calculate the sustainable growth rate for Amgen. Can Amgen keep growing at 18 percent per year indefinitely? (See www.amgen.com.)

16.11 Engineering Support Systems supplies electronic equipment and logistic services for U.S. and other nation's armed forces (www.engineeredsupport.com). Using Table 16.3, describe the profit model used by this firm. Determine the firm's return on total capital and return on equity.

CHAPTER 17

The Financial Plan

Budgets are not merely affairs of arithmetic, but in a thousand ways go to the root of prosperity of individuals, the relation of classes, and the strength of kingdoms.
William E. Gladstone

Entrepreneurs build a financial plan to determine the economic potential for their venture. This plan provides an estimate of the potential of the venture. Of course, any estimate is based on a set of assumptions regarding sales revenues and costs. Using the best available information and their intuition, entrepreneurs calculate the potential profitability of the venture. Furthermore, they need to determine the flow of cash monthly to identify the cash investments that will be required over a two- or three-year period. Also, an income statement and a balance sheet are required to demonstrate profitability and liquidity.

Using the estimates of sales, the venture team can determine the number of units it needs to sell to breakeven. Furthermore, they can calculate several measures of profitability that demonstrate the return provided by their venture for investors. The best ventures grow sales consistently and provide positive cash flow and profit early in their life. ∎

17.1 BUILDING A FINANCIAL PLAN

A sound business plan as previously discussed is based on a solid vision and business model. The business model is an expression of the theory of the business in the form of a vision and story. This story also needs to make sense financially. The business model tells a story about the customer and the value proposition that leads to revenue and profit. To create this value for the customer, a new firm needs to build a financial plan that describes the expected revenues, cash flows, profits, and investments necessary to achieve them. The purpose of any business is to create value for its customers and to generate a return on investment for its owners. A financial plan provides an estimate of projected cash flow and return on investment.

To create a financial plan, entrepreneurs must clearly state their assumptions about sales and costs. What resources will it take, over what timeframe, to achieve expected sales and profitability? The calculation of cash flows is based on a set of assumptions, which we will call the **base case,** that portrays the most likely outcomes. It may be prudent to also determine the outcome of a situation in which the expectations are not realized as expected, called the **pessimistic case. Cash flow** is the amount of cash flowing into or out a firm during a specific period. It is arrived at by subtracting the amount paid out in dividends from the net profit and then add back noncash expenses such as depreciation (see Table 17.10 for a glossary of accounting and financial terms).

The entrepreneurs' goal is to develop a solid set of financial projections that will include a pro forma income statement. **Pro forma** means provided in advance of actual data. Pro forma statements are forecasts of financial outcomes. The creation of a set of financial projections starts with a sales forecast based on a set of assumptions regarding the customer and sales growth. Then the calculation of projected sales over a two or three year period can be developed. This is step 1, as shown in Table 17.1. The second step is to state the assumed costs of doing business in the timeframe described in step 1. In step 2, the costs associated with the projected sales can be calculated. Step 3 is to calculate the expected income and cash flow forecast over the timeframe based on a set of assumptions regarding the timing of sales and receipts as well as payables to vendors and others. The final step is to calculate the balance sheet on an annual basis for the two- or three-year period. The balance sheet at the starting point of the new venture will need to be described by stating the assumed starting investments and required assets.

The cash flows, assets, balance sheet, and revenue projections are all interconnected through linkages. Accounting items are classified into "accounts" according to their nature, translated into monetary units, and organized in statements (see Table 17.10 for a glossary of terms). The basic accounting formula is

$$\text{Assets} = \text{liabilities} + \text{equity}$$

where assets are what the company owns and liabilities are the amounts it owes to another person or entity. Equity is the company's net worth (book value) expressed as

TABLE 17.1 Four steps to building a financial plan.

1. Sales forecast

- Timeframe—two or three years.
- Assumptions about sales per customer, number of customers, and growth rate of sales.
- Calculation of the sales forecast.

2. Costs forecast

- Assumptions about the costs of doing business in the specified time frame.
- Calculation of the costs associated with the projected sales of step 1.

3. Income and cash flow forecast

- Assumptions about the timing of cash receivables and payables specified in the time frame.
- Calculation of the income and cash flow associated with the projected sales and costs on a monthly basis over the timeframe.

4. Balance sheet

- Assumptions about the starting value of cash and assets.
- Calculated based on the income and cash flows from step 3.

$$\text{Equity} = \text{assets} - \text{liabilities}.$$

Equity is the ownership of the firm, usually divided into certificates called common or preferred shares of stock. The assets and liabilities are linked to income and expenses, as shown in Figure 17.1. Assets are used to generate income, and liabilities require expenses such as rent, payments, or return of loans. Book value is the firm's net worth and is often called accounting value. Market value is share price times the number of issued shares. Note that book value is not equal to market value, which is the perceived value of the firm given its growth potential.

The financial plan is critical to the evaluation of the business model of the new venture. With sound assumptions, the projected results will help in evaluating the venture and its financial viability. The resulting financial plan is only as good as the quality of the assumptions. One reason forecasting models are so fallible is that they rely on the assumptions that the user chooses to input [Riggs, 2004].

New firms should select two or three parameters of the business that display the greatest impact on the cash flows of the business. Examples are sales growth rate and new customer acquisition rate. They should then test the changes in sales as each of these parameters is changed. For example, a software firm, which licenses its products, will examine the potential range for its growth rate of licenses sold.

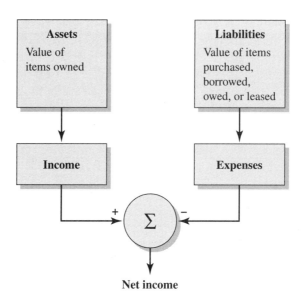

FIGURE 17.1 Assets generate income and liabilities lead to expenses. Net income is income minus expenses.

17.2 SALES PROJECTIONS

Sales projections will normally be developed for a two- or three-year period on a monthly basis. The sales forecast for a new venture is often the weakest link in the financial plan. Since the new venture has not actually obtained sales, the firm can only work with assumptions based on inadequate information.

In this chapter, consider a fictitious new venture, named e-Travel. It seeks to sell travel guide books (e-books) directly to readers who will download the guides via the Internet and read them on their e-book reader, such as a laptop computer or a handheld device. Short books such as these travel guides operate on a pull model, since customers order them only when they wish to read them. This is opposite of the usual push model, in which the publisher prints the book and then tries to find a purchaser. For travelers, an e-reader is easier to use than several heavy books. Using keywords such as "pizza restaurant Denver," the reader can use the search function to get the information quickly.

The new venture, named e-Travel, needs to build a financial plan. The first step is to build the sales projections. e-Travel has created a network of the best authors of travel guides. These authors have signed publisher agreements and have provided e-Travel with electronic guides to over 500 cities, regions, and leisure and recreation destinations throughout the world. All these guides are written following a common format, and key search words are identified.

e-Travel expects to sell 1,200 guides starting in the third month of operation. The price per guide is $15, paid by credit card when the guide is ordered online. Based on market research, the expected growth rate of sales is 10 percent per month. The pessimistic growth rate is 1 percent per month. The sales projections for the expected growth rate are shown in Table 17.2 for a three-year period. Assuming a 10 percent growth rate per month, sales amount to $3 million in year 3.

17.3 COSTS FORECAST

To determine the expected costs of doing business, the new venture team must examine their needs for facilities, equipment, and employees. Our example business, e-Travel, will need an office, computers, software, and office furniture. The authors will constantly update the information in their guides and will receive royalties at 12 percent of net revenues paid on the fifteenth day of the month following the sale generated by their guides. The costs of doing business will include salaries, marketing, and communication costs, and normal office utilities and supplies. The cost assumptions for e-Travel are summarized in Table 17.3. Every new venture needs to construct a set of assumptions at a similar level of detail.

17.4 INCOME STATEMENT

The income statement reports the economic results of a firm over a time period. The income statement is calculated as shown in Figure 17.2 [Maher, Stickney, and Weil, 2004]. The income (profit and loss) statement for e-Travel depicts the venture's expected performance over a period of time—in this case, three years. The sales, costs, and profits or losses are shown monthly. The purpose of the income statement is to show how much profit or loss is generated. Due to the online nature of e-Travel, it has no cost of goods sold. The profit and loss statement is shown in Table 17.4. The firm is profitable by the fifth month and shows a profit of $19,954 for the first year (for the base case).

17.5 CASH FLOW STATEMENT

The cash flow statement shows the actual flow of cash into and out of the venture. The cash flow statement tracks when the venture actually receives and spends the cash. A venture with positive cash flow can continue to operate without new debt or equity capital. If a cash flow statement reveals projected negative cash in some period, it will be necessary to plan for a new capital infusion. We define cash flow as the sum of retained earnings minus the depreciation provision made by a firm [Maher, Stickney, and Weil, 2004].

A growing business needs cash to operate. The detailed cash flow process is shown in Figure 17.3. Firms calculate their cash on hand at the end of each month. Therefore,

$$TC(N+1) = (CF - Disbursements) + TC(N)$$

TABLE 17.2 Sales projections for the expected growth rate (10 percent per month).

Year 1

Month	1	2	3	4	5	6	7	8	9	10	11	12	Year total
Units	0	0	1,200	1,320	1,452	1,597	1,757	1,933	2,126	2,339	2,573	2,830	19,127
Price per unit	$15	$15	$15	$15	$15	$15	$15	$15	$15	$15	$15	$15	
Sales dollars	$0	$0	$18,000	$19,800	$21,780	$23,955	$26,355	$28,995	$31,890	$35,085	$38,595	$42,450	$286,905

Year 2

Month	1	2	3	4	5	6	7	8	9	10	11	12	Year total
Units	3,113	3,424	3,766	4,143	4,557	5,013	5,514	6,065	6,672	7,339	8,073	8,880	66,559
Price per unit	$15	$15	$15	$15	$15	$15	$15	$15	$15	$15	$15	$15	
Sales dollars	$46,695	$51,360	$56,490	$62,145	$68,355	$75,195	$82,710	$90,975	$100,080	$110,085	$121,095	$133,200	$998,385

Year 3

Month	1	2	3	4	5	6	7	8	9	10	11	12	Year total
Units	9,768	10,745	11,820	13,002	14,302	15,732	17,305	19,036	20,940	23,034	25,337	27,871	208,892
Price per unit	$15	$15	$15	$15	$15	$15	$15	$15	$15	$15	$15	$15	
Sales dollars	$146,520	$161,175	$177,300	$195,030	$214,530	$235,980	$259,575	$285,540	$314,100	$345,510	$380,055	$418,065	$3,133,380

TABLE 17.3 Cost assumptions for e-Travel.

- Author royalties: 12 percent of net revenues paid on the fifteenth of the month following the sale

- Credit card: 1 percent to credit card providers, paid electronically as sale is processed

- Office rent: $1,500 per month

- Physical assets (computers, furniture, etc.): $48,000 per year; four-year life

- Depreciation of equipment (monthly):

Year 1	Year 2	Year 3
$1,000	$2,000	$3,000

- Salaries (monthly):

	Year 1	Year 2	Year 3
President	$4,000	$ 5,500	$ 6,500
Vice president	4,000	5,000	6,000
Administrative manager	1,500	3,000	3,500
Total:	$9,500	$13,500	$16,000

- Social Security taxes and other benefits: 15 percent of total salaries

- Marketing (monthly):

Year 1	Year 2	Year 3
$2,000	$2,500	$3,000

- Utilities, supplies, travel, communication (monthly):

Year 1	Year 2	Year 3
$2,000	$3,000	$4,000

- Interest expense: $1,000 per month ($100,000 loan at 12 percent per year; interest paid monthly; principal to be paid at end of five years)

- Income taxes: 30 percent of income before taxes

where TC(N+1) is the cash at the end of month (N+1), TC(N) is the total cash at the end of month (N), and CF is the cash flow for the month.

The cash flow statement for e-Travel is provided in Table 17.5. It is assumed that the founders invest $140,000 in cash and obtain a bank loan of $100,000 secured by their personal assets. This $240,000 is estimated to be required to cover the initial purchase of long-term assets such as computers and furniture, as shown in month 1 of year 1 in Table 17.5. The initial investment of $240,000 can be considered an equity investment since the loan is personally guaranteed by the two founders. Under the base case assumption of 10 percent growth in revenues each month, the cash flow quickly becomes positive.

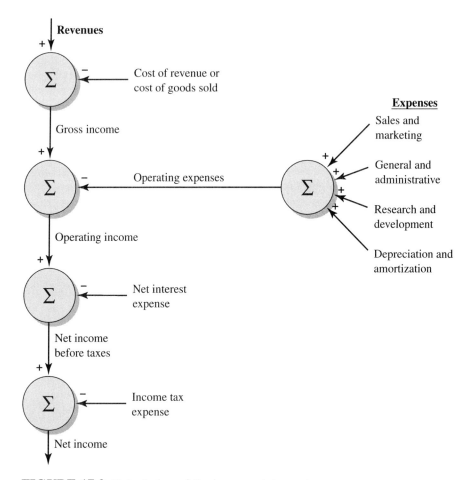

FIGURE 17.2 Calculation of the income statement.

17.6 BALANCE SHEET

The new venture team should prepare a balance sheet at the opening of the business and for the end of each year. The balance sheet depicts the conditions of the business by displaying the assets, liabilities, and owners' equity of the business [Maher, Stickney, and Weil, 2004]. The format for a balance sheet is shown in Figure 17.4 for a business at the end of the year.

The balance sheet for e-Travel is shown in Table 17.6. The balance sheet shows the assets, such as cash, equipment, furniture, and accumulated depreciation. The liabilities are the loan payable and the royalties to the authors. Owners' equity consists of contributions of $140,000 and retained earnings. Table 17.6 shows the balance sheet at the end of month 1, year 1, year 2 and year 3. The balance sheet provides evidence of the financial strength of e-Travel.

TABLE 17.4 Profit and loss statement.

Year 1

Month	1	2	3	4	5	6	7	8	9	10	11	12	Year total
Revenues	$0	$0	$18,000	$19,800	$21,780	$23,955	$26,355	$28,995	$31,890	$35,085	$38,595	$42,450	$286,905
Expenses:													
Author royalties	0	0	2,160	2,376	2,614	2,875	3,163	3,479	3,827	4,210	4,631	5,094	34,429
Credit card charges	0	0	180	198	218	240	264	290	319	351	386	425	2,871
Marketing	2,000	2,000	2,000	2,000	2,000	2,000	2,000	2,000	2,000	2,000	2,000	2,000	24,000
Depreciation	1,000	1,000	1,000	1,000	1,000	1,000	1,000	1,000	1,000	1,000	1,000	1,000	12,000
Interest	1,000	1,000	1,000	1,000	1,000	1,000	1,000	1,000	1,000	1,000	1,000	1,000	12,000
Office rent	1,500	1,500	1,500	1,500	1,500	1,500	1,500	1,500	1,500	1,500	1,500	1,500	18,000
Salaries	9,500	9,500	9,500	9,500	9,500	9,500	9,500	9,500	9,500	9,500	9,500	9,500	114,000
Social Security and benefits	1,425	1,425	1,425	1,425	1,425	1,425	1,425	1,425	1,425	1,425	1,425	1,425	17,100
Utilities, supplies, travel, communication	2,000	2,000	2,000	2,000	2,000	2,000	2,000	2,000	2,000	2,000	2,000	2,000	24,000
Profit (loss) before income tax	(18,425)	(18,425)	(2,765)	(1,199)	523	2,415	4,503	6,801	9,319	12,099	15,153	18,506	28,505
Income tax (credit)	(5,528)	(5,528)	(830)	(360)	157	725	1,351	2,040	2,796	3,630	4,546	5,552	8,551
Net profit (loss)	($12,897)	($12,897)	($1,935)	($839)	$366	$1,690	$3,152	$4,761	$6,523	$8,469	$10,607	$12,954	$19,954

Year 2

Month	1	2	3	4	5	6	7	8	9	10	11	12	Year total
Revenues	$46,695	$51,360	$56,490	$62,145	$68,355	$75,195	$82,710	$90,975	$100,080	$110,085	$121,095	$133,200	$998,385
Expenses:													
Author royalties	5,603	6,163	6,779	7,457	8,203	9,023	9,925	10,917	12,010	13,210	14,531	15,984	119,805
Credit card charges	467	514	565	621	684	752	827	910	1,001	1,101	1,211	1,332	9,985
Marketing	2,500	2,500	2,500	2,500	2,500	2,500	2,500	2,500	2,500	2,500	2,500	2,500	30,000
Depreciation	2,000	2,000	2,000	2,000	2,000	2,000	2,000	2,000	2,000	2,000	2,000	2,000	24,000
Interest	1,000	1,000	1,000	1,000	1,000	1,000	1,000	1,000	1,000	1,000	1,000	1,000	12,000
Office rent	1,500	1,500	1,500	1,500	1,500	1,500	1,500	1,500	1,500	1,500	1,500	1,500	18,000
Salaries	13,500	13,500	13,500	13,500	13,500	13,500	13,500	13,500	13,500	13,500	13,500	13,500	162,000
Social Security and benefits	2,025	2,025	2,025	2,025	2,025	2,025	2,025	2,025	2,025	2,025	2,025	2,025	24,300
Utilities, supplies, travel, communication	3,000	3,000	3,000	3,000	3,000	3,000	3,000	3,000	3,000	3,000	3,000	3,000	36,000
Profit before income tax	15,100	19,158	23,621	28,542	33,943	39,895	46,433	53,623	61,544	70,249	79,828	90,359	562,295
Income tax	4,530	5,747	7,086	8,563	10,183	11,969	13,930	16,087	18,463	21,075	23,948	27,108	168,689
Net profit	$10,570	$13,411	$16,535	$19,979	$23,760	$27,926	$32,503	$37,536	$43,081	$49,174	$55,880	$63,251	$393,606

(continued on next page)

TABLE 17.4 (continued)

Year 3

Month	1	2	3	4	5	6	7	8	9	10	11	12	Year total
Revenues	$146,520	$161,175	$177,300	$195,030	$214,530	$235,980	$259,575	$285,540	$314,100	$345,510	$380,055	$418,065	$3,133,380
Expenses:													
Author royalties	17,582	19,341	21,276	23,404	25,744	28,318	31,149	34,265	37,692	41,461	45,607	50,168	376,007
Credit card charges	1,465	1,612	1,773	1,950	2,145	2,360	2,596	2,855	3,141	3,455	3,801	4,181	31,334
Marketing	3,000	3,000	3,000	3,000	3,000	3,000	3,000	3,000	3,000	3,000	3,000	3,000	36,000
Depreciation	3,000	3,000	3,000	3,000	3,000	3,000	3,000	3,000	3,000	3,000	3,000	3,000	36,000
Interest	1,000	1,000	1,000	1,000	1,000	1,000	1,000	1,000	1,000	1,000	1,000	1,000	12,000
Office rent	1,500	1,500	1,500	1,500	1,500	1,500	1,500	1,500	1,500	1,500	1,500	1,500	18,000
Salaries	16,000	16,000	16,000	16,000	16,000	16,000	16,000	16,000	16,000	16,000	16,000	16,000	192,000
Social Security and benefits	2,400	2,400	2,400	2,400	2,400	2,400	2,400	2,400	2,400	2,400	2,400	2,400	28,800
Utilities, supplies, travel, communication	4,000	4,000	4,000	4,000	4,000	4,000	4,000	4,000	4,000	4,000	4,000	4,000	48,000
Profit before income tax	96,573	109,322	123,351	138,776	155,741	174,402	194,930	217,520	242,367	269,694	299,747	332,816	2,355,239
Income tax	28,972	32,797	37,005	41,633	46,722	52,321	58,479	65,256	72,710	80,908	89,924	99,845	706,572
Net profit	$67,601	$76,525	$86,346	$97,143	$109,019	$122,081	$136,451	$152,264	$169,657	$188,786	$209,823	$232,971	$1,648,667

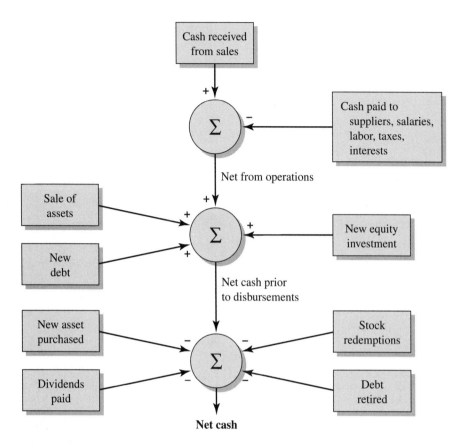

FIGURE 17.3 Cash flow process.

17.7 RESULTS FOR A PESSIMISTIC GROWTH RATE

Any new venture needs to plan for the likely case and prepare for the worst case. For e-Travel, we will assume the pessimistic case occurs when the sales grow at a rate of only 1 percent per month. The summary of the results for the pessimistic case are shown in Table 17.7. Table 17.7a shows the sales projections for the first three years. Sales for year 3 are $284,010 for the pessimistic case, while they were estimated at $3,133,380 for the expected case. Table 17.7b shows the profit and loss statement for the pessimistic case. Note that the firm is not profitable in the pessimistic case.

   We show the cash flow statement for the pessimistic case in Table 17.7c. Notice that the ending cash balance turns negative in month 1 of year 3. The company would need a cash infusion in month 1 of year 3 to continue operating.

TABLE 17.5 Cash flow statement.

Year 1

Month	1	2	3	4	5	6	7	8	9	10	11	12	Year total
Operating activities													
Net profit (loss)	($12,897)	($12,897)	($1,935)	($839)	$366	$1,690	$3,152	$4,761	$6,523	$8,469	$10,607	$12,954	$19,954
Add: Depreciation	1,000	1,000	1,000	1,000	1,000	1,000	1,000	1,000	1,000	1,000	1,000	1,000	12,000
Add: Increase in royalties payable			2,160	216	238	261	288	316	348	383	421	463	5,094
Cash flow from operations	(11,897)	(11,897)	1,225	377	1,604	2,951	4,440	6,077	7,871	9,852	12,028	14,417	37,048
Investing activities													
Purchase of long-term assets	(48,000)												(48,000)
Financing activities													
Bank loan	100,000												100,000
Owners' cash contributions	140,000												140,000
Increase (decrease) in cash	180,103	(11,897)	1,225	377	1,604	2,951	4,440	6,077	7,871	9,852	12,028	14,417	229,048
Beginning cash balance	0	180,103	168,206	169,431	169,808	171,412	174,363	178,803	184,880	192,751	202,603	214,631	0
Ending cash balance	$180,103	$168,206	$169,431	$169,808	$171,412	$174,363	$178,803	$184,880	$192,751	$202,603	$214,631	$229,048	$229,048

Year 2

Month	1	2	3	4	5	6	7	8	9	10	11	12	Year total
Operating activities													
Net profit	$10,570	$13,411	$16,535	$19,979	$23,760	$27,926	$32,503	$37,536	$43,081	$49,174	$55,880	$63,251	$393,606
Add: Depreciation	2,000	2,000	2,000	2,000	2,000	2,000	2,000	2,000	2,000	2,000	2,000	2,000	24,000
Add: Increase in royalties payable	509	560	616	678	746	820	902	992	1,093	1,200	1,321	1,453	10,890
Cash flow from operations	13,079	15,971	19,151	22,657	26,506	30,746	35,405	40,528	46,174	52,374	59,201	66,704	428,496
Investing activities													
Purchase of long-term assets	(48,000)												(48,000)
Increase (decrease) in cash	(34,921)	15,971	19,151	22,657	26,506	30,746	35,405	40,528	46,174	52,374	59,201	66,704	380,496
Beginning cash balance	229,048	194,127	210,098	229,249	251,906	278,412	309,158	344,563	385,091	431,265	483,639	542,840	229,048
Ending cash balance	$194,127	$210,098	$229,249	$251,906	$278,412	$309,158	$344,563	$385,091	$431,265	$483,639	$542,840	$609,544	$609,544

Year 3

Month	1	2	3	4	5	6	7	8	9	10	11	12	Year total
Operating activities													
Net profit	$67,601	$76,525	$86,346	$97,143	$109,019	$122,081	$136,451	$152,264	$169,657	$188,786	$209,823	$232,971	$1,648,667
Add: Depreciation	3,000	3,000	3,000	3,000	3,000	3,000	3,000	3,000	3,000	3,000	3,000	3,000	36,000
Add: Increase in royalties payable	1,598	1,759	1,935	2,128	2,340	2,574	2,831	3,116	3,427	3,769	4,146	4,561	34,184
Cash flow from operations	72,199	81,284	91,281	102,271	114,359	127,655	142,282	158,380	176,084	195,555	216,969	240,532	1,718,851
Investing activities													
Purchase of long-term assets	(48,000)												(48,000)
Increase (decrease) in cash	24,199	81,284	91,281	102,271	114,359	127,655	142,282	158,380	176,084	195,555	216,969	240,532	1,670,851
Beginning cash balance	609,544	633,743	715,027	806,308	908,579	1,022,938	1,150,593	1,292,875	1,451,255	1,627,339	1,822,894	2,039,863	609,544
Ending cash balance	$633,743	$715,027	$806,308	$908,579	$1,022,938	$1,150,593	$1,292,875	$1,451,255	$1,627,339	$1,822,894	$2,039,863	$2,280,395	$2,280,395

Balance sheet of ABC Corporation, 31 December 200X

Current assets (liquid in less than a year)	{	Cash and equivalents Accounts receivable Inventories	Current liabilities (payable in less than a year)	{	Accounts payable Accounts expenses Short-term debt
Fixed assets	{	Property, plant, and equipment (minus depreciation)	Long-term liabilities	{	Bonds issued Bank loans
Other assets	{	Intangibles (minus depreciation) Investment securities	Shareholders' equity	{	Common stock Additional paid-in capital Retained earnings

Total assets = total liabilities + shareholders' equity

FIGURE 17.4 Format for a balance sheet.

To be a successful venture, e-Travel needs to attain a sales growth rate that exceeds 4 percent per month over the first two years. This figure can be determined by modifying the spreadsheet calculation with various growth rates.

17.8 BREAKEVEN ANALYSIS

In the initial stages of building a financial plan, it is useful to know when a profit may be achieved. **Breakeven** is defined as when the total sales equals the total costs. Total sales (R) are

$$R = Q \times P$$

where q = number of units sold and P = price per unit. Total cost (TC) is

$$TC = FC + VC$$

where FC = total fixed costs and VC = variable costs. Thus, breakeven is the volume of sales (Q) at which the venture will neither make a profit nor incur a loss. Sales in excess of the volume of sales needed to cover costs will result in a profit.

Total fixed costs are $221,100 for e-Travel in year 1, and variable costs are 13% of sales, since royalty and credit card costs are 12% and 1%, respectively. Then, to determine Q, we have

$$R = TC$$

$$R = \$221,100 + (0.13 \times R)$$

or

$$0.87R = \$221,100$$

TABLE 17.6 Balance sheet.

End of month 1 of year 1			End of year 1		
Assets			**Assets**		
Cash		$180,103	Cash		$229,048
Equipment and furniture		48,000	Equipment and furniture		48,000
Accumulated depreciation		(1,000)	Accumulated depreciation		(12,000)
Total assets		$227,103	Total assets		$265,048
Liabilities			**Liabilities**		
Loan payable		$100,000	Royalties payable		$5,094
			Loan payable		100,000
Owners' equity			**Owners' equity**		
Owners' contributions		140,000	Owners' contributions		140,000
Retained earnings (deficit)		(12,897)	Retained earnings		19,954
Total owners' equity		127,103	Total owners' equity		159,954
Total liabilities and owners' equity		$227,103	Total liabilities and owners' equity		$265,048

End of year 2			End of year 3		
Assets			**Assets**		
Cash		$609,544	Cash		$2,280,395
Equipment and furniture		96,000	Equipment and furniture		144,000
Accumulated depreciation		(36,000)	Accumulated depreciation		(72,000)
Total assets		$669,544	Total assets		$2,352,395
Liabilities			**Liabilities**		
Royalties payable		$15,984	Royalties payable		$50,168
Loan payable		100,000	Loan payable		100,000
Owners' equity			**Owners' equity**		
Owners' contributions		140,000	Owners' contributions		140,000
Retained earnings		413,560	Retained earnings		2,062,227
Total owners' equity		553,560	Total owners' equity		2,202,227
Total liabilities and owners' equity		$669,544	Total liabilities and owners' equity		$2,352,395

TABLE 17.7A Sales projections for the pessimistic growth rate (1 percent per month).

Year 1		Year 2		Year 3	
	Year total		Year total		Year total
Units	12,550	Units	16,804	Units	18,934
Price per unit	$15	Price per unit	$15	Price per unit	$15
Sales dollars	$188,250	Sales dollars	$252,060	Sales dollars	$284,010

TABLE 17.7B Profit and loss statement for the pessimistic growth rate (1 percent per month).

Year 1		Year 2	
	Year total		Year total
Revenues	$188,250	Revenues	$252,060
Expenses:		Expenses:	
Author royalties	22,590	Author royalties	30,246
Credit card charges	1,883	Credit card charges	2,521
Marketing	24,000	Marketing	30,000
Depreciation	12,000	Depreciation	24,000
Interest	12,000	Interest	12,000
Office rent	18,000	Office rent	18,000
Salaries	114,000	Salaries	162,000
Social Security and benefits	17,100	Social Security and benefits	24,300
Utilities, supplies, travel, communication	24,000	Utilities, supplies, travel, communication	36,000
Profit (loss) before income tax	(57,323)	Profit before income tax	(87,007)
Income tax (credit)	0	Income tax	0
Net profit (loss)	($57,323)	Net profit (loss)	($87,007)

Year 3	
	Year total
Revenues	$284,010
Expenses:	
Author royalties	34,079
Credit card charges	2,840
Marketing	36,000
Depreciation	36,000
Interest	12,000
Office rent	18,000
Salaries	192,000
Social Security and benefits	28,800
Utilities, supplies, travel, communication	48,000
Profit before income tax	(123,709)
Income tax	0
Net profit (loss)	($123,709)

TABLE 17.7C Cash flow statement for the pessimistic growth rate (1 percent per month).

Year 1

Month	1	2	3	4	5	6	7	8	9	10	11	12	Year total
Operating activities													
Net profit (loss)	($18,425)	($18,425)	($2,765)	($2,609)	($2,452)	($2,295)	($2,138)	($1,982)	($1,812)	($1,643)	($1,473)	($1,304)	($57,323)
Add: Depreciation	1,000	1,000	1,000	1,000	1,000	1,000	1,000	1,000	1,000	1,000	1,000	1,000	12,000
Add: Increase in royalties payable	2,160	22	21	22	21	22	23	24	23	24			2,362
Cash flow from operations	(17,425)	(17,425)	395	(1,587)	(1,431)	(1,273)	(1,117)	(960)	(789)	(619)	(450)	(280)	(42,961)
Investing activities													
Purchase of long-term assets	(48,000)												(48,000)
Financing activities													
Bank loan	100,000												100,000
Owners' cash contributions	140,000												140,000
Increase (decrease) in cash	174,575	(17,425)	395	(1,587)	(1,431)	(1,273)	(1,117)	(960)	(789)	(619)	(450)	(280)	149,039
Beginning cash balance	0	174,575	157,150	157,545	155,958	154,527	153,254	152,137	151,177	150,388	149,769	149,319	0
Ending cash balance	$174,575	$157,150	$157,545	$155,958	$154,527	$153,254	$152,137	$151,177	$150,388	$149,769	$149,319	$149,039	$149,039

Year 2

Month	1	2	3	4	5	6	7	8	9	10	11	12	Year total
Operating activities													
Net profit (loss)	($8,234)	($8,064)	($7,895)	($7,712)	($7,529)	($7,346)	($7,164)	($6,981)	($6,798)	($6,615)	($6,432)	($6,237)	($87,007)
Add: Depreciation	2,000	2,000	2,000	2,000	2,000	2,000	2,000	2,000	2,000	2,000	2,000	2,000	24,000
Add: Increase in royalties payable	23	23	24	25	25	25	26	25	25	25	25	27	298
Cash flow from operations	(6,211)	(6,041)	(5,871)	(5,687)	(5,504)	(5,321)	(5,138)	(4,956)	(4,773)	(4,590)	(4,407)	(4,210)	(62,709)
Investing activities													
Purchase of long-term assets	(48,000)												(48,000)
Increase (decrease) in cash	(54,211)	(6,041)	(5,871)	(5,687)	(5,504)	(5,321)	(5,138)	(4,956)	(4,773)	(4,590)	(4,407)	(4,210)	(110,709)
Beginning cash balance	149,039	94,828	88,787	82,916	77,229	71,725	66,404	61,266	56,310	51,537	46,947	42,540	149,039
Ending cash balance	$94,828	$88,787	$82,916	$77,229	$71,725	$66,404	$61,266	$56,310	$51,537	$46,947	$42,540	$38,330	$38,330

(continued on next page)

TABLE 17.7C (continued)

Year 3

Month	1	2	3	4	5	6	7	8	9	10	11	12	Year total
Operating activities													
Net profit	($11,416)	($11,220)	($11,024)	($10,829)	($10,633)	($10,424)	($10,216)	($10,007)	($9,799)	($9,589)	($9,380)	($9,172)	($123,709)
Add: Depreciation	3,000	3,000	3,000	3,000	3,000	3,000	3,000	3,000	3,000	3,000	3,000	3,000	36,000
Add: Increase in royalties payable	27	27	27	27	27	29	29	29	29	28	29	29	337
Cash flow from operations	(8,389)	(8,193)	(7,997)	(7,802)	(7,606)	(7,395)	(7,187)	(6,978)	(6,770)	(6,561)	(6,351)	(6,143)	(87,372)
Investing activities													
Purchase of long-term assets	(48,000)												(48,000)
Increase (decrease) in cash	(56,389)	(8,193)	(7,997)	(7,802)	(7,606)	(7,395)	(7,187)	(6,978)	(6,770)	(6,561)	(6,351)	(6,143)	(135,372)
Beginning cash balance	38,330	(18,059)	(26,252)	(34,249)	(42,051)	(49,657)	(57,052)	(64,239)	(71,217)	(77,987)	(84,548)	(90,899)	38,330
Ending cash balance	($18,059)	($26,252)	($34,249)	($42,051)	($49,657)	($57,052)	($64,239)	($71,217)	($77,987)	($84,548)	($90,899)	($97,042)	($97,042)

Therefore,

$$0.87 \, (Q \times \$15) = \$221{,}100$$

or

$$Q = 16{,}943$$

Therefore, after selling about 17,000 e-Travel guides, the firm is profitable.

17.9 MEASURES OF PROFITABILITY

The shareholders of a venture are interested in profitability of the firm. The primary measure for the investor is return on invested capital, also called **return on investment** (ROI). The ROI for a venture is

$$\text{ROI} = \frac{\text{net income}}{\text{investment}}$$

where the income is distributed or allocated to the investor. As a firm grows, it may not actually distribute cash to its investors for some period. In that case, it is retaining earnings and using the retained cash earnings as investment capital. Then the retained earnings are added to the original equity investments to yield the owners' equity [Riggs, 2004].

Net income and owners' equity can provide the ratio called **return on equity** (ROE). ROE is calculated as

$$\text{ROE} = \frac{\text{net income}}{\text{owners' equity}}$$

The return on equity for e-Travel for year 2 for the projected base case is

$$\text{ROE} = \frac{\$393{,}606}{\$553{,}560} \times 100\% = 71.1\%$$

The return on investment can be calculated at the time of distribution of cash or when the common stock is priced in a public market. If the ownership held by the original investors in e-Travel can be sold at the end of year 3 for $720,000, the multiple (M) achieved by the investor group is

$$M = \frac{\$720{,}000}{\$240{,}000} = 3.0$$

The annual compound return over the three years is 44.2 percent since $(1.442)^3 = 3.0$. Therefore, the annual return on investment (ROI) is

$$\text{ROI} = 44.2\%$$

Note that we designate the investment as $240,000 since that original investment is the total investment by the founders. We consider a loan countersigned personally as equivalent to an equity investment by the founders.

TABLE 17.8 Profit and loss statement for AgraQuest as projected in its original business plan.

	Profit and loss statement				
	1996	**1997**	**1998**	**1999**	**2000**
Sales	550	2,375	4,250	8,900	16,500
Income	(2,338)	(3,094)	(2,768)	(1,693)	4,284
Average shares outstanding	7,002	9,462	10,972	11,722	11,722

Note: All figures are in thousands of dollars or shares.

TABLE 17.9 Projected investment requirements as provided in AgraQuest's original business plan.

	Projected investment requirements				
	1996	**1997**	**1998**	**1999**	**2000**
Capital required	6,500	5,200	8,600	10,200	0

Note: All figures are in thousands of dollars.

17.10 AGRAQUEST

The original business plan for AgraQuest, dated May 5, 1995, requested start-up financing of $1.1 million for equipment, $2.5 million for operations, and $2.9 million for cash reserves—a total initial investment of $6.5 million. It also projected an initial public offering after five years. The projected sales and income for the first five years is shown in Table 17.8. The projected investment needs for the five years are provided in Table 17.9.

AgraQuest was unable to meet the expected results, and sales grew to only $6.4 million in 2003. The two assumptions that caused the projections to be unrealistic were: 1) contract screening and natural molecules sold and 2) product development schedule. AgraQuest's plan assumed that screening of molecules for other firms and sale of molecules to other firms would result in $2.4 million in 1997 and $4.3 million in 1998. None of these revenues were realized. Furthermore, it projected sales of its natural products would be $2.9 million in 1999 and $6.5 million in 2000. These sales were severely delayed because the assumption was that products would be approved by the Environmental Protection Agency in 18 months, while it actually took 36 months. Furthermore, the pipeline of products developed more slowly than planned. These problems demonstrate the fragility of assumptions for any business plan.

17.11 SUMMARY

The entrepreneurial team builds their financial plan to determine the economic potential for their venture and demonstrate it to potential investors. The plan uses projected figures based on the underlying assumptions of the business venture.

This plan shows the profit and loss statement and the cash flow statement, which can be used to draw up the balance sheet. Monthly figures are used for the first year or two and quarterly figures for the next two or three years. Furthermore, a calculation of the sale of the required number of units for breakeven will be useful. The best new ventures are able to grow sales consistently and show positive cash flow and profit early in their life.

> **Principle 17**
> A sound financial plan demonstrates the potential for growth and profitability for a new venture and is based on the most accurate and reliable assumptions available.

17.12 EXERCISES

17.1   A new venture is launched with an initial investment (cash on hand) of $80,000 and generates sales of $40,000 each month. It has monthly operating costs of $36,000. The firm purchases equipment costing $30,000 each month for the first four months. Calculate the return on investment at the end of 12 months. Determine if the cash on hand remains positive at the end of each month. What, if any, investment is required and when?

17.2   A software firm has fixed costs of $800,000 and variable costs of $12 per unit. Calculate the breakeven quantity (Q) when each unit sells for $50. If the firm sells 50,000 units in a year, what is its profit for that year? Assume the tax rate for the firm is 20 percent.

17.3   Viscotech Inc. is planning to enter the field of electro-optical systems for automated optical inspection and detection of defects in manufacturing components and modules. The projected financial revenues and income are shown in the table below. The firm plans to start with an equity investment of $10,000 and a five-year loan of $500,000. Determine the cash on hand at the end of each year. Also, determine the return on equity and the return on investment for each year.

Viscotech projections.

	Year 1	Year 2	Year 3	Year 4	Year 5
Revenues	1,500	3,400	5,900	10,600	15,400
Income after tax	(500)	(100)	200	400	600
Depreciation	250	300	350	400	400
Average shareholders' equity	1,000	700	600	800	1,400
Long-term debt	500	450	300	200	100

Note: All figures are in thousands of dollars.

17.4 A new firm, Sensor International, is preparing a plan based on its new device to be used in a security network. The cost of manufacturing, marketing, and distributing a package of six sensors is 14 cents, and the price to the distributor is 68 cents. The firm calculates its one-time fixed costs at $121,000. Determine the number of units required for breakeven.

17.5 Sensor International sells 300,000 packages in its first year and 400,000 in its second. If the investor's original investment was $100,000, determine the return on investment in year 1 and year 2 for the firm. Assume a tax rate of 20 percent.

17.6 Superconductor Inc. is planning to start on January 1, 2005, with an initial investment of $200,000 from its founder team. It projects sales of $12,000 in February 2005 and a growth rate of sales of 10 percent per month for the foreseeable future (at least two years). It expects to receive payment upon delivery for its unique devices. Its costs are $18,000 for its first month of operation, and it plans for a growth rate of expenses at 2 percent per month. Prepare a cash flow statement for the first 24 months of the firm, and determine if and when an additional investment would be required.

17.7 A new medical device technology company has fixed costs of $420,000 and variable costs per unit of $3,100. Competitor analysis shows that the price for a comparable product ranges from $6,500 to $9,300. The goal of the firm is to attain profitability this year while increasing its market share next year. What price should the firm select? What is the break-even quantity for the selected price?

17.8 Reconsider the firm described in Exercise 17.7 when it is determined that fixed costs have declined to $300,000 and the firm has determined by studies that it can only expect to sell 60 units this year. What price should it select to ensure selling the 60 units profitably?

TABLE 17.10 Brief glossary of accounting and financial terms.

Assets—The value of what the company owns.

Balance sheet—The financial statement that summarizes the assets, liabilities, and shareholders' equity at a specific point in time.

Book value—The net worth of the firm, calculated by total assets minus intangible assets (patents, goodwill) and liabilities.

Cash flow—The sum of retained earnings minus the depreciation provision made by the firm.

Depreciation—The allocation of the cost of a tangible, long-term asset over its useful life. The reduction of the value of asset from wear and tear.

Discount rate—The rate at which future earnings or cash flow are discounted because of the time value of money.

Dividends—A distribution of a portion of the net income of a business to its owners.

Earnings per share—The ratio of net income to shares of stock outstanding.

Equity—The firm's net worth (book value).

Financial statement—A report summarizing the financial condition of a business. It normally includes a balance sheet and an income statement.

Income statement—A financial statement that summarizes revenues and expenses.

Liabilities—The amounts owed to other entities.

Net income—Total income for the period less total expenses for the period.

Pro forma—Provided in advance of actual data.

Retained earnings—Represents the owner's claim on the earnings that have not been paid out in dividends.

Return on investment—The ratio of net income to investment.

Return on revenues—The ratio of net income to revenues.

Return on stockholders' equity—Net income divided by average stockholders' equity.

Revenues—Sales after deducting all returns, rebates, and discounts.

Statement of cash flows—The statement that summarizes the cash effects of the operating, investing, and financing activities for a period of time.

Working capital—Current assets minus current liabilities.

CHAPTER 18

Sources of Capital

Capital is to the progress of society what gas is to a car.
James Truslow Adams

Entrepreneurs can estimate the capital required for their new business by reviewing the financial projections they prepare using the methods detailed in Chapter 17. In examining the projections and the cash flow statement, it becomes clear what capital will be needed and when. The entrepreneurs can provide some of the required capital, and friends and family can help by investing in the new business. Most businesses that expect to grow to a significant scale will need outside capital investments from professional investors. Typically, several stages of investment will be required over the life of the business.

This chapter addresses the task of attracting investors to a new business and creating an investment offering that will meet the firm's needs and the investors' requirements for an attractive return. In this chapter, we describe the funds that may be available from friends and family, individual investors, and professional investors, as well as banks.

Several stages of investment may be required, and it must be determined what percentage ownership is offered to the investor. This determination is based on the valuation of the new business at each stage. With the mutual agreement of the investor and the new firms, the terms of the deal will be recorded and the arrangement completed.

An alternative to an equity investment is debt financing from a bank or other financial institution. Furthermore, a line of credit can help finance short-term cash flow requirements.

Many firms use an initial public offering (IPO) to raise additional growth capital and to offer early investors a means of harvesting the value created in an emerging firm. Preparation for an IPO can be an important task for a firm with solid growth potential. ∎

18.1 FINANCING THE NEW VENTURE

The financial projections for a new venture, as described in Chapter 17, provide the entrepreneur with an estimate of the cash flow over the first two or three years. From the cash flow statement, the entrepreneurial team can determine the requirement for financial capital. Some new firms are cash positive almost immediately, and the entrepreneurs themselves can provide the necessary start-up funds. On the other hand, if it takes one year or more for the cash flow to become positive, a sizable investment may be necessary. Most high-tech firms take one to three years to become cash flow positive.

A new venture may require capital to purchase fixed assets such as computers and buildings. Furthermore, a new business needs capital to operate while building a customer base. **Working capital** is the capital used to support a firm's normal operations and is defined as current assets minus current liabilities. Financial capital is necessary to permit a new venture to purchase assets and provide working capital. As a firm grows, its needs for financial capital will normally increase. Choosing the right sources of capital for a business can be as important as choosing the team members and the location of the business. This decision will influence the future of the firm.

Finding the capital they need can be a difficult and time-consuming task for entrepreneurs. Getting and completing an investment agreement can take three to 9 months or longer. For many entrepreneurs, it may be wise to first secure a few customers and then seek investors. Finding the right financial backer takes a good business plan and time. It will be necessary to continuously tell the venture's story and answer myriad questions. However, revealing proprietary information understandably makes entrepreneurs uneasy. The chance that important information will leak to competitors is real.

Many investors take a long time reviewing the plan and interviewing the team, only to turn down the proposal in the end. The entrepreneur must assume a tentative deal will not be consummated and keep looking for investors even when one investor is seriously interested. While it's tempting to end the hard work of finding money, continuing the search not only saves time if the deal falls through but also strengthens the negotiating position.

Financial capital for new ventures is available, but the key is knowing where to look. Entrepreneurs must do their homework before attempting to raise money for their ventures. Understanding which sources of funding are best suited to the various stages of a company's growth and then learning how those sources operate are essential to success.

The issue of how much money to seek is difficult to resolve. The entrepreneurs want the investors to provide all the money necessary before positive cash flow. However, most investors want to divide their investment into several milestone-based stages. Furthermore, most investors will be wary of the pro forma projections and tend to accept only the pessimistic projection or worse. The investors attempt to factor uncertainty into their calculations, while the entrepreneur is more certain of the projections.

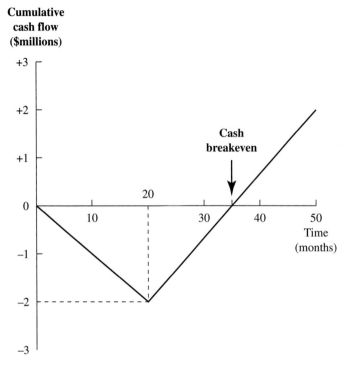

FIGURE 18.1 Idealized cash flow diagram for a new enterprise.

An idealized cash flow diagram is shown in Figure 18.1. This new enterprise has a burn rate of $100,000 per month in the first 20 months. It starts generating positive cash flow in the twenty-first month and reaches cumulative cash breakeven in the thirty-fifth month. This firm would require an investment of at least $2 million.

Uncertainty of venture outcomes can lead to a wide range of estimates of results. Breaking a firm's development into several stages can help investors build confidence in the firm over time. However, staged investments require the entrepreneur to raise funds several times—a potentially long and costly effort.

Often the two parties, investors and entrepreneurs, possess asymmetrical information. The investors may know more about the industry, while the entrepreneurs may know more about the venture. Furthermore, it may be difficult for the investors to determine an appropriate value for the intellectual property of the new venture. What is a patent worth? Often one finds out in court. Furthermore, how does one value the team's capabilities? Finally, the dynamics of the marketplace are difficult to assess. In addition, the marketplace provides variable valuation multiples as the market mood changes. These four factors that lead to different perceptions of the investors and the entrepreneurs are summarized in Table 18.1 [Gompers and Lerner, 2001].

**TABLE 18.1 Five factors that lead to the different perceptions
of investors and entrepreneurs.**

■ Uncertainty of projected outcomes.	■ Dynamics of the industry and the financial marketplace.
■ Asymmetrical information.	■ Concentration of wealth in a venture by entrepreneurs while investors will have a diversified portfolio.
■ Assigning a value to intellectual property and intangibles.	

The process of securing investment capital requires that the effects of the
first four factors shown in Table 18.1 be reduced in the negotiation stage. This
can be done through a full discussion of the risks, clear goals, and value of intellectual property and the leadership team. Also, the investors must build confidence that the entrepreneurs can properly manage the firm within the dynamics
of the marketplace. Finally, it should be recognized by both parties that the entrepreneurs will have their financial wealth concentrated in the venture, while the
investors will have a set of diversified investments. The goal of the entrepreneurs
is to find investors who examine the factors in Table 18.1 and eventually align
their view of these factors with those of the entrepreneurs. These investors can
be called *aligned investors.* Entrepreneurs are advised to find aligned investors
since they will be responsive to changing needs and provide required flexibility.

Entrepreneurs who are seeking financing send a credible signal to potential
investors through their willingness to invest in the business. Investors may be attracted to entrepreneurs who are willing to invest at least one-half of their private wealth in the venture [Ogden, Jen, and O'Connor, 2003].

18.2 VENTURE INVESTMENTS AS REAL OPTIONS

Investors make capital investments in opportunities for future cash returns. Professional investors often speak of an opportunity as an option. Opportunities are
options, which may be defined as rights but not obligations to take some action
in the future. Investments in new ventures can be viewed as investments in opportunities with an uncertain outcome. The outcome of an investment in a new
venture is highly uncertain. However, an investment in any new venture may create unforeseen opportunities that can be exploited in the future. A sound practice
in the early stages of developing a business is to keep a number of options open
by committing investments only in stages while exploring multiple business
paths. Once uncertainty has been reduced to a tolerable level and widespread
consensus exists within the organization on an appropriate path, the full commitment to that path can be made. Investors who hold options are given the right
to make a decision now or at a later date. They can exercise that right to either
proceed to the next step or cut their losses and decide to cease investing.

In financial terms, an option is simply the right to purchase an asset at some
future date and at a predetermined price. A **real option** is the right to invest in
(or purchase) a real asset—a new start-up firm at a future date.

Christopher Columbus developed a plan to seek out exotic spices and develop a spice trade to Europe. He was an Italian sea captain who submitted, without success, a proposal to the king of Portugal in 1484 to reach
Asia by sailing west. Over the next four years, he made inquiries of many
European courts. Finally, in 1492, after years of fruitless effort, he received the support of Queen Isabella and the Court of Madrid. The court
invested 1.4 million maravedis (currency of the time), and Columbus invested 250,000 maravedis, mostly obtained from friends and family. The
contract called for his designation as "admiral" and receipt of one-eighth
of the profits from of all gold, silver, gems, and spices produced or mined
in "his dominions."

Columbus returned to Spain in March 1493 with some gold and a
large amount of information about how to get to the "New World." Queen
Isabella had purchased a real option on the discovery of gold and spices.
As a result of Columbus's discovery, she decided to exercise the option
and send Columbus, Pizzaro, and Cortez to the New World to find and
bring back fabulous wealth to Spain.

Intellectual capital (knowledge) can be transformed to economic capital to
increase or create cash flows as well as strategic capital to exploit new opportunities. We state that economic capital is the intrinsic value of the series of cash
flows. The strategic capital is the option value (OV).

We may state broadly that the value of an investment in a new venture (V)
is

$$V = IV + OV$$

where IV = intrinsic value and OV = option value.

Net present value (NPV) is the present value of the future cash flow of a
venture discounted at an appropriate rate (r). The intrinsic value of a venture is
the net present value (NPV) of the venture using a discount rate (r), equal to the
expected return for the venture.

The net present value of a series of cash flow (c_n) is

$$NPV = \sum_{n=0}^{N} \frac{c_n}{(1+r)^n} \qquad (18.1)$$

where n = 0, 1, 2, $\cdots$ N. For example, the NPV for a new firm over the first two
years might be

$$NPV = -100,000 + \frac{65,000}{(1+r)} + \frac{35,000}{(1+r)^2}$$

where r = 0.15, the discount rate for this firm. The initial cash flow is negative
since an investment of $100,000 was required at n = 0. Then we may calculate
NPV as

$$NPV = -\,100,000 + 65,000\ (0.870) + 35,000\ (0.756)$$
$$= -\,100,000 + 83,000$$
$$= \$ - 17,000$$

The intrinsic value (IV) is equal to NPV for this case. However, the investor has an option to reinvest in the firm after the first two years. Typically, investors establish options by making an initial investment in a venture, which grants them an option to invest again later. As information flows over time, the uncertainty is reduced, and later-stage investments can be seen as less risky.

The value of an option (OV) is a function of four factors: the life of the option (T), the volatility (or uncertainty) of the price of the underlying asset (σ), the discount rate (r), and the level of the exercise price (X), relative to the current price of the venture (P). Clearly, having a longer period of time in which to decide whether or not to exercise an option increases the likelihood that the value of the firm will in the future exceed the exercise price. The higher the degree of uncertainty about the future value of the venture, the more one should be willing to pay for the option. The uncertainty can be represented by the standard deviation of the firm's price (σ) over the period (T). If the firm is new, use the standard deviation for comparable firms.

The value of an option increases with the relative value of the current stock price at the initial investment (P) to the exercise price of the option (X). The ratio P/X increases as X declines. The value of an option also increases with the discount rate (r), since higher discount rates make the option more valuable due to high discounting of future cash flows. The four factors contributing to the value of a real option are summarized in Table 18.2.

The value of the option is based on the four factors T, σ, P/X, and r. Then the option value is

$$OV = f(T, \sigma, P/X, r)$$

which increases as all four factors increase. The option value can be calculated using the Black-Scholes formula [Boer, 2002]. An option calculator is available at www.blobek.com/black-scholes.html. Another valuation model for options is the binomial model which uses a decision tree format. [Copeland 2004]

As a first approximation, we can use a linear approximation for OV as follows:

$$OV = k_1 T + k_2 \sigma + k_3 (P/X) + k_4 r$$

TABLE 18.2 Value of a real option based on four factors.

- Increases with the level of uncertainty measured by the standard deviation (σ).
- Increases with the length of time (T) the person holding the option has to decide whether or not to exercise it.
- Increases with the ratio of the current stock price (P) to the exercise price (X). The ratio is P/X.
- Increases with the discount rate (r).

where k_i are unspecified constants. For high relative values of the four factors, the option value will be significant and may exceed the NPV for the first few years of a high-risk venture.

Let us again consider the hypothetical case discussed in the preceding paragraphs. An initial investment of $100,000 was made with an option to invest in a second round of investments after two years. Thus, consider the case where the period (T) is relatively long, the standard deviation of the firm's value high, and the discount rate relatively high. Also, if the current stock price of the firm at the time of the initial investment is low, say $10, and if the exercise price is preset at $5, then the option value is quite high. Perhaps an investor could estimate an option value of $50,000. Then the value of the investment is

$$V = IV + OV$$
$$= -\$17,000 + 50,000 = +\$33,000$$

Often, the value of an investment is largely its real option value.

Genentech was founded in 1976 by entrepreneur Robert Swanson and biochemist Herbert Boyer to use gene-splicing technology to develop new pharmaceuticals. Genentech, devoid of profit, went public in 1980, raising $35 million. With little cash flow in sight, Genentech was valued solely on its potential to exploit its intellectual property and introduce important new pharmaceuticals. Genentech became profitable in 1993, and its total market value was $38 billion in 2004.

Investments that appear overly risky from a purely financial view may be viable once the opportunities for future action are taken into account. However, as soon as an option no longer promises to provide future value, it should be abandoned. Entrepreneurs tend to allow unpromising ventures to drag on far too long. Entrepreneur teams become imbued with a collective belief that their venture will succeed, a conviction that overcomes skepticism [Carr, 2002].

18.3 SOURCES OF CAPITAL

Many sources of financial capital exist for a new business, as listed in Table 18.3. There are two types of capital: equity and debt. **Equity capital** represents the investment by a person in ownership through purchase of the stock of the firm. The holders of equity shares are called stockholders. **Debt capital** is money that a business has borrowed and must repay in a specified time with interest. An example is a leased vehicle. Debt capital usually does not include any ownership interest in the new firm.

Equity financing of a venture at formation usually involves funds provided by founders, friends, and family. Lenders and equity investors expect the entrepreneur team to invest a significant amount of their own capital in the new business.

TABLE 18.3 **Sources of capital.**

■ Founders	■ Leasing companies
■ Family and friends	■ Established companies
■ Small business investment companies	■ Public stock offering
■ Small Business Innovation Research grants	■ Government grants and credits
■ Wealthy individuals (angels)	■ Customer prepayments
■ Venture capitalists	■ Pension funds
■ Banks	■ Insurance companies

1. Founding stage
• Entrepreneurial team begins with a vision, business model, and strategy

2. Seed stage
• Initial financial capital

3. Growth stage
• Growth capital required

4. Harvest stage
• IPO or acquisition provides return to investors and founders

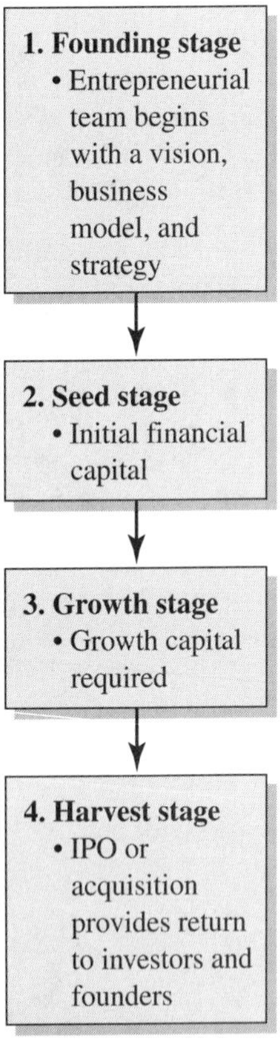

FIGURE 18.2
Four financial steps in building a successful firm.

If an entrepreneur will not invest in a risky venture, why should the passive investor? Usually, a new business can only obtain debt capital after some period in the marketplace and success evidenced by a good balance statement. Investments from family and friends are an excellent source of seed capital and can get a start-up far enough along to attract money from private investors or venture capital companies. Family and friends are often willing to invest because of their relationships with one or more of the founders. However, family and friends should receive and review all the financing documents. Furthermore, they should be able to afford losing their investment and be comfortable with the risk.

In 1995, Mike and Jackie Bezos invested $300,000 in their son Jeff's start-up, Amazon.com. Today, those shares are worth many times what they paid for them. New ventures can approach professional investors who may invest in very promising, high-growth ventures. Wealthy individuals will also invest in new ventures. They are often called **angels,** and it is estimated that angels personally invest in 20,000 firms annually. **Venture capitalists** are professional managers of investment funds. They normally invest in about 2,000 new firms annually. The number of investments made by these two groups increases in good economic conditions. Angels and venture capitalists have invested the same total amount in recent years.

The Small Business Administration (SBA) will help fund start-ups as well as provide advice on funding sources. Small Business Innovation Research grants are available to technology startups for the development of innovative technologies (see www.sba.gov/sbir).

Almost all new enterprises firms are not in a position to seek their early-stage financial capital through an initial public offering on the public stock market. Only the most qualified and experienced teams with a outstanding opportunity are able to make an initial public offering in the early stages of building a business. A spin-off of an established company may be able to make an IPO.

The four financial steps for building a successful firm are summarized in Figure 18.2. Often, a firm can start with seed capital from its founders and friends. However, most technology firms require significant capital in the growth phase and turn to professional or wealthy investors.

18.4 BOOTSTRAP AND SEED FINANCING

The initial funds used to launch the new firm are usually called **seed capital.** The first round of capital needs may be limited, and the funds may be readily available. Launching a start-up with modest funds from the entrepreneurial team, friends, and family is often called **bootstrap financing.** To bootstrap a venture means to start a firm by one's own efforts and to rely solely on the resources available from oneself, family, and friends. For many ventures with modest potential returns to investors, it is best to attract investors who are known to the entrepreneurs. Professional investors can only back a small fraction of the number of firms that start up each year. They seek to invest in large opportunities with high rates of return that have defensible competitive advantages, well-defined business plans, and well-known, proven founders.

An advantage of bootstrap financing is the ability to make some mistakes and yet keep going. Professional investors can hinder entrepreneurs from following the try-it, fix-it approach required in the uncertain environments in which start-ups flourish. Entrepreneurs who are unsure of their markets or who don't have the experience to deal with investor pressure are better off without other people's capital, even if they can somehow get investors to overlook their limited credentials and experience.

Bootstrap entrepreneurs often start with a modest business plan and a limited opportunity and look for a quick route to breakeven and positive cash flow. Many entrepreneurs underestimate the marketing costs entailed in overcoming customer inertia and conservatism, especially with respect to new, unproven products.

Bootstrap companies start small and build their experience and know-how as they go. Eventually, these modest beginnings can turn into large successes as the firms find new opportunities. For example, Princeton Review was launched to compete with local private tutors of uneven quality. Eventually, the firm found its place on the national scene and competed with the Kaplan chain. Actual figures are difficult to obtain, but up to 75 percent of start-ups are bootstrap, self-financed firms. Self-financed new firms concentrate on the sales activity to bring cash flow into the business.

Many businesses fit the model of a bootstrap opportunity. They keep costs low, seek out markets that competitors are ignoring, and build the business one step at a time. Entrepreneurs often must fund a significant portion of their new business start-up since investors and lenders are reluctant to provide the required capital [Quadrini, 2002]. These funding constraints cause the entrepreneur to accumulate assets that can be used in a start-up.

Pierre Omidyar launched AuctionWeb on Labor Day, 1995, while he was still employed full-time at a software firm. An experienced software developer, Omidyar was intrigued by the opportunities to build a business on the Internet, and his vision was to provide a "perfect market" for buyers and sellers on an Internet auction site [Cohen, 2002]. He wrote the program for the site over the Labor Day weekend using his personal website provided by his home Internet service provider. The best domain name he could find was eBay.com. Throughout the fall of 1995, AuctionWeb had hosted thousands of auctions. By February 1996, Omidyar's Internet service provider started charging him a commercial rate for his website, so he started charging a small fee for each sale. Starting in February, AuctionWeb was profitable. By April, AuctionWeb took in $5,000 in sales. In June, when revenues doubled to $10,000, Omidyar decided to quit his day job. He quickly attracted a friend, Jeff Skoll, to join him, first as a consultant and then full time by August 1996. By late 1996, they moved out of Omidyar's house to an office building in Campbell, California. By October 1996, AuctionWeb had a total of four employees when it hosted 28,000 auctions that month. AuctionWeb was dedicated to thriftiness and controlling costs. By January 1997, the site had hosted 200,000 auctions, and Auction Web was projecting revenues of over $4 million for 1997. Early in 1997, Omidyar and Skoll wrote their first business plan and started looking for investors. By June 1997, Benchmark Capital, a venture capital firm, paid $5 million for 21.5 percent of the company, which then changed its name to eBay.

Bootstrap companies usually follow five rules: 1) start small and probe the market, 2) learn from your customer and adjust the business model, 3) adjust the revenue and profit engine, 4) keep costs to a minimum, and 5) start expanding the company, once the new venture starts growing, while keeping the cost curve below the revenue curve.

A useful indicator of the value of the quality of a new venture proposal for funding is the actual proportion of the lead entrepreneurs' wealth that is committed to the new venture [Prasad, Vozikis, and Bruton, 2001]. Typically, bootstrap start-ups can take care of the first (seed) round of the company. If the firm starts to grow, it may need investments from professional investors. The advantages and disadvantages of bootstrap financing are listed in Table 18.4.

> Tom Siebel earned an MBA and an MA in computer science from the University of Illinois. After graduation, he joined Oracle in 1982. By 1992, Siebel and Pat House had founded Siebel Systems. For the first 18 months, everyone worked for no salary and received equity shares. Siebel stated:
>
> > This was never about making money. It was never about going public; it was never about the creation of wealth. This was about an attempt to build an incredibly high-quality company. [Malone, 2002]

TABLE 18.4 Advantages and disadvantages of bootstrap financing.

Advantages	Disadvantages
■ Low pressure on valuation	■ Unable to fund growth phase
■ Easy terms on ownership	■ Lack of funding commitment for future
■ Control by founders	■ Loss of advice from professional investors
■ Little time spent on finding investors	

18.5 ANGELS

Angels are wealthy individuals, usually experienced entrepreneurs, who invest in business start-ups in exchange for equity in the new ventures. The term *angel* for an investor in a new firm was originally used for a person who backs a new theater production on Broadway. Angels are people who share the vision of the new venture and provide support, advice, and money. In a sense, angels can provide wings for lifting a new creation. Angels often have personal experience and interest in the industry that the start-up is entering. Angel investing is a fast-growing segment of the new business financing industry, and it is often ideal for start-ups that have outgrown the capacity of investments from friends and family but are still too small to attract the interest of venture capital companies. For example, after raising the money to launch Amazon.com from family and friends, Jeff Bezos turned to angels, attracting $1.2 million from a dozen angels. Angels invest about $16 billion a year in early-stage U.S. companies.

Angels often invest because they understand the industry and are attracted by the opportunity as well as the potential return. They may place less emphasis on an early harvest strategy and enjoy working with new entrepreneurs. Angels serve as investors, advisers, and mentors for the new ventures they support. They help new entrepreneurs create and refine a business model, find top talent, build business processes, test their ideas in the marketplace, and attract additional funding. Angels tend to invest close to home and limit their investments to early-stage companies. Most of the new ventures they fund were recommended to them by a business associate or an angel group. A summary list of criteria for investments by angels is provided in Table 18.5.

Angels can be helpful investors, but sometimes they can be overbearing and have a negative impact. Choosing the right angel is important.

In some geographical regions, angels join together to form groups of angels. These groups work together to screen investment opportunities. For example, the Band of Angels meets monthly in Silicon Valley to hear a few presentations by new venture start-ups (see www.bandangels.com). Angel groups are established in many cities (see Appendix C for a partial list).

In 1976, Steve Wozniak and Steve Jobs designed the Apple I computer in Jobs's bedroom and built the prototype in Jobs's house. To start the company,

TABLE 18.5 Criteria for angel investments.

- Within the industry in which the angel has experience.
- Located within a few hours' driving distance of the angel.
- Recommended by trusted business associates.
- Entrepreneurs with attractive personal characteristics such as integrity and coachability.
- Good market and growth potential.
- Seeking an investment of $100,000 to $1 million that offers minority ownership of about 40 percent.

Jobs sold his Volkswagen and Wozniak sold his Hewlett-Packard calculator, which raised $1,300. With that capital and credit from local electronic suppliers, they set up their first production line. Jobs met Mike Markkula, a former marketing manager at Intel and wealthy angel who invested $91,000 in cash and personally guaranteed a bank line of credit for $250,000. Jobs, Wozniak, and Markkula each held one-third ownership of Apple Computer [Young, 1988]. Markkula, the angel investor, became chairman of the company in 1977.

The founders of Google, Sergey Brin and Larry Page, approached Sun Microsystems cofounder Andy Bechtolsheim in 1998, who wrote a $100,000 check after a 15-minute pitch describing the new business. By 1999, the venture capital firms of Kleiner Perkins and Sequoia also invested.

Angels are usually available through referrals from their colleagues. However, a referral isn't something you ask for. A referral is a favor that you earn. You can ask for a referral, but it can be awkward and difficult.

18.6 VENTURE CAPITAL

Venture capital is a source of funds for new ventures that is managed by investment professionals on behalf of the investors in the venture capital fund. The people who manage the venture capital fund are called venture capitalists. These funds typically invest in new ventures with high potential returns. The private venture capital firm is seeking equity participation in these high-potential firms. Each year, about 2,000 to 3,000 new ventures receive funding from venture capital. The venture capital firm engages in careful screening and due diligence before investing. It brings in other venture capital investors in a financing round and prepares strict contracts and restrictions (called term sheets) with the new venture. Venture capital firms are usually interested in technology ventures with high potential.

A typical venture capital firm will have enough investments so as to diversify its portfolio. Each venture capital firm will have several partners who are experienced, full-time investors. Their knowledge of finance and technology enables them to judge the potential investment. Normally, venture capital investing is staged financing. Thus, new information at each stage enables the venture capitalist to make good decisions. The new ventures are held to staged goals, milestones, and deadlines for achieving these milestones. A typical investment from a venture capital firm is $1 million to $3 million in the first financing round, with a total commitment of $5 million to $13 million over several stages. Many entrepreneurs start up a new business to gain independence, only to find that they have a new set of partners—the venture capitalists acting as members of the board of directors.

A venture capital portfolio of 10 to 15 new ventures may achieve an overall annual return of 30 percent. Perhaps one-half of the new ventures fail or provide a low return. Fortunately, two or three new ventures provide an overall annual return of 50 to 100 percent. Therefore, venture capital firms are looking for new ventures that potentially can return at least 40 or 50 percent annually. Obviously, these candidates must be firms with high-potential growth.

TABLE 18.6 Investment stages.

1. Seed or start-up stage: Complete the team, formalize the plan, complete initial arrangements. Financial capital from angels, friends, and family.

2. Development stage (series A): Product development and prototype, ready for launch. Financial capital from venture capital funds.

3. Growth stage (series B or C, and others as required): Launch and growth phase. Financial capital from venture capital firms.

4. Competitive or maturity stage (initial public offering): Mature firm in a competitive context. Financial capital from offerings in the public equity markets.

The four investment requirements used by the venture capitalists in the selection process are: 1) the industry is well-known to them; 2) the amount of the investment is greater than $1 million; 3) the company is at the appropriate stage of progress; and 4) the potential return is 40 percent or more annually. The track record of the team is also critically reviewed [Gompers and Sahlman, 2002].

Venture capitalists prefer staged or phased financing. The four stages are shown in Table 18.6. Venture capital is normally available in the development and growth stages, with greatest emphasis on the development stage. Venture capital money is limited-term money [Zider, 1998]. Typically, a venture capital firm wants to harvest (realize) its return on investment within five to seven years. The harvest is usually facilitated through an initial public offering in the public equity markets or an acquisition by an established company.

New enterprises that can build up to large businesses can require large capital investments as they grow. Amazon raised $8 million from venture capitalists and $54 million from its 1997 IPO. The largest capital infusion for Amazon was a $1.25 billion debt offering in 1999.

The venture capitalists expect the new business to achieve certain outcomes, often called milestones, at the end of a funding stage before proceeding to the next stage. Examples of the milestones for each stage are shown in Table 18.7. Each milestone functions as a miniplan about each stage for the venture. For example, a milestone may state that a working prototype will be available in six months.

TABLE 18.7 Milestones for each stage of funding.

Stage	Milestone = expected outcome
Seed	Formation of initial team and completion of business plan
Series A	Product development completed
Series B	Product test and customer acceptance proven
Series C	Launch product into market

A special type of venture capital is available for social benefit firms or those that promise to create products that lead to sustainable resources or environments. For example, the Silicon Valley Social Venture Fund, known as SV2, is dedicated to funding organizations that facilitate social change (www.sv2.org). The Social Enterprise Alliance offers good contacts (www.se-alliance.org).

Venture capital funds concentrate on attractive, disruptive, and high-growth industries. They concentrated on computers and biotechnology in the 1980s and communications and the Internet in the 1990s. Areas of concentration in the 2000s include nanotechnology, biomedical devices, genomics, and communications infrastructure. By investing in emerging industries, venture capitalists hope to build great companies in important industries while reaping large rewards.

Venture capitalists carry out one or more of four functions: 1) they obtain finance capital for start-ups, 2) they evaluate projects for other participants in the venture, 3) they provide expertise for the development of the firm, and 4) they serve as the central coordinator for all the participants involved during a firm's infancy.

In a typical deal at the development stage, a group of two or three venture capitalists will invest $5 million to $10 million in exchange for 40 percent to 60 percent preferred-equity ownership. The preferred class of stock provides the venture capitalists with preference over common stock held by founders, family, friends, and other first-stage investors. They will hold a liquidation preference on rights to assets. Furthermore, venture capitalists seek voting rights over key decisions such as the sale of the firm or the timing of the IPO. They often require seats on the board of directors.

Venture capital is high-risk and high-return capital available to high-potential firms. Venture capitalists plan for a 40 percent or more annualized return on their investment with built-in protections and controls over the new venture. The structure of most venture capital deals favors the venture capitalist and may place the entrepreneur at a disadvantage. However, the venture capital firms bring risk money, industry knowledge and contacts, and a pathway to a public stock offering. The risk-and-reward profiles for various types of investments are shown in Figure 18.3. Venture capitalists work at the high risk–high return end of the profile.

Venture capital investments in new start-ups and emerging ventures averaged $25 billion per year over the decade of the 1990s and peaked at $106 billion in 2000 during the Internet dot.com and telecommunications boom. Assuming $10 billion per year for new start-ups is available in the future and 3,000 ventures are funded, the average investment is $3.3 million per new venture. The venture capital investments in the United States for 1995 to 2003 are shown in Table 18.8. Investments outside the United States are about equal to the U.S. investments.

Venture capitalists get good rates of return by buying shares of private companies early and then helping management use that cash to turn the start-ups into businesses with growing revenues, profitability, and cash flow. Since they want to get really good returns, the big winners generally have to earn 10 times the

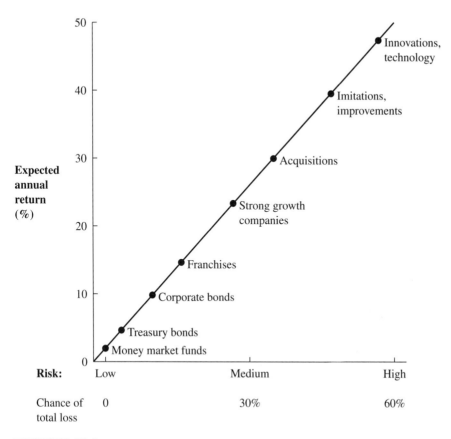

FIGURE 18.3 Risk and reward profile for various investments.

TABLE 18.8 U.S. venture capital investments for 1995–2003.

Year	Number of companies	Number of deals	Number of investors	Amount ($millions)
1995	1,555	1,869	694	7,630.76
1996	2,090	2,610	818	11,506.79
1997	2,550	3,181	1,030	14,772.27
1998	2,992	3,689	1,187	21,244.44
1999	4,453	5,603	1,825	54,348.39
2000	6,388	8,066	2,502	105,839.58
2001	3,812	4,603	2,024	40,625.89
2002	2,549	3,031	1,698	21,268.29
2003	2,286	2,784	1,430	18,364.99

Source: MoneyTree Survey by PricewaterhouseCoopers/Thomson Venture Economics/NVCA

TABLE 18.9 Characteristics of an attractive venture capital investment.

■ Potential to become a leading firm in a high-growth industry with few competitors.	■ Outstanding opportunity.
	■ Founders' capital invested in the venture.
■ Highly competent and committed management team and high human capital (talent).	■ Recognizes competitors and has a solid competitive strategy.
■ Strong competitive abilities and a sustainable competitive advantage.	■ A sound business plan showing how cash flow turns positive within a few years.
■ Viable exit or harvest strategy.	■ Demonstrated progress on the product design and good sales potential.
■ Reasonable valuation of the new venture.	

original investment money in four to five years. A multiple of ten times is equivalent to a 58.5 percent annual return over the five years. If a venture capitalist invests $5 million in the early stage of a company, his or her ownership stake has to be worth $50 million in five years. If the venture capitalist owns one-half of the company, the firm must be worth $100 million after five years.

The characteristics of a good venture capital deal from the venture capitalists' view and the venture founders' view are summarized in Table 18.9. A large opportunity in a fast-growing industry led by a very competent, experienced team is attractive to venture capitalists. With a good venture capitalist partner, the founders of the new venture can realize their dream of a well-capitalized venture that can make a big difference in their industry. At the same time, the founding team must recognize that venture capital has a high cost of potential loss of ownership, control, and even the vision of the venture. A list of venture capital firms is available from the National Venture Capital Association (www.nvca.org).

TABLE 18.10 Five steps for a venture capital deal.

1. Determine the amount of cash needed and its use.

2. Locate appropriate venture capital investors and secure a referral to them.

3. Determine which risks are to be reduced in this financing round:

   ■ Team risk—people

   ■ Capital risk—financial

   ■ Technology risk—innovation

   ■ Market risk—industry and competitors

4. Agree on valuation and ownership structure.

5. Agree on a contract (term sheet) describing the deal and its terms.

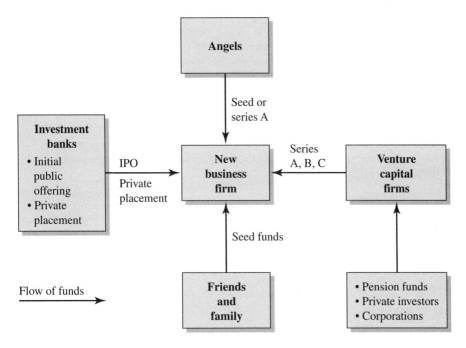

FIGURE 18.4 Potential for the flow of funds to a new business firm.

New and emerging business can use the five steps outlined in Table 18.10 to secure venture capital. Every one of these steps takes time, and the whole process may take three to 12 months. Given the extent of technology markets, there is no scarcity of good ideas. What is scarce are experienced, competent, and committed entrepreneurs. Investors strive to find the best entrepreneurial teams to invest in.

The overall investment process for a new firm can be portrayed as shown in Figure 18.4. The seed round of financing may be achieved with friends and family. Angels may supply the series A round of financing. Venture capitalists will often supply the funds in series B and C. Then investment bankers will facilitate an IPO or large private placement.

Fred Smith used his family money to found Federal Express (FedEx) in 1973. He leased several planes and built a 25-city network. He knew additional funding would depend on a solid start, so he tested the system for two weeks, shipping empty packages cross-country. In mid-April 1973, he opened for business. To expand his network, Smith turned to venture capitalists. He got his venture capital because the FedEx system worked and he had his own money in it. By 1975, FedEx was profitable. FedEx sold stock to the public with an IPO in 1978.

18.7 CORPORATE VENTURE CAPITAL

Large companies such as Intel, Microsoft, and Cisco Systems have engaged in investing in external start-ups [Chesbrough, 2002]. **Corporate venture capital** is the investment of corporate funds in start-up firms that are not part of the corporation. In this case, established corporations are acting as venture capitalists. However, the corporate venture capitalist in many cases is looking for new ventures that can offer synergies between itself and the new venture. Another approach is for the corporate venture capitalist to make an investment to exploit its industry knowledge and then generate a high return on investment. For example, Dell Ventures has made several investments to make high returns.

An example of a strategic investment is one of many made by Microsoft in start-up companies that could help advance its Internet businesses. Other firms, such as Intel, make corporate venture capital investments in firms making products complementary to their own. Another possible corporate investment strategy is in start-ups that may be valuable to the corporation's future operations. A summary of the four forms of corporate venture capital investments is provided in Figure 18.5.

Many start-ups can benefit from corporate venture capital, since corporations can make good partners and the cost of the capital can be less than that from regular venture capitalists. However, corporate venture capitalists can take longer to close a deal. Nevertheless, corporate venture capital may increase the credibility of the new venture and may not exert as much control as regular venture capitalists. Perhaps the most important benefit of corporate venture capital

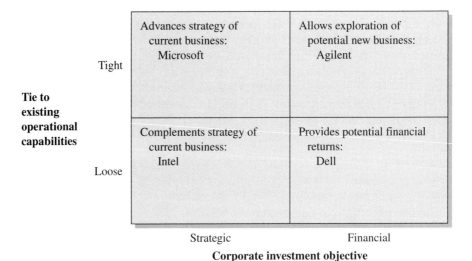

FIGURE 18.5 Four forms of corporate venture capital.

Source: Burgelman, Christensen, and Wheelwright, 2004.

is the potential for a strong partnership between the corporate venture capitalists and the new venture that strategically couples the strength of the large corporation with the innovation of the start-up [Mason and Rohner, 2002].

Entrepreneurs need to be cautious since corporate venture capitalists often invest to be aware of potential competitors and may not have their interests aligned with those of the start-up.

Intel's venture capital arm was, in 2002, by far the most active investor of any kind in technology companies that received venture capital. Intel invested $200 million in more than 100 companies in 2002 [Clark, 2003a]. Intel has made 800 investments in the decade 1992 to 2002, accounting for a total of $8 billion. Intel Capital's main mission is to nurture technologies that could stimulate demand for Intel's chip businesses.

18.8 VALUATION

The **valuation rule** is the algorithm by which an investor such as an angel or venture capitalist assigns a monetary value to a new venture. For many operating businesses, determining the net present value (NPV) is the best method for selecting alternative projects. A new enterprise, however, has uncertain pro forma cash flows, and investors find it difficult to use projected cash flows as reliable measures. Even if potential cash flows can be estimated, how does an investor decide what is a fair price for a share of ownership in a new firm?

The value of a company is equal to the present value of all dividends or cash disbursements paid now or later. A new firm has no historical results to use to project future cash flows. Furthermore, fads and social dynamics play a role in the determination of the value of a firm. Thus, determining the value of a start-up is difficult.

Discounted cash flow is a method of calculating the present value of a future stream of cash flow based on discounting back future flows at the end of a number of years using a discount rate (r) (see Equation 18.1). Let us start with the discounted cash flow rules for the valuation of a new venture, ABC Inc., with projected cash flows as shown in Table 18.11. Since the firm is not yet operating, these projections are subject to wide uncertainty. With a potential investment in the first year, what percentage of the firm's ownership should the investor require? What is the discount rate for this calculation?

TABLE 18.11 Projected cash flow and profit for ABC Inc.

Year	1	2	3	4	5	6
Sales	0	1,000	2,500	5,000	8,000	10,000
Profit	−600	−10	400	650	1,000	1,200
Cash flow	−1,100	0	500	1,200	1,500	1,800

Note: All figures in thousands of dollars.

The **discount rate** is the rate (r) at which future earnings or cash flow is discounted because of the time value of money. The discount rate (r) for a firm is its cost of capital. Therefore, the discount rate for investors will be their expected return on investment. Thus, the new enterprise may propose using a discount rate of 15 percent, while the investors may demand a return of 30 percent or more. McNulty and colleagues [2002] have shown the real cost of capital of a biotech start-up is about 35 percent. Also, for the cash flow calculation, we need to have an estimate of the cash flow for year 6 and later. It is unreasonable to project into year 6 and later since the estimates become less reliable for later years. Thus, investors will be reluctant to use the discounted cash flow method for valuation of a firm that is yet to provide any reliable cash flow.

Venture capitalists or angels will want a harvest of the value of the firm by an IPO, acquisition, or buyout of their share of ownership by year 5. Realistically, the IPO or acquisition may not happen until year 6 or 7. Examining the projections for ABC Inc. shown in Table 18.11, a valuation can be determined for year 5 using a method favored by venture capitalists and angels.

The **new venture valuation rule** uses the projected sales, profit, and cash flow in a target year (N) and the projected earning growth rate (g) for five years after year N. The investment by the investor (I) is made at the beginning of year 1. The investor requires an annual return (gain) on investment of G for N years. Thus, investors expect a capital return after N years of $(1 + G)^N$ times their original investment I. Therefore,

$$CR = (1 + G)^N \times I$$
$$= M \times I$$

where CR is the capital return and M is the multiple of the investment. Thus, if an investor invests $1.1 million in the series A stage and the expected annual return on investment, G, is 45 percent over a five-year period, then

$$M = (1 + 0.45)^5 = 6.41$$

Therefore,

$$CR = M \times I = 6.41 \times 1.1 = 7.05$$

The percentage ownership (PO) demanded by the investor will then be

$$PO = \frac{CR}{MV} \times 100\%$$

where MV is the expected market value of the new venture in year N.

To calculate the expected market value in year 5, we may use the price-to-earnings ratio (PE) of comparative firms to estimate the market value in year N. Then,

$$MV = PE \times EN$$

where EN is the earnings in year N. In this case for year 5, EN = $1,000,000 (Table 18.11). The comparative PE ratio is obtained by looking at the industry

PE ratios while accounting for the expected growth rate of earnings over the ensuing years. In this case, we might expect a growth rate of earnings of 20 percent for several years following year 5. Then, examining the industry data, we estimate the appropriate PE ratio is 16. Therefore, we have

$$MV = 16 \times \$1,000,000$$
$$= \$16,000,000$$

Then the required percentage ownership is

$$PO = \frac{CR}{MV} \times 100\%$$
$$= \frac{7.05}{16.00} \times 100\% = 44.0\%$$

The market value can also be calculated using a price-to-sales ratio (PS) for comparative firms. If comparative firms have a PS = 2.3, then the market value would be MV = PS × S. The market value for ABC Inc. is

$$MV = PS \times S$$
$$= 2.3 \times \$8,000,000 = \$18,400,000$$

Then the percentage ownership required by the investor would be

$$PO = \frac{CR}{MV} \times 100\% = \frac{7.05}{18.40} \times 100\% = 38.3\%$$

Using these calculations, the investor may reasonably expect to receive 40 to 50 percent of the firm's ownership in a series A investment. Given the uncertain nature of the sales and profit projections, the valuation of a firm is a function of the potential of the firm to achieve a big success in a short time. This simple example assumes no more stages of investment are required in the five-year period.

Netscape was formed by Marc Andreesen, co-author of the University of Illinois Mosiac Web browser, and Jim Clark, founder and former CEO of Silicon Graphics. As an angel investor, Clark invested $3 million in Netscape in May 1994. Clark, a respected entrepreneur, then offered a $6.4 million investment opportunity (at $2.25 per share) for 15 percent of the firm to Kleiner Perkins, a premier venture capital firm. Netscape's sales rocketed to $16 million in the first six months of operations [Lewis, 2000]. Just before the IPO, Clark owned 30.0 percent of the firm and Andreesen owned 12.3 percent. The multiple (M) they achieved in 17 months on their seed shares was 37.3. Kleiner Perkins achieved a multiple of 12.4 on its first-stage investment in just 13 months. Netscape was the first high-profile venture of the Internet boom companies and reaped large rewards. A summary of the valuation of Netscape (price per share) is shown in Table 18.12.

When a firm makes an offering, it raises an investment (INV). The firm hopes to set its pre-money (before the investment) value as PREMV. Then

$$\text{post-money value} = \text{pre-money value} + \text{investment}$$

TABLE 18.12   Netscape valuation at four stages.

Stage	Date	Price per share	New investment ($ millions)
Seed	4/94	$0.75	$3.1
Series A	7/94	$2.25	$6.4
Series B	6/96	$9.00	$18.0
IPO	8/96	$28.00	$160.0

When post-money value is POSMV, we have

$$POSMV = PREMV + INV$$

The percentage of the company sold to the investors is

$$\frac{INV}{POSMV} \times 100\%$$

Consider a firm, EZY Inc., with a series of investments as shown in Table 18.13. A set of venture capital investors invests in the firm in the series A round at 90 cents per share and owns 40.0 percent of the firm after the purchase. At the next stage, series B, the firm has missed its milestones, and the investors demand a reduced share price of 50 cents per share. This round (or stage) of financing is called a *down round,* since the price per share goes down. As a result, the venture capitalists receive a significant increase in their ownership in the firm.

The experience of FedEx in its early days of venture capital investments is shown in Table 18.14. Note the down round in September 1974.

The timing of staged investments is a critical issue for the CEO and CFO of a new company that has not yet achieved breakeven. This company is using cash from an earlier investment to cover its negative cash flow. We define **burn rate** as cash in minus cash out, on a monthly basis. Thus, if a firm has $800,000 cash in the bank and has a burn rate of $100,000 per month, it will run out of cash in eight months unless it can change its burn rate to zero.

Some companies spend money unwisely, so a good rule is to keep the burn

TABLE 18.13   Financial Stages of EZY Inc.

Stage	Investors	Price per share	Ownership by FFF*	Ownership by venture capitalists
Seed	Founders Friends Family	$0.10	100%	—
Series A	Venture capital group	$0.90	60.0%	40.0%
Series B	Venture capital group	$0.50	34.0%	66.0%

*Founders, friends, and family.

TABLE 18.14 Venture capital financing of FedEx.

	Date	Investment ($ millions)	Share price ($ per share)
Series A	September 1973	$12.25	—
Series B	March 1974	$6.40	$7.34
Series C	September 1974	$3.88	$0.63
IPO	1978	—	$6.00

Source: Gompers and Lerner, 2001.

rate as low as possible. The alternative is to raise new rounds of new financing, if that is possible.

> A great opportunity matched with experienced entrepreneurs often leads to success, but not always. Steve Papermaster and Frank Moss formed Agillion in December 1999 in Austin, Texas. Papermaster had founded several firms and was formerly CEO of Tivoli Systems. Agillion offered software to small businesses to provide Web pages for them and charged $29.95 per month for its software. It needed hundreds of thousands of small customers to be profitable. It had pro forma projections showing profitability within two years. By July 2002, Agillion filed for bankruptcy. Its burn rate was $65,900 a day—spending $67 million over its life of 33 months. In that same period, its sales totaled $146,947 [Hawkins, 2002]. As one of its former employees said, "It was too good to be true."

18.9 TERMS OF THE DEAL

The terms of the investment transaction are critical to the entrepreneur. The issues that concern professional investors are provided in Table 18.15. Clearly, trust and integrity are necessary between investors and the founding entrepreneurs. The venture capitalist willing to pay the highest price is not necessarily

TABLE 18.15 Issues to be resolved within the terms of the deal.

■ Percent ownership for the investor group.	■ Type of security.
■ Timing of investment.	■ Reservation of ownership for employees (stock option pool).
■ Control exerted by investor.	■ Antidilution provisions.
■ Vesting periods for ownership by the entrepreneur team.	■ Milestones of achievement, if there are multiple stages or steps to the investment.
■ Rights to require an IPO and registration rights.	■ Stock option plans.

the person whom the entrepreneur will want most in the deal. Another venture capitalist who is not willing to pay quite as much may be a better partner in growing the business.

Professional investors will normally want **preferred stock,** which has claims on dividends and assets before common stock owners. With all the factors involved in completing a financing deal, the new venture needs to engage an attorney to review any terms of the deal and perhaps represent the new venture in negotiating the deal.

The terms of any deal should reflect the likelihood that the firm will require more capital at a later time. Many deals make it difficult to raise capital later at an attractive price. Often the deal is written on an assumption that everything works out as planned—an unlikely outcome. The plan for the future should include reasonable methods of obtaining new capital infusions. Often the provisions for protecting the investors in the current deal will be onerous if another capital infusion is required.

The terms of the deal should address the means of achieving potential return and the allocation of risk between investor and new venture. It may be better to lower the price to the investor and get the investor to share the future risk with the new venture. The investors are looking for protection of their investment, and the new venture is seeking a capital infusion but needs to retain the right to pursue future capital infusions. Excessive protection clauses need to be traded away for lower prices for stock purchased by the investor. If possible, the investor should pay less for ownership and share the risk with the new enterprise.

18.10 DEBT FINANCING

New ventures with sales and cash flow can consider recurring short-term or long-term debt financing. Debt provides financial leverage to a firm and enables the firm to increase its return on equity. The principle of financial leverage works as long as a firm's earnings are larger than the interest charged for the borrowed money. Of course, if the firm's net earnings should drop below the interest cost of borrowed money, the return on owners' equity will decrease. Thus, most new ventures avoid financial leverage until they achieve stable growth.

Debt financing usually is easier to arrange, and often cheaper, than equity. Any profitable firm can borrow money if it's willing to pay high enough interest rates. Borrowers have the advantage of not giving up ownership or control of the firm—unless they can't pay the interest. Also, the tax deduction for interest paid cuts the effective cost of debt.

Using debt exposes a company to another kind of risk. If it doesn't show enough profit to cover debt payments, its existence may be in peril.

Banks and the Small Business Administration will lend to qualified firms. Asset-backed borrowing may be available to new ventures. Leasing equipment is a form of borrowing. Venture leasing, while expensive, may be cheaper than giving away more equity. An asset leaseback is also a way to secure cash by giving up an asset such as a building or equipment.

Many new ventures arrange for a line of credit, a form of short-term borrowing, to be used as needed. The firm pays a fee for the right to access borrowed funds as needed. Sometimes a new venture can secure a bank loan, which will be guaranteed by the Small Business Administration. These asset-based loans can be favorable to the new venture, if needed.

Small Business Administration loans in 2001 amounted to $10 billion to about 50,000 borrowers. Under the program, the federal agency doesn't actually make the loan—banks do, mostly—but it guarantees about 75 percent of the loaned amounts. And it covers losses on the guaranteed portions from a combination of fees charged to banks and taxpayer funds.

18.11 INITIAL PUBLIC OFFERING

The first public equity issue of stock made by a company is referred to as an **initial public offering** (IPO). The newly issued stock is sold to all interested investors of the general public in a cash offer. In the United States, the IPO is a sale of a portion of the company to the public by filing with the Securities and Exchange Commission (SEC) and listing its stock on one of the stock exchanges. The offer is managed via a financial intermediary, an investment bank, which aids in the sale of the securities. The investment bank performs such services as formulating the method of issuance, pricing, and selling the new securities.

Determining the offering price is a difficult task, but new issues are normally priced somewhat below their intrinsic value to ensure a sufficient number of new buyers. The total cost to the firm for issuance of new securities can be up to 10 percent of the funds raised in a typical offering raising $50 million.

The new venture firm has three possible reasons to issue an IPO: 1) raise new capital, 2) liquidity, and 3) image or brand. Many fast-growing firms will need large capital infusions of more than $30 million, and the public market is suitable for these larger amounts. Second, liquidity—the ability to easily convert ownership to cash—is facilitated by the IPO. The third reason is to help build brand reputation by fastening public ownership of the new venture firm. These advantages of issuing an IPO are summarized in Table 18.16.

TABLE 18.16 Advantages and disadvantages of issuing an IPO.

Advantages	Disadvantages
■ Raising new capital with the possibility of later, additional offerings.	■ Offering costs and effort required.
■ Liquidity: Ability to convert ownership to cash, potential of harvest for investors and founders.	■ Disclosure requirements and scrutiny of operations.
	■ Perceived pressures to achieve short-term results.
■ Visibility: Build brand and reputation.	■ Possible loss of control to a majority shareholder.

Entrepreneurs and early investors can obtain a harvest of the value they have created by issuing an IPO. The timing of the IPO is critical since the IPO market can be very volatile. The IPO market was favorable in the period 1998 to 2000 and very unfavorable in the period 2001 to 2003.

The disadvantages of issuing on IPO include: 1) offering costs, 2) disclosure and scrutiny, and 3) perceived short-term pressures. These disadvantages are listed in Table 18.16. For small offerings, less than $25 million, the costs can amount to 15 percent of the offering. The time required to prepare all the documents can also be onerous to a small, emerging firm. The disclosure and scrutiny can be burdensome. Any misstep can cause havoc with the company's share price. Furthermore, many new firms may find the pressure for reports of quarterly earnings improvements difficult to satisfy.

Ben & Jerry's Ice Cream built its business to $4 million by 1983 and wanted to grow over the next few years, so it needed to raise new funds for working capital and a new plant. It turned to its community in Vermont, where it was well known, and arranged a public stock offering for Vermont residents only. With a goal of raising $750,000, it offered stock at a minimum of 12 shares at $10.50 each, for a total of $126. This plan offered to sell 17.5 percent of the firm for $750,000, implying a postmoney value of the firm of $4.29 million. Nearly 18,000 Vermont households bought stock, and about one-third bought the minimum number of shares (12). A year after the Vermont offering, Ben & Jerry's followed up with a national IPO to raise additional expansion funds [Cohen and Greenfield, 1997].

Determining the appropriate offering price is the most important thing the lead investment bank must do for an initial public offering. The issuing firm faces a potential cost if the offering price is set too high or too low. If the issue is priced too high, it may be unsuccessful and be withdrawn. If the issue is priced below the true market price, the issuer's existing shareholders will experience an opportunity loss.

The IPO marketing process includes a road show. Road shows involve the lead underwriters and key firm managers marketing the firm to prospective investors (institutional investors) via presentations in major cities and one-on-one meetings with mutual fund managers.

Amazon.com was founded in July 1994 and opened for business in July 1995. It reached sales of $16 million for the quarter ending March 31, 1997, but was operating at a loss. Jeff Bezos hoped to take the company public to raise additional funds and to build a public recognition for it. The total investment from Bezos, angels, and Kleiner Perkins amounted to approximately $9 million. Amazon selected Deutsche Morgan Grenfell (DMG) as its investment bankers in February 1997 and began the process of preparing the necessary documents for submission to the Securities and Exchange Commission (SEC). Amazon went public with an IPO of 3 million shares at a price of $18 each. Its sales per share, pre-IPO, were approximately $1.80. Thus, the firm went public at a price-to-sales ratio of 10. The firm increased its sales from $148 million in 1997 to $1.5 billion in 2002, but was unprofitable in those years.

If a firm intends to eventually issue an IPO when the market is favorable, it is wise to work from the beginning to be in position to qualify its IPO. Thus, a firm planning to go public needs to meet all the regulatory requirements in place when creating the prospectus. This means having audited financial statements, a complete management team, a sustainable competitive advantage, and an independent board of directors.

The prospectus or selling document is a part of the information provided to the SEC for approval. The information in the prospectus must be presented in an organized, logical sequence and an easy-to-read, understandable manner to obtain SEC approval. Some of the most common sections of a prospectus include the cover page, prospectus summary, description of the company, risk factors, use of proceeds, dividend policy, capitalization, dilution, selected financial data, the business, management, and owners, and the financial statements.

The cover page of the prospectus for Netflix's IPO in 2002 is shown in Figure 18.6a. The table of contents for the Netflix prospectus is shown in Figure 18.6b. The summary page, the offering page, and the summary financial data are provided in Figures 18.6c, d, and e, respectively. At the time of the IPO, the business was not yet profitable, but the company was narrowing its losses. Netflix raised $82.5 million from the offering of 5.5 million shares. Netflix used the cash for operations and to pay off debt of $14 million.

Its 2001 revenues of $76 million reflected real demand, much of it in the San Francisco Bay Area. Nationwide, subscribers totaled 456,000 in 2001, up from 292,000 the year before. Netflix's financial picture includes its relatively low costs.

The average annual number of IPOs over the period 1971 to 2001 was 300. The number of IPOs and the proceeds from these IPOs are shown for the period 1971 to 2001 in Table 18.17. The average proceeds of an IPO in the 1990s were $100 million.

The market for IPO issuance is cyclical. Thus, the ability of a company to go public is a function of timing as well as its financial performance. There are several reasons for going public. First, it's one way to obtain the financial resources needed to grow. Second, it may do great things for the firm's reputation. Third, it's an excellent tool for recruiting employees. Fourth, and probably most important, it provides liquidity and an avenue for eventually cashing out of the company. However, only a few firms are able to go public, even in good times. A firm needs to be in an industry with favor in the market and be able to show a good financial story of growth and profitability. The firm also needs to be able to

TABLE 18.17 IPO offerings for selected years in the United States.

Year	1971	1976	1981	1986	1991	1996	2001
Number of issues	253	40	347	693	365	864	111
Total proceeds ($ billions)	1.1	0.3	3.1	17.7	16.3	56.1	36.3

5,500,000 Shares

Common Stock

This is Netflix, Inc.'s initial public offering of common stock. We are selling all of the shares.

Prior to the offering, no public market existed for the shares. The common stock has been approved for quotation on the Nasdaq National Market under the symbol "NFLX."

Investing in our common stock involves risks that are described in the "Risk Factors" section beginning on page 5 of this prospectus.

	Per Share	Total
Public offering price	$15.00	$82,500,000
Underwriting discount	$1.05	$5,775,000
Proceeds, before expenses, to Netflix, Inc.	$13.95	$76,725,000

The underwriters may also purchase up to an additional 825,000 shares from us at the public offering price, less the underwriting discount, within 30 days from the date of this prospectus to cover overallotments.

Neither the Securities and Exchange Commission nor any state securities commission has approved or disapproved of these securities or determined if this prospectus is truthful or complete. Any representation to the contrary is a criminal offense.

The shares will be ready for delivery on or about May 29, 2002.

Merrill Lynch & Co.

Thomas Weisel Partners LLC

U.S. Bancorp Piper Jaffray

The date of this prospectus is May, 2002.

FIGURE 18.6A

TABLE OF CONTENTS

You should rely only on the information contained in this prospectus. We have not, and the underwriters have not, authorized any other person to provide you with different information. If anyone provides you with different or inconsistent information, you should not rely on it. We are not, and the underwriters are not, making an offer to sell these securities in any jurisdiction where the offer or sale is not permitted. You should assume that the information appearing in this prospectus is accurate only as of the date on the front cover of this prospectus or other date stated in this prospectus. Our business, financial condition, results of operations and prospects may have changed since that date.

Through and including June 16, 2002 (the 25th day after the date of this prospectus), all dealers effecting transactions in these securities, whether or not participating in this offering, may be required to deliver a prospectus. This is in addition to the dealers' obligation to deliver a prospectus when acting as underwriters with respect to their unsold allotments or subscriptions.

Netflix is a registered trademark and Netflix.com, CineMatch and Mr. DVD are trademarks of Netflix, Inc. Each trademark, trade name or service mark of any other company appearing in this prospectus belongs to its holder.

FIGURE 18.6B

SUMMARY

This summary highlights information contained elsewhere in this prospectus. You should read the entire prospectus carefully, including "Risk Factors" and our financial statements and the notes to those financial statements appearing elsewhere in this prospectus before you decide to invest in our common stock.

Our Company

We are the largest online entertainment subscription service in the United States providing more than 600,000 subscribers access to a comprehensive library of more than 11,500 movie, television and other filmed entertainment titles. Our standard subscription plan allows subscribers to have three titles out at the same time with no due dates, late fees or shipping charges for $19.95 per month. Subscribers can view as many titles as they want in a month. Subscribers select titles at our Web site (www.netflix.com) aided by our proprietary CineMatch technology, receive them on DVD by first-class mail and return them to us at their convenience using our prepaid mailers. Once a title has been returned, we mail the next available title in a subscriber's queue. In 2001, our total revenues were $75.9 million, and our net loss was $38.6 million. For the three months ended March 31, 2002, our total revenues were $30.5 million, and our net loss was $4.5 million. As of March 31, 2002, we had an accumulated deficit of $141.8 million.

In 2001, domestic consumers spent more than $29 billion on in-home filmed entertainment, representing approximately 78% of total filmed entertainment expenditures, according to Adams Media Research. Consumer video rentals and purchases comprised the largest portion of in-home filmed entertainment, representing $21 billion, or 71% of the market in 2001, according to Adams Media Research.

The home video segment of the in-home filmed entertainment market is undergoing a rapid technology transition away from VHS to DVD. The DVD player is the fastest selling consumer electronics device in history, according to DVD Entertainment Group. In September 2001, standalone set-top DVD player shipments outpaced VCR shipments for the first time in history, and this trend continued throughout the remainder of 2001. At the end of 2001, approximately 25 million U.S. households had a standalone set-top DVD player, representing an increase of 91% in 2001, according to Adams Media Research. Adams Media Research estimates that the number of U.S. households with a DVD player will grow to 69 million in 2006, representing approximately 62% of U.S. television households in 2006.

Our subscription service has grown rapidly since its launch in September 1999. We believe our growth has been driven primarily by our unrivalled selection, consistently high levels of customer satisfaction, rapid customer adoption of DVD players and our increasingly effective marketing strategy. We primarily use pay-for-performance marketing programs and free trial offers to acquire new subscribers. In the San Francisco Bay area, where the U.S. Postal Service can make one- or two-day deliveries from our San Jose distribution center, approximately 2.8% of all households subscribe to Netflix.

Our proprietary CineMatch technology enables us to create a customized store for each subscriber and to generate personalized recommendations which effectively merchandise our comprehensive library of titles. We provide more than 18 million personal recommendations daily. In April 2002, more than 11,000 of our more than 11,500 titles were selected by our subscribers.

We currently provide titles on DVD only. We are focused on rapidly growing our subscriber base and revenues and utilizing our proprietary technology to minimize operating costs. Our technology is extensively employed to manage and integrate our business, including our Web site interface, order processing, fulfillment operations and customer service. We believe our technology also allows us to maximize our library utilization and to run our fulfillment operations in a flexible manner with minimal capital requirements.

FIGURE 18.6C

The Offering

Common stock offered by Netflix ....   5,500,000 shares

Common stock to be outstanding after the offering ......   20,648,074 shares

Use of proceeds ....   We estimate that our net proceeds from this offering will be approximately $74.7 million. We intend to use the net proceeds for:

- repayment of approximately $14.1 million of indebtedness under our subordinated promissory notes, including accrued interest as of May 22, 2002; and

- general corporate purposes, including working capital.

Risk factors ....   See "Risk Factors" and other information included in this prospectus for a discussion of factors you should carefully consider before deciding to invest in shares of our common stock.

Nasdaq National Market symbol..................................   NFLX

Unless we indicate otherwise, all information in this prospectus: (1) assumes no exercise of the overallotment option granted to the underwriters; (2) assumes the conversion into common stock of each outstanding share of our preferred stock, which will occur automatically upon the completion of this offering; (3) is based upon 15,148,074 shares outstanding as of May 22, 2002, including shares to be issued to certain studios immediately prior to this offering based on our capitalization as of May 22, 2002; (4) gives effect to a one-for-three reverse stock split effected in May 2002; and (5) excludes:

- 4,352,472 shares of common stock issuable upon the exercise of stock options outstanding as of May 22, 2002, with a weighted average exercise price of $3.11 per share, of which 1,162,022 were vested as of May 22, 2002, and 988,608 shares of common stock available for future option grants under our 1997 Stock Plan and 2002 Stock Plan, as of May 22, 2002;

- 7,017,962 shares of common stock issuable upon exercise of warrants with a weighted average exercise price of $3.20 per share; and

- 583, 333 shares of common stock reserved for issuance under our 2002 Employee Stock Purchase Plan.

FIGURE 18.6D

Summary Financial and Other Data

The summary financial data below should be read together with "Management's Discussion and Analysis of Financial Condition and Results of Operations" and the consolidated financial statements and the related notes included elsewhere in this prospectus.

	Year Ended December 31,			Three Months Ended March 31,	
	1999	2000	2001	2001	2002
	(in thousands, except per share data)				
Statement of Operations Data:					
Total revenues	$ 5,006	$ 35,894	$ 75,912	$ 17,057	$30,527
Gross profit (loss)	633	11,033	26,005	(1,120)	15,369
Operating loss	(30,031)	(57,557)	(37,227)	(20,417)	(4,054)
Net loss	(29,845)	(57,363)	(38,618)	(20,598)	(4,508)
Net loss per common share:					
Basic and diluted	$ (21.41)	$ (40.57)	$ (21.15)	$ (12.26)	$ (2.20)
Pro forma—basic and and diluted[1]			$ (2.74)		$ (0.30)
Supplemental pro forma[2]			$ (2.60)		$ (0.26)
Number of shares used in computing per common share amounts:					
Basic and diluted	1,394	1,414	1,086	1,680	2,047
Pro forma—basic and and diluted[1]			14,099		14,834
Supplemental pro forma[2]			14,532		15,701

	As of March 31, 2002		
	Actual	Pro Forma[1]	Pro Forma As Adjusted[3]
	(in thousands)		
Balance Sheet Data:			
Cash and cash equivalents	$ 15,671	$ 15,671	$ 76,471
Working capital (deficit)	(9,547)	(9,547)	51,253
Total assets	44,740	44,740	105,540
Long-term debt, less current portion	4,117	4,117	959
Redeemable convertible preferred stock	101,830	—	—
Stockholders' equity (deficit)	(90,872)	10,958	74,916

	Year Ended December 31,			Three Months Ended March 31,	
	1999	2000	2001	2001	2002
	(in thousands)				
Other Data:					
EBITDA[4] (unaudited)	$(21,223)	$(28,179)	$ (1,716)	$(3,600)	$ 3,583
Adjusted EBITDA[5] (unaudited)	(24,405)	(43,860)	(13,722)[6]	(8,012)[6]	666
Number of subscribers (unaudited)	107	292	456	303	603
Net cash provided by (used in):					
Operating activities	$(16,529)	$(22,706)	$ 4,847	$(2,805)	$ 6,505
Investing activities	(19,742)	(24,972)	(12,670)	(4,087)	(5,798)
Financing activities	49,408	48,375	9,059	(927)	(1,167)

(1) The pro forma balance sheet data, pro forma net loss per share—basic and diluted, and pro forma number of shares—basic and diluted give effect to the conversion of all outstanding shares of our preferred stock into shares of common stock automatically upon completion of this offering.

(2) The supplemental pro forma net loss per share—basic and diluted gives effect to the assumed repayment of our subordinated promissory notes as of July 11, 2001 with the proceeds from the offering for the shares solely sold to repay these subordinated promissory notes.

(3) The pro forma as adjusted column gives effect to the sale of 5,500,000 shares of common stock offered by us at the initial public offering price of $15.00 per share and the application of the net proceeds from the offering, after deducting underwriting discounts and commissions and estimated offering expenses, including repayment of our subordinated promissory notes.

(4) EBITDA consists of operating loss before depreciation, amortization of intangible assets, amortization of DVD library, non-cash charges for equity instruments granted to non-employees, gains or losses on disposal of assets and stock-based compensation. EBITDA provides an alternative measure of cash flow from operations. You should not consider EBITDA as a substitute for operating loss, as an indicator of our operating performance or as an alternative to cash flows from operating activities as a measure of liquidity. We may calculate EBITDA differently from other companies.

(5) Adjusted EBITDA consists of EBITDA less amortization of DVD library. Adjusted EBITDA provides an alternative measure of cash flow from operations. You should not consider Adjusted EBITDA as a substitue for operating loss, as an indicator of our operating performance or as an alternative to cash flows from operating activities as a measure of liquidity. We may calculate Adjusted EBITDA differently from other companies.

(6) Adjusted EBITDA for the year ended December 31, 2001 and for the three months ended March 31, 2001 has been "normalized" to reflect DVD library amortization as if a one-year amortizable life had been used beginning as of January 1, 2000 instead of July 1, 2001. As more fully discussed in Note 1 to the Notes to Financial Statements, on January 1, 2001, we revised our DVD library amortization policy from an accelerated method using a three-year life to the same accelerated method over a one-year life.

FIGURE 18.6E

TABLE 18.18 Conditions that favor an IPO.

■ Market value greater than $200 million.	■ Profit margins greater than 14 percent.
■ Sales greater than $100 million.	■ Return on capital greater than 14 percent.

sell a large enough number of shares to net at least $50 million so that its total costs of the IPO process remain reasonable.

While a fast-growing, high-impact company may benefit from an IPO, it is wise to consider the burdens of public ownership. Aggressive regulation, stringent record keeping requirements, and a fickle public market can be significant to smaller firms. Firms with a market value of less than $200 million may wish to avoid these burdens and remain private. Furthermore, firms with a market value of less than $300 million are often thinly traded and illiquid. The conditions required to go public are summarized in Table 18.18.

18.12 AGRAQUEST

The original AgraQuest business plan dated May 5, 1995, provided the profit and loss statement shown in Table 17.8. It also provided a table of projected investment requirements, shown in Table 17.9. These projections were optimistic but consistent with the growth in interest in biotech firms and their potential to become large, important companies. The original business plan called for an IPO after five years.

Pam Marrone showed the plan to numerous venture capitalists over an 18-month period. Eventually, in February 1997, she found a team of investors interested in opportunities in sustainable agriculture. These venture capitalists as a group invested $3.2 million in the firm for about 46 percent of the firm. Subsequent stages of venture capital were attracted over the years, as shown in Table 18.19. Note the down round (F), where the price per share dropped significantly.

TABLE 18.19 Sequence of venture capital rounds for AgraQuest.

Year	Round	Amount raised ($ thousands)	Share price ($)	Percent owned by venture capital firms
1995	Seed—Founders	50	0.025	—
1995	Seed—Family and friends	450	0.35	—
1997	A—Venture capital	3,200	1.05	46.1%
1998	B—Venture capital	6,000	1.70	59.6%
1999	C—Venture capital	7,100	2.35	66.4%
April 2000	D—Venture capital	7,200	3.20	72.3%
December 2000	E—Venture capital	15,000	5.00	76.1%
2003	F—Venture capital	9,550	1.45	88.0%

As a result, after round F, the venture capital firms owned about 88 percent of the firm. Marrone, the original founder, owned only 5 percent of the firm as a result of all the venture capital rounds.

The initial public offering market was favorable in 2000, and AgraQuest prepared an IPO document and SEC registration to raise $75 million with a share price of $11 to $13. The lead underwriter (investment banker) was Merrill Lynch. The prospectus was completed and filed in August 2001. However, the market for IPOs disappeared with the September 11, 2001, terrorist attacks. As a result, an expenditure of $1 million on underwriting costs was lost as the IPO filing was withdrawn in April 2002.

18.13 SUMMARY

The entrepreneur leaders of a new start-up firm create a set of financial projections that can be used to estimate the cash investments that will be needed as well as when they will be needed. Using that information, the entrepreneurs seek out investment capital. For the first or seed round, they may rely on their own funds and investments from friends and family. Eventually, most technology firms will need a significant investment from wealthy individuals called angels or professional investors called venture capitalists.

Several stages of investment may be the best means of acquiring investments based on performance milestones for each stage. A real option is the right to purchase an asset at a future date. Thus, venture capital investors often use staged investing with milestones to exercise their investment opportunity with a start-up. For big-impact start-ups, it is the future value of the potential of the new firm that is most attractive.

Using venture capital valuation methods, a start-up's value is established and an agreement on the division of ownership may be obtained. As the expected growth of revenues and profitability are achieved, the firm and the investors may wish to exercise an option to sell. This could be an initial public offering (IPO) or the acquisition of the start-up by a larger firm in order to harvest the value that has been created by the partnership of investors, founders, and employees.

Principle 18
Many kinds of sources for investment capital for a new and growing enterprise exist and should be compared and managed carefully.

18.14 EXERCISES

18.1   Viscotech Inc. is described in Exercise 17.3. Determine the percentage ownership an angel group may demand for investing the $1 million sought by Viscotech at the start of year 1. If Viscotech is unable to obtain the bank loan of $500,000, it will need an equity investment of $1.5 million. What ownership percentage will the angel group demand for this investment? Assume the annual interest payment on the loan was planned to be 10 percent of the principal.

18.2   Glenn Owens's attractive technology start-up requires $10 million to launch. Projections show earnings of $10 million and sales of $80 million in the fifth year. The venture capital firm expects a return of 50 percent per year for the five-year period prior to an IPO. What ownership portion should the venture capitalist expect to receive?

18.3   DGI, a new firm in formation, has developed a set of projections shown in the table below. The expected and pessimistic cases are shown. The harvest of the firm is planned for the fifth year. The firm is seeking an initial investment of $1 million before launching in year 1 as well as a commitment for $1 million at the end of year 2 for expansion. Acting as an adviser to a venture capital firm, prepare an offer of investment to submit to DGI. Assume the PE ratio in DGI's industry is 15.

18.4   A venture capitalist is considering investing $4 million in a series B financial round. This investor requires a 50 percent return on her money over the five-year period to IPO. The expected sales grow at 60 percent per year for the first five years and reach $22 million in the fifth year. What percentage of the firm will this investor demand? Assume that the investor will value the firm in the range of four to six times sales.

18.5   The CEO of an early-stage software company is seeking $5 million from venture capitalists. The reasonable projected net income of $5 million in year 5 can be valued at a PE of 20. Furthermore, the sales in year 5 are projected at $25 million. What share of the company would the venture capitalists require if their anticipated rate of return is 50 percent? The company has 1 million shares outstanding before the venture capitalists purchase shares. What price per share should the venture capitalists pay?

18.6   A venture capitalist is interested in investing $6 million in a new venture. He estimates that at the end of year 4, the firm will generate $30 million in revenues and $8 million in profit. The term sheet will

Revenues and net income for the expected case and the pessimistic case for DGI.

Year	1	2	3	4	5
Expected revenues	0.84	2.82	5.44	8.35	11.55
Expected net income	0.18	1.25	2.67	4.17	5.86
Pessimistic revenues	0.42	1.41	2.72	4.18	5.78
Pessimistic net income	(0.11)	0.26	0.77	1.25	1.81

Note: All figures in millions of dollars.

call for an IPO at the end of year 4. He demands a return of 40 percent per year. What percentage equity in the firm should he require when comparable ratios in the industry are PE of 18 and price-to-sales of 3.0?

18.7   Apollo Group (symbol APOL) is a for-profit educational firm (www.apollogrp.com). The firm was started in 1981 and had its IPO in December 1994. Its earnings growth rate was 38 percent per year for 1998 to 2003. Apollo's system of schools had 187,500 students in 2003, with about 40 percent participating as online students. Examine the balance sheet and profit and loss statement. Determine the return on capital for this firm.

18.8   RedEnvelope is an online gift retailer that went public in 2003 (www.redenvelope.com). The firm had sales of $70 million in 2003. The firm went public via the OpenIPO process using bids for shares (see www.openipo.com). Examine the OpenIPO process and the RedEnvelope offering. What was the postmoney valuation of the firm? Was this valuation reasonable?

18.9   Rfco is attempting to commercialize its radio frequency chip for future cell phones [Lohr, 2003]. Increasingly, cell phones are a combination of radios and computers equipped with tiny cameras, video screens, and music players. Rfco is planning to design and manufacture integrated circuits to support cell phone functions that enable a "world phone." Rfco raised $16.5 million in a first round of financing from four venture capital firms (see www.rfcosemi.com). Estimate the potential market for Rfco chips, and determine what share of Rfco should have been sold to the four venture capital firms in 2003.

18.10   Consider a new firm in the nanotechnology field that seeks a second round of financing. This year, it has revenues of $2 million and projects profitability of $200,000 next year on revenues of $3 million. It is raising $1 million from a new set of investors. What share of the company should it offer to the new investors? Assume it can increase profits at a rate of 25 percent per year over the next five years.

CHAPTER 19

Presenting the Plan and Negotiating the Deal

Leadership involves remembering past mistakes, an analysis of today's achievements, and a well-grounded imagination in visualizing the problem of the future.
Stanley C. Allyn

The creators of a new enterprise need to tell their story about the future of their business. A good story will motivate investors, employees, and partners. A well-told narrative plan, one that demonstrates a novel solution to an important problem, can be inspirational and motivational. Establishing credibility and trust through presentations of the new venture's plan may lead to an investment.

A short presentation of the plan (called a pitch) can help interest investors in seeing a more complete presentation. The 10-slide pitch, often delivered in an investor's office or at a venture fair, may create the necessary interest from an investor.

The negotiation of a deal with an investor is an important part of the process. One can cement the relationship or destroy it through the negotiation process. Often agreements with terms contingent upon performance may be appropriate. The integrated story and the business plan should show how the business solution would be profitable within a reasonable period. The investors are interested in a favorable return. They also want to sense that they will be partners with trustworthy, capable entrepreneurs. ■

19.1 THE NEW VENTURE STORY

Stories are an integral part of the process by which founders start and build new ventures, acquire needed resources, and generate new wealth [Lounsbury and Glynn, 2001]. Storytelling is applicable to corporate new ventures as well as independent new ventures. Stories play a critical role in the process of the emergence of a new business. Stories that are told by or about a new venture define it in ways that can lead to favorable interpretations of the social benefits and wealth-creating possibilities of the venture, thus enabling resources to flow to the new enterprise. Furthermore, stories can legitimize new ventures, thus helping to build acceptance of them. A good entrepreneurial story attracts financial and human resources as well as builds industry acceptance. An attractive story can help build support for a new venture. Successful entrepreneurs must shape interpretations of the nature and potential of their new venture for those who may supply needed resources.

A **story** is a narrative of factual or imagined events. Stories depict a course of challenge, plan, actions, and outcomes in a manner similar to a plan. A good story and business plan define an opportunity, a concept, cause and effect, and an outcome all held together in a holistic way. Stories use plot lines and twists to capture the imagination and interest of the listener. The story of the new venture tells the ultimate goal of the venture, the ideological challenge, and the means of achieving the goal. The creation of an appealing and coherent story may be a useful form of communication for entrepreneurs in the attempt to attract interest and support for their idea and plan. To be effective, the content of the story must align with the interests and background of the listener. A well-crafted story about a new venture emphasizes the goals and merits of the venture.

As founder, Jim Clark provided the story for both Netscape and Healtheon as mentioned earlier when recounting to Michael Lewis [2000]:

> It dawned on Clark that the food chain of capitalism was missing a link, and that, if he summoned the nerve to hoist himself up, he could be that link. And that if he didn't have the nerve to do so he would make a mockery of his entire remarkable climb. . . . His role in the valley was clear: *he was the author of the story.* He was the man with the nerve to invent the tale in which all the characters—the engineers, the [venture capitalists], the managers, the bankers—agreed to play the role he assigned to them. And if he was going to retain his privilege of telling the stories, he had to make sure that the stories had happy endings.

In constructing a legitimate identity for a new venture, the storyteller tries for a balance of alignment with existing challenges and the potential for distinctiveness. The investor may be drawn by the credibility of the story and the storyteller to seriously consider investment.

New firms should not rely solely on a pure bullet-point presentation. A list of bullet points reduces a set of issues to a few points but provides little fabric or motivation. Bulleted lists are often generic in meaning and leave challenges and relationships unspecified. Furthermore, they often leave critical assumptions

TABLE 19.1 Three stages of the story.

1. **Set the stage:** Define the current situation, the current players, and the opportunities coherently and clearly. Life and business are in a delicate balance.

2. **Introduce the dramatic conflict:** An inciting incident or need throws life out of balance. Describe the challenges and opportunities, and the need for a coherent plan to proceed toward success and a new balance.

3. **Reach a resolution:** Portray a coherent plan describing how the new venture can overcome the obstacles and succeed by following the plan.

unstated. A good story includes all the challenges, relationships, and assumptions in the very fabric of the narrative. A presentation can use some bullet-point slides but should emphasize the story it wants to tell.

The new venture story consists of three elements, as shown in Table 19.1 [Shaw, Brown, and Bromiley, 1998]. The first step is the setting of the stage by describing the current conditions of the industry, society, and the existing relationships and opportunities. Next, the storyteller introduces the dramatic conflict by describing the challenges confronting the venture, the need for a plan to overcome the challenges, and the critical obstacles and issues. Finally, in the third step, the storyteller describes the plan to overcome the obstacles, secure the necessary resources, and move on to success.

Jim Clark contracted an illness in late 1995 that gave him personal experience with the health care industry, its bureaucracy, and resulting paperwork demands. He responded with a concept: The patient would have a password and a digital record, and the doctor would use the Internet for billing forms with no hassle for both parties. He drew a diagram of the four players in the health care industry, as shown in Figure 19.1. He then placed Healthscape in the center of the diagram as the solution, via the Internet, for the entire industry. Clark told a story of critical importance to our nation, described the dramatic challenges and opportunities, and portrayed the solution: his new venture, Healthscape (later called Healtheon). With his compelling story, Clark went on to raise millions of dollars to build Healtheon into a new company, which later was sold to WebMD [Lewis, 2000].

A well-told narrative plan that shows a difficult situation and a novel solution leading to improved market conditions can be galvanizing. When listeners can locate themselves in the story, their sense of commitment and involvement is enhanced. By conveying a powerful impression of the process of succeeding, narrative plans can motivate and mobilize resources and investors.

Stories help entrepreneurs deliver direction and inspiration more powerfully than a logical argument. Facts convey information, and stories convey meaning. Entrepreneurs need facts to back up their plans and a story to convey the goals and meaning of the venture. [Gargiulo 2002]

Like a good book or movie, a story needs an "inciting incident," or turning point, and a good ending. With the help of colleagues and allies, the entrepreneur turns challenges and obstacles into new opportunities and an outcome desired by all.

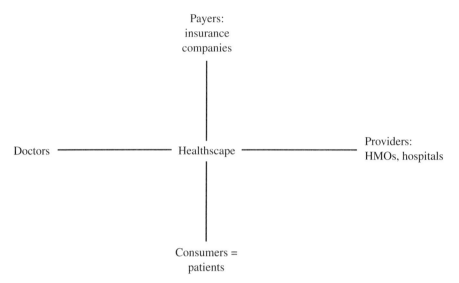

FIGURE 19.1 Healthscape diagram.

Robert Johnson created an innovative plan to produce television programs targeted to black viewers. Johnson met with several investors and told them his story of the opportunity to serve black viewers with TV programs that resonated with their values and experiences. The power of this story generated significant financial and cable channel resources. As a result, Black Entertainment Television (BET) was launched on January 8, 1980 (see www.bet.com).

A great story well told will pack enough power to be memorable. We can forget bullet points, but a great story is unforgettable. It displays the struggle of expectation and hope versus reality and challenge [McKee, 2003]. An example of an entrepreneur's story follows:

My father and I were very close. In 1999, he exhibited congestive heart disease that appeared to be untreatable. While in the hospital for further tests, he died in the middle of the night. These tests were inadequate and failed my father and me. I have found and licensed a patent for a new blood test, but the Food and Drug Administration has been slow to respond, and the inventor has gone on to other issues. Our early tests have shown good results for a low-cost test that can illuminate paths to a cure.

Our firm, Heartease, needs $1 million to complete our FDA certification. We have the data and proof of the test's efficacy. We need you on our team with your insights and money. Together, we can save your father or mother.

19.2 THE SHORT VERSION OF THE STORY

Often entrepreneurs have a chance to make their case to a potential investor or ally. The short version of the new venture story is often called the **elevator pitch,** which gets its name from the 2-minute opportunity to tell a story during an elevator ride. A short version of the venture story quickly demonstrates that the entrepreneurs know their business and can communicate it effectively. The secret of strong short stories is in grabbing the attention of listeners, convincing them with the promise of mutual benefit, and setting the stage for follow-up. The storyteller should speak in terms the listeners can relate to. A good storyteller communicates with the passion that comes from knowing that this opportunity may never come again.

The goal of the short story is to get approval to proceed to the next step, where the entrepreneur can tell the longer version of the story and secure new colleagues, allies, and investors. The short version of the venture story starts with an introduction, moves into a description of the opportunity, and then describes the potential benefits of the new venture. New venture leaders need to connect to sources of resources by using their short story to gain links to new allies. Chance meetings will offer opportunities to entrepreneurs to make a brief case for their venture. A prepared short version of the venture story can be a powerful door opener.

A pitch can start with a captivating question such as:

"If IKEA can provide baby-sitting, why can't movie theaters?"

The short version of the story should convey the vision of the venture. In the case of Genentech, it is: "We discover and make biotechnology pharmaceutical products to reduce or overcome the effects of cardiovascular, pulmonary, and cancer diseases." An important vision can provide the inspiration for the venture and its short story.

> As the founder of Intuit, Scott Cook describes his venture in this way: Homemakers need to pay the family's bills. They hate the hassle of bill collection and payment. They need a software computer program to quickly and easily pay their bills. Other programs are too slow and too hard to learn. Our solution is a fast, easy-to-use program with no instruction books needed. So the bill payer needs Quicken!

19.3 THE PRESENTATION

The new venture team will be expected to verbally present their business plan to investor groups, angels, talent, allies, and suppliers. The purpose of this meeting will be to persuade them to cooperate, support, and participate in the new venture. Effective persuasion is a negotiating and learning process that leads to a shared vision. Formerly, people thought of persuasion as a simple process: state

TABLE 19.2 Four-step method of persuasion.

■ Establish credibility with the investor, ally, client, or talent.	■ Offer solid, compelling evidence to support the plan.
■ Frame the goals of a new venture to be consistent with the goals of the investors or allies. Describe the unique benefits of the venture.	■ Build a good relationship with the investor or ally.

the position or plan, outline the supporting arguments, and then ask for the action or deal sought after. Today, most would-be investors and allies want to engage in a dialogue with the new venture team. The entrepreneur outlines the venture and invites feedback and alternative solutions. Persuasion involves compromise and the development of relationships with the investors and other participants.

New enterprises must be able to sell their story to potential investors and clients. Selling ideas is intrinsically difficult. Clients and investors are naturally risk-averse when it comes to large projects that call for large investments with payoffs that are many years in the future. They are even more risk-averse when the projects do not originate from within their own organization.

Effective persuaders first establish credibility. Second, they frame their goals in a way that establishes common ground with their listener. Third, they offer solid evidence to support their plan. Finally, they build a good relationship with their potential investors or allies. This four-step method of persuasion is summarized in Table 19.2 [Conger, 1998].

Credibility and trust is established through experience over time. Thus, the shortest path to credibility is with someone the entrepreneurs already know. If the person is new to the entrepreneurs, their track record and references will be important. Their expertise and experience in the venture industry will be significant. The expertise required for the venture must exist on the venture team. It may be useful to review the concept of influence and persuasion as described in Section 11.10.

The business plan must appeal strongly to the potential investor or ally. Framing the unique benefits of the new venture so that they match the goals of the investor or ally is critical. The next step is to provide solid evidence supporting the business plan. Here, a vivid story or analogy will help to bring the plan alive. Finally, it is important to build a good relationship with the investor or ally. In this step, the entrepreneurs demonstrate their commitment to the plan and show some of their passion for the project.

A good presentation captures the listeners and lets them respond to the problem the entrepreneurs propose to solve. A four-part "pitch" is described in Table 19.3.

In presenting the business plan, it is also useful to answer the nine questions provided in Table 19.4. Of course, there will be other issues, but these nine are almost always part of a presentation. It is helpful to rehearse the presentation

TABLE 19.3 Four-part pitch.

1. This is a painful problem that customers need to solve.	3. This company has a profitable and proven solution for this problem.
2. Many customers who have this painful problem have money to spend on alleviating the pain.	4. Management has implemented and planned effectively in the past, and they can execute well in the future.

with an audience of a few trusted colleagues or friends who can respond with suggestions.

Trust is the basis of nearly every enduring business relationship. The presentation of a business plan is a vehicle for winning trust from potential investors and allies. Trust, confidence, and relationships are built over time. Speaking about the venture's goals needs to invoke a positive response from the listener. What difference will it make if the venture is successful? Will people live better lives or enjoy new alternatives?

Most presentations to venture capitalists and other investors or allies should be based on about 10 slides. Each slide should have no more than 30 words. This will keep the listener interested and the presenters focused on their key points. A sample 10 slide presentation is outlined in Table 19.5. Ten slides can be used for a 20-minute presentation with time for discussion.

In any presentation, the speaker should convey a sense of urgency about the problem and a strong commitment to make the solution robust. Good speakers highlight the unique benefits on slides 3 and 4.

Listeners are swayed by the quality of the ideas but even more by evidence that the presenters are creative and innovative. The investors or new team members seek to be part of a creative collaboration. The listeners look for passion and evidence that the proposed solution is a big change or discontinuity. Furthermore, the goal is to engage the listeners so that they become part of a creative collaboration. The best outcome is that the presenters successfully project themselves as

TABLE 19.4 Nine questions to answer in the business plan presentation.

1. What is the product and what problem is it solving?
2. What are the unique benefits of the product?
3. Who is the customer?
4. How will it be distributed and sold?
5. How many people will buy it in the first and second years?
6. How much will it cost to design and build the product?
7. What is the sales price?
8. When will you break even?
9. Who are the key team members and how are they qualified to build this business?

TABLE 19.5 A sample ten-slide presentation.

1. Company name, presenter name, contact information.
2. Description of the problem: the need and the market.
3. Solution: the product and its key benefits.
4. Business model and profitability.
5. Competition and strategy.
6. Technology and related processes.
7. Marketing and sales plans.
8. Leadership team and prior experience.
9. Financial projections summary.
10. Current status and funds required.

creative types and get their listeners to view themselves as creative collaborators in the process of building the new venture [Elsbach, 2003].

19.4 NEGOTIATING THE DEAL

After the presentation of the business plan and follow-on discussions, an investor or ally may be confident about the potential venture but hold different expectations about it. Thus, the investor and the venture team may have different views regarding the valuation for the firm and the appropriate terms of the deal. A good deal is one that fairly meets the needs of the new venture while enabling the relationship between the firm and the investor to flourish in the future. Thus, the pricing and terms of the deal must be balanced with the future of the relationship. If possible, the new venture needs to have alternatives to closing a bad deal. With a good alternative, the new venture can walk away from a bad deal. A new venture should understand its own interests and its own no-deal alternatives or options. Price, control, and ownership percentage are usually key factors for negotiation.

Negotiating a fair deal is a skill that can be learned. Most entrepreneurs have limited experience negotiating a fair deal with investors. **Negotiation** may be defined as a decision-making process among interdependent parties who do not share identical preferences. Consider an example of a manager and an employee negotiating a pay raise for the employee. They are interdependent but have different preferences for the outcome. The employee wants a raise, and the manager wants improved performance.

The best negotiation should produce an efficient, wise agreement and not damage the relationship between the parties. A wise or good agreement is one that meets the legitimate interests of both parties, resolves conflicts fairly, and is durable [Fisher and Ury, 1991]. One should try to avoid parties locking into a position, but rather reach for common ground.

TABLE 19.6 Four principles of negotiations.

1. Focus on describing the problem (task or deal) and take the people out of the discussion.

   Goal: All participants are solving the problem.

2. Focus on the interests of the parties, not their original positions.

   Goal: Each party states what they seek and the associated goals.

3. Generate a variety of options or possibilities that advance the interests of the parties.

   Goal: Several real solutions.

4. Create a final deal based on fair and objective standards.

   Goal: Real, measurable standards.

Source: Fisher and Ury, 1991.

The process of negotiating a good deal for all concerned can be based on four principles, as summarized in Table 19.6. Try to take personalities out of the discussion and avoid locking into positions. Try to get everyone working for a fair deal. Talk about the interests and goals of each party and avoid taking rigid positions. Then generate several possible solutions that advance the interests of all parties. Finally, select the best solution with measurable outcomes.

Often negotiations will stall, and it may be best to try to reshape the scope and sequence of the negotiation. One or more of the parties can scan widely to identify elements outside of the deal currently on the table that might create a more favorable structure. For example, they may introduce new parties and terms to the deal and try to satisfy all parties [Lax and Sebenius, 2003].

Investors tend to have goals based on return on investment and time horizons to receiving this return. Entrepreneurs tend to have goals based on growth, success, and achievement, as well as return on investment. Both parties need to help generate some good options so that there is room to adjust the deal. Finally, the parties select a deal that gives them a fair solution to their needs. This deal should include measurable outcomes and adjustments in ownership or other factors if the agreed-to outcomes are not realized.

When making an investment, a venture capitalist considers three forms of risk: the market risk (establishing customers for the product), the technological risk (the extent to which the technologies or concepts are well developed and not threatened by potential competitors), and the management risk (the team's technical and leadership competencies to develop the new firm). The venture capitalist wants to know which risks will be reduced by their investment in this stage.

Differences regarding valuation and ownership usually reflect differences in estimations of the future performance of the new venture. Therefore, it may be wise to develop an agreement that includes terms that are contingent on the outcome of designated measures or events. Using contingent terms in a contract enables the parties to bet on the future rather than argue about it. An agreement with contingences could include terms on such measures as revenues, profits, or number of customers achieved in an agreed-upon period. The actual outcome could lead to an agreed-upon ownership percentage. For example, the investor agrees to an ownership split of 70-30 (firm versus investor) if the goals and milestones are met but will reserve shares for a readjusted split of 60-40 if they are not met. Investors and new ventures can come to very different conclusions about many kinds of future events, such as sales and market shares as well as competitors' moves. Whenever such a difference exists, so does an opportunity to craft a contingent contract that both sides believe to be in their best interests. A contingent contract results in the two parties sharing the risk.

A common way to reach agreement about ownership is to offer a deal with warrants to tie ownership to actual performance. A warrant is a long-term option to acquire common shares, usually at a nominal price. For example, an investor may receive a warrant to receive an agreed-upon number of shares if certain performance levels are not achieved by a certain date [Smith and Smith, 2004].

A factor that can complicate negotiations is the matter of dilution of ownership by the founder group. Investors will usually seek an *antidilution* clause to protect themselves from dilution of their ownership percentage. This antidilution clause will usually be triggered by a lower price (down round) in a subsequent financing round. All the terms of an investment agreement, often called a *term sheet,* should be reviewed by the new venture firm's attorney. A term sheet is a funding offer from a capital provider. It lays out the amount of an investment and the conditions under which the investors expect the entrepreneur to work using their money. The key is to remember that it's just an offer, and the entrepreneur can counter that offer and negotiate all the terms before finally accepting the funds. An excellent reference on these matters is *The Entrepreneur's Guide to Business Law,* by Constance Bagley and C. E. Dauchy (Boston: South-Western, 2002).

19.5 CRITICAL ISSUES FOR THE BUSINESS PLAN

We discussed the development of the business plan in Section 8.9 and the presentation of the business plan in Section 19.3. After the business plan is presented to a few potential investors, their suggestions and criticisms may require adjusting the business model of the firm or other parts of the business plan. Perhaps it is necessary to revise the product to make it more compelling. Offering a product that is "nice to have" for a customer is different than offering a product that the customer "must have." It is nice to have vitamins available, but when one has a headache, an aspirin or Advil is a "must have." Is the product a complete solution to the customer's problem, or is it only a part of the solution?

Investors will ask: Are the people committed to the business, and is the opportunity large in potential? Can we identify and reduce the risks? Will there be a way we can harvest our return on investment? Is the estimated growth rate attractive?

The plan in written form and the verbal presentation should hold together as an integrated whole. Use of outdated or incorrect data will leave doubts for investors. Unsubstantiated assumptions can also hurt a plan. Does the plan include an honest recognition of the competitors?

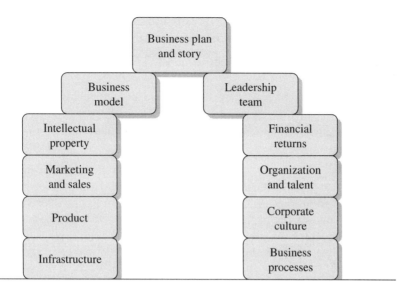

FIGURE 19.2 Integral nature of the business plan and the firm.

The business plan and its associated story can be viewed as the keystone of the business arch, as shown in Figure 19.2. All the elements of the business come together to form a business whole.

It will help to validate a business plan if the new venture has paying customers at the time of presentation. It is even better if the firm is about to become profitable. Investors need reasons to invest. They want to see advance orders, a letter of intent, or a customer list. That shows proof of customers and how they value the firm's product.

In 1968, Gordon Moore and Robert Noyce left Fairchild Semiconductor to found Intel. They brought along Andy Grove and several other colleagues. At Intel, they saw an opportunity to make a silicon transistor and later an integrated circuit. Moore and Noyce were leaders in their field and knew Arthur Rock, a San Francisco venture capitalist. They asked him one day if he could raise $3 million to start Intel. Rock had secured commitments for the $3 million investment by that evening. Moore and Noyce were well-known and could command a good deal for their new business. Most entrepreneurs are not as fortunate.

If the leaders of a new venture do not have a big, proven network and background, they will need to work hard to find and satisfy investors. They must convince them that their business is a one-time chance to get involved with something important that will exploit a big opportunity.

19.6 AGRAQUEST

The search for venture capital was long and hard for Pam Marrone. She pitched her business plan to over 200 venture capital firms. Her plan called for a new firm, AgraQuest, entering a big market and creating a new solution. The market for pesticides is $28 billion and is dominated by large chemical companies. AgraQuest had a solution that used natural pesticides. Her story was:

1. Pesticides and herbicides help farmers to keep their yields up and avoid disastrous pestilence.

2. Chemical pesticides are harmful to people and the environment.

3. AgraQuest can readily develop natural biopesticides that protect people and the environment, and allow farmers to use them right up to the harvest.

Eventually, Marrone was introduced to the Investors Circle (www. investorscircle.net) by Calvert Social Ventures, and she presented at a venture fair in Chicago in May 1996. This led to interest from Rockefeller Ventures and its eventual investment in AgraQuest.

Marrone learned to give a passionate presentation and capture the listener's attention. Most investors, however, are wary of the agricultural market and declined to invest. The all-natural solution had been tried before and had failed.

One of her angel investors in the seed round, James Schlindwein (former CEO of Sara Lee and Sysco), started introducing her to agricultural venture investors. Marrone found her angel investor to be her best mentor and coach.

Eventually, Marrone found an interested set of social responsibility venture capital firms. They were dedicated to the natural solution and saw AgraQuest as a company that could succeed. It was a well-led company with a good process for discovering new biopesticides. AgraQuest received $3.2 million in venture capital from a group of social venture capitalists in 1997. By 2003, the company had received a total of $48 million in venture capital.

19.7 SUMMARY

The potential for a successful new business can be communicated through a short pitch, a formal presentation, and a written business plan. Most entrepreneurs will need the skills to communicate their vision and solution in all three forms. Through these presentations of the entrepreneurs' story, the potential investor, new employee, or ally learns to understand the opportunity and recognize the competencies of the team members.

As investors become interested in the new enterprise, negotiations about valuation and performance milestones will commence. Conducting negotiations that retain and enhance the rapport with the investor is essential. The entrepreneur is negotiating until the moment of execution of an agreement and a transfer of funds. All the negotiations continuously address issues surrounding product, team, processes, business model, and intellectual property.

Principle 19

The creation and communication of a compelling story about a venture and the resulting skillful negotiations to close a deal with investors are critical to all new enterprises.

19.8 EXERCISES

19.1 Quick Stores, a new drug store chain, has decided that free-standing stores in high-traffic areas are its best strategy. It also features drive-through pickup of pharmaceuticals. Develop a story for Quick Stores that can be used with potential investors.

19.2 A new firm, mail-stamps.com, intends to sell postage stamps over the Internet. Write a short elevator pitch for this start-up.

19.3 A new firm develops and distributes electronic games for personal computers. These games teach children to read, recognize symbols, and perform mathematics. This new venture needs $1 million to launch a nationwide campaign for its products. Prepare a short story that will persuade a venture capitalist to support the firm.

19.4 Nuance Communications spun off from SRI International in April 2000. An IPO was issued on April 13, 2000, selling 4.5 million shares at $17 (see www.nuance.com). Nuance develops natural-voice interface software that enables access to content and services from the telephone. Prepare a story for this firm that could be used in attracting distributors and alliances.

19.5 Charmed Technology is spin-off from the MIT Media Lab that offers wearable Internet products and technologies (www.charmed.com). Write a story that depicts this venture and its opportunity to build an important business.

19.6 IBM is creating a business based on its genetic research tools [Harmon, 2003]. It develops hardware and software for genetics research and development companies. Create a story that communicates this opportunity.

19.7 Business plan contests offer an opportunity for entrepreneurs to present their business plans. One such contest is offered by the National Institute for Entrepreneurship (see www.venturebowl.com). Visit the site, study the winning presentations, and prepare a brief report on a presentation that interests you.

19.8 As the CEO of a new technology venture, you and your team have set a valuation for your firm of $10 million (pre-money) and found a willing venture capital firm. The venture capital firm, however, has set a valuation of $6 million. Revenues next year are projected to be about $6 million, and the firm will be profitable next year. Identify a negotiation approach for achieving a reasonable compromise valuation.

CHAPTER 20

Leading a Technology Venture to Success

Well done is better than well said.
Benjamin Franklin

C reating a business plan for a new enterprise is important, but implementing the plan successfully is essential. Execution of a plan is a discipline for connecting strategy with reality by aligning goals and the firm's people to achieve the desired results. Execution is about turning a concept into a great business.

New businesses move from start-up to growth to maturity in stages. Managing a new business through these stages requires different skills and organizational arrangements. Start-ups need to plan for having the right people in the right positions as they grow.

Organizations, like people, need to learn and adapt to change. Organizing for recognizing and responding to challenges can build resilience in a start-up firm. The ability to adapt to change may be a firm's only truly sustainable advantage. Furthermore, to achieve long-term success, a firm needs an ethical base for action. ■

20.1 EXECUTION OF THE BUSINESS PLAN

Once the new venture has secured the necessary resources, the firm proceeds to the **implementation** phase by carrying out or putting into effect the elements of the business plan. Another common term for the implementation is execution, which is a system of getting things done. **Execution** is a discipline for meshing strategy with reality, aligning the firm's people with goals, and achieving the results promised [Bossidy and Charan, 2002]. Often, the unique difference between a successful company and its competitors is the ability to execute its plan.

Execution is not just tactics; rather, it is competency and an associated system built into a firm's goals and culture. Both Dell Computer and Gateway sell personal computers directly to customers, but Dell requires one-fifth the working capital needed to generate a million dollars of sales than does Gateway. Both companies sell build-to-order PCs, but Dell's asset velocity is five times that of Gateway. Asset velocity is defined as the ratio of sales dollars to net assets, which include plant and equipment, inventories, and working capital.

Execution is the process of determining how well a firm is performing, acting to improve performance, following through, and ensuring accountability. Execution is following through on the strategy of the business plan. Any firm with a sound business model and strategy is still only as good as the implementation of the strengths of the model and strategy. In preparing the plan, all the team members have built their expectations and strategies into an agreed-upon, coherent road map. Thus, the team knows what needs to be done. The next step is to agree on how it is going to happen. Who does what and by when? The team sets short-term goals and priorities, and then assigns tasks to individuals. Rewards and recognition are linked to on-time performance. Missing deadlines is costly. Therefore, setting realistic deadlines is important. The success of a new enterprise can be attributed to the execution skills of many team players rather than the decision-making skills of an omniscient entrepreneur. Six questions that can be effectively used to achieve solid implementation are provided in Table 20.1.

The new venture needs team members with a wide set of capabilities so that one person can be responsible for several tasks. If necessary, additional people may be engaged to accomplish unique or difficult tasks. These tasks flow seamlessly from the firm's strategy. The team describes where it wants to be by a certain time and then divides the tasks among its members. It is critical to have a realistic assessment of the effort and time required for major tasks. The elements of an operating plan are tasks, milestones, and objectives. The leadership team

TABLE 20.1 Six questions for implementation.

1. **Why** is the objective a priority?	4. **Who** is on the team and will be accountable?
2. **What** is the action and the expected outcome?	5. **When** will the activity be completed?
3. **How** will the action be achieved?	6. **Where** will it be accomplished?

will need to make trade-offs between tasks and goals so that the operating plan is realistic. A specific, written operating plan will help the firm to move forward efficiently. Also, reviews of accomplishments and the operating plan will help keep the team on task.

Execution is fundamentally about turning a concept into a business. For example, after a short period, the firm may find its prototype has defects or the sales channel is not as attractive as originally thought. Every new venture runs into trouble before reaching fruition. Then it is time to reconfirm the vision and recommit to the execution task. This requires persistence and follow-through. While admitting mistakes, the team's focus must stay on the long-term strategy. For example, in 1998, when fraud started to increase on eBay, the online auction firm created an antifraud campaign. Nonpaying bidders would receive only one warning before receiving a 30-day suspension from bidding. At the same time, eBay offered free insurance through Lloyd's of London. As a result, eBay became more successful.

Execution is hard work. The setting of goals and deadlines should be the task of those who need to accomplish the work. The establishment of priorities is a big part of sound execution. Tasks can be divided into "must do," "should do," and "nice to do," with the priority kept on the "must do" activities as much as possible. The use of measurable goals and lists of necessary tasks and deadlines is very helpful. Cyrus Field tried four times to lay a telegraph cable across the Atlantic Ocean to link the United States and Britain. Using a new steamship, the *Great Eastern,* he succeeded in 1866 after nine years of effort. The saga was a dramatic example of Thomas Edison's maxim that "genius is 1 percent inspiration, 99 percent perspiration."

Consider the highly competitive computer-aided design software business and two solid competitors, Mentor Graphics and Cadence Design (see www.mentor.com and www.cadence.com). Mentor and Cadence offer software for chip design at competitive prices. With aggressive competition and demanding customers, execution and details are everything. A firm's survival depends on repeat business. Knowing what details are more important than others can make a big difference. Few competitive advantages cannot be quickly imitated. Therefore, a new firm needs to out-execute its competitors. It must continually underpromise and overdeliver.

An emerging new venture often needs help on operational issues as it moves toward growth. One source of help is the new firm's suppliers and customers. They have in-depth capabilities that can often be accessed by a new firm. Often these large firms want the new firm to succeed and will lend a hand on tough issues.

During the initial period following launch of the firm, one of the primary goals will be to build and grow revenues. One key measure that can help in the early stages of a business is the ratio of revenues to expenditures plus assets employed, called a business index (BI), where

$$BI = \frac{\text{revenues}}{\text{expenses} + \text{assets}}$$

TABLE 20.2 Seven steps to building great companies.

1. **Leadership:** Leaders are ambitious for the firm, possess strong will and resolve, desire sustained results, and retain personal humility.

2. **People:** Choose the right people, put them in the right positions, create a road map for success, and communicate it to everyone.

3. **Success:** Unwavering faith the firm will prevail in the long run, confront the realities and facts, and respond.

4. **Organizing principle:** Act on your passion, competencies, and economic engine to create a core principle for the business.

5. **Culture:** Build a culture of discipline where everyone is responsible for results; stay focused.

6. **Technology:** Select a technology application that will accelerate the firm's momentum.

7. **Momentum:** Build momentum slowly and consistently for the long run by constantly creating entrepreneurial projects.

Source: Collins, 2001.

The goal of a business is to steadily increase the BI ratio by growing revenues faster than expenses and assets.

As a new firm grows, one useful measure of sound execution is the sales-per-employee ratio. A successful technology company will have at least $200,000 sales per employee. For example, the emerging firm Network Appliance has about $500,000 sales per employee (www.netapp.com).

Great companies execute flawlessly. They deliver products that *consistently* meet customer expectations. Furthermore, they empower customer representatives on the frontline to respond to varying customers' needs. The goal is to achieve almost perfect operational execution by constantly improving processes, training staff, and eliminating inefficiencies. An example of a firm with great execution is United Technologies [Joyce, Nohria, and Roberson, 2003].

Jim Collins [2001] describes seven steps to building a great business, as summarized in Table 20.2. If the new enterprise can implement most of these very well, it will have a good chance for success.

Jack Welch, retired CEO of General Electric, is perhaps the best-known executive of 1985 to 2000. He advocated many principles and methods of execution, such as "work-out" and "bullet-train speed." The key to Welch's popularity is his plain but powerful rhetoric about the critical issues of the firm [Lowe, 2001]. He insisted that each operating unit be number one or two in its market. While GE is a very large company, it retains today the atmosphere of entrepreneurial activity. General Electric is one of the "great companies" and meets all seven criteria for Collins's great companies.

20.2 STAGES OF A BUSINESS

A new business venture is expected to grow over time, normally following an S-curve, as shown in Figure 20.1. The five stages of a firm are: start-up, take-off, growth, slowing growth, and maturity. During the start-up period, the firm is organizing itself, accumulating the necessary resources, and launching its product. The second stage is take-off, when revenues start to grow. The growth stage is often most profitable. Eventually, the firm's growth slows in the slow-growth period. Finally, the firm reaches the mature phase. Figure 20.1 shows a high-growth trajectory and a slow-growth trajectory for two businesses. The high-growth firm experiences a growth rate of 40 percent per year or better. The slow-growth firm will have a growth rate of 10 percent per year.

Technological ventures needing a long start-up period for developing their product have to show success and raise their funds in stages. They continually must demonstrate credibility and engender trust in the investor community. The CEO and leadership of the venture must execute a creative and truthful strategy that unites the interests of the investors and the employees [Kleiner, 2003b].

Moving from start-up to take-off may engender serious stresses as management practices that were appropriate for a smaller size and earlier time no longer work and are scrutinized by frustrated managers. High-growth ventures encounter

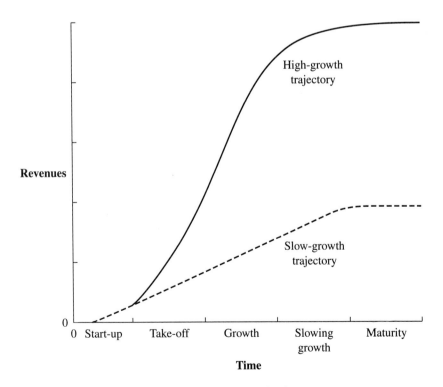

FIGURE 20.1 Growth trajectories for two businesses.

fluid situations where rapid change occurs and chaos appears. New products are released and marketing plans are in flux. New hires join the firm and decision-making is too slow. With high growth rates, new capital infusions may be necessary. Long working days become common and the potential for employee burnout increases. As ventures progress through stages, the original founders may depart and are replaced by management professionals. Fast-growing firms are more likely to replace the founders. Founders are more likely to remain, however, if they hold a sizable percentage ownership of the firm [Boeker and Karichalil, 2002].

A new venture on a slow-growth trajectory will enjoy a less-demanding workweek and fewer competitors but may be faced with limited profitability and access to capital.

TABLE 20.3 Stages of a business.

Start-up	Goals:	Design, make and sell; exploit the opportunity
		Create the product and go to market
		Build the core business
		Emphasis on creativity
		Informal communication
Take-off	Goals:	Efficiency of operations
		Refine and strengthen the business model
		New procedures introduced
		Invest in quality and customer service
		Communication becomes formal
		Emphasis on direction from leaders
Growth	Goals:	Expansion of revenues and market share
		Expand product lines
		Hierarchy in place
		Move to decentralized structure
		Emphasis on delegation of tasks and responsibilities
Slowing growth	Goals:	Consolidation and renewal of the firm
		Leadership required
		New initiatives needed
		Manage working capital
		Emphasis on coordination and renewal
Maturity	Goals:	New innovation
		Strong culture and history
		Development of personnel seeking new opportunities
		Emphasis on collaboration and renewal

As a new firm enters the take-off phase, the need for additional capital, re-sources, and employees will lead to more regularized processes and increasingly formalized communication. The take-off phase requires management skills and budgeting, accounting, and purchasing capabilities. By the time a firm enters the growth phase, the company moves toward decentralization and delegation of tasks. At this time, the firm may add midlevel managers for such tasks as pur-chasing, fulfillment, and sales. In the phase called "slowing growth," the chal-lenge of flat revenues calls for new innovation and entrepreneurial leadership with an emphasis on renewal. The factors of the stages of a business are sum-marized in Table 20.3.

When the business grows, the founding team is incredibly busy. Rapid growth puts an enormous strain on them. The business outgrows its production facilities and management capabilities. Typically, the management crunch hits in the third or the fourth year. That is when firms tend to outgrow their management base with quality falling, delivery dates missed, and customers not paying on time. As the firm enters the growth phase, the leadership team needs to ask: What does the business need at this stage? Founding CEOs tend to depart when their firm reaches a rapid growth phase and needs strong managerial expertise [Boeker and Karichalil, 2002].

During the growth phase, competition heats up. Most products that are technology-based experience competition driven by technology, as shown in Figure 20.2 [Hirsh, Hedlund, and Schweizer, 2003]. As technologies like fuel

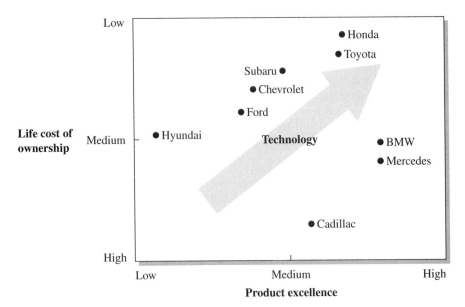

FIGURE 20.2 Technology drives most competition toward excellence and lower life-cycle cost of ownership. In this figure, automobiles are used to illustrate the principle.

cells and hybrid engines improve, customers demand better performance with lower life-cycle cost of ownership. Firms that respond to these demands will continue to succeed.

Founded in 1982, Sun Microsystems experienced stresses of rapid growth and myriad challenges in 1984 that showed that the first CEO, Vinod Khosla, was not the person for the next stage of the firm's life. It was clear that Khosla was a great visionary, but he lacked some leadership skills to handle the growth phase. Within a few months, the board of directors urged Khosla to step down. Khosla left the company soon afterward, and his co-founder, Scott McNealy, took over as CEO.

Entrepreneurs often are unable to make the transition to the growth mode by becoming executives of a high-growth firm. The habits and skills that make entrepreneurs successful can undermine their ability to lead larger organizations. Entrepreneurs tend to focus on details and tasks in the early stage of a firm, as they should. As the firm grows, a leader needs to work on leading a larger, more complex organization. However, entrepreneurs can learn to grow with their firm by developing their relationship, network, and strategic capabilities and by moving from a task orientation to a coordinating approach [J. Hamm, 2002].

As the firm approaches the slowing growth phase, a move toward hierarchy may be appropriate. Hierarchy helps us to handle the complexity of large organizations. Furthermore, people envision career ladders, readily understand the system, and identify with one subunit. Hierarchical structures provide rewards of power and status, and may be the best for managing complex activities in large firms. The leader strives to keep the best of the small firm—empowerment, teamwork, and shared leadership—while accepting the benefits of hierarchy in the large firm [Levitt, 2003].

Often an analogy helps us understand the management of transitions. We can envision the early-state company in the start-up and take-off as a jazz band playing in a jam session with wonderful improvisations. The jazz band has fewer than 20 members, and all the members know each other and seamlessly take their turn leading. Several players can play several instruments. When a firm grows to more than 50 people, it starts to shift to acting like an orchestra with its separate sections—strings, wind instruments, and percussion. The orchestra needs one coordinating leader, called a conductor. He or she has a score they follow and describes a strategy the orchestra will follow for each piece of music. They act as one—as should a growing firm.

Cisco Systems's revenue growth between 1995 and 2000 was 53 percent per year, primarily through acquisitions of small companies in exchange for Cisco stock. By 2002, Cisco's growth stalled, and the goal became managing expenses. The motto for the old Cisco was: faster, more sales. The motto for the new Cisco is: slower, better, profitable. From mid-1999 to late 2000, Cisco doubled its payroll from 22,000 to 44,000 employees. In 2001, the growth abruptly ended when businesses stopped buying. Telecommunications companies discovered they had massively overbuilt and ordered too much equipment from Cisco.

Revenue fell for the first time in Cisco's history. By the summer of 2001, sales plunged one-third from their level six months earlier. By 2003, Cisco was cutting 8,500 workers [Thurm, 2003].

With a talented leadership team, a new venture can navigate the challenges of growth and the demand for initiatives as needs change. Throughout the evolution of the firm, the challenge is to contain and reduce costs, improve operating margins, manage demand and capacity, and continuously innovate.

An example of a successful move from the take-off to the growth stage is that of eBay. With a successful site and venture capital firm investment, eBay began looking in 1997 for a capable manager to take it into the growth phase. It found Meg Whitman, an experienced manager with solid marketing credentials. Whitman arrived in 1998 and immediately worked on the execution of an IPO for September 1998. In the first quarter of 1998, more than $100 million in goods changed hands on eBay, and the company's revenues exceeded $3 million a month [Cohen, 2002]. eBay sold about 9 percent of the company and raised $62.8 million. The IPO price was $18 per share, which jumped to $53 on the opening of the market. Whitman built a brand that promised trustworthy trading for buyer and seller. She turned eBay into a powerful auction firm with about $1.9 billion in revenues and a profit margin of 20 percent in 2003.

In 1991, Motorola founded a spin-off called Iridium LLC to build a set of 66 low-orbit satellites that would allow subscribers to make phone calls from any global location. In 1998, the company launched its service, charging $3,000 for a handset and $3 per minute for calls. By 1999, Iridium filed for bankruptcy. As the CEO of Iridium said [Finkelstein, 2003]:

> We're a classic MBA case study in how not to introduce a product. First, we created a marvelous technological achievement. Then we asked how to make money on it.

Downturns and recessions happen in every industry. The response of an emerging business to these tests can lead to renewal and success or disarray and failure. Few companies have soared as high, sunk as low, and struggled to survive as the software maker Novell. Founded in 1983, it has a formidable competitor, Microsoft, and has experienced large swings in revenues and profitability. In the early 1980s, Novell pioneered the market for network operating systems. In the early 1990s, Novell missed the shift to the Internet and lost market share. Drifting through the 1990s, Novell tried for a renewal with Eric Schmidt as leader. Trying to move out of the slowing growth phase, Schmidt attempted to renew the innovation of the firm. By 1999, Novell launched new software for networks and the Internet. However, Novell revenues remained flat from 1997 to 2002, and profitability remained elusive.

By the time a company reaches maturity with 4,000 stores and 1.4 million employees, there is a good chance it will have lost its entrepreneurial zeal. Not so with Wal-Mart, with its mission of goods for the masses. Jim Collins [2003] states that Wal-Mart has built a consistent, growing company through its cultlike culture as well as its commitment to everyday low prices. Wal-Mart's discipline is: "Never think of your company as great, no matter how successful it becomes."

Managing a downturn is as challenging as managing a period of fast growth. In a recession, customers are slow to pay their bills, and suppliers become weak. Furthermore, the availability of new capital dries up. If possible, an emerging new venture can reformulate a positive agenda for managing through the down period by renewing and tightening its strategy while avoiding overreacting. An economic downturn is an opportunity to clean the slate and get back to economic reality. Every downturn is a chance to rebuild the core business. Contrary to conventional wisdom, downturn winners avoid diversification. Focus makes sense, along with renewal of the business [Rigby, 2001]. With a focus on the core business and a renewed strategy, the firm can see beyond the bad times. While managing costs, the firm prepares for the next upturn. If a firm has the resources, selected acquisitions may be a wise step for the future.

Making it through a downturn isn't easy, and there's no ready path to success. Companies that successfully handle a downturn refocus on their core business and renew their strategy. They maintain a long-term view and strive to earn the loyalty of employees, suppliers, and customers. Coming out of the downturn, they maintain momentum in their business to stay ahead of the competition [Rigby, 2001].

Randy Komisar [2000] describes a start-up as requiring three types of CEOs at successive stages of its development. He uses descriptors in terms of dogs. The first CEO of a start-up is the "retriever." This CEO assembles the core team and the product to fit the original vision, and proceeds to access the necessary resources. The second CEO is the "bloodhound" who must sniff out a trail and find the right market and profitable customers. The third CEO is the "husky" who executes well and pulls the established firm steadily forward.

Another organizational issue flowing through the stages of a company's life is the matter of executive succession. As the firm moves through the stages, it often must change its CEO. As needs change, the board of directors and the investors ask whether the incumbent has the skills to manage the firm through and into the next stage. Fewer than 40 percent of founder CEOs make it past the second round of venture capital financing [Bailey, 2003d]. Venture capitalists, like major league team managers, often wait too long to replace their CEO or starting pitcher.

Successful executive succession can lead to superior organizational performance [Dyck, 2002]. Succession planning is necessary for any new venture

TABLE 20.4 Four factors of executive succession and the relay race analogy.

1. **Sequence**	Ensuring the successor has the appropriate skills and experience to lead the organization in the next stage. The executive in the start-up stage should have strong entrepreneurial skills, while the executive in the take-off phase should have good organizational skills.1.
2. **Timing**	Ensuring the leadership baton is passed in a timely and expeditious way from incumbent to successor.
3. **Baton-passing technique**	Methods used for passing the baton. Ensuring that the baton is handed off as expected and the incumbent lets go of the power.
4. **Communication**	Harmonious cooperation and clear communication between incumbent and successor.

as it moves from one stage to another. The four factors of a good succession are sequence, timing, technique, and communication, as described in Table 20.4. Using the analogy of a relay race, there is a positive relationship between successful passing of the leadership baton and organizational performance. Succession is facilitated when the incumbent and the successor have a shared understanding of the timing and technique of the hand-off.

20.3 THE ADAPTIVE CORPORATION

Successful entrepreneurs know a great deal about what kind of priorities matter to their customers at particular historical junctures. Fashion, self-expression, status, community, and control are important factors in different periods of a customer's life. Entrepreneurs have a deep knowledge of their customers and products, and create meaningful brands and a range of organizational capabilities that consistently deliver on the promises of their brands. Furthermore, they learn from their experience and make rapid adjustments [Koehn, 2001].

No business plan survives its ultimate collision with reality. Changes in the marketplace and competition require any firm to react to change. Entrepreneurs must determine whether the assumptions on which their organization was built match the current reality.

One of the biggest tasks for leaders in growing firms is to repeatedly mobilize their companies to change to meet new opportunities and competitive challenges. The leadership team needs to reinvent its strategy through a process of continuous renewal in a time of constant change; this process is called strategic learning [Pietersen, 2002]. A key capability of leaders in adaptive organizations is the capacity to adapt. These leaders do not get stopped by tough challenges they encounter but learn new lessons and go on. Aldous Huxley [1990] stated it as: "Experience is not what happens to a man. It is what a man does with what happens to him." **Strategic learning** is a cyclical process of adaptive learning using four steps: learn, focus, align, and execute. This adaptive process of learning and executing, if done well, may be one of a company's sustainable competitive advantages. Challenging discontinuities are new technologies, globalization, the Internet, deregulation, convergence, and channel disintermediation. A firm's strategy defines how it will respond to its challenges. Thus, the leadership team needs to focus the firm's resources on the best opportunities in the shifting context of the business world.

A **learning organization** captures, generates, shares, and acts on knowledge by revising its strategy as new knowledge becomes available. This type of firm is an **adaptive enterprise**—one that changes its strategy or business model as the conditions of the marketplace require.

In a small start-up consisting of 10 to 20 people with shared values and objectives, an informal process of developing renewed strategies will suffice. As the firm grows, it needs to continue to renew its strategy as conditions in the competitive marketplace require. At that time, the ability to adapt becomes a required organizational capability. The leadership team needs to learn from its experiences, adjust its strategy, and execute those changes. Learning to deal with change, discontinuity and uncertainty and adapt in a timely way needs to become a skill for leaders of new ventures. [Buchanan, 2004]

The effective management of risk is critical to success. Assessing the risks associated with new initiatives can help managers make adjustments to mitigate these risks. Dell Computer is always asking what could go wrong and considering ways to mitigate the downside. The characteristics of successful CEOs are shown in Table 20.5.

The learning organization, as described in Section 9.3, uses a learning process or cycle as shown in Figure 20.3. The goal of the learning process is to generate new strategies in a cycle of renewal. The first step consists of a situation analysis (see Section 4.3) of the competitive marketplace, industry dynamics, and the firm's strengths and weaknesses. The outcome of this step is insight into the issue and alternative responses. The second step is to redefine the vision, mission, strategy, and adjusted business model (see Section 4.5). The outcome of this step is a statement of the performance, resource, and capability gaps in

TABLE 20.5 Characteristics of successful CEOs.

Prone to failure	More likely to succeed
1. Arrogant, hubris	1. Humble, open-minded
2. Overconfident with their answers to problems	2. Realistic, learning, always challenging answers
3. Underestimate major obstacles and risks	3. Examine carefully potential negative consequences and all risks
4. Rely on what worked in the past	4. Challenge every decision and look for wise changes and opportunities to learn

Source: Finkelstein and Mooney, 2003.

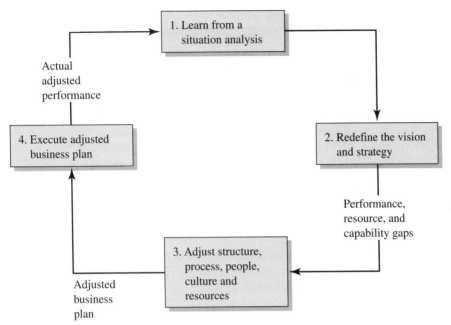

FIGURE 20.3 Strategic learning cycle of a learning organization.
Source: Adapted from Pietersen, 2002.

terms of the desired strategy and the actual reality. The third step is to adjust the structure, process, people, and culture of the firm to work toward the new strategy. The outcome of this step is an adjusted business plan. The fourth step is to execute the newly adjusted business plan. The outcome of the fourth step is the actual adjusted performance. After a period of actual performance, the learning cycle starts again. The firm continues to learn and adjust as it repeats the strategic learning cycle.

Effective learning involves continuing to ask the key questions about customers, competitors, capabilities, resources, and profitability. Entrepreneurs should not rely on excessive overconfidence and overcommit up-front resources, hindering their willingness to learn and adjust. The best degree of confidence lies at the level of willingness to decide and move ahead with the expectation that new knowledge will help the firm learn and adjust [Simon and Houghton, 2003].

Corporations formerly built to last like pyramids are now more like temporary arrangements. With a changing landscape, the adaptive organization has become a reality. CEOs and their firms fail when they fail to execute their strategy. The culture of the firm, if widely understood and shared, can help the firm's people to execute the strategy. With a strong culture and less formal direction, employees can take ownership over their actions and execute well. New firms can recruit, train, and reward people to take responsibility for their actions [Chatman and Cha, 2003].

Most organizations face all kinds of unpredictable challenges that collectively place huge demands on people's creativity and imaginations. Resilience is the ability to recover quickly from setbacks. **Resilience** is a skill that can be learned and increased. A resilient organization acts on its learning and possesses a staunch acceptance of reality, a set of strongly held positive values, and a powerful ability to adapt [Coutu, 2002]. The firm has a clear, undistorted sense of reality about its competitive position. Then it needs the ability to make meaning out of difficult challenges. Resilient leaders build a new, improved vision of the future. Value systems of resilient firms evoke meaning and noble purpose. The third factor is the ability to make a future from what is available—a kind of inventiveness or ability to improvise a solution. Companies that survive regard improvisation as a core skill [Coutu, 2002]. Highly resilient organizations have the ability to read a weak signal of a problem and respond to it. Action tempered by reflection is the best method of responding to change and information [Coutu, 2003]. One way to identify threats and weaknesses is to bring in outsiders to test the resilience of the firm and its processes.

George Eastman and Henry Strong created the Eastman Kodak firm in 1880 based on a new photographic technology process for preparing film negatives. The company set out to become the best gelatin-coated dry glass plate photography company and sold shares to the public in 1884. Eastman sought to create a mass-market camera and developed a strategy based on continuous innovation and learning. Eastman hired technical graduates of the University of Rochester and MIT to continuously innovate. The first Kodak, placed on the market in 1888, was a simple, handheld box camera containing a 100-exposure roll of paper stripping film. The entire camera was sent back to the manufacturer for developing, printing, and reloading when the film was used up. In 1889, Eastman introduced transparent film. By 1900, Eastman introduced the Kodak Brownie with the motto: "You push the button and we do the rest." In 1914, he traveled to Boston to visit Harvard Business School, where he hired Marion Folsom as a manager who added business capabilities and leadership to the firm. By 1927, Eastman Kodak had a virtual monopoly on the photographic industry in the United States. For half a century, Kodak, a "learning company," continuously adapted its strategy and products to build the most powerful brand in photography [Tedlow, 2001].

Successful start-up teams have a business plan, but they are willing to adapt it as needed. Flexcar (www.flexcar.com) was started by Neil Peterson in Seattle to provide a time-sharing plan for automobile users. He thought that customers would use cars provided by his time-sharing company rather than buy their own car. This approach was popular in Europe, but Peterson found that people in Seattle still wanted their own car. He quickly learned to switch his marketing campaign to steer it toward universities and businesses, where the business model and sales pitch made more sense [Thomas, 2003b].

A current example of an adaptive corporation is Starbucks. The 6,000-unit chain exhibits an uncanny ability to learn, innovate, and change. In late 2001, Starbucks introduced a prepaid debit card for customers. By late 2003, it had issued 14 million of these cards, which accounted for 15 percent of all transactions. Furthermore, Starbucks has continually adjusted its menu, increasing offerings and innovations.

> Philippe Kahn, co-founder of Borland, described adaptivity as improvisation [Malone, 2002]:
>
> I don't know what an entrepreneur is, but to me it's the difference between a jazz musician and a classical musician. I think a classical musician is a kind of guy who's gonna work for a big company. A jazz musician's gotta work in a small band and know how to improvise. I think that's really the analogy, that's really the difference.

It is wise to build an adaptive learning corporation from a new venture's inception. Competitive advantage depends, in large part, on the ability of organizations to constantly change and reinvent themselves. They accomplish this through building and rebuilding a shared vision and team learning [Gabor, 2000]. To a great extent, the iterative, adaptive process of Figure 20.3 is the only sustainable advantage for a firm in a dynamic economy.

20.4 ETHICS

Life is filled with difficult ethical challenges. **Ethics** are a set of moral principles for good human behavior. Ethics provides the rules for conducting activities in a manner acceptable to society. **Moral principles** are concerned with goodness (or badness) of human behavior and usually are provided as rules and standards of human behavior. Thus, a common moral rule would be: do not lie. Of course, such rules are subject to interpretation, thus, the concept of the white lie.

Ethics is concerned with doing the right (moral) thing. Society also establishes laws to guide actions. For example, U.S. law states that bribes and kickbacks are illegal. Laws are subject to interpretation, and paying a fee for sales help may be legal, while paying a bribe is illegal.

The success of new ventures, either profit or nonprofit, depends on winning against competitors. The competitive marketplace can put pressure on the entrepreneur to act unethically. The business leader finds it difficult to be fair to others without sacrificing customers or profits. The troubles of Enron, WorldCom, and Arthur Andersen show the poor practices that arise when competitive pressures win out over ethical principles.

Ethical conduct may reach beyond the law, since the law is inadequate for every task. Doing the right thing is an undefined but helpful standard. One moral goal would be to tell the truth. Thus, a businessperson would try to provide full and truthful information about his or her product or service. Telling the truth is a critical part of integrity, and integrity is the basis for reputation. Thus, firms, at the least, find it in their interest to be truthful. Fortunately, good ethics and self-interest usually coincide, since most firms want to develop and maintain a high reputation [Beauchamp and Bowie, 2001].

Integrity can be defined as truthfulness, wholeness, and soundness. It can be described as the consistency of our words and our actions or our character and our conduct. A corporate model of integrity is based on ethical principles embedded in the corporate culture so that all stakeholders can conduct business to attain mutual benefits [Kaptein and Wempe, 2002]. Factors that impede ethical decisions are lack of openness about decisions and the self-interests of individuals. While firms have clear-cut business objectives such as profitability, they must consider them subordinate to ethical values. A firm's integrity cannot be sacrificed to short-term gain. The firm's moral compass points the way. The spotlight is on the CEO and his or her integrity.

Arthur Andersen, an accounting firm, went from the motto "think straight, talk straight" to "lie, cheat, and shred" [Toffler, 2003]. Andersen found it difficult to be truthful as an auditor of a firm when it sought the client's lucrative consulting business. We know that the lack of truth and the collapse of integrity can lead to terrible outcomes, as illustrated in the Enron case of 2002. On Enron, Robert Bryce and M. Ivins write [2002]:

> Enron failed because its leadership was morally, ethically and financially corrupt. Whether the question was accounting or marital fidelity, the executives who inhabited the 50th floor at Enron's headquarters became incapable of telling the truth, to the Securities and Exchange Commission, to their spouses, or to their employees. That corruption permeated everything they did, and it spread through the company like wildfire.

Perhaps the best ethical principle is [McLemore, 2003]:

> Steer clear of trouble by spotting it well in advance and by acting honorably, conscientiously, and nobly.

MiniScribe was a Longmont, Colorado, producer of disk drives that found itself in trouble when IBM canceled major purchasing contracts. When actual sales failed to materialize, the CEO badgered and bullied MiniScribe executives to meet quarterly revenue targets, no matter what it took. Executives turned to cooking the books. Their "cooking" activities included counting raw inventory as finished goods, creating false inventory, and grossly overstating actual shipments.

As MiniScribe's sales and profits continued to tumble, the pressures on the firm increased. At one point, executives rented a private warehouse. Over a weekend, staffers and spouses packed bricks in disk drive shipping boxes, then shipped pallet-loads of them to "BW," a fake customer. To pull off the brick shipping plan, they created a custom computer program called "Cook Book." Over $4 million worth of "very hard" disks were shipped utilizing Cook Book.

> Once the fraud was exposed, MiniScribe's stock plummeted, and investors lost hundreds of millions of dollars, but not before the same executives, in an example of insider trading, sold most of their shares at a healthy profit. The CEO and CFO served time in prison.
>
> Source: www.tactics.com.

As one becomes an entrepreneurial leader, the pressures to win at any cost will become powerful. To be a good team player, one may be asked to cut corners. McLemore [2003] wisely suggests you say, "I don't feel comfortable doing that." Of course, the risk of loss of your position is real, but you can repeat, without judgment, "I am uncomfortable doing that."

The rise and fall of Enron was based on a partnership between the financial division of Enron and the investment bankers who put together the deals. "Enron loves these deals," wrote a Chase banker in 1998, "as they are able to hide funded debt from their equity analysts" (McLean, 2003). On Wall Street, investment bankers call their innovative structured-finance arrangements their "technology." In investment banking, the ethic for many was: "Can you get the deal? If you can, and you're not likely to be sued or jailed, it's a good deal."

McLemore [2003] suggests two tests for difficult actions: 1) Are things so distorted that you lose sleep over them? and 2) Could you live with the newspaper reporting your actions tomorrow? If difficult issues arise, perhaps it is best to discuss the issues with a knowledgeable and trustworthy friend. Can you live with the action? Ernest Hemingway wrote in his novel, *Death in the Afternoon:* "What is moral is what you feel good after, and what is immoral is what you feel bad after." Several tools for acting ethically are provided in Table 20.6.

> Entrepreneurs should plan an ethical foundation for the firm so that they can build integrity and reputation. Maintenance of integrity is critical to the long-run success of any firm. Intel CEO Craig Barret promotes a "3M" philosophy to help leaders make decisions with the right amount of integrity and ethics. The three Ms—manager, media, and mother—represent the three constituencies with whom leaders should be comfortable sharing their decision. Only if the leaders expect that their manager, the media, and their mother will all approve of their decision should they proceed with their chosen course of action.
>
> Source: Barrett, 2003.

20.5 AGRAQUEST

AgraQuest prepared a business plan that stated that the U.S. Environmental Protection Agency would certify and permit its natural biopesticides within one year. The actual time to approve Serenade was two years, and AgraQuest's

TABLE 20.6 Tools for acting ethically in tough situations.

- Maintain involvement in a variety of activities and with a variety of people. This will help avoid being pressured to go along with what everybody around you says is acceptable behavior.

- Create a personal board of directors consisting of people you admire and who possess admirable values. If difficult situations arise, call on them for advice.

- Keep a cash reserve of six months to a year of salary. This will allow you to escape further involvement with an unethical firm and seek other options.

- Increase your defenses to negative forces of influence and persuasion by reading a book on the subject such as Cialdini, 2001.

- Apply the "front, left-hand-side page of the Wall Street Journal" test. Would your actions change if you knew they would be exposed publicly someday?

- Write down your personal core values for later use. They can provide a helpful reference in difficult times to remind you of what is important in your life.

- Take a break before making a decision. When feeling pressured, ask for time to leave the room and gather your thoughts.

second product, Sonata, took 30 months for approval. Furthermore, the firm developed a marketing campaign for its first product that was fundamentally flawed. As a result of poor execution and slow adaptation to challenges and difficulties, AgraQuest has always been late meeting its milestones.

On the other hand, the implementation of the research and development strategy and processes has been relatively flawless. It is the golden goose, but it is slow in gestating the products. AgraQuest is a science company with a poor execution record in product certification, marketing, and sales. What could have AgraQuest achieved if it followed the iterative process of Figure 20.3 early in its life?

20.6 SUMMARY

The implementation of a creative, well-defined business plan is essential to the success of a new enterprise. Good execution depends on the logical alignment of the firm's strategy with its goals and the efforts of its people. Turning a concept into a successful reality depends on goals, deadlines, teamwork, and focus on achieving the desired outcomes. Choosing the right people for the right jobs and helping them see where to go and how to get there are critical elements of building a great company. With the right people, a great strategy and a sound road map a start up can strive to achieve an outstanding execution of the plan.

A new business grows from fledgling start-up to growth to maturity in stages. Managing a new business through the stages requires varying skills and organizational arrangements. Start-ups need to have talented, multiskilled people in place as they move through the stages of growth.

Emerging firms are constantly subject to challenge and change. Organizing a firm for resilient response to these challenges calls for an adaptive corporation. The ability to adapt to change may be a firm's only sustainable advantage. Furthermore, a firm needs to sustain its ethical principles through difficult times.

Principle 20
The ability to continuously and ethically execute a business plan and adapt that plan to changing conditions provides a firm with a sustainable competitive advantage.

20.7 EXERCISES

20.1 Southeby's and Christie's are the two largest upscale auction houses. Both enjoyed a growing business in the boom years of the late 1990s. In 2000, both firms were accused of price fixing. The Sherman Antitrust Act was passed in 1890 to control the power of trusts and monopolists. In 1995, both firms announced they would charge a fixed, nonnegotiable sliding-scale commission on the sales price [Stewart, J. 2001]. Is this the age-old tactic of price fixing? What constitutes legal pricing policies versus illegal price fixing?

20.2 Your emerging new company is selling a high-priced software system to newspapers and magazines. Each sale amounts to $100,000 or more. Your firm is scheduled to deliver a system next week to one of your best customers. However, your chief technical officer has just told you that they have found a major software error that will take two to three weeks to fix. You are counting on the sale within this month so that you can meet payroll and pay all your delinquent bills. Your CFO suggests you ship the system now and send in a team later to fix the error. Your CTO wants to fix it first and then ship. What should you do?

20.3 In 1995, Nancy Evans and Candice Carpenter created iVillage.com. Their mission was to build virtual communities for grown-ups to call home on the Internet. As Carpenter put it, "What we're trying to do is create extremely targeted communities you want to be in right away." Carpenter's previous job was in the brick-and-mortar media corporation, QVC, where she had been in charge of starting up a new upscale home shopping channel. iVillage's website began with the idea of creating a community that would represent the "Internet for the rest of us." Carpenter's repertoire in the media industry was very useful when it came to raising funds for the venture. Her reputation helped iVillage raise a total of $67 million through the first four rounds of venture capital fundraising while under the skepticism of being an East Coast Internet start-up.

iVillage purposely chose to become the ultimate "noncontent" provider, which meant its visitors' contributions and interactions are what gives the site its value, not editorials written by in-house experts. After the first three years of operation, its motto was refined to "Join our community of smart, competent, real women today." iVillage made its target audience educated female baby-boomers, causing the firm to balance its site between personalization and community [Levere, 2002].

iVillage stock price tripled upon its IPO in the spring of 1999. In 2000, Douglas McCormick took over as Carpenter's assistant in the business and in just two months was declared the executive officer. Douglas had been a key player in starting the Lifetime cable channel and has a lot of experience in the cable industry. In February 2001, iVillage paid $27 million for its main competitor, Women.com.

Examine the history of iVillage and try to find the gaps in execution by this firm. Visit ivillage.com and determine if the firm has a future. What stage is it in presently? Has it adapted its business plan to reality?

20.4   Applebee's International Inc. opened its first restaurant in 1986, and by 2003, it had 1,500 restaurants in 49 states and eight countries. Applebee's is the world's largest chain in the "casual dining" category and is known for its excellent execution. The average restaurant has annual sales of $3 million, and the firm has a plan for growing to 2,500 restaurants. Any restaurant operates on the basis of sound execution and service. Applebee's plan is based on 1) menu selection, 2) quality, 3) personal service, 4) convenience of location, 5) a family-friendly atmosphere, and 6) price-value relationship. Visit an Applebee's restaurant and rate it on execution on all six factors.

20.5   Capstone Technology is a developer, assembler, and supplier of microturbine technology (www.capstoneturbine.com). In 1998, it was the first company to offer a proven power source using microturbine technology. Its primary customers are in the on-site power production and hybrid-electric vehicle markets. Recently, revenues have been declining, and the firm has not achieved profitability. Study this company and recommend changes that can lead to profitability.

20.6   United Technologies operates in diversified industries that include aircraft engines, elevators, aerospace, and electronic controls (www.utc.com). The firm has exhibited steady growth based on its execution capabilities. Describe the execution capabilities of this firm using the format of Table 20.2.

20.7   In May 2003, Zipcar of Boston decided it was time to bring in new funding to reach profitability (see www.zipcar.com). However, the willing investors insisted on replacing the CEO and the board of directors. Examine Zipcar's subsequent progress in terms of execution and need at different stages of the life of this company.

20.8 Your cash-strapped company is bidding for a badly needed contract. As the bid deadline nears, an employee of your nearest competitor pays you a visit. He says he'll provide details of your competitor's bid in return for the promise of a job in six months, after the dust has settled. You know your competitor can survive losing this contract; you can't. Unfortunately, hiring a new employee will mean someone who currently works for you will have to go. Even so, is this an offer you can't refuse?

20.9 You are an officer of a struggling company that has just negotiated a life-saving merger. Secrecy is a must; a leak will nullify the deal. You know that a longtime employee whose job will be eliminated in the merger is on the verge of buying an expensive new house. Do you tell the employee to hold off and thereby raise suspicions and rumors among other employees, or do you pretend everything's normal, even though doing so will most likely result in financial ruin for the employee?

20.10 Your new firm is considering offering one of two health benefit options. One is more complete but also more costly than the other. Should you ask your employees to all accept the lower-cost option? Should you explain the benefits of both plans? If you do, most people will prefer the better plan. What should you do?

20.11 The CFO of Tyco International is reported to have told the chief human resources officer to pay him and the CEO a large bonus in 2000. When she questioned the CFO about documentation of approvals of these multimillion-dollar bonuses, he reportedly said, "Why don't you trust me? I've never lied to you?" [Maremont, 2003] How would you respond? What is a good way to handle such difficult situations?

20.12 As a sales representative of your firm, you often meet with new businessowners. Recently, one business founder told you he fired an employee just after she became ill with cancer. What would you say to him, if anything? Would you continue to do business with this firm?

20.13 While attending an industry meeting, you find a file left behind by one of the participants from another firm. Opening the file, you find a plan for a new product that will compete with your product. Should you read it and return it later? Should you copy it and show it to your boss?

20.14 Select an example of a white-collar (business) crime such as the Enron case and describe what happened. How could have this crime been avoided?

REFERENCES

Aaker, David. 2001. *Developing Business Strategies.* 6th ed. New York: Wiley and Sons.

Aaker, David, and E. Joachimsthaler. 2000. *Brand Leadership.* New York: Free Press.

Aaker, David, V. Kumar, and G. S. Day. 2001. *Marketing Research.* New York: Wiley & Sons.

Abboud, Leila, and S. Hensley. 2003. "New Prescription for Drug Makers." *Wall Street Journal* (14 October), p. A1.

Adler, Carlye. 2002. "Would You Pay $2 Million for This Franchise?" *Fortune* (1 May).

Adner, Ron, and D. A. Levinthal. 2002. "The Emergence of Emerging Technologies." *California Management Review* (Fall), pp. 50–66.

Afuah, Allan. 1998. *Innovation Management.* New York: Oxford University Press.

Afuah, Allan, and C. Tucci. 2001. *Internet Business Models and Strategies.* New York: McGraw-Hill.

Agarwal, Rajshree, M. B. Sarkar, and R. Echambadi. 2002. "The Conditioning Effect of Time on Firm Survival." *Academy of Management Journal* 5:971–94.

Ahuja, Gautam, and C. M. Lampert. 2001. "Entrepreneurship in the Large Corporation." *Strategic Management Journal* 22:521–43.

Aiello, Robert, and M. Watkins. 2000. "The Fine Art of Friendly Acquisition." *Harvard Business Review* (December), pp. 101–16.

Akst, Daniel. 2003. "The Very Model of a Modern Modular House." *Wall Street Journal* (29 May), p. D8.

Albrinck, Jill, J. Hornery, D. Kletter, and G. Neilson. 2002. "Adventures in Corporate Venturing." *Strategy and Business* 22:119–29.

Alsop, Ronald. 2002. "Perils of Corporate Philanthropy." *Wall Street Journal* (8 June), p. B1.

Alvarez, Sharon, and J. B. Barney. 2001. "How Entrepreneurial Firms Can Benefit from Alliances with Large Partners." *Academy of Management Executive* 15(1):139–48.

Anders, George. 2003. *Perfect Enough.* New York: Penguin Putnam.

Antoncic, B., and R. Hisrich. 2001. "Intrapreneurship: Construct Refinement and Cultural Validation." *Journal of Business Venturing* 16:495–527.

Arndt, Michael. 2002. "Whirlpool Taps Its Inner Entrepreneur." *Business Week Online* (7 February).

Astebro, Thomas. 1998. "Basic Statistics on the Success Rate and Profits for Independent Inventors." *Entrepreneurship Theory and Practice* (Winter): 41–48.

Bagley, Constance, and C. E. Dauchy. 2002. *The Entrepreneur's Guide to Business Law.* Boston: South-Western.

Bailey, Jeff. 2003a. "In Fast Food, Bigger Doesn't Ensure Success." *Wall Street Journal* (18 May), p. B6.

Bailey, Jeff. 2003b. "A Restaurant's Turnaround Is All in the Details." *Wall Street Journal* (20 May), p. B3.

Bailey, Jeff. 2003c. "Can You Teach Someone to Be an Entrepreneur?" *New York Times* (4 September), p. 32.

Bailey, Jeff. 2003d. "For Investors, Founders Are Short-Term CEOs." *Wall Street Journal* (21 October), p. A24.

Bailey, Jeff. 2002. "When Selling Your Company Means Letting Go of the Dream." *Wall Street Journal* (10 December), p. 86.

Baker, Wayne. 2000. *Achieving Success Through Social Capital.* San Francisco: Jossey-Bass.

Balachandra, R., M. Goldschmitt and J. Friar, 2004, "The Evolution of Technology Generations," IEEE Trans. on Engineering Management, February, 3–12.

Barkema, Harry, J. Baum, and E. Mannix. 2002. "Management Challenges in a New Time." *Academy of Management Journal* 5:916–30.

Barney, Jay. 2002. *Gaining and Sustaining Competitive Advantage,* 2d ed. Upper Saddle River, N.J.: Prentice-Hall.

Barney, Jay. 2001. "Is the Resource Based View a Useful Perspective for Strategic Management Research? Yes." *Academy of Management Review* (January): 41–56.

Baron, James, and M. T. Hannan. 2002. "Organizational Blueprints for Success in High-Tech Start-Ups." *California Management Review* 3:18–24.

Baron, Robert, and G. D. Markman. 2003. "Beyond Social Capital." *Journal of Business Venturing* 18:41–60.

Barrett, Craig. 2003. Address at the AEA/Stanford Executive Institute, Stanford University, Palo Alto, Calif. (August 14).

Barta, Patrick, and C. Binkley. 2003. "U.S. Travelers' Satisfaction Rises." *Wall Street Journal* (21 May), p. D2.

Barthelemy, Jerome. 2003. "The Seven Deadly Sins of Outsourcing." *Academy of Management Executive* 2:87–100.

Bartlett, Christopher, and S. Ghoshal. 2002. "Building Competitive Advantage through People." *MIT Sloan Management Review* (Winter), pp. 34–41.

Batstone, David. 2003. *Saving the Corporate Soul.* San Francisco: Jossey-Bass.

Batten, Frank. 2002a. "Out of the Blue and into the Black." *Harvard Business Review* (April), pp. 112–19.

Batten, Frank. 2002b. *The Weather Channel.* Boston: Harvard Business School Press.

Bauman, Adam. 2003. "More Companies Are Routing Calls via Internet." *New York Times* (1 September), p. 14.

Baumol, William. 2002. *The Free Market Innovation Machine.* Princeton, N.J.: Princeton University Press.

Baumol, William J. 1993. *Entrepreneurship, Management and the Structure of Payoffs.* Cambridge: MIT Press.

Beatty, Jack. 2001. *Colossus—How the Corporation Changed America.* New York: Broadway Books.

Beauchamp, Tom, and N. E. Bowie. 2001. *Ethical Theory and Business,* 6th ed. Upper Saddle River, N.J.: Prentice-Hall.

Bechky, Beth. 2002. "Gofers, Gaffers and Grips: Coordination and Role Enactment in Film Production." Working paper, University of California, Davis.

Beckwith, Harry. 2003. *What Clients Love.* New York: Warner Books.

Bellow, Adam. 2003. "In Praise of Nepotism." *Atlantic Monthly* (July), pp. 98–105.

Bennis, Warren. 1994. *On Becoming a Leader.* Reading, Mass.: Addison-Wesley.

Bennis, Warren, and R. J. Thomas. 2002a. "Crucibles of Leadership." *Harvard Business Review* (September), pp. 39–45.

Bennis, Warren, and R. J. Thomas. 2002b. *Geeks and Geezers.* Boston: Harvard Business School Press.

Berk, Christina C. 2003. "A Reliable Vending Machine?" *Wall Street Journal* (12 November), p. B4.

Berstein, Peter. 1996. *Against the Gods.* New York: Wiley & Sons.

Best, Michael H. 2001. *The New Competitive Advantage.* New York: Oxford University Press.

Best, Michael H. 1990. *The New Competition.* Cambridge: Harvard University Press.

Bhargava, Hement. 2003. "Contingency Pricing for Information Goods." *Journal of Management Information Systems* (Fall): 115–38.

Bhide, Amar. 2000. *The Origin and Evolution of New Business.* New York: Oxford University Press.

Biggart, Nicole. 1989. *Charismatic Capitalism.* Chicago: University of Chicago Press.

Bird, Barbara. 2003. "Learning Entrepreneurship Competencies." *International Journal of Entrepreneurship Education* 1:203–27.

Birkinshaw, Julian. 2003. "The Paradox of Corporate Entrepreneurship." *Strategy and Business* 30:46–57.

Birley, Sue. 2002. "Universities, Academics, and Spinout Companies: Lessons from Imperial." *International Journal of Entrepreneurship Education* 1:133–53.

Black, J. Stewart, and H. B. Gregersen. 2002. *Leading Strategic Change.* Upper Saddle River, N.J: Prentice-Hall.

Boeker, Warren, and R. Karichalil. 2002. "Entrepreneurial Transitions." *Academy of Management Journal* 3:818–26.

Boer, Peter. 2002. *The Real Options Solution.* New York: Wiley & Sons.

Bolino, Mark, W. Turnley, and J. Bloodgood. 2002. "Citizenship Behavior and the Creation of Social Capital." *Academy of Management Review* 4:505–22.

Bossidy, Larry, and Ram Charan. 2002. *Execution.* New York: Crown.

Bosworth, Michael. 1995. *Solution Selling.* New York: McGraw-Hill.

Boulding, William, and M. Christen. 2001. "First Mover Disadvantage." *Harvard Business Review* (October), pp. 20–21.

Bower, Joseph. 2001. "Not All M&As Are Alike and That Matters." *Harvard Business Review* (March), pp. 93–101.

Bradley, Bill, P. Jansen, and L. Silverman, 2003. "The Nonprofit Sector's $100 Billion Opportunity." *Harvard Business Review* (May), pp. 94–103.

Brandenburger, Adam, and Barry Nalebuff. 1997. *Co-opetition.* New York: Currency Doubleday.

Brandt, Richard, and T. Weisel. 2003. *Capital Instincts.* New York: Wiley & Sons.

Breeden, Richard. 2002. "Entrepreneurship Stabilizes in the U.S. after Drop." *Wall Street Journal* (19 November), p. B7.

Brenner, D., and S. Lyshevski. 2002. *Handbook of Nanoscience, Engineering and Technology.* Boca Raton, Fla.: CRC Press.

Bronscomb, Lewis, F. Kodama, and R. Florida. 1999. *Industrializing Knowledge.* Cambridge: MIT Press.

Brown, David. 2002. *Inventing Modern America.* Cambridge: MIT Press.

Brown, John Seely, and Paul Duguid. 2001. "Creativity versus Structure: A Useful Tension." *MIT Sloan Management Review* (Summer), pp. 93–94.

Brown, John Seely, and Paul Duguid. 2000. *The Social Life of Information.* Boston: Harvard Business School Press.

Brown, Ken. 2003. "A Banking Success Story." *Wall Street Journal* (11 May), p. C1.

Brown, Shona, and K. M. Eisenhardt. 1998. *Competing on the Edge.* Boston: Harvard Business School Press.

Brush, Candida, P. G. Greene, and M. M. Hart. 2001. "From Initial Idea to Unique Advantage." *Academy of Management Executive* 15(1):64–78.

Bryant, Keith, and Henry Dethloff. 1985. *A History of American Business.* Upper Saddle River, N.J.: Prentice-Hall.

Bryce, Robert, and M. Ivins. 2002. *Pipe Dreams: Greed, Ego and the Death of Enron.* New York: Public Affairs–Perseus.

Buchanan, Mark, 2004. "Power Laws and the New Science of Complexity Management," Strategy and Business, 34:71–79.

Buffett, Warren. 2003. Berkshire Hathaway Annual 2002 Report Letter to Shareholders, Omaha, Neb.

Bunnell, David. 2000. *Making the Cisco Connection.* New York: Wiley & Sons.

Burgelman, Robert A. 2002. *Strategy as Destiny.* New York: Free Press.

Burgelman, Robert A., C. M. Christensen, and S. C. Wheelwright. 2004. *Strategic Management of Technology and Innovation.* Burr Ridge, Ill.: McGraw-Hill Irwin.

Bygrave, William, and M. Minniti. 2000. "The Social Dynamics of Entrepreneurship." *Entrepreneurship Theory and Practice* (Spring): 25–36.

Carlassare, Elizabeth. 2001. *DotCom Divas.* New York: McGraw-Hill.

Carlisle, Tamsin. 2002. "Ballard Technology Isn't Engine of Growth." *Wall Street Journal* (27 December), p. B4.

Carr, Nicholas. 2003. "IT Doesn't Matter." *Harvard Business Review* (June), pp. 41–49.

Carr, Nicholas. 2002. "Unreal Options." *Harvard Business Review* (December), p. 22.

Chakravorti, Bhaskar, 2004. "The New Rules for Bringing Innovations to Market," *Harvard Business Review* (March), pp. 58–67.

Chandler, Alfred, and J. W. Cortada. 2003. *A Nation Transformed.* New York: Oxford University Press.

Chandler, Gaylen, et al. 2002. "Initial Size and Membership Change in Emerging and New

Venture Teams." Presented at Academy of Management Annual Conference, Denver.

Chatman, Jennifer, and S. E. Cha. 2003. "Leading by Leveraging Culture." *California Management Review* (Summer), pp. 20–33.

Chatterjee, Sayan. 1998. "Delivering Desired Outcomes Efficiently." *California Management Review* (Winter), pp. 78–94.

Chen, David. 2003. "Leaping Forward Online." *New York Times.com* (25 October).

Chesbrough, Henry. 2003. *Open Innovation.* Boston: Harvard Business School Press.

Chesbrough, Henry. 2002. "Making Sense of Corporate Venture Capital." *Harvard Business Review* (March), pp. 90–99.

Christensen, C. M., and M. Overdorf. 2000. "Meeting the Challenge of Disruptive Change." *Harvard Business Review* (March), pp. 66–77.

Christensen, Clayton. 2002. "The Rules of Innovation." *Technology Review* (June) 21–28.

Christensen, Clayton. 1999. *Innovation and the General Manager.* Burr Ridge, Ill.: McGraw-Hill Irwin.

Christensen, Clayton, and M. E. Raynor. 2003. *The Innovator's Solution.* Boston: Harvard Business School Press.

Christensen, Clayton, M. Raynor, and M. Verlinden. 2001. "Skate to Where the Money Will Be." *Harvard Business Review* (November), pp. 73–81.

Christensen, Clayton, and Richard Tedlow. 2000. "Patterns of Disruption in Retailing." *Harvard Business Review* (January), 36–42.

Christensen, Clayton, M. Verlinden, and G. Westerman. 2002. "Disruption, Disintegration and the Dissipation of Differentiability." *Industrial and Corporate Change* 5:955–993.

Churchill, Neil, and J. W. Mullins. 2001. "How Fast Can Your Company Afford to Grow?" *Harvard Business Review* (May), pp. 135–42.

Cialdini, Robert. 2001. "Harnessing the Science of Persuasion." *Harvard Business Review* (October), pp. 72–79.

Cialdini, Robert. 1993. *Influence.* New York: Morrow.

Clark, Don. 2003a. "Intel's John Miner to Be President of Investment Unit." *Wall Street Journal* (18 April), p. C5.

Clark, Don. 2003b. "Renting Software Online." *Wall Street Journal* (3 June), p. B1.

Clemons, Eric, and J. A. Santamaria. 2002. "Maneuver Warfare." *Harvard Business Review* (April), pp. 57–65.

Cohen, Adam. 2002. *The Perfect Store.* Boston: Little, Brown and Company.

Cohen, Ben, and Jerry Greenfield, 1997. Ben and Jerry's Double Dip, New York: Simon and Schuster.

Cohen, Don, and L. Prusak. 2001. *In Good Company.* Boston: Harvard Business School Press.

Collins, James, and William Lazier. 1992. *Beyond Entrepreneurship.* Upper Saddle River, N.J.: Prentice-Hall.

Collins, James, and J. Porras. 1996. "Building Your Company's Vision." *Harvard Business Review* (September), pp. 65–77.

Collins, Jim. 2003. "Bigger, Better, Faster." *Fast Company* (June), pp. 74–76.

Collins, Jim. 2001. *Good to Great.* New York: Harper Collins.

Conger, Jay. 1998. "The Necessary Art of Persuasion." *Harvard Business Review* (May), pp. 85–95.

Copeland, Tom, and Peter Tufano, 2004. "A Real-World Way to Manage Real Options," *Harvard Business Review* (March), pp. 90–99.

Corstijens, Marcel, and J. Merrihue. 2003. "Optimal Marketing." *Harvard Business Review* (October), pp. 114–21.

Coughlan, Anne, and K. Grayson. 1998. "Network Marketing Organizations." *International Journal of Research in Marketing* 15:401–26.

Courtney, Hugh. 2001. *20-20 Foresight.* Boston: Harvard Business School Press.

Coutu, Diane. 2003. "Sense and Reliability." *Harvard Business Review* (April), pp. 84–90.

Coutu, Diane. 2002. "How Resilience Works." *Harvard Business Review* (May), pp. 46–55.

Covey, Stephen, and A. R. Merrill, 1996. First Things First. New York: Free Press.

Covin, Jeffrey, D. P. Slevin, and M. B. Heeley. 1999. "Pioneers and Followers." *Journal of Business Venturing* 15:175–210.

Crawford, Fred, and Ryan Matthews. 2001. *The Myth of Excellence.* New York: Crown Business.

Cross, Rob, and L. Prusak. 2002. "The People Who Make Organizations Go—or Stop." *Harvard Business Review* (June), pp. 105–12.

Curtis, Wayne. 2002. "The Iceberg Wars." *Atlantic Monthly* (March), pp. 88–100.

Dahan, Ely, and V. Srinivasan. 2000. "The Predictive Power of Internet-Based Product Concept Testing." *Journal of Product Innovation Management* 17:99–109.

Davenport, Thomas, and J. Glaser. 2002. "Just in Time Delivery Comes to Knowledge Management." *Harvard Business Review* (July), pp. 107–11.

Davenport, Thomas, and L. Prusak. 1998. *Working Knowledge.* Boston: Harvard Business School Press.

Davidsson, Per. 2002. "What Entrepreneurship Research Can Do for Business and Policy Practice." *International Journal of Entrepreneurship Education* 1:5–24.

Davis, Julie L., and S. S. Harrison. 2001. *Edison in the Boardroom.* New York: Wiley & Sons.

Davis, Stan, and Christopher Meyer. 2000. *Future Wealth.* Boston: Harvard Business School Press.

Dees, J. Gregory, J. Emerson, and P. Economy. 2002. *Strategic Tools for Social Entrepreneurs.* New York: Wiley & Sons.

DeLong, Thomas, and V. Vijayaraghavan. 2003. "Let's Hear It for B Players." *Harvard Business Review* (June), pp. 96–102.

Dembo, Ron, and Andrew Freeman. 1998. *Seeing Tomorrow.* New York: Wiley & Sons.

DeMeyer, Arnold, C. H. Loch, and M. T. Pich. 2002. "Managing Project Uncertainty." *MIT Sloan Management Review* (Winter), pp. 60–67.

Demos, Nick, S. Chung, and M. Beck. 2002. "The New Strategy and Why It Is New." *Strategy and Business* 25:15–19.

Deutschman, Alan. 2000. *The Second Coming of Steve Jobs.* New York: Broadway Books.

Devaraj, Sarv, and R. Kholi. 2002. *The IT Payoff.* Upper Saddle River, N.J.: Prentice-Hall.

Dhar, Ravi. 2003. "Hedging Customers." *Harvard Business Review* (May), pp. 86–92.

Diamond, Jared. 2000. "The Ideal Form of Organization." *Wall Street Journal* (12 December), p. A17.

Dorf, Richard C. 2004. *Handbook of Engineering Tables.* Boca Raton, Fla.: CRC Press.

Dorf, Richard C. 2001. *Technology, Humans and Society: Toward a Sustainable World.* San Diego: Academic Press.

Dorf, Richard. 1999. *The Technology Management Handbook.* Boca Raton, Fla.: CRC Press.

Douglas, Evan, and D. Shepard. 2002. "Self-Employment as a Career Choice." *Entrepreneurship, Theory and Practice* (Spring): 81–89.

Douglas, Evan, and Dean Shepard. 1999. "Entrepreneurship as a Utility Maximizing Response." *Journal of Business Venturing* 15:231–51.

Downes, Larry, and C. Mui. 1998. *Unleashing the Killer App.* Boston: Harvard Business School Press.

Doz, Yves, and Gary Hamel. 1998. *Alliance Advantage.* Boston: Harvard Business School Press.

Drucker, Peter. 2002. *Managing in the Next Society.* New York: St. Martin's Press.

Drucker, Peter F. 1995. *Managing in a Time of Great Change.* New York: Penguin Books.

Drucker, Peter. 1993. *Innovation and Entrepreneurship.* New York: Harper Collins.

Dyck, Bruno, et al. 2002. "Passing the Baton: The Importance of Sequence, Timing, Technique and Communication in Executive Succession." *Journal of Business Venturing* 17:143–62.

Dye, Renee. 2000. "The Buzz of Buzz." *Harvard Business Review* (November), pp. 139–46.

Dyson, James, and R. Uhlig. 2001. *A History of Great Inventions.* New York: Carroll and Graf.

Economist. 2003a. "Beyond the Nanotype." (March 15), pp. 23–24.

Economist. 2003b. "Pots of Promise." (May 24), pp. 69–71.

Economist. 2003c. "The Big Easy." (May 31), p. 57.

Economist. 2003d. "Who Gets Eaten and Who Gets to Eat." (July 12), pp. 61–63.

Economist. 2003e. "The Promise of a Blue Revolution." (August 9), pp. 19–21.

Economist. 2002. *Pocket World in Figures.* London: Economist Books.

Eisenhardt, Kathleen, and D. N. Sull. 2001. "Strategy as Simple Rules." *Harvard Business Review* (January), pp. 106–16.

Elias, Stephen, and R. Stim. 2003. *Patent, Copyright and Trademark,* 6th ed. Berkeley: Nolo Press.

Elsbach, Kimberly. 2003. "How to Pitch a Brilliant Idea." *Harvard Business Review* (September), pp. 40–48.

Enriquez, Juan, and Ray Goldberg. 2000. "Transforming Life, Transforming Business: The Life-Science Revolution." *Harvard Business Review* (March), pp. 96–104.

Erikson, Truls. 2002. "Entrepreneurial Capital." *Entrepreneurship, Theory and Practice* (Spring): 275–90.

Ettenberg, Elliott. 2002. *The Next Economy.* New York: McGraw-Hill.

Fahey, Liam, and R. M. Randall. 1998. *Learning from the Future.* New York: Wiley & Sons.

Farrell, Diana. 2003. "The Real New Economy." *Harvard Business Review* (October), pp. 105–12.

Fasser, Yetim, and D. Brettner. 2002. *Management for Quality in High Technology Enterprises.* New York: Wiley & Sons.

Fast Company, 1997, "John Doerr's Startup Manual," (March), p. 82.

Ferrary, Michel. 2003. "The Gift Exchange in the Social Networks of Silicon Valley." *California Management Review* (Summer), pp. 120–36.

Fine, Charles, et al. 2002. "Rapid Response Capability in Value Chain Design." *MIT Sloan Management Review* (Winter), pp. 69–75.

Finkelstein, Sydney. 2003. *Why Smart Executives Fail.* New York: Portfolio Penguin.

Finkelstein, Sydney, and A. Mooney. 2003. "Not the Usual Suspects." *Academy of Management Executive,* 2:101–12.

Fisher, Lawrence M. 2003. "Symantec's Strategy-Based Transformation." *Strategy and Business* 30:81–89.

Fisher, Lawrence. 2002. "Yves Doz: The Thought Leader Interview." *Strategy and Business* 29:115–23.

Fisher, Robert, and W. Ury. 1991. *Getting to Yes.* New York: Penguin.

Fleming, Lee, and Olar Sorenson. 2001. "The Dangers of Modularity." *Harvard Business Review* (September), pp. 20–21.

Florida, Richard. 2002. *The Rise of the Creative Class.* New York: Basic Books.

Fogel, Robert W. 2000. *The Fourth Great Awakening and the Future of Egalitarianism.* Chicago: University of Chicago Press.

Forelle, Charles. 2003. "Gillette, Schick Battle, Blades Drawn." *Wall Street Journal* (12 November), p. B4a.

Forethought. 2001. "Inside Boeing's Big Move." *Harvard Business Review* (October), pp. 22–23.

Frank, Robert H. 2002. *Microeconomics and Behavior,* 5th ed. New York: McGraw-Hill.

Freiberg, Kevin, and Jackie Freiberg. 1997. *Nuts!* New York: Broadway Books.

Friedman, Thomas L. 1999. *The Lexus and the Olive Tree.* New York: Farrar, Straus, and Giroux.

Friedman, Thomas L., 2004. "Giving Young Arabs Hope for Jobs." World Economic Forum, Davos, www.azstarnet.com/sn/vote/7377.php.

Gabor, Andrea. 2000. *The Capitalist Philosophers.* New York: New York Times Books.

Gaffney, John. 2001. "How Do You Feel About a $44 Tooth Bleaching Kit?" *Business 2.0* (October), pp. 126–27.

Galor, Oden, and O. Moav. 2002. "National Selection and the Origin of Economic Growth." *Quarterly Journal of Economics* (November): 1133–91.

Gargiulo, Terrence, 2002. Making Stories, Westport, CT: Quorum.

Garrett, E. M. 1992. "Branson the Bold." *Success* (November), p. 22.

Garvin, David. 1993. "Building a Learning Organization." *Harvard Business Review* (July), pp. 78–91.

Gatewood, Elizabeth. 2001. "Busting the Stereotype." *Kelley School Business Magazine* (Summer), pp. 14–15.

Gersick, Kelin, and John Davis. 1997. *Life Cycles of a Family Business.* Boston: Harvard Business School Press.

Gerstner, Louis. 2002. *Who Says Elephants Can't Dance?* New York: Harper Collins.

Ghemawat, Pankaj. 2003. "The Forgotten Strategy." *Harvard Business Review* (November), pp. 76–84.

Gibson, Elizabeth, and A. Billings. 2003. *Big Charge at Best Buy.* Palo Alto, Calif.: Davies-Black.

Girard, Kim. 2000. "Pandesic's Failed Union." *Business 2.0* (September), pp. 16–18.

Gittell, Jody H. 2003. *The Southwest Airlines Way.* New York: McGraw-Hill.

Gladwell, Malcolm. 2001. "Smaller." *New Yorker* (26 November), pp. 60–68.

Gladwell, Malcolm. 2000. *The Tipping Point.* Boston: Little, Brown.

Godiesh, Orit, and James Gilbert. 2001. "Transforming Corner-Office Strategy into Frontline Action." *Harvard Business Review* (May), pp. 73–79.

Goldenberg, Jacob, et al. 2003. "Finding Your Innovation Sweet Spot." *Harvard Business Review* (March), pp. 120–29.

Goleman, Daniel, R. Boyatzis, and A. McKee. 2002. *Primal Leadership.* Boston: Harvard Business School Press.

Goleman, Daniel, R. Boyatzis, and A. McKee. 2001. "Primal Leadership: The Hidden Driver of Great Performance." *Harvard Business Review* (December), pp. 44–51.

Gomes, Lee. 2003. "Is Antitenor Plan by Priceline Founder Genius or Just Goofy?" *Wall Street Journal* (30 June), p. B1.

Gompers, Paul, and J. Lerner. 2001. *The Money of Invention.* Boston: Harvard Business School Press.

Gompers, Paul, and W. A. Sahlman. 2002. *Entrepreneurial Finance.* New York: Wiley & Sons.

Gordon, John S. 2002. *A Thread across the Ocean.* New York: Walker.

Gosling, Jonathan, and H. Mintzberg. 2003. "The Five Minds of a Manager." *Harvard Business Review* (November), pp. 54–63.

Gossage, Bobbie. 2003. "Tabling Benefits." *Inc.* (June), pp. 46–48.

Gourville, John, and D. Soman. 2002. "Pricing and the Psychology of Consumption." *Harvard Business Review* (September), pp. 91–96.

Grabowski, Robert, L. Navarro-Serment, and P. K. Khosla. 2003. "An Army of Small Robots." *Scientific American* (November), pp. 63–67.

Graebner, Melissa. 2004. "Momentum and Serendipity: How Acquired Leaders Create Value in the Integration of Firms." 25:*Strategic Management Journal.*

Graebner, Melissa, and K. Eisenhardt. 2002. "The Other Side of the Story." Working paper, Stanford University, Palo Alto, Calif.

Green, Heather. 2003. "Companies That Really Get It." *Business Week* (25 August), p. 144.

Greene, Patricia, C. G. Brush, and M. M. Hart. 1999. "The Corporate Venture Champion." *Entrepreneurship Theory and Practice* (Spring): 103–22.

Grimes, Ann. 2003. "Powerful Connections." *Wall Street Journal* (30 October), p. B1.

Grove, Andy. 2003. "Churning Things Up." *Fortune* (11 August), pp. 115–18.

Gruley, Bryan. 2003. "Michigan Potboiler." *Wall Street Journal* (14 February), p. A1.

Gumpert, David E. 2002. *Burn Your Business Plan.* Needham, Mass.: Lauson Publishers.

Gundry, Lisa, and H. Welsch. 2001. "The Ambitious Entrepreneur." *Journal of Business Venturing* 16:453–70.

Haddad, Charles. 2003. "FedEx and Brown Are Going Green." *Business Week* (4 August), pp. 60–62.

Hamel, Gary. 2001. "Revolution versus Evolution: You Need Both." *Harvard Business Review* (May), pp. 150–56.

Hamel, Gary. 2000. *Leading the Revolution.* Boston: Harvard Business School Press.

Hamm, John. 2002. "Why Entrepreneurs Don't Scale." *Harvard Business Review* (December), pp. 110–15.

Hamm, Steve. 2003. "Borders Are So 20th Century." *Business Week* (22 September), pp. 68–72.

Hammer, Michael. 2001. *The Agenda.* New York: Crown Business.

Handy, Charles. 1999a. *The New Alchemists.* London: Hutchinson.

Handy, Charles. 1999b. *The Hungry Spirit.* New York: Broadway Books.

Hardy, Quentin. 2003. "All Eyes on Google." *Forbes* (26 May), pp. 100–10.

Hargadon, Andrew. 2003. *How Breakthroughs Happen.* Boston: Harvard Business School Press.

Hargadon, Andrew, and Y. Douglas. 2001. "When Innovations Meet Institutions." *Administrative Science Quarterly* 46 (September): 476–501.

Hargadon, Andrew, and A. Fanelli. 2002. "Action and Possibility." *Organization Science* 13(3):290–302.

Hargadon, Andrew, and R. I. Sutton. 2000. "Building an Innovation Factory." *Harvard Business Review* (June), pp. 157–66.

Harmon, Amy. 2003. "I.B.M. Looks to Genetics to Map a New Business." *New York Times* (25 August).

Harry, Mikel, and R. Schroeder. 2000. *Six Sigma.* New York: Currency Doubleday.

Harzberg, Friderick. 2003. "How Do You Motivate Employees?" *Harvard Business Review* (January), pp. 87–92.

Hastie, Reid, and R. M. Dawes. 2001. *Rational Choice in an Uncertain World.* Thousand Oaks, Calif.: Sage.

Hawkins, Lori. 2002. "Agillion's Brief, Fast Life." *Austin-American Statesman* (3 February), p. 3.

Heifetz, Ronald, and D. Laurie. 2001. "The Work of Leadership." *Harvard Business Review* (December), pp. 131–40.

Helft, Miguel. 2002. "Fashion Fast Forward." *Business 2.0* (May), pp. 61–66.

Henderson, Rebecca, and Kim Clark. 1990. "Architectural Innovation." *Administrative Science Quarterly* 35:9–30.

Herrick, Thaddeus. 2003. "Glass Breaks Out." *Wall Street Journal* (17 July), p. B1.

Hill, Charles W. L. 2003. *Global Business,* 2d ed. New York: McGraw-Hill.

Hill, Charles W., and Gareth R. Jones. 2001. *Strategic Management,* 5th ed. Boston: Houghton Mifflin.

Hill, Charles, and F. Rothaermel. 2003. "The Performance of Incumbent Firms in the Face of Radical Technological Innovation." *Academy of Management Review* 28:257–74.

Hill, Michael, R. Ireland, S. Camp, and D. Sexton. 2002. *Strategic Entrepreneurship.* Malden, Mass.: Blackwell.

Hirsh, Evan, S. Hedlund, and M. Schweizer. 2003. "Reality Is Perception—The Truth about Car Brands." *Strategy and Business* (Fall): 20–25.

Hitt, Michael A., R. D. Ireland, S. M. Camp, and D. L. Sexton. 2001. "Entrepreneurial Strategies for Wealth Creation." *Strategic Management Journal* 22:479–91.

Hoang, Ha. 2003. "Is Your Company Over-Allianced?" *Strategy and Business* 30:9–11.

Hof, Robert. 2002. "Turning Rust into Gold." *Business Week* (10 June), p. 102.

Holcombe, Randall G.. 2001. "The Invisible Hand and Economic Progress." *Entrepreneurial Inputs and Outcomes* 13:281–326.

Holmes, Stanley. 2003. "For Starbucks, There's No Place Like Home." *Business Week* (9 June), p. 48.

Holt, Douglas. 2003. "What Becomes an Icon Most?" *Harvard Business Review* (March), pp. 43–49.

Horngren, Charles, G. Foster, and S. Datar. 2003. *Cost Accounting.* Upper Saddle River, N.J.: Prentice-Hall.

Hotchner A. E., and P. Newman. 2003. *Shameless Exploitation in the Pursuit of Common Good.* New York: Doubleday.

Hounshell, David. 1985. *From the American System to Mass Production.* Baltimore: Johns Hopkins University Press.

Howe, Jeff. 2002. "The Next Wave." *Wired* (January), pp. 109–12.

Hoover, Gary. 2001. *Hoover's Vision.* New York: Texere.

Huntington, Tom. 2003. "The Gimmick That Ate Hollywood." *Invention and Technology* (Spring): 34–45.

Hutchinson, Bill. 2003. "Art Behind the Scenes." *Santa Fean* (June), pp. 43–45.

Huxley, Aldous. 1990. *The Perennial Philosophy.* New York: Harper Collins.

Iansiti, Marco, and Roy Levien, 2004. "Strategy as Ecology." *Harvard Business Review* (March), pp. 68–78.

Ibarra, Hermina. 2002. "How to Stay Stuck in the Wrong Career." *Harvard Business Review* (December), pp. 40–47.

Ittnev, C. D., and D. F. Larcker. 2003. "Coming Up Short on Nonfinancial Performance." *Harvard Business Review* (November), pp. 88–96.

Jackman, Jay, and M. H. Strober. 2003. "Fear of Feedback." *Harvard Business Review* (April), pp. 101–7.

Jakle, John, and K. A. Sculle. 1999. *Fast Food— Roadside Restaurants in the Automobile Age.* Baltimore: Johns Hopkins University Press.

Jakle, John A., K. A. Sculle, and J. S. Rogers. 1996. *The Motel in America.* Baltimore: Johns Hopkins University Press.

Jassawalla, Avan, and H. C. Sashittal. 2002. "Cultures that Support Product Innovation Processes." *Academy of Management Executive* (August): 42–54.

Jonietz, Erica. 2003. "Biotech + Medicine." *Technology Review* (October): 72–80.

Joyce, William, N. Nohria, and B. Roberson. 2003. *What Really Works.* New York: Harper Collins.

Julien, Pierre Andre, and C. Pamangalahy. 2003. "Competitive Strategy and Performance of Exporting SMEs." *Entrepreneurship Theory and Practice* (Spring): 227–45.

Katila, Riitta. 2002. "New Product Search over Time." *Academy of Management Journal* 5:995–1010.

Kanter, Rosabeth Moss. 2003. "Leadership and the Psychology of Turnarounds." *Harvard Business Review* (June), pp. 58–67.

Kaplan, Robert, and D. Norton. 2000. *The Strategy Focused Organization.* Boston: Harvard Business School Press.

Kaptein, Muel, and J. Wempe. 2002. *The Balanced Company.* New York: Oxford University Press.

Karlgaard, Rich. 2003. "The Cheap Decade." *Forbes* (31 March), p. 37.

Kash, Rick. 2002. *The New Law of Demand and Supply.* New York: Doubleday.

Kellner, Tomas. 2002. "One Man's Trash." *Forbes* (4 March), pp. 96–98.

Kessler, Eric, and Paul Bierly. 2002. "Is Faster Really Better? An Empirical Test of the Implications of Innovation Speed." *IEEE Transactions on Engineering Management* (February): 2–12.

Khurana, Rakesh. 2002a. *Searching for a Corporate Savior.* Princeton, N.J.: Princeton University Press.

Khurana, Rakesh. 2002b. "The Curse of the Superstar CEO." *Harvard Business Review* (September), pp. 60–66.

Killian, Erin. 2003. "Butter 'Em Up." *Forbes* (9 June), pp. 175–76.

Kim, W. Chan, and R. Mauborgne. 2003. "Tipping Point Leadership." *Harvard Business Review* (April), pp. 60–69.

King, Martin Luther. 1963. *Strength to Love.* Minneapolis: Fortress Press.

King, Tom. 2002. "Hollywood Previews." *Wall Street Journal* (26 April), p. W9.

Kipling, R. 1902. *Just So Stories.* New York: Morrow.

Kirchhoff, Bruce A. 1994. Entrepreneurship and Dynamic Capitalism. New York: Praeger.

Kirkman, Bradley, et al. 2002. "Five Challenges to Virtual Team Success." *Academy of Management Executive* 3:67–79.

Kirkpatrick, David. 2003. "Brainstorm 2003." *Fortune* (27 October), pp. 187–90.

Kirkpatrick, David. 2002. "The Online Grocer Version 2.0." *Fortune* (25 November), pp. 217–26.

Klein, Alec. 2003. *Stealing Time.* New York: Simon and Schuster.

Klein, Mark, and A. Einstein. 2003. "The Myth of Customer Satisfaction." *Strategy and Business* 30:8–9.

Kleiner, Art. 2003a. "Are You in with the In Crowd?" *Harvard Business Review* (July), pp. 86–92.

Kleiner, Art. 2003b. "Making Patient Capital Pay off." *Strategy and Business* (Fall): 26–30.

Koehn, Nancy. 2001. *Brand New.* Cambridge: Harvard Business School Press.

Kogut, Bruce. 2003. *The Global Internet Economy.* Cambridge: MIT Press.

Kolbert, Elizabeth. 2003. "The Car of Tomorrow." *New Yorker* (11 August), pp. 30–40.

Komisar, Randy. 2000. *The Monk and the Riddle.* Boston: Harvard Business School Press.

Kotler, Philip. 2001. *A Framework for Marketing Management.* Upper Saddle River, N.J.: Prentice-Hall.

Kotler, Philip, N. Roberto, and N. Lee. 2002. *Social Marketing.* Thousand Oaks, Calif.: Sage.

Kotter, John P., and D. S. Cohen. 2002. *The Heart of Change.* Boston: Harvard Business School Press.

Kouzes, James, and B. Z. Posner. 2002. *Leadership Challenge.* San Francisco: Jossey-Bass.

Krajewski, Lee, and L. Ritzman. 2002. *Operations Management,* 6th ed. Upper Saddle River, N.J.: Prentice-Hall.

Kramer, Roderick. 2003. "The Harder They Fall." *Harvard Business Review* (October), pp. 58–66.

Kramer, Roderick. 2002. "When Paranoia Makes Sense." *Harvard Business Review* (July), pp. 62–69.

Kuemmerle, Walter. 2002a. "Home Base and Knowledge Management in International Ventures." *Journal of Business Venturing* 17:99–122.

Kuemmerle, Walter. 2002b. "A Test for the Fainthearted." *Harvard Business Review* (May), pp. 4–8.

Labovitz, George, and V. Rosansky. 1997. *The Power of Alignment.* New York: Wiley & Sons.

Landers, Peter. 2003. "With New Patent, Mayo Clinic Owns a Cure for Sniffles." *Wall Street Journal* (30 April), p. A1.

Landes, David S. 1998. *The Wealth and Poverty of Nations.* New York: Norton.

Langley, Monica, 2003. "In Tough Times Head to Warren Buffet." *Wall Street Journal* (14 November), p. A1.

Lapre, Michael, and L. N. Van Wassenhove. 2002. "Learning across Lines." *Harvard Business Review* (October), pp. 107–11.

LaSalle, Diana, and T. A. Britton. 2003. *Priceless.* Boston: Harvard Business School Press.

Lashinsky, Adam. 2003. "Meg and the Machine." *Technology Review* (1 September), pp. 70–78.

Lavallo, Dan, and D. Kahneman. 2003. "Delusions of Success." *Harvard Business Review* (July), pp. 56–63.

Lax, David A., and J. K. Sebenius. 2003. "3-D Negotiation: Playing the Whole Game." *Harvard Business Review* (November), pp. 65–74.

Lax, Eric. 1985. "Banking on Biotech Business." *New York Times* (22 December).

Lechler, Thomas. 2001. "Social Interaction: A Determinant of Entrepreneurial Team Venture Success." *Small Business Economics* 16:263–78.

Leibovich, Mark. 2002. *The New Imperialists.* Paramus, N.J.: Prentice-Hall.

Leifer, Richard, et al. 2000. *Radical Innovation.* Boston: Harvard Business School Press.

Leonard, Devin. 2003. "Songo in the Key of Steve." *Fortune* (12 May), pp. 53–62.

Leonard-Barton, Dorothy. 1995. *Wellsprings of Knowledge.* Boston: Harvard Business School Press.

Leung, Shirley. 2003. "How Come There's No McDonald's of Chinese Food?" *Wall Street Journal* (23 January), p. A1.

Leung, Shirley. 2002. "Armchairs, TVs and Espresso—Is It McDonald's?" *Wall Street Journal* (30 August), p. A1.

Levere, Jane. 2002. "Pop Goes the Pop-up Ads." *New York Times* (29 July), p. C4.

Levesque, Moren, Dean Shepard, and Evon Douglas. 2002. "Employment or Self-Employment: A Dynamic Utility-Maximizing Model." *Journal of Business Venturing* 17:189–210.

Levitt, Harold. 2003. "Why Hierarchies Thrive." *Harvard Business Review* (March), pp. 96–102.

Lewis, Michael. 2003. *Moneyball.* New York: Norton.

Lewis, Michael. 2000. *The New, New Thing.* New York: Norton.

Lilien, Gary, and A. Rangaswamy. 2003. *Marketing Engineering.* Upper Saddle River, N.J.: Prentice-Hall.

Lodish, Leonard, H. Morgan, and A. Kallianpur. 2001. *Entrepreneurial Marketing.* New York: Wiley & Sons.

Lohr, Steve. 2003a. "Silicon Valley Hikes Wireless Frontier." *New York Times* (7 April).

Lohr, Steve. 2003b. "A Once and Present Innovator." *New York Times* (6 May), p. C1.

Lohr, Steve. 2003c. "Oldest Living Startup Tells All." *New York Times* (10 November).

Longaberger, Dave. 2001. *Longaberger.* New York: Harper Collins.

Lord, Michael, S. W. Mandel, and J. D. Wager. 2002. "Spinning out a Star." *Harvard Business Review* (June), pp. 115–21.

Lounsbury, Michael, and M. Glynn. 2001. "Cultural Entrepreneurship: Stories, Legitimacy and the Acquisition of Resources." *Strategic Management Journal* 22:545–64.

Loveman, Gary. 2003. "Diamonds in the Data Mine." *Harvard Business Review* (May), pp. 109–13.

Low, Murray B., and E. Abrahamson. 1997. "Movements, Bandwagons and Clones: Industry Evolution and the Entrepreneurial Process." *Journal of Business Venturing* 12:435–57.

Lowe, Robert A. 2001. "The Role and Experience of Start-ups in Commercializing University Inventions." *Entrepreneurial Inputs and Outcomes* 13:189–222.

Lucier, Chuck, and J. Dyer. 2003. "Creating Chaos for Fun and Profit." *Strategy and Business* 30:14–20.

Luhnow, David, and C. Terhurie. 2003. "A Low-Budget Cola Shakes up Markets South of the Border." *Wall Street Journal* (27 October), p. A1.

Lynn, Gary, and Richard Reilly. 2002. *Blockbusters.* New York: Harper Collins.

MacMillan, Ian. 2003. "Global Gamesmanship." *Harvard Business Review* (May), pp. 62–71.

Magretta, Joan. 2002. *What Management Is.* New York: Free Press.

Maher, Michael, C. P. Stickney, and R. L. Weil. 2004. *Managerial Accounting,* 8th ed. Cincinnati: Southwestern.

Majumdar, Sumit. 1999. "Sluggish Giants, Sticky Cultures and Dynamic Capability Transformation." *Journal of Business Venturing* 15:59–78.

Malik, Om. 2003. *Broadbandits.* Hoboken, N.J.: Wiley & Sons.

Malone, Michael. 2002. *Betting It All: The Entrepreneurs of Technology.* New York: Wiley & Sons.

Mangalindan, Mylene. 2003. "Rising Clout of Google." *Wall Street Journal* (16 July), p. A1.

Mangalindan, Mylene. 2002. "For Bulk E-Mailer, Pestering Millions Offers Path to Profit." *Wall Street Journal* (13 November), p. A1.

Mangalindan, Mylene, and S. L. Hwang. 2001. "Insular Culture Helped Yahoo! Grow, But Has Now Hurt It in the Long Run." *Wall Street Journal* (9 March), p. A1.

Mankiw, N. G. 2000. *Principles of Economics,* 2d ed. Dallas: South-Western.

Manla, Markku, T. Keil, and S. A. Zahra. 2003. "Corporate Venture Capital and Recognition of Technological Discontinuities." Proceedings of Academy of Management Annual Meeting (August), Seattle.

Marcus, Bernie, and Arthur Blank. 1999. *Built from Scratch.* New York: Crown Books.

Maremont, Mark. 2003. "Tyco Ex-Officer Says Swartz Made Admission on Bonuses." *Wall Street Journal* (30 October), p. B1.

Markman, Gideon, D. Balkin, and R. A. Baron. 2002. "Inventors and New Venture Formation." *Entrepreneurship Theory and Practice* (Winter): 149–65.

Martin, Roger L. 2002. "The Virtue Matrix." *Harvard Business Review* (March), pp. 69–75.

Martin, Roger, and M. C. Moldoveanu. 2003. "Capital versus Talent." *Harvard Business Review* (July), pp. 36–41.

Mason, Heidi, and T. Rohner. 2002. *The Venture Imperative.* Boston: Harvard Business School Press.

McCall, Morgan. 1998. *High Flyers.* Boston: Harvard Business School Press.

McCuen, Jess. 2003. "Failure of Genius." *Inc.* (August), pp. 90–95.

McElroy, Mark. 2003. *The New Knowledge Management.* Boston: Elsevier.

McGrath, James, F. Kroeger, M. Traem, and J. Rocken Haeuser. 2001. *The Value Growers.* New York: McGraw-Hill.

McKee, Robert. 2003. "Storytelling That Moves People." *Harvard Business Review* (June), pp. 51–55.

McKenzie, Ray. 2001. *The Relationship-Based Enterprise.* New York: McGraw-Hill.

McLean, Bethany, and P. Elkind. 2003. *The Smartest Guys in the Room.* New York: Portfolio.

McLemore, Clinton. 2003. *Street-Smart Ethics.* Louisville, Ky.: Westminster John Knox Press.

McMullen, Jeffrey, and Dean Shepard. 2002. "Regulatory Focus and Entrepreneurial Intention." Presentation at the Academy of Management Meeting (August).

McNulty, James, et al. 2002. "What's Your Real Cost of Capital?" *Harvard Business Review* (October), pp. 114–21.

Melicher, Ronald, and C. Leach. 2003. *Entrepreneurial Finance.* Boston: Southwestern.

Melnyk, Steven, and M. Swink. 2002. *Value-Driven Operations Management.* New York: McGraw-Hill.

Merrick, Amy. 2003. "How Gingham and Polyester Rescued a Retailer." *Wall Street Journal* (9 May), p. A1.

Meyer, Christopher. 1998. *Relentless Growth.* New York: Free Press.

Mezias, John, and W. H. Starbuck. 2003. "What Do Managers Know, Anyway?" *Harvard Business Review* (May), pp. 16–17.

Michael, Steven C. 2003. "First Mover Advantage through Franchising." *Journal of Business Venturing* 18:61–80.

Mickle Thwart, John, and A. Wooldridge. 2003. *The Company.* New York: Modern Library.

Miles, Morgan, and J. Covin. 2002. "Exploring the Practice of Corporate Venturing." *Entrepreneurship, Theory and Practice* (Spring): 21–40.

Minniti, Maria, and W. Bygrave. 2001. "A Dynamic Model of Entrepreneurial Learning." *Entrepreneurship Theory and Practice* (Spring): 5–16.

Mintzberg, Henry, B. Ahlstrand, and J. Lampel. 1998. *Strategy Safari.* New York: Free Press.

Mintzberg, Henry, J. Lampel, J. B. Quinn, and S. Ghoshal. 2003. *The Strategy Process,* 4th ed., Upper Saddle River, N.J.: Prentice-Hall.

Modis, Theodore. 1998. *Conquering Uncertainty.* New York: McGraw-Hill.

Mokyr, Joel. 2003. *The Gifts of Athena—Historical Origins of the Knowledge Economy.* Princeton, N.J.: Princeton University Press.

Montague, Tonda. 2001. "Southwest Airlines Turns 30." *Southwest Airlines Spirit,* June: pp. 58–69.

Moore, Geoffrey. 2002. *Crossing the Chasm.* New York: Harper Collins.

Moore, Geoffrey. 2000. *Living on the Fault Line.* New York: Harper Collins.

Moore, Geoffrey. 1999. *Inside the Tornado.* New York: Harper Collins.

Morris, Michael, and D. F. Kuratko. 2002. *Corporate Entrepreneurship.* Orlando, Fla.: Harcourt.

Murtha, Thomas, S. Lenway, and J. Hart. 2001. *Managing New Industry Creation.* Stanford, Calif.: Stanford University Press.

Nalebuff, Barry, and Ian Ayres. 2003. *Why Not?* Boston: Harvard Business School Press.

Nambisan, Satish. 2002. "Designing Virtual Customer Environments for New Product Development." *Academy of Management Review* 27(3):392–413.

Nellore, Rajesh. 2001. "The Impact of Supplier Visions on Product Development." *Journal of Supply Chain Management* (Winter): 27–36.

Nohvia, Nitin, W. Joyce, and B. Roberson. 2003. "What Really Works." *Harvard Business Review* (July), pp. 43–51.

Norman, Donald. 1998. *The Invisible Computer.* Cambridge: MIT Press.

Norman, Patricia. 2001. "Are Your Secrets Safe? Knowledge Protection in Strategic Alliances." *Business Horizons* (November), pp. 51–60.

Northouse, Peter G. 2001. *Leadership,* 2d ed. Thousand Oaks, Calif.: Sage.

Ogden, Joseph, F. Jen, and P. O'Connor. 2003. *Advanced Corporate Finance.* Upper Saddle River, N.J.: Prentice-Hall.

Orenstein, Susan. 2003. "The Love Algorithm." *Business 2.0* (August), pp. 117–21.

Packard, David. 1995. *The HP Way.* New York: Harper Collins.

Paine, Lynn S. 2003. *Value Shift.* New York: McGraw-Hill.

Palmer, Jay. 2003. "Reflected Glory." *Barron's* (19 May), p. 31.

Paydarfar, David, and William Schwartz. 2001. "An Algorithm for Discovery." *Science* (6 April), p. 13.

Perlow, Leslie, and S. Williams. 2003. "Is Silence Killing Your Company?" *Harvard Business Review* (May), pp. 52–58.

Perlow, Leslie, G. Okhuysen, and N. P. Repenning. 2002. "The Speed Trap." *Academy of Management Journal* 5:931–55.

Perseus. 2002. *Business: The Ultimate Resources.* Cambridge: Perseus.

Peters, Tom, and R. Waterman. 1982. *In Search of Excellence.* New York: Harper & Row.

Petroski, Henry. 2003. *Small Things Considered: Why There Is No Perfect Design.* New York: Knopf.

Pfeffer, Jeffrey, and R. Sutton. 2000. *The Knowing-Doing Gap.* Boston: Harvard Business School Press.

Pietersen, Willie. 2002. *Reinventing Strategy.* New York: Wiley & Sons.

Plitt, Jane R. 2000. *Martha Matilda Harper and the American Dream.* Syracuse, N.Y.: Syracuse University Press.

Pogue, David. 2003. "For TiVo and Replay, New Reach." *New York Times* (29 May), p. 86.

Porter, Michael. 2001, "Strategy and the Internet." *Harvard Business Review* (March), pp. 63–78.

Porter, Michael E. 1998. *On Competition.* Boston: Harvard Business School Press.

Porter, Michael, and S. Stern, 2001, "Innovation: Location Matters." *MIT Sloan Management Review* (Summer), pp. 28–36.

Post, James, L. E. Preston, and S. Sachs. 2002. "Managing the Extended Enterprise." *California Management Review* (Fall), pp. 6–20.

Pottrack, David, and T. Pearce. 2000. *Clicks and Mortar.* San Francisco: Jossey-Bass.

Prahalad, C. K., and A. Hammond. 2002. "Serving the World's Poor, Profitably." *Harvard Business Review* (September), pp. 48–57.

Prahalad, C. K., and V. Ramaswamy. 2000. "Co-opting Customer Competence." *Harvard Business Review* (January), pp. 79–87.

Prasad, Dev, G. Vozikis, and G. Bruton. 2001. "Commitment Signals in the Interaction between Business Angels and Entrepreneurs." *Entrepreneurial Inputs and Outcomes* 13:45–69.

Preston, John T. 2002. "Success Factors in Technology-Based Entrepreneurship: An MIT Perspective." *International Journal of Entrepreneurship Education* 1(2):277–94.

Prusak, Laurence, and D. Cohen. 2001. "How to Invest in Social Capital." *Harvard Business Review* (June), pp. 86–94.

Quadrini, Vincerizo. 2001. "Entrepreneurial Financing, Savings and Mobility." *Entrepreneurial Inputs and Outcomes* 13:71–94.

Quinlan, Mary Lou. 2003. *Just Ask a Woman.* Hoboken, N.J.: Wiley & Sons.

Quinn, James B., J. J. Baruch, and K. A. Zien. 1997. *Innovation Explosion.* New York: Free Press.

Rappaport, Alfred et al., 2001. Expectations Investing, Boston: Harvard Business School Press.

Reichheld, Frederick. 2001. "Lead for Loyalty." *Harvard Business Review* (July), pp. 76–84.

Reiss, Bob. 2000. *Low Risk, High Reward.* New York: Free Press.

Ridgway, Nicole. 2003. "Something to Sneeze at." *Forbes* (21 July), pp. 102–4.

Ries, Al, and Jack Trout. 2001. *Positioning: The Battle for Your Mind.* New York: McGraw-Hill.

Rigby, Darrell. 2001. "Moving Upward in a Downturn." *Harvard Business Review* (June), pp. 99–105.

Rigby, Darrell, and C. Zook. 2002. "Open Market Innovation." *Harvard Business Review* (October), pp. 80–89.

Riggs, Henry. 2004. *Financial and Economic Analysis for Engineering and Technology Management,* 2d ed. Hoboken, N.J.: Wiley & Sons.

Robbins-Roth, Cynthia. 2000. *From Alchemy to IPO.* Cambridge, Mass.: Perseus.

Roberts, Michael, and Diana Gardner, 2000. Advanced Inhalation Research. Harvard Business School, Case 899292.

Rogers, Everett. 1995. *Diffusion of Innovations,* 4th ed. New York: Free Press.

Rohlfs, Jeffrey. 2001. *Bandwagon Effects in High Technology Industries.* Cambridge: MIT Press.

Rogers, Paul, T. Holland, and D. Haas. 2002. "Value Acceleration: Lessons from Private-Equity Masters." *Harvard Business Review (*June), pp. 94–101.

Roman, Kenneth. 2003. How to Advertise, 3rd ed. New York: Thomas Dunne.

Roos, Goran, A. Bainbridge, and K. Jacobsen. 2001. "Intellectual Capital as a Strategic Tool." *Strategy and Leadership* 4:21–26.

Ross, Stephen, R. Westerfield, and J. Jaffe. 2002. *Corporate Finance.* New York: McGraw-Hill Irwin.

Rothaermel, Frank, and D. Deeds. 2004. "Exploration and Exploitation Alliances in Biotechnology." *Strategic Management Journal* (Winter): 100–21.

Ruef, Martin. 2002. "Strong Ties, Weak Ties and Islands." *Industrial and Corporate Change* 3:427–49.

Ryan, Rob. 2002. *Smartups.* Ithaca, N.Y.: Cornell University Press.

Ryans, Adrian, R. More, D. Barclay, and T. Deutscher. 2000. *Winning Market Leadership.* New York: Wiley & Sons.

Sadler, Robert. 2000. "Corporate Entrepreneurship in the Public Sector." *Australian Journal of Public Administration* (June): 25–43.

Sahlman, William. 1999. *The Entrepreneurial Venture.* Boston: Harvard Business School Press.

Santoli, Michael. 2003. "Dowagers in Distress." *Barron's* (10 March), p. 19.

Sanyal, Rajib. 2001. *International Management.* Upper Saddle River, N.J.: Prentice-Hall.

Sarker, M. B., R. Echambadi, and J. Harruson. 2001. "Alliance Entrepreneurship and Firm Market Performance." *Strategic Management Journal* 22:701–11.

Sathe, Vijay. 2003. *Corporate Entrepreneurship.* New York: Cambridge University Press.

Sawhney, Mohan, and J. Zabin. 2001. *The Seven Steps to Nirvana.* New York: McGraw-Hill.

Schmetter, Bob. 2003. *Leap.* New York: Wiley & Sons.

Schoemaker, Paul. 2002. *Profiting from Uncertainty.* New York: Free Press.

Schrage, Michael. 2002. "Ease of Learning." *Technology Review* (December), p. 23.

Schrage, Michael. 2001. "Playing around with Brainstorming." *Harvard Business Review* (March), pp. 149–54.

Schrage, Michael. 2000. *Serious Play.* Boston: Harvard Business School Press.

Schultz, Howard. 1997. New York: Hyperion.

Schumpeter, Joseph. 1984. *Capitalism, Socialism and Democracy.* New York: Harper Torchbooks.

Schwartz, Evan. 2002. *The Last Lone Inventor.* New York: Harper Collins.

Secrest, Meryle. 2001. *Somewhere for Me.* New York: Knopf.

Seeley Brown, John, and P. Duguid. 2002. *The Social Life of Information.* Boston: Harvard Business School Press.

Selden, Larry, and G. Colvin. 2003a. "M&A Needn't Be a Loser's Game." *Harvard Business Review* (June), pp. 70–79.

Selden, Larry, and G. Colvin. 2003b. "What Customers Want." *Fortune* (7 July), pp. 122–25.

Selden, Larry, and G. Colvin. 2002. "Will This Customer Sink Your Stock?" *Fortune* (30 September), pp. 127–32.

Shane, Scott. 2001. "Technological Opportunities and New Firm Creation." *Management Science* (February): 205–20.

Shaw, George Bernard. 1903. *Man and Superman.* New York: Penguin.

Shaw, Gordon, R. Brown, and P. Bromiley. 1998. "Strategic Stories: How 3M Is Rewriting Business Planning." *Harvard Business Review* (May), pp. 41–50.

Shepard, Dean. 2003. "Learning from Business Failure." *Academy of Management Review* 2:318–28.

Shepard, Dean, Evan Douglas, and Mark Shanley. 2000. "New Venture Survival." *Journal of Business Venturing* 15:393–410.

Shepard, Dean, R. Ettenson, and A. Crouch. 2000. "New Venture Strategy and Profitability." *Journal of Business Venturing* 15:449–67.

Shepard, Dean, and N. F. Krueger. 2002. "An Intentions-Based Model of Entrepreneurial Teams." *Entrepreneurship Theory and Practice* (Winter): 167–85.

Shepard, Dean, and M. Levesque. 2002. "A Search Strategy for Assessing a Business Opportunity." *IEEE Transactions on Engineering Management* (May): 140–54.

Shepard, Dean, and Mark Shanley. 1998. *New Venture Strategy.* Thousand Oaks, Calif.: Sage Publications.

Sheth, Jagdish, and R. Sisodia. 2002. *The Rule of Three.* New York: Free Press.

Shrader, Rodney, and Mark Simon. 1997. "Corporate versus Independent New Ventures." *Journal of Business Venturing* 12:47–66.

Shulman, Seth. 2003. "The Vision Thing." *Technology Review* (May): 75.

Shuman, Jeffrey. 2001. *Collaborative Communities.* Chicago: Dearborn.

Silverstein, Michael J. 2004. *Trading Up.* Hoboken, N.J.: Wiley & Sons.

Simon, Mark, and S. M. Houghton. 2003. "The Relationship between Overconfidence and the Introduction of Risky Products." *Academy of Management Journal* 2:139–49.

Slywotzky, Adrian. 2002. *The Art of Profitability.* New York: Warner Books.

Slywotzky, Adrian. 1996. *Value Migration.* Boston: Harvard Business School Press.

Slywotzky, Adrain, et al. 1999. *Profit Patterns.* New York: Random House.

Slywotzky, Adrian, and David Morrison. 2000. *How Digital Is Your Business?* New York: Crown Business.

Slywotzky, Adrian, and Richard Wise. 2002. "The Growth Crisis and How to Escape It." *Harvard Business Review* (July), pp. 73–83.

Smillie, Dirk. 2003. "Marine Dream." *Forbes* (10 November), pp. 187–90.

Smith, Janet, and R. L. Smith. 2004. *Entrepreneurial Finance,* 3d ed. Hoboken, N.J.: Wiley & Sons.

Snyder, Bill. 2003. "LEDL Lamps Light the Way for Social Entrepreneurs." *Stanford Business* (November): 14–16.

Sonnenfeld, Jeffrey. 2002. "What Makes Boards Great." *Harvard Business Review* (September), pp. 106–12.

Southwick, Karen. 1999. *High Noon.* New York: Wiley & Sons.

Spekman, Robert, and L. Isabella. 2000. *Alliance Competence.* New York: Wiley & Sons.

Spulker, Daniel. 2004. *Management Strategy.* Burr Ridge, Ill.: McGraw-Hill.

Stein, Nicholas. 2003. "America's Most Admired Companies." *Fortune* (3 March), pp. 81–90.

Stein, Nicholas. 2001. "Yes, We Have No Profits." *Fortune* (26 November), pp. 26–28.

Sterman, John D. 2000. *Business Dynamics.* New York: McGraw-Hill.

Sternberg, Robert, L. A. O'Hara, and T. I. Lubart. 1997. "Creativity as Investment." *California Management Review* (Fall), pp. 8–21.

Stevenson, Howard, et al. 1999. *New Business Ventures and the Entrepreneur,* 5th ed. Burr Ridge, Ill.: McGraw-Hill Irwin.

Stevenson, Seth. 2003. "How to Beat Nike." *New York Times Magazine* (5 January), pp. 29–33.

Stewart, James. 2001. "Bidding War." *New Yorker* (15 October), pp. 42–50.

Stewart, Thomas A. 2001. *The Wealth of Knowledge.* New York: Currency Books.

Stipp, David. 2003. "Speed Reading Your Genes." *Fortune* (1 September), pp. 150–54.

Stires, David. 2001. "The Best Thing Since Sliced Bread." *Fortune* (29 October), p. 38.

Stone, Florence. 2002. *The Oracle of Oracle.* New York: Amacom.

Stringer, Kortney. 2003. "How Do You Change Consumer Behavior?" *Wall Street Journal* (17 March), p. R6.

Sull, Donald. 2003. "Managing by Commitments." *Harvard Business Review* (June), pp. 82–91.

Surowiecki, James. 2003. "EZ Does It." *New Yorker* (8 September), p. 36.

Sutton, John. 2001. *Technology and Market Structure.* Cambridge: MIT Press.

Sutton, Robert. 2002. *Weird Ideas That Work.* New York: Free Press.

Sutton, Robert. 2001. "The Weird Rules of Creativity." *Harvard Business Review* (September), pp. 94–103.

Swank, Cynthia. 2003. "The Lean Service Machine." *Harvard Business Review* (October), pp. 123–29.

Swanson, Robert A. "Cofounder of Gerentech," an Oral History. Bancroft Library U of Ca. Berkeley 2001.

Sweetman, Bill. 2003. "E-Plane." *Popular Science* (December), p. 38.

Szulanski, Gabriel, and S. Winter. 2002. "Getting It Right the Second Time." *Harvard Business Review* (January), pp. 62–69.

Tabarrok, Alexander. 2002. *Entrepreneurial Economics.* New York: Oxford University Press.

Tallman, Stephen, and K. Fladmoe-Lindquist. 2002. "Internationalization Globalization and Capability Strategy." *California Management Review* (Fall), pp. 110–34.

Tapscott, Don, D. Ticoll, and A. Lowy. 2000. *Digital Capital.* Boston: Harvard Business School Press.

Taub, Eric. 2003. "DVDs Meant for Buying but Not for Keeping." *New York Times* (21 July).

Taylor, Suzanne, and K. Schroeder. 2003. *Inside Intuit.* Boston: Harvard Business School Press.

Tedeschi, Bob. 2003. "End of the Paper Chase." *Business 2.0* (March), p. 64.

Tedlow, Richard S. 2001. *Giants of Enterprise.* New York: Harper Collins.

Teitelman, Robert. 1989. *Gene Dreams.* New York: Basic Books.

Tennant, Geoff. 2002. *Design for Six Sigma.* Burlington, Vt.: Gower.

Thaler, Linda, and R. Koval. 2003. *Bang!* New York: Currency.

Thomas, Paulette. 2003a. "Owner's Contacts Make Business a Success." *Wall Street Journal* (18 February), p. B9.

Thomas, Paulette. 2003b. "Entrepreneur's Biggest Problems." *Wall Street Journal* (17 March), p. R1.

Thomas, Paulette. 2003c. "Two Women's Drive Opens a Closed Door." *Wall Street Journal* (13 May), p. B7.

Thompke, Stefan. 2001. "Enlightened Experimentation." *Harvard Business Review* (February), pp. 48–52.

Thompke, Stefan, and D. Reinertsen. 1998. "Agile Product Development." *California Management Review* (Fall), pp. 8–28.

Thompke, Stefan, and Eric von Hippel. 2002. "Customers as Innovators." *Harvard Business Review* (April), pp. 74–81.

Thompson, Leigh. 2003. "Improving the Creativity of Organizational Work Groups." *Academy of Management Executive* (February): 96–109.

Thompson, Nicholas. 2003. "Netflix Uses Speed to Fend off Wal-Mart Challenge." *New York Times* (29 September).

Thurik, Roy, S. Wennekers, and L. M. Uhlaner. 2003. "Entrepreneurship and Economic Performance." *International Journal of Entrepreneurship Education* 1:157–79.

Thurm, Scott. 2003. "A Go-Go Giant of the Internet Age, Cisco Is Learning to Go Slow." *Wall Street Journal* (7 May), p. A1.

Thurm, Scott. 2002. "Cisco Details the Financing for Its Start-Up." *Wall Street Journal* (12 March), p. A3.

Toffler, Barbara Ley. 2003. *Final Accounting.* New York: Broadway Books.

Ullman, David. 2003. *The Mechanical Design Process.* New York: McGraw-Hill.

Ulrich, Karl, and S. Eppinger. 2004. *Product Design and Development,* 3d ed. New York: McGraw-Hill.

Ulwick, Anthony. 2002. "Turn Customers Input into Innovation." *Harvard Business Review* (January), pp. 91–97.

Van den Ende, Jan, and N. Wijaberg. 2003. "The Organization of Innovation and Market Dynamics." *IEEE Transactions on Engineering Management* (August): 374–82.

Van Pragg, C. M., and J. S. Cramer. 2001. "The Roots of Entrepreneurship and Labor Demand: Individual Ability and Low Risk Aversion." *Economica* (February): 45–62.

Vogelstein, Fred. 2003a. "Mighty Amazon." *Fortune* (26 May), pp. 61–74.

Vogelstein, Fred. 2003b. "24/7 Customer." *Fortune* (24 November), p. 212.

Vonderembse, Mark, and G. White. 2004. *Operations Management.* Hoboken, N.J.: Wiley & Sons.

Waaser, Ernest et al., 2004. "How You Slice It: Smarter Segmentation for Your Sales Force." *Harvard Business Review* (March), pp. 105–10.

Weightman, Gavin. 2003. *The Frozen-Water Trade.* New York: Hypercon.

Welch, Jack. 2002. *Jack: Straight from the Gut.* New York: Warner.

Wellman, Barry. 2001. "Computer Networks as Social Networks." *Science* (14 September), pp. 2031–33.

Wennekers, Sander, L. M. Uhlaner, and R. Thurik. 2002. "Entrepreneurship and Its Conditions." *International Journal of Entrepreneurship Education* 1:25–64.

Wild, John, K. R. Subramanyam, and R. Haloey. 2003. *Financial Statement Analysis.* New York: McGraw-Hill.

Windham, Laurie. 2000. *The Soul of the New Consumer.* New York: Allworth Press.

Winer, Russell. 2001. "A Framework for Customer Relationship Management." *California Management Review* (Summer), pp. 89–104.

Winer, Russell. 2000. *Marketing Management.* Upper Saddle River, N.J.: Prentice-Hall.

Wingfield, Nick, and K. Lundegaard. 2003. "eBay Is Emerging as Unlikely Giant in Used-Car Sales." *Wall Street Journal* (7 February), p. A1.

Wood, Robert C., and G. Hamel. 2002. "The World Bank's Innovation Market." *Harvard Business Review* (November), pp. 104–12.

Woolley, Scott, and Q. Hardy. 2003. "The Wonderful World of Wirelessness." *Forbes* (1 September), pp. 94–102.

Wu, Amy. 2003. "A Specialty Food Store with a Discount Attitude." *New York Times* (27 July).

Yakura, Elaine. 2002. "Charting Time: Timelines as Temporal Boundary Objects." *Academy of Management Journal* 5:956–70.

Young, Jeffrey S. 1988. *Steve Jobs—The Journey Is the Reward.* Glenview, Ill.: Scott-Foresman.

Zacharakis, Andrew, G. D. Meyer, and J. DeCastro. 1999. "Differing Perceptions of New Venture Failure." *Journal of Small Business Management* (July): 1–14.

Zahra, Shaker, D. O. Neubaum, and M. Huse. 2000. "Entrepreneurship in Medium Size Companies." *Journal of Management* 5:947–76.

Zider, Bob. 1998. "How Venture Capital Works." *Harvard Business Review* (December), pp. 131–39.

Zimmerman, Monica, and G. J. Zeitz. 2002. "Beyond Survival: Achieving New Venture Growth by Building Legitimacy." *Academy of Management Review* 3:414–31.

Business Plans

EZGUARD LIFE SCIENCES

Asha Nayak, M.D., Ph.D.
Nick Mourlas, Ph.D
Chris Eversull, B.S., M.A.
Kurt Grote, B.S.
Frederick Winston, B.S., M.S.

Spring 2002

I. Executive Summary

1. Clinical Need

Physicians are increasingly performing interventional procedures to treat common vascular diseases that have caused vessels to narrow or occlude. Untreated, these vascular conditions lead to strokes and heart attacks, which in 2002 cost more than $50B per year to treat (U.S.). Despite physicians' intention to re-establish blood flow to vital organs, their interventional methods are frequently complicated by the liberation of clots and debris. This results in embolization (blockage of distant blood vessels) and subsequent permanent brain/cardiac damage with costly clinical consequences.

2. EZGuard

EZGuard has envisioned a novel two-component, disposable device and method to safely, effectively, and easily overcome the tragic side effects of embolization. Because of the pressing need physicians have for a safe and effective interventional solution and the superiority of our product, our expert advisors concur that our technology should demand a substantial fraction of the more than $500M per year (and rapidly growing) distal protection market. Although other solutions have been developed, physicians are frustrated by the technical problems and safety concerns associated with these devices and are using them out of dire necessity and for lack of a more optimal solution. EZGuard's solution uniquely meets *every* customer criteria for an improved distal production product, as determined by a careful review of the literature and discussions with vascular surgeons, interventional cardiologists, and neurologists.

3. The Competition and EZGuard's Competitive Advantage

There is considerable competition in the distal protection market. Currently, one FDA (Food and Drug Administration) approved device (PercuSurge) is commercially available and several others are under development. EZGuard is positioned to capture substantial market share in this large and growing marketplace based on its expected *superior clinical efficacy and safety*. The fact that EZGuard's solution is *faster* and *simpler* to use than existing devices should further accelerate its rapid adoption. Because our technology differs significantly from that of competing devices, we are confident that our intellectual property (IP) will enable us to create a highly competitive business in this space. Our IP will be generated under experienced counsel, who will aid us in writing and filing our patents to best exclude new competition from emerging.

Unlike competing solutions, EZGuard's technical success does not rely on delivery of tools distal to the diseased vessel, making it readily compatible with many more interventional tools and anatomies than competing devices. We will distinguish ourselves from our competition by providing a multipurpose device which can be used in combination with different stent systems and interventional tools, rather than ally ourselves with a single stent manufacturer at the outset.

Our predecessors have thankfully paved the FDA 510(k) approval pathway and are aggressively petitioning for reimbursement from medical insurers. Their data has convinced the medical field of the need for distal protection, such that it is now established as standard-of-care for saphenous vein graft (SVG) revascularization. While aware of their potential threats, we are grateful to our competitors for these critical and costly contributions.

4. The Market

The initial market consists of an approximately 250,000 SVG recanalizations per year, an established market with existing approval and reimbursement. At $2,000 per distal protection device (in 2002), this constitutes a $500M per year market. In addition, we expect carotid stenting to be approved for stroke prevention by mid-to-late 2003. The FDA approval of carotid stenting will quickly add another approximately 168,000 procedures per year (patients currently receive surgical carotid endarterectomy [CEA], which will be replaced by the less invasive stenting procedure), and will grow to include those patients who currently are not candidates for surgical CEA, but who will be candidates for carotid stenting. Therefore, we predict the number of carotid stent procedures to rise from 168,000 in its first year of approval to a 500,000 per year plateau by its fourth year. At $2,000 per device, carotid stenting represents a $320M per year market in 2004, increasing to a $1B per year market by 2007. Our aim is to get our device approved as soon as safely possible, so that even if it must be used off-label, it can be available to physicians as they are becoming more familiar with carotid stenting. Our technology has a secure place in SVG recanalization procedures and is uniquely timed to anticipate the emergence of carotid stenting—a procedure that presents an even greater opportunity on the horizon.

5. Financial Strategy

We conservatively estimate two years for technology development and FDA 510(k) approval. Two years is a standard timeline for achieving these milestones for low technical and therapeutic risk medical devices similar to ours. We have assumed a conservative 0.2 percent market penetration within the first year, ramping up to 15 percent market penetration by FY5. This financial growth will yield an operating income of $230M by FY5. We will require an initial investment of approximately $5M to prototype and provide preclinical proof of concept and an additional $10M during FY2 to obtain FDA approval and launch the product to market. Fiscal years (FY) begin upon funding in Summer 2002.

6. The Team

EZGuard's management team is an energetic, highly qualified, and experienced group of friends who work very well together. Among us are two physicians and three engineers with considerable experience in the early stage development of medical devices. We are all completing our training this year and are eager to launch our first medical device company. Our advisory board comprises a panel of highly successful medical entrepreneurs who are providing guidance on financial, legal, and regulatory issues. We look forward to their continued involvement and plan to formally add members to our team as the company's needs grow.

II. Business Plan

1. Clinical Need

Stroke is the leading cause of long-term disability in the United States, resulting in health care costs of over $43B per year. Because of the grave consequences of stroke for which there is currently no reliable treatment, intense efforts are being focused on prevention. Several conditions have been identified as treatable risk factors for stroke, including carotid stenosis, a progressive, age-related narrowing of the carotid arteries, which supply blood to the brain. Carotid stenosis is a complex vascular condition in which the carotid artery is narrowed due to the accumulation of abnormal tissue (plaque, lipid, calcium, and/or clot).

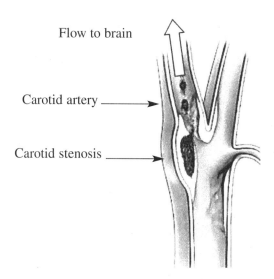

Flow to brain

Carotid artery

Carotid stenosis

FIGURE 1 Illustration of a flow restricting stenosis in the carotid artery.

Studies have shown that treating carotid stenosis significantly reduces the risk of future stroke. For years, carotid stenosis has been treated by specialist vascular surgeons who bypass or re-open existing carotid arteries in an open surgical procedure under general anesthesia.

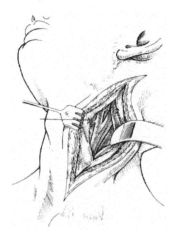

FIGURE 2 Illustration of carotid endarterectomy (CEA), an open surgical procedure used to treat carotid stenosis.

However, because of the high surgical risk of this procedure, and the fact that many patients are poor candidates for surgery due to other medical problems, a new nonsurgical approach has been developed called carotid stenting. During this procedure, the carotid artery is accessed from within (via a catheter placed in a peripheral artery), angioplastied (dilated with a balloon), and buttressed open by an implantable metal stent. Carotid stenting is showing promising results in large, multicenter clinical trials, and most specialists in this field agree that this method will be in widespread use by 2003 or 2004.[5,8] However, one of the greatest criticisms of carotid stenting has been that debris (clot and fat) is liberated during the procedure, which results in occlusion of distant vessels and iatrogenic stroke *during* this preventative procedure. Adequate "distal protection" would ensure the safety of this procedure and is something physicians desperately need.[5]

Despite the disadvantage of debris liberation, physicians are compelled by the many advantages of this nonsurgical approach and continue to support carotid stenting trials with great enthusiasm. However, because the distal protection technologies that have emerged are ridden with technical problems (described below), distal protection remains a largely unmet clinical need. Several cardiologists have indicated that they would embrace a device that provided full distal protection without compromising blood flow to the brain or threatening the integrity of the blood vessels involved.[3] Such a device would allow them to more confidently perform carotid stenting procedures and to treat more high-risk patients.

2. Market Analysis

An estimated 2.5 million Americans suffer from carotid stenosis that is severe enough to warrant prophylactic treatment for the prevention of stroke.[13] Worldwide, another 2.5 million patients may be candidates. Currently, 168,000 surgical endarterectomies are performed per year in the U.S. alone.[13] These patients would be immediate candidates for carotid stenting as soon as the technique is approved through clinical trials. At $2,000 per distal protection device, this represents a $336M per year U.S. opportunity. Additionally, many patients who need treatment of carotid stenosis are not included in this number because they are too ill to undergo surgery or do not live near a hospital staffed by a specialty vascular surgeon who performs carotid endarterectomy. These patients (more than 100,000 per year) would comprise an even larger stenting market (an additional $200M per year U.S.). Lastly, as the population ages and greater efforts are made to diagnose carotid stenosis, these numbers are expected to rise even further.

The market predictions above are supported by a number of independent sources. First, the industry has demonstrated a cost tolerance of $2,000, even for distal protection devices that only partially address customer needs (as described below). A 2000 First Union Securities analyst report predicts a $592M per year (U.S.) plus $162M per year (international) market for distal carotid protection devices.[7] A 2001 Frost and Sullivan report predicts a $275M per year U.S. beachhead market as carotid stenting replaces surgical endarterectomy.

Carotid stenting is only one of several medical procedures that warrant meticulous distal protection. Other procedures include interventions in the neurovasculature (treatment of aneurysms, arteriovenous malformations, vascular tumors) and elsewhere (treatment of saphenous vein graft [SVG] occlusions in patients after coronary bypass surgery, renal artery stenosis, and stenoses of the arteries of the arms and legs). These procedures are dramatically increasing in number as tools are emerging to access previously untreatable diseases. The need for distal protection is so pressing that once a device is FDA approved for one of these indications, physicians routinely use it off-label for the other applications, as evidenced by the PercuSurge GuardWire Protection System, the first and only FDA approved distal protection device to date. Although approved for SVG procedures, it is used in many other contexts.

Of the applications listed above, SVG recanalization is the most attractive entry point market because of its *current* established presence (175,000 cases per year [U.S.] and 75,000 cases per year [international], creating a distal protection market of nearly $500M per year) and the feasibility of rapid clinical trials in this area. SVG recanalization is a procedure that is already reimbursed and distal protection is considered a required standard-of-care. Physicians are equally frustrated with existing devices in this area as they are in carotid applications. Since SVG occlusion is a late complication of coronary artery bypass surgery, the market is a by-product of surgeries performed during the 1990s. As nonsurgical therapies develop for coronary artery disease, fewer saphenous vein grafting surgeries are projected in upcoming years. Nevertheless, the SVG occlusion market will remain predictably at the size stated above into the first decade of the twenty-first century because of the 7–10 year delay in emergence of SVG occlusion after surgery. As discussed below, our approach will be to enter the market with our SVG application, targeting the growing carotid stenting market in a most timely fashion as it emerges from clinical trials and gains FDA approval.

3. Existing Methods

Several companies have attempted solutions to the problem of distal protection. To date, no single device provides the complete, safe, and simple-to-use protection that physicians need. An introduction to the major competing devices follows:

Balloon Occlusion

PercuSurge, Inc. was the first to recognize the clinical need for distal protection and addressed it with a low-profile distal balloon system called the GuardWire. It is currently the only distal protection device approved by the FDA; although only FDA-approved for *cardiac* applications, it is being routinely used off-label in carotid and other interventions. It sells for $1,800.

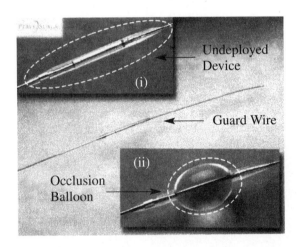

FIGURE 3 Medtronic's PercuSurge Balloon for distal carotid protection: as delivered (i) and after deployment (ii).

The procedure requires a special hollow GuardWire to be passed through the narrowing of the vessel and inflation of an occlusive balloon distal to the lesion. The balloon obstructs normal blood flow through the vessel, creating a barrier for debris, which can be suctioned away through a proximal suction catheter. Although physicians are pleased with PercuSurge's reduction in embolization rates (16% without protection, 9% with the PercuSurge system), they are nervous about using this device because distal blood flow is *completely* stopped for the duration of the procedure.[3] Especially in the brain and heart, this occlusion can be very poorly tolerated. With time, the distal tissues die, and thus, physicians feel tremendously pressured to complete the procedure quickly while using this device. Further, recent studies show that the initial passage of the GuardWire across the unprotected lesion results in liberation of significant amounts of debris.[9] The EZGuard solution overcomes this problem by providing distal protection even during insertion of the first tools across the lesion.

Filter Solutions

In an attempt to better preserve distal *flow* during protection, second generation (filter) devices are currently being evaluated in clinical trials. AngioGuard and FilterWire are large-profile umbrellalike devices, which are inserted (again past an unprotected lesion) into place, allowing blood to flow through the filters but trapping small debris, which cannot pass through the filter's pores. Originally the devices were very bulky with 100μm pores. Later versions have a decreased device profile and pore size of 80μm. These filters, however, leave much to be desired in that (1) it is often difficult to get a large filter past a narrow occlusion, (2) there is no distal protection while the filter is being inserted across the lesion, (3) the filter can fill with debris and lead to complete occlusion, (4) its metal edges can traumatize the distal vessel and cause spasm, dissection, or perforation of these critical vessels, (5) small debris (< 80 μm, which can cause significant stroke) can still pass through the filter,[10] and (6) snug apposition of the filter against the vessel wall is very hard to ensure, and thus, a channel often persists for unprotected flow of debris to the distal vessels. Although the filter companies are aggressively trying to address these concerns, many of them are inherent to the filter concept, and an optimal solution with these devices is thought to be unlikely.

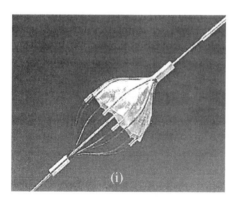
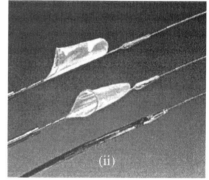

FIGURE 4 The AngioGuard (i) and Filter Wire (ii) feature utilize mechanical filtration through a porous mesh (80–100μm pores).

Flow Reversal

Most recently, ArteriA, Inc. proposed a flow-reversal method, which employs an elaborate system of balloons and catheters to reverse the flow of blood across the lesion so that any debris that is liberated during the procedure flows directly into an external catheter where it can be filtered before being returned to the body via the venous system. While clinicians initially embraced the general concept behind this idea, the technology has since met with many technical hurdles. First, only a fraction of patients are candidates for safe flow reversal, which requires an intact Circle of Willis (a highly anastomotic connection of arteries supplying the brain, which is notoriously variable in humans and is only determined at the time of the procedure by cerebral arteriogram). Second, an elaborate system of catheters must be set up, including an extra-corporeal bypass tract (Figure 5), which routes reversed blood back into the body after debris has been filtered out. This cumbersome setup is believed to limit the acceptance of this technology. Further, for the many stenotic lesions that occur in the common carotid artery, this technique would require withdrawal of a balloon past a fully deployed stent. Thus, this limits the application of the flow-reversal technique to stenoses entirely within the *internal* carotid artery and not those that extend through the carotid bifurcation into the common carotid artery.

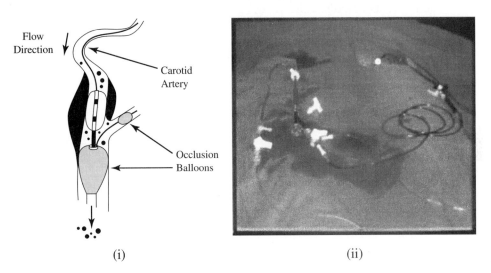

FIGURE 5 The ArteriA solution: schematic of occlusion balloons and resulting flow direction (i); and photograph the ArteriA external arterial-venous shunt in clinical use (ii).

4. Regulatory Pathway

PercuSurge, Inc. paved the regulatory pathway, acquiring 510(k) FDA approval in 2001 for use of its GuardWire product in cardiac procedures. Their approval was based on an 800 patient randomized multicenter trial comparing patients who had saphenous vein graft (SVG) revascularization procedures with and without occlusive balloon protection.[2] Because of the striking reduction in stroke risk with this device, future distal protection trials will require products to work as well as the current solution. In other words, our trial will need to show "noninferiority" versus PercuSurge. The cost to complete such a trial is estimated at $4.5M, based on a 600 patient enrollment at $7,500 per patient. The FDA has allowed approval under the 510(k) pathway for distal protection devices with predicate technology pre-dating 1976. Because our device meets these criteria, we are confident that the FDA will not require premarket approval (PMA), a process feared by most medical device companies because it requires a trial that is usually 10-fold greater in cost.

5. Competition and Existing IP

While there is a variety of intellectual property (IP) surrounding distal arterial protection, it has been carefully examined (under counsel of Dr. Makower) to ensure that our approach does not infringe on any existing solutions. As our technology involves a directionally inflating balloon and a suction system engineered for in-line filtration, special attention was paid to inventions that described these features. Patent diagrams of the most closely related inventions are shown below. These patents describe balloons designed to segmentally deliver *stents* of variable length and suction systems that rely on a *proximal occlusion balloon.* Thus, our inventions are unique in purpose, design, and method. A disclosure has been filed with Stanford's Office of Technology and Licensing (OTL), and a patent application is being prepared.

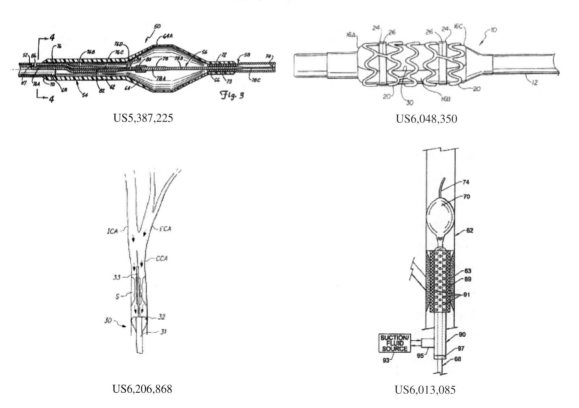

US5,387,225

US6,048,350

US6,206,868

US6,013,085

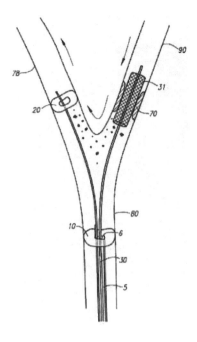

US6,146,370

6. Target Customers

Interventional cardiologists currently perform carotid stenting and many other procedures (described above), which would benefit from distal protection methods. Discussions with several trend-setting cardiologists have revealed the universal acceptance of this need and the growing frustration with existing technologies. Historically in this industry, devices that show superior clinical outcomes or procedural performance are met with rapid adoption. Product loyalties and barriers to change are much smaller than in other industries.

7. Solution Requirements

Discussions with several interventional cardiologists and a detailed analysis of the literature has rendered the following list of customer requirements for the ideal solution:

- Removes debris ($50\mu m$[10]) during high-risk periods
- Preserves distal artery flow
- Low profile (deliverable through 6-7 French system)
- Flexible, maneuverable
- Simple to use
- Cost < $2,500

Unlike all competitor devices, EZGuard uniquely addresses all of these customer needs.

8. The EZGuard Solution

EZGuard employs an innovative suction system, which collects debris delivered to it by a novel directing balloon. It allows easy-to-use intermittent suction to be turned on and off by the physician during periods of high embolic risk while preserving full distal blood flow during the majority of the procedure. The special balloon effectively creates a distal occlusion while debris is 'milked' toward the suction catheter during stenting. The suction system incorporates a physician-operated bi-directional catheter with an in-line filter, which collects debris when suction is on. When suction is reversed, the debris-containing filter lapses off-line so that a continuous stream of filtered blood can re-enter the carotid circulation. This solution meets all of the requirements listed above while addressing the problems of all previous devices. Of note, a recent report in the *Archives of Neurology* supports the idea that there are short high-risk periods during carotid stenting and suggests that protection targeted at these times would be an optimal way of balancing prevention of embolic events and assurance of adequate distal blood flow.[9]

9. Business Strategy

Industry Characteristics
As evidenced by many predicate devices in interventional cardiology, the physician customers in this market have a tolerance for high-cost devices, which are demonstrated to improve clinical outcomes. High gross profit margins are the norm, with the most dramatic example being the coronary stent (costing approximately $0.50 to manufacture and selling for $1,500.00). The culture of cardiologists is to embrace new technologies; changes are rapid in the capture and loss of market share.

Risk Assessment
Our up-front development costs are high due to expensive clinical trials. Fortunately, predicate devices have limited the requirements (of trial size, clinical endpoints, etc.) for FDA approval. Our approach will be to attain FDA approval for EZGuard on the basis of clinical "noninferiority" versus the current industry standard (PercuSurge). Noninferiority status requires a shorter, smaller clinical trial than proving "superiority." Clinician acceptance will be won on abundant preclinical data, which demonstrate that features of EZGuard promote improved clinical outcomes and procedural safety, and on expert testimonial. Expert members of our advisory board concur that the theoretical benefits of EZGuard are most compelling. They have expressed unanimous support of our technology.

As we hope to build a self-sustainable company that does not depend on acquisition, our vision is a long-term one. We accept the up-front financial risks of funding an expensive clinical trial because it is the necessary cost of an enormous opportunity. Once approved for one medical application (SVG occlusions), much smaller trials (which can be federally and academically sponsored) will drive the use of our device in many other applications (renal artery stenosis, peripheral occlusive disease, etc.). Thus, a one-time clinical trial cost will avail a multitude of ongoing opportunities.

Unique Challenges

Although many other applications are envisioned, EZGuard is currently designed for use in carotid stenting and SVG recanalization. Presently, approval (and reimbursement) for carotid stenting awaits the completion of a large randomized multi-center controlled (CREST) trial, anticipated to end in early 2003. Experts expect it to replace other (surgical) therapeutic modalities and become the standard-of-care by 2004 or 2005. Furthermore, because the procedure is much less invasive than surgery, many more patients (who are currently not treated due to ineligibility for surgery) will be treated using carotid stenting. Because of the uncertainties aforementioned, it is difficult to confidently predict the rate and extent of market development and penetration.

Addressing Technology and Adoption Challenges

We plan to face the technical and adoption challenges of our technology with a highly focused but flexible approach: (1) We will design our distal protection technology so that it can be readily applied to either SVGs or carotid arteries. (2) We will initially rely on SVG recanalization (an existing, well-established, and reimbursed market) as our entryway to FDA approval. SVG trials are smaller, shorter, and cheaper than carotid trials, and would allow our technology a faster track to market. Once on the market, physicians have historically used devices in multiple areas where a theoretical benefit is appreciated. (3) We are aiming to reach the market just as CREST trial results are paving the way for FDA approval for carotid stenting. Availability of our device at this time would allow off-label usage of our superior technology for carotid stenting even before a full-scale clinical trial using EZGuard is performed in the carotids. This strategy will also support ongoing sales in the large and growing carotid market while such a trial is being run. Superior results in the SVG market should attract the interest of the NIH [National Institutes of Health] and academic medical centers, making external funding of this trial more probable. (4) If CREST results are unable to win approval for carotid stenting, our device will still sustain a profitable company in the $500M per year SVG recanalization market. Our focus will then be steered in the direction of other applications (renal artery stenosis, peripheral artery occlusions) before significant funds have been unnecessarily lost to carotid stenting trials. (5) Lastly, we will explore the highly risky but potentially enormous role for our device in the treatment of acute myocardial infarction (heart attack). If successful, the ability to treat acute myocardial infarction would powerfully position us with both a superior technology and a unique (and extremely large market) indication, making us an invaluable partner for the large stent manufacturers who currently do not possess indications for this prized application.

Development Plan

Year 1—Engineers with expertise in vascular biology and design of catheter-based technologies will be hired to iterate and reiterate our prototype system until it meets the specific end-user requirements expressed by our expert advisors. A manufacturing engineer will ensure cost effectiveness and scalability are factored into the design process, and our legal consultant will ensure that design changes continue to respect preexisting IP. The major goals of our first year will be to complete a final prototype ready for preclinical testing and a quality IP portfolio.

Year 2—Additional engineers will be hired to support small-scale manufacturing of our devices for preclinical trials. Clinical trial design for SVG occlusions will be completed by Q1-2003 and manufacturing of sufficient devices will be completed by Q3-2003. We will enroll patients from late Q3-2003 through Q3-2004. With clinical endpoints of the trial limited to 30 days of follow-up, the trial should be complete by Q4-2004. Meanwhile, trend-setting cardiologist advocates will demonstrate the EZGuard system in live cases at high-profile meetings such as the Transcatheter Cardiovascular Therapeutics (TCT) Meeting, Arizona Heart Meeting, and American College of Cardiology meetings to increase awareness and convey endorsement of our technology. A CEO will be hired during year two to lead the company toward upcoming sales.

Year 3—Goals for this year will be to complete our clinical trial, obtain FDA and overseas regulatory approval, and scale up manufacturing efforts to meet anticipated customer demands. A sales force will be developed to support sales beginning this year; the sales force will target 10 key high-volume medical centers. Interest from potential acquirers will be entertained, especially from stent companies with existing international sales and marketing presence.

Year 4—Full-scale manufacturing will meet U.S. and international customer demands. An international sales force will be built or contracted, depending on expected adoption rates in different countries. With expensive clinical trials complete, more R and D and legal efforts will go into carving out new opportunities (outside of the carotids and SVGs) for the EZGuard system. A CFO will be hired early in the year to poise the company for acquisition/IPO.

Year 5—Full-scale manufacturing will continue to meet customer demands. The sales and marketing force will be tailored to meet unique customer needs on both demographic and clinical grounds. Having reached profitability, the company may present an attractive candidate for acquisition by a large stent manufacturer. Development of new applications for the EZGuard device will continue.

Years 6–8—Continued courtship of potential acquirers and development of new clinical applications, while full-scale manufacturing and sales are supported. Although it is difficult to predict directions this far in the future, we expect to develop a viable stand-alone company that has an effective platform for distal protection that can be customized for several different applications, each with an enormous market. FDA approval in one application (SVGs and/or carotids) will allow off-label use of the device in a number of another applications.

Throughout this process, disruptive technologies in medicine and health care will be followed, and new opportunities will be addressed in due course. We hope to foster a culture that focuses on reaching milestones while maintaining the flexibility to react to changes in the health care environment quickly and effectively.

Entry to Market

By providing a product that addresses a well-established clinical need in a far superior way than all existing approaches, we expect EZGuard to quickly capture significant market share. By involving the trend-setting cardiologists on our expert panel in all phases of our design, development, and testing process, we anticipate their continued support of our product. Their testimonials at industry-wide meetings and courses they teach to cardiologists around the country will drive acceptance of our device. Historically, this method has led to the success of many devices in this industry. An international sales force will supplement this process.

10. Financial Analysis

Our projections conservatively estimate two years for technology development and acquiring FDA 510(k) approval. Two years is a standard timeline for achieving these milestones for low technical and therapeutic risk medical devices similar to ours. We have assumed a 0.2 percent market penetration within the first year, ramping up to 15 percent market penetration by FY5. This financial growth will yield an operating income of $230M by FY5. We will require an initial investment of approximately $5M to prototype and provide preclinical proof of concept, and an additional $10M during FY2 to obtain FDA approval and launch the product to market. See Appendix A for details.

11. The Team

Asha Nayak, MD, PhD Neuroscience—Stanford Innovation Fellow. Practicing internist with PhD in neuroscience; detailed understanding of human brain anatomy, vasculature, and function; contacts in clinical medicine and neuroscience. Has observed many carotid stenting and saphenous vein graft occlusion revascularization procedures and understands their challenges.

Nick Mourlas, MS Applied Physics, PhD Electrical Engineering— Stanford Innovation Fellow; extensive research and industry-consulting experience in the design, development, and testing of biosensor technologies.

Chris Eversull, BS, MA—Stanford Medical Scholars Fellow, Stanford medical student; five years' experience developing cardiovascular technologies as R and D Engineer for Thomas Fogarty's Bacchus Vascular and Medtronic-AneuRx, resulting in five patents; expertise includes prototyping, device testing, manufacturing, and applying for FDA approval.

Kurt Grote, BS, MD in 6/02—Stanford medical student; 1996 Olympic gold medalist; part-time consultant for Piper-Jaffray on medical technologies; research experience in ophthalmology.

Frederick Winston, BS, MS in 6/02—Mechanical engineer and chemist with prototyping experience; former full-time Venture Capital Analyst at U.S. Bancorp Piper Jaffray Ventures.

12. Expert Advisors

Josh Makower, MD, MBA—CTO, Transvascular Inc. Extensive large corporate (Pfizer, Inc.) and start-up company (ExploraMed, TVI) operating experience in cardiovascular medical devices.

Paul Yock, MD—Professor of Cardiovascular Medicine and Director of the Biomedical Technology Innovation Program, Stanford University. Serial medical entrepreneur; inventor of the rapid exchange (RX) catheter system in widespread use today.

Fred St. Goar, MD—Interventional Cardiologist at El Camino Hospital, investigator in carotid stenting trials, founding member of medical device companies including HeartPort and E-Valve.

James Joye, DO—Interventional Cardiologist, Director of Peripheral Vascular Interventions at the Cardiovascular Institute of El Camino Hospital, investigator in carotid stenting trials, instructor who runs carotid training courses for cardiologists, founder of CryoVascular Technologies.

Gregory Robertson, MD—Interventional Cardiologist, Sequoia Hospital. Consultant to Lumend, Inc. and Fox Hollow Technologies. Participant in numerous preclinical and clinical trials for development of cardiac devices.

Thomas Hinohara, MD—Interventional Cardiologist, Sequoia Hospital. Consultant to Lumend, Inc. and Fox Hollow Technologies. Participant in numerous preclinical and clinical trials for development of cardiac devices.

Howard Holstein, JD—Expert in FDA regulatory affairs with more than 20 years' experience advising medical device companies.

III. Selected References

1. Ackerstaff, R., et al., "Association of Intraoperative Transcranial Doppler Monitoring Variables with Stroke from Carotid Endarterectomy," *Stroke* (August 2000):1817–23.
2. "510(k) Summary" Application for GuardWire System Approval in Saphenous Vein Graft Occlusions, PercuSurge, Inc., 5/25/01.
3. Discussions with Thomas Hinohara, MD, and Gregory Robertson, MD, Interventional Cardiologists, Sequoia Hospital, Redwood City, California; James Joye, DO and Fred St. Goar, MD, Interventional Cardiologists, The Cardiovascular Institute, El Camino Hospital, Los Altos, California.

4. Discussions with Greg Albers, MD, and David Tong, MD, Neurologists, Stroke Center, Stanford University Medical Center.

5. Fasseas, P., et al., "Distal Protection Devices During Percutaneous Coronary and Carotid Interventions," *Current Controlled Trials in Cardiovascular Medicine* 2(6):286–91.

6. "Guidance for Industry and FDA: Guidance for Neurological Embolization Devices," U.S. Department of Health and Human Services, Food and Drug Administration, 8/13/99.

7. Jay, A., and Robins, J., "Interventional Neuroradiology, Stroke, and Carotid Artery Stenting: The Opportunity for Medical Devices in the Brain Comes of Age," First Union Securities Analyst Reports, 8/10/00.

8. Ohki, T., and Veith, F., "Carotid Artery Stenting: Utility of Cerebral Protection Devices," *Journal of Invasive Cardiology* 13(1):47–55.

9. Orlandi, G., et al., "Characteristics of Cerebral Microembolism during Carotid Stenting and Angioplasty Alone," *Archives of Neurology* 58(9):1410–13.

10. Rapp, J., et al., "Atheroemboli to the Brain: Size Threshold for Causing Acute Neuronal Cell Death," *Journal of Vascular Surgery* 32(1): 68–76.

11. Tannenbaum, Larry, "Corporate Financial Planning—Business Plan Construction" lecture, CFO and Senior Vice President, Metrika, Inc., 4/4/01.

12. Tabet, S., et al., "Screening for Asymptomatic Carotid Artery Stenosis," *Guide to Clinical Preventive Services, Cardiovascular Disease,* 2d ed., U.S. Preventive Services Task Force, 1996.

13. Wein, T., and Bornstein, N., "Stroke Prevention: Cardiac and Carotid-Related Stroke," *Neurologic Clinics* 18(2).

Appendix A—Assumptions

We predict capturing 0.2 percent of the SVG and carotid distal protection markets by FY3, ramping exponentially up to 15 percent by FY5. To do this, we will target key clinical centers both in the U.S. and internationally. These centers will have at least two physicians who perform on average 1.4 procedures per week. We expect to capture 50 percent of the sales at our target centers based on our superior technology. We expect to have one salesperson per center in FY3 and each salesperson covering five centers by FY5. This means a growth in sales force to about 200 salespeople covering 980 centers by FY5. (There are 2,100 interventional labs and 19,500 interventional cardiologists in the U.S. alone). We predict that our SVG clinical trial will require 600 patients enrolling over an 18 month period at $7,500 per patient. These numbers are conservative and reflect prior clinical trial experience in this area. Salaries and benefits are estimated at $250,000/yr (engineers), $200,000/yr (sales staff), $300,000/yr (part-time legal counsel), $600,000/yr (CEO), $400,000/yr (VP), and $70,000/yr (admin).

Appendix A—Financial Projections

Price	$ 2,000	
COGS	$ 200	
Patients	US	618,000
	International	300,000
	TOTAL:	918,000

		FY1	FY2	FY3	FY4	FY5
Revenue	United States	$ –	$ –	$ 942,484	$13,218,792	$185,400,000
	International	$ –	$ –	$ 457,516	$ 6,416,889	$ 90,000,000
	TOTAL:	$ –	$ –	$ 1,400,000	$19,635,682	$275,400,000
Units Sold	United States	0	0	471	6,609	92,700
	International	0	0	229	3,208	45,000
	TOTAL:	0	0	700	9,818	137,700
COGS		$ –	$ –	$ 140,000	$ 1,963,568	$ 27,540,000
Op. Expense	Sales & Marketing	$ –	$ 200,000	$ 2,000,000	$ 3,147,332	$ 39,342,857
	General & Admin	$ 470,000	$ 470,000	$ 470,000	$ 1,540,000	$ 2,680,000
	Research & Devt	$ 1,250,000	$ 2,500,000	$ 2,500,000	$ 2,500,000	$ 2,500,000
	Regul/Clinical	$ –	$ 1,500,000	$ 3,000,000	$ –	$ –
	Legal	$ 300,000	$ 300,000	$ 300,000	$ 300,000	$ 300,000
	TOTAL:	$ 1,720,000	$ 4,670,000	$ 7,970,000	$ 7,187,332	$ 44,522,857
Operating Income		$(1,720,000)	$(4,670,000)	$(6,570,000)	$12,448,350	$230,877,143
Capital Expenses		$ (500,000)	$(1,000,000)			
Taxes		$ –	$ –	$ –	$ 4,979,340	$ 92,350,857
Cash Flow		$(2,220,000)	$(5,670,000)	$(6,570,000)	$ 7,469,010	$138,526,286

I-MOS SEMICONDUCTORS

Kailash Gopalakrishnan
Adam Wegel
Rajit Marwah
Tod Sacerdoti

Autumn 2002

I. Executive Summary

Introduction

I-MOS Inc. (I-MOS), a semiconductor intellectual property start-up, has developed a disruptive transistor technology that offers a 1000x reduction in static power dissipation and a 30 percent increase in chip performance. This technology solves a problem that has plagued the semiconductor industry for the past 30 years: how to increase the density of transistors on a chip without exponentially increasing the heat generated. The I-MOS solution not only dramatically reduces static power consumption and increases chip performance, but it does so without increasing semiconductor fabrication costs. In fact, this solution is fully compatible with all the existing tools and processes currently used in semiconductor fabrication.

I-MOS will commercialize this technology by licensing it to integrated device manufacturers (e.g., Intel, IBM, Motorola), fabless semiconductor companies (e.g., Xilinx, Qualcomm, Nvidia), and semiconductor foundries (e.g., TSMC, UMC, Chartered). I-MOS has already initiated discussions with Intel, Xilinx, and TSMC, and all three have expressed an interest in licensing this technology.

Market Opportunity and Solution

The semiconductor industry is expected to grow from $140 billion in 2002 to $240 billion by 2005.[1] The industry is currently plagued with the problem of how to increase the density of transistors on a chip without exponentially increasing the heat generated. This exponential increase in the "Static Leakage Problem" occurs because at the same time transistor threshold voltages have been scaled down substantially, escalating the leakage per transistor, the number of transistors per chip has dramatically increased. A significant opportunity exists for a high-performance, cost-effective, and scalable solution at the transistor level that addresses the Static Leakage Problem.

I-MOS has developed a disruptive transistor technology that reduces the static power dissipation by 1000x and provides a 30 percent increase in chip performance without increasing semiconductor manufacturing costs. The I-MOS solution is cost-effective, by working in conjunction with standard CMOS

---

[1] Goldman Sachs, *Technology: Semiconductors Industry Primer 2002.*

processes, and increases manufacturing efficiencies, due to lower device variations. Two-thirds of the semiconductor industry, $93 billion in 2002 or $160 billion by 2005, can benefit from the I-MOS technology. However, I-MOS will initially target the segments of the industry that are particularly power and performance sensitive, including wireless devices, consumer electronics, graphics, and networking products. These segments currently represent approximately $50 billion today or $86 billion by 2005.

Business Model

The semiconductor intellectual property (IP) industry was an $890 million industry in 2002 and is growing at 25 percent annually.[2] I-MOS will contribute to this growth rate by utilizing a fabless nonexclusive IP-licensing business model in which we outsource all manufacturing and sales of the semiconductors incorporating our technology. Along with our IP, we also provide manufacturers with design libraries, SPICE models, and layout modifier software tools that seamlessly port existing chip designs to incorporate the new I-MOS technology.

We will work with our foundry partners to get our technology up and running in mainstream processes, and optimized for performance and static power, for every chip generation. We license our technology, on a nonexclusive basis, to integrated device manufacturers, fabless semiconductor companies and semiconductor foundries. We derive multiple revenue streams from each licensing arrangement: the manufacturers pay I-MOS an upfront license fee and a per chip royalty fee, and the foundries pay a per wafer fee.

Technology

The I-MOS technology was invented and patented by I-MOS Chief Technology Officer Kailash Gopalakrishnan under the guidance of Dr. James Plummer, Dean of the Stanford School of Engineering. I-MOS will have an exclusive license on the technology from the Stanford Office of Technology Licensing. Our technology has two applications: (1) it reduces static power dissipation by 1000X, has comparable dynamic power and has up to 30 percent increase in performance, or (2) it can reduce static power dissipation by 1000X, have comparable performance and reduce dynamic power by 20 percent over existing CMOS technology. These technological advancements are achieved by using breakdown voltage modulation, which reduces the subthreshold slope, resulting in much higher ON currents for performance and much lower OFF currents for lower static power than CMOS.

Current Status and Milestones

We have extensive device modeling and simulation results, and have demonstrated proof of concept with a successful initial run in silicon. Silicon-based prototypes were fabricated in the Stanford Nanofabrication Facility and have validated the I-MOS design.

---

[2] UBS Warburg, Global Equity Research, *At the Core of IT,* May 2002.

■ **Milestone One:** We are planning to build silicon, germanium, and strained silicon, submicron devices to validate the scaling behavior of the I-MOS device shown in our simulations. We expect to have these experiments completed by June of 2003, coinciding with our first round of funding.

■ **Milestone Two:** Our funding will allow us to demonstrate the I-MOS design in simple circuits, understand trade-offs, determine device characterization, study scaling properties, and sign up our first fab customer. We anticipate completing this milestone and raising our second round of funding by September of 2004.

■ **Milestone Three:** We intend on demonstrating the I-MOS design on large SRAM test vehicle circuits, showing reliability, optimizing for performance and power, and signing up our first beta customers.

Financial Projections (millions)

	Yr 1	Yr 2	Yr 3	Yr 4	Yr 5	Yr 6	Yr 7
Revenue	$0	$0	$8	$19	$37	$73	$142
Net Income (AT)	($5)	($9)	($6)	($1)	$14	$18	$37
After Tax Margin	NA	NA	–68%	–6%	23%	25%	26%
Revenue Growth	NA	NA	NA	125%	98%	96%	95%

II. Market Opportunity

Summary:

■ The semiconductor industry is enormous and growing.

■ The semiconductor IP industry is experiencing rapid growth.

■ The Static Leakage Problem has become a major concern for semiconductor companies.

The semiconductor industry is enormous and growing

The worldwide market for semiconductor devices is large and growing, driven by the demand for electronic systems that are dependent on semiconductors. According to Gartner Research, the 2002 worldwide semiconductor market is $140 billion and is expected to grow to $240 billion by 2005.

Historically, the demand for semiconductor devices was met by vertically integrated semiconductor manufacturers, who designed, manufactured and tested their own products, in their own facilities, with their own tools. In the 1990s, the process for design and manufacturing grew in complexity and the cost of developing manufacturing facilities reached unmanageable levels, driving a disaggregation of the semiconductor industry. This disaggregation created a wave of growth of new semiconductor companies including fabless semiconductor chip designers, semiconductor equipment and tools vendors, and third-party semi-conductor manufacturers, or foundries.

The semiconductor IP industry is experiencing rapid growth

In recent years, semiconductor design has increased further in complexity and geometries have rapidly decreased to meet market demand. Access to cutting-edge technology has become a competitive weapon for semiconductor manufacturers who are now forced to compete in an increasingly dynamic market. In the last year alone, the top 10 semiconductor companies spent over $12B on internal research and development. As semiconductor research has become more specialized, companies have begun to look to the outside for the latest technologies as a predictable way to generate R and D. These industry trends have accelerated the disaggregation of the semiconductor industry, fueling a new wave of growth in the market for semiconductor IP licensing companies. In 2001, semiconductor companies spent nearly $1 billion on IP licensing, and this is expected to grow at over 40 percent.[3] The total available market for IP licensing includes memory, microprocessors, microcontrollers, microperipherals, DSPs, ASICs and custom chips, representing two-thirds of the semiconductor industry, or $93 billion in 2002 and $160 billion by 2005.

The Static Leakage Problem has become a major concern for semiconductor companies

For the past 30 years, semiconductor companies have been plagued with the problem of how to increase the density of transistors on a chip without exponentially increasing the heat generated. In the past, this "Static Leakage Problem" has been addressed by creating distance between the individual transistors in order to remove the concentration of heat or by finding ways to rapidly cool the chip when the heat is generated. In recent years, however, new geometries of semiconductors have been reduced at an increasing rate and the transistor threshold voltages, the voltage between the on and off positions at which the transistor is just considered on, have been scaled down substantially. As a result, the Static Leakage Problem per transistor has increased exponentially. In addition, the number of transistors per chip has dramatically increased. These factors are contributing to an exponentially increasing Static Leakage Problem.

> "The [static leakage problem] is so bad that if current trends continue, future chips could reach 2,000 watts/cm^2, equivalent to a nuclear reactor . . . designers will have to trade off some performance to reduce power."—Shekhar Borkar, director of the Circuit Research Lab at Intel

The industry has long been aware of the Static Leakage Problem, but it has been perceived to be the result of a fundamental principle of thermodynamics and,

---

[3] UBS Warburg, Global Equity Research, *At the Core of IT,* May 2002.

therefore, an impossible problem to solve. Semiconductor companies have addressed the problem by employing a variety of circuit workarounds, with varying degrees of success. One workaround used by circuit designers is to switch the power supply off from the circuits so that no static power may be dissipated when computing is done. This method is fundamentally inefficient due to the difficulty in determining which circuits are computing and which are not computing in a semiconductor. In addition, in a lot of high-density memory circuits such as SRAM and DRAM, it is not physically possible to switch the circuits off because these memories lose information when the power supply is switched off. In addition, the design complexity and cost of development increase significantly.

A more modern workaround proposed for future generation of chips is to design transistors with different threshold voltages on the same chip. For critical paths on the circuit, the high-speed (low threshold voltage) transistor may be used, but for the rest of the circuit the low speed (high threshold voltage) transistors would be used. Although this appears to be a convenient solution, it still results in a substantial reduction in performance. In addition, this solution introduces design complexity because circuit and layout designers have to design their circuits keeping two or more different transistors in mind and placing high-performance transistors judiciously in critical paths only. Every new transistor with a new threshold voltage necessitates additional process steps and new masks, which can increase the cost of making chips significantly. The design tools needed to accomplish these steps simply do not exist at this point.

> Intel Corp. plans to incorporate "sleep transistors" onto future-generation microprocessors to push clock frequencies higher and help tame the worsening leakage current that threatens high-speed processor designs.—Article in the *EE Times* on June 14, 2002

Several major market trends will exacerbate the Static Leakage Problem in the future. Static leakage per transistor has increased from 1 pA/μm in the 1990s to tens of nA/μm today and is projected to increase to many μA/μm by 2010 for high performance transistors. In addition, the number of transistors per chip is projected to increase well beyond the billion-transistor mark within the next few years. Coinciding with the technology market trends is the rapidly increasing demand for low power and high performance wireless devices with long battery lives, multimedia applications functionality, and many additional capabilities. Businesses and consumers are consistently demanding performance increases in laptops, PDAs, digital cellular phones, digital cameras, MP3 players, and other wireless products. Semiconductor companies are limited in their ability to deliver these performance improvements unless the Static Leakage Problem is addressed. I-MOS believes that the current market trends have created a significant opportunity for a high-performance, cost-effective, and scalable solution at the transistor level that addresses the exponentially increasing Static Leakage Problem.

"We're scaling our gate thickness on our devices to less than 10 angstroms for future generations. The leakage to that is just unbelievably high. So we're looking furiously for new materials throughout the industry."
—Jeffrey Welser, IBM project manager for advanced CMOS

III. Solution

> **Summary:**
> - Dramatically reduced static power dissipation
> - Increased chip performance
> - Cost-effectiveness
> - Increased manufacturing efficiencies

I-MOS has developed a disruptive transistor technology that reduces the static power dissipation by 1000x and provides a 30 percent increase in chip performance without increasing semiconductor manufacturing costs. I-MOS is focusing the application of its revolutionary technology on the Static Leakage Problem and, specifically, in chips manufactured in strained silicon and germanium, where the technology is most effective.

Since the transistor is the fundamental building block of a semiconductor, the I-MOS technology has a wide range of applications. However, the semiconductor markets that drive wireless devices such as laptops, PDAs, and cell phones, are most affected by the Static Leakage Problem. Imagine laptop batteries that lasted through multiple plane flights or PDAs that could play movies—I-MOS technology will push semiconductor driven products to new levels of performance. The key benefits of the I-MOS solution are:

Dramatically reduced static power dissipation

I-MOS technology reduces the overall static power dissipation of the chip by solving the problem at the fundamental level—by reducing the static leakage current per transistor. The I-MOS technology reduces the static leakage of the transistor 1000x less than standard CMOS transistors.

Increased chip performance

The I-MOS technology enables customers to achieve up to a 30 percent increase in performance over CMOS semiconductors with comparable levels of dynamic power. Alternatively, if dynamic power is the main concern of our customer, the I-MOS technology can deliver a 20 percent reduction in dynamic power over CMOS with a comparable level of performance.

Cost-effectiveness

Application of the I-MOS technology does not significantly impact manufacturing costs. The I-MOS technology is designed to work in conjunction with standard CMOS processes and the only necessary manufacturing modification is to selectively replace standard transistors from the design with the I-MOS transistors.

Increased manufacturing efficiencies

Typically, design changes that are necessary for higher performance result in a reduction in semiconductor yields. Due to our unique technology, our customers can achieve both higher performance and a higher yield, thereby generating a lower cost of goods sold. This manufacturing efficiency stems from the fact that there are certain variations from wafer to wafer, die to die and within the die itself. These variations cause certain transistors to be leakage prone and others to be slower. I-MOS devices have much lower variations as compared to standard CMOS devices, and this guarantees significantly lower development time, a shorter learning curve and higher yields.

IV. Business Model and Strategy

Summary:
- Complete technology development
- Develop strategic relationships with foundries
- Enable products in large and rapidly growing wireless markets
- Generate revenue through a combination of upfront license fees and royalties
- Build out our patent portfolio

Our goal is to establish the I-MOS transistor technology as the standard for high-performance semiconductors in the market. The broad application of this technology has led us to pursue a capital-efficient, IP-licensing business model in which we neither manufacture nor sell semiconductors incorporating our technology. We license the I-MOS technology, on a nonexclusive basis, to integrated semiconductor manufacturers, foundries and fabless semiconductor companies. This business model is scalable, has low fixed and variable costs, has high switching costs for our customers and is synergistic as we grow, in that it allows us to make interproduct sales and upgrade customers on relative products.

The disruptive transistor technology that I-MOS has developed can add value to approximately two-thirds of manufactured semiconductors. We will initially target the segments of the industry that are particularly power and performance sensitive, including: wireless devices, consumer electronics, graphics, and networking products. These segments represent approximately $50 billion in 2002 or $86 billion by 2005.

The I-MOS business model and strategy are designed to establish the I-MOS transistor as the standard transistor for high-performance semiconductors in the market and is based on achieving the following milestones.

Complete technology development

I-MOS has developed a revolutionary transistor technology that solves the Static Leakage Problem in the lab. We have extensive device modeling and simulation results that show this technology works and have done an initial run in silicon to show the proof of concept. In order to demonstrate this technology and get it up and running in the marketplace, we must first study scaling properties, demonstrate circuits, understand trade-offs and device characterization, optimize for speed and get our LMII layout tool completed.

Develop strategic relationships with foundries

I-MOS must develop strong relationships with foundries today in order to incorporate our technology into their mainstream processing when technology development is complete. Foundries benefit from this relationship because they will be able to provide the most current transistor technology to their customers and therefore increase margins. Our value proposition to the foundries is powerful due to the minimal costs associated with bringing our technology on board. The strength of our relationship with the foundry will influence our ability to derive royalty fees from their customers as well, a common practice in the industry. An important element of our strategy may be developing strong relationships with the fabless semiconductor companies in order to create pull-through demand to influence the adoption of the I-MOS technology.

Enable products in large and rapidly growing wireless markets

I-MOS is initially targeting the largest and most rapidly growing segments of the wireless device market for our transistor technology. We believe these segments are the most likely to adopt this new technology because we enable performance advantages that can help them compete in the marketplace. These large and high-growth wireless markets include cell phones, PDAs, portable MP3 devices, digital cameras and handheld video games. Target customers in these markets include Broadcom, Texas Instruments, Qualcomm, SST Corp., Cirrus Logic, and others.

Generate revenue through a combination of upfront license fees and royalties

We intend on generating revenue from three types of companies—foundries, semiconductor manufacturers, and fabless semiconductor companies. Initially, we will work with foundries to bring our technology into their mainstream processes as we prove our technology and demonstrate scalability. After completing development in the foundries, we will license our technology to both fabless and integrated semiconductor companies who will generally pay a multi-million dollar upfront license fee to I-MOS for the broad use of our technology in their semiconductors. These fees may hinge on specific milestones for deliverables from I-MOS or the production of chips by the company. Initially, most engineering costs associated with bringing a customer on board will be consumed by I-MOS, but as we develop our customer base this will become a revenue generator for the Company. We hope to use this up-front fee structure as a capital efficient way to fund future development.

Both foundries and licensees will be responsible for paying I-MOS royalties on sales occurring throughout the life of the I-MOS technology being licensed. Foundries will pay I-MOS a per wafer fee for all wafers used in production. Fabless licensees will pay I-MOS a per chip fee, usually corresponding to a percentage of the cost of the chip. Integrated manufacturers will pay a per chip fee as a percentage of cost as well, but the fee structure will be slightly higher. Royalty rates will typically range from 2.5 percent–5 percent of the average selling price (ASP) of the chip; however, a wide variety of factors will impact the royalty rate including size of the up-front fee, number of chips in production, and many others.

Build out our patent portfolio

A key element of the I-MOS strategy is the development and protection of our intellectual property. We will aggressively pursue both offensive and defensive patents surrounding the core I-MOS transistor patent and will maintain a focus building out our patent portfolio.

Case Study #1: Xilinx Inc.

Xilinx Inc., founded in 1984 and headquartered in San Jose, California, is the leading supplier of complete programmable logic solutions, including advanced integrated circuits, software design tools, predefined system functions delivered as cores, and unparalleled field engineering support. Xilinx is the world-leader in FPGA-based products and its PLD solutions have more than 50 percent market share. Xilinx currently uses the most advanced 130 nm UMC process and it has 250M transistors in its high-end product 2VP125 (about 5X greater than Pentium IV). In an FPGA, a good fraction of transistors are unused in the design process but all the transistors contribute to static leakage. We talked to an advanced circuit researcher from Xilinx, who told us "Static power dissipation is our biggest problem scaling forward because our customers always want higher performances. We would be very interested in looking at new technologies that can help us alleviate this problem. Reducing static power would also help us to expand our market by replacing ASIC's in cellphones."

Case Study #2: Intel Inc.

Intel Inc. is the world leader in semiconductor manufacturing technology and supplies the computing and communications industries with chips, boards, systems, and software building blocks that are the "ingredients" of

(continued on next page)

(continued from page 501)

computers, servers and networking and communications products. Intel's products are always at the cutting edge of technology: its highest-end Pentium IV processors contain more than 50M transistors, and Intel has devoted a significant fraction of its R and D budget in device and circuit related solutions to the problems of static power dissipation. The director of the Circuit Research Lab at Intel noted, "The static leakage problem is so bad that if current trends continue, future chips could reach 2,000 watts/cm^2, equivalent to a nuclear reactor." We talked to an advanced process-integration group manager at Intel who informed us that "Undoubtedly, the static leakage problem seems to be one of the biggest roadblocks for deep submicron transistor scaling. We are constantly looking for new solutions that can push us to a better position on the static power versus performance curve."

V. Sales and Marketing Plan

Summary:
- Use direct technical sales force to target fabs and end users
- Offer support as well as technology

Our sales and marketing activities will be primarily focused on establishing and maintaining licensing arrangements with foundries, semiconductor manufacturers, and fabless semiconductor companies. Our sales strategy is to pursue targeted customers through the combination of our direct sales force and our strategic alliances. Since we generally license our technology to foundries that, in turn, sell products incorporating these technologies to fabless semiconductor companies, the foundry alliances we form will be an integral part of our sales strategy. In establishing these alliances, we seek partners who we believe will allow us to grow the overall market for our I-MOS transistor technology.

Use direct technical sales force to target fabs and end users

Our marketing activities will also be directed toward the end product manufacturers. Through targeted advertising and co-marketing efforts, we will be focused on increasing the awareness of the I-MOS technology and generating interest from potential end customers. We believe these efforts will create demand for our product from the product manufacturers, which will create pull-through demand to the semiconductor manufacturers who will end up licensing our technology.

Offer support as well as technology

Our goal will be to establish ourselves as an important partner to our customers and develop relationships both at the executive level and at the engineering level.

We anticipate dedicating substantial resources toward marketing to customers, implementing our technologies and supporting our customers. This will include on-site engineering support, other technical support, trade shows, and more traditional marketing activities. We believe a close working relationship with our customers will allow us to identify new product areas and technologies to focus on for our future research and development efforts.

VI. Technology Overview

Summary:
- CMOS technology is thermodynamically limited to 60 mV/decade.
- I-MOS transistor physics achieves 5 mV/decade.
- No other solution offers comparable performance.
- Technology roadmap developed for future products and services.

CMOS technology is thermodynamically limited to 60 mV/decade

The scaling of transistor feature sizes has demanded that chip supply voltages be continually shrunk in order to keep the dynamic power dissipation within tolerable limits for reliable operation. Reducing this supply voltage, while maintaining transistor performance, requires the threshold voltage of the transistors to be reduced in accordance with the supply voltage. One good metric of how much a transistor leaks in the "off" state is the subthreshold slope which describes the rate of change of the current with respect to the voltage below the threshold. In conventional transistors, this subthreshold slope is limited by basic thermodynamic principles to 60 mV/decade, which means that the transistor current changes by one order of magnitude for every 60 mV change in the voltage. Therefore, reducing the threshold voltage increases the leakage current of the transistor exponentially. In addition, chip temperatures have steadily increased in the past decade, thereby dramatically increasing the leakage current as well. The number of transistors per chip has doubled every two years for the past 30 years, and all these factors contribute to a dramatic rise in the static power dissipation in chips. These levels of static power dissipation are unacceptable in most applications and may impose fundamental limitations on transistor scaling itself.

I-MOS transistor physics achieves 5 mV/decade

I-MOS has developed a disruptive transistor technology (Impact Ionization MOS, or I-MOS) that reduces static power dissipation by 1000X over conventional transistors and delivers up to 30 percent enhancement in performance. This dramatic reduction in static power dissipation stems from a *fundamental breakthrough in transistor physics*. Conventional CMOS transistors work by modulating the charge in the channel by using a process called drift-diffusion controlled by a gate terminal. The subthreshold slope in CMOS-based systems is thus thermodynamically limited to kT/q or 60 mV/decade at room temperature. The I-MOS works by avalanche breakdown voltage modulation in specifically

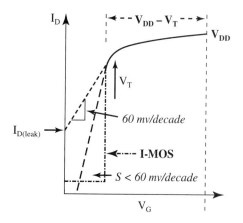

FIGURE 1 Comparison of the subthreshold slope of conventional transistors and the I-MOS and its implications on leakage.

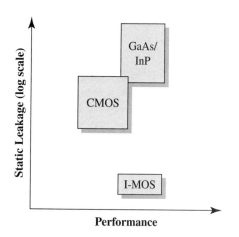

FIGURE 2 Comparison between the static power and performance trade-offs for CMOS, GaAs, and I-MOS transistors over all process corners.

designed p-n junction diodes. The avalanche breakdown process is an extremely abrupt and fast process. This built-in gain mechanism amplifies the nonlinearity of the system resulting in a subthreshold slope much lower than kT/q (~ 5 mV/decade or lower) (Fig. 1). We believe that I-MOS is the only technology that offers this huge reduction in static power and is the only semiconductor device in the world that has reduced the subthreshold slope below the thermodynamic limit of kT/q.

The 30 percent enhancement in performance is obtained by a combination of a number of factors. As mentioned earlier, the impact ionization process is an extremely steep and fast process and enhances the drive current in a transistor by about 20 percent over CMOS. In addition, the avalanche process causes a reduction in the swing at the output of the transistors by about 20 percent, which increases transistor speed. This enhancement in transistor performance is significant considering that it takes the industry 2–3 years to scale to a new transistor technology to enhance performance by 30 percent.

No other known future technology offers comparable performance

GaAs and InP based transistors provide high performance but also suffer from high static leakage. I-MOS is the only solution that can provide both higher levels of performance and dramatically lower static power. In addition, GaAs and InP are new materials and have more complex and expensive fabrication processes, which would likely require an expensive redesign of all current fabrication facilities. In contrast, the basic I-MOS concept works in any material. Our modeling work has shown that the best material for the I-MOS transistor would be strained silicon or germanium. IBM, Intel, AMD, TSMC, UMC, and the rest of the semiconductor industry, have announced their intentions to introduce strained silicon in mass production before 2005.

The static power dissipation problem is a lot worse especially if one considers all the process corners under which one would like the chip to operate. The worst-case process corner typically operates with 100 times more static power dissipation than the typical case. Since these levels of static power are unacceptable, this would end up directly affecting the yield. In addition to reducing this static power, the I-MOS also has about three times lower variability as compared to CMOS. This can improve the yield significantly for chips incorporating the I-MOS technology. A comparison is illustrated in Fig. 2.

Technology roadmap developed for future products and services

Our core competency is the transistor technology related IP and the design tools that we offer to incorporate this I-MOS technology in any chip design. Typically, any transistor technology has three phases in its lifetime. The initial phase lasts for a couple of years and requires substantial research and development on the technology, process optimization and reliability testing of the transistor. After this phase, the technology is used in high performance chips and other designs that really need the extra computing power and reduced static power. In the final phase, which lasts for about three years, the technology is used in lower-end chips that are designed for reduced cost. We estimate that the I-MOS transistors that we introduce at every technology node would go through similar phases.

After doing extensive research and development from 2003–2005 on the 0.080 μm technology node, we plan to introduce the I-MOS design into production around 2005. We also plan to introduce advanced generations of I-MOS transistors in 2007, 2010, 2013 and 2016 for the 0.065 μm, 0.045 μm, 0.032 μm, and 0.022 μm technology nodes corresponding to the roadmap outlined by the International Technology Roadmap for Semiconductors (ITRS).

In addition, we envision expanding to have direct products in the future because the fundamental technology has wide ranging implications in a number of diverse fields. After having proven the technology in a number of diverse markets, we plan to introduce SRAM- and DRAM-based memory products along the way in parallel to pursuing licensing to other semiconductor markets. Furthermore, the I-MOS technology also has significant applications in the power semiconductor industry including power amplifiers, synchronous rectifiers, ESD protection, and we plan to introduce products that satisfy the needs of those markets as well.

One more significant breakthrough in computing stems from the optoelectronic based applications of the I-MOS transistor. I-MOS provides a good device template in photodetector applications and can provide a gain-bandwidth product that is about a 1,000 times higher than conventional CMOS/GaAs based detectors. This can solve a very critical problem associated with clock distribution on chip. Optics, which has been proposed as an alternative to conventional techniques of electrical clock distribution in any ASIC or microprocessor chip, are becoming increasingly difficult because of increasing chip areas, rising clock frequency, and increased numbers of latches. The proprietary optoelectronic I-MOS technology is the only solution that can provide optoelectronic detectors for optically clocking integrated circuits. We plan to license our proprietary optical I-MOS solutions for on-chip clocking and chip-to-chip signalling to all partners, fabs and customers who need this technology for increased scalability in their high frequency designs. Fig. 3 summarizes all the products that I-MOS plans to offer and their rough time frames. We believe that this rich variety of products in both the IP licensing and product space will strongly position the company to scale upwards in accordance with Moore's law well into the twenty-first century.

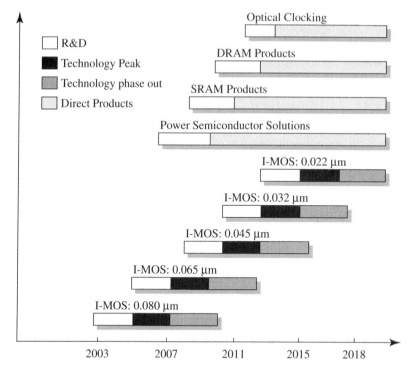

FIGURE 3 Future products based on the proprietary I-MOS technology.

VII. Technology Implementation

Summary:
- Compatible with existing fab processes
- Partner with fabs to implement technology
- Provide tools and support in addition to technology

Compatible with existing fab processes

I-MOS is fully compatible with all of the current manufacturing tools and processes used for making strained silicon transistors. The footprint (or the area) occupied by the I-MOS transistor is exactly the same as the area occupied by conventional CMOS transistors for the same generation. In other words, there is no extra cost to be incurred in implementing the I-MOS technology. Most of the differences in the I-MOS device and structure are absorbed in the lithography masks and in the implant and anneal steps that are used in processing MOS transistors. We expect that the worst-case scenario would be the addition of a couple more masks for extra implant / spacer steps in the process of optimizing the device. One more distinct advantage of the I-MOS transistor is that its higher speed over CMOS permits us to go backward one technology generation for the same level of performance and much reduced static power. This permits us to cut down on the ever-increasing costs of manufacturing transistors and makes I-MOS chips much cheaper to fabricate than current CMOS chips for lower power and comparable or higher levels of performance and integration.

In low power chips for portable applications, in addition to reducing static power, the I-MOS could also be tweaked so that higher performance may be traded off for reduced dynamic or active power. We envision that we may be able to reduce the dynamic power by 20 percent or more for comparable levels of performance over CMOS transistors.

Circuit design using the I-MOS technology is only marginally different from conventional circuit design using CMOS transistors, and the slight difference stems from the fact that the output voltage range of a logic block containing the I-MOS transistor does not span the entire supply voltage range. This reduced swing at the output helps enhance performance and reduce dynamic power dissipation but comes at the expense of reduced noise margin, which is used to differentiate logic levels "1" and "0" from each other. We believe that this penalty can be easily accomodated because of lower spread in the I-MOS transistor thresholds. I-MOS also holds intellectual property rights for the circuit design methodology using the I-MOS transistor.

Partner with fabs to implement technology

The implementation strategy would be to do joint development with the foundries and have the technology in the base-line process of the fabs. I-MOS will work with the foundries to understand the scalability of the transistor and develop simple circuit modules incorporating the I-MOS transistor as part of their first year milestones. We will then develop high-density and large-array SRAM test vehicles to understand the performance and static-power trade-off in these devices. We will also be optimizing the device specifications to maximize performance and minimize standby power. As stated earlier, the I-MOS technology is fundamentally compatible with all the existing tools and processes in IC fabrication. The I-MOS fabrication would use the conventional mask sets and would lithographically modify the ion-implantion and spacer steps that the I-MOS device would recieve. This is evidenced by the fact that conventional transistors are n-p-n while the I-MOS is a specially designed p-i-n. This is necessary for implementing the breakthrough physics that fundamentally alters the device characteristics. The conventional transistors, also fabricated simultaneously on the chip, would be processed as they are normally done. Therefore, it is possible to fabricate I-MOS devices and conventional CMOS transistors on the same chip. In the process of the optimizing device characteristics for performance and reliability, we anticipate the need to add up to two more mask steps. This is relatively inexpensive, considering that most conventional IC processes today require about 33–40 mask steps. In addition, we believe that with the wide-spread acceptance of this technology and with process optimizations, we may be able to eliminate the low power (high V_T) and the low standby power (high V_{DD}, high

V_T) transistors on the layout and cut down on the number of mask sets by about 5–10. Since a new scaled technology is introduced every 2–3 years, I-MOS would work with the foundries for developing a scaled I-MOS technology for every generation. I-MOS would then jointly work with these foundries to license this technology to their customers.

Provide tools and support in addition to technology

In addition to the I-MOS transistor related intellectual property, we also offer design tools, standard SPICE models for these transistors, and layout related tools for incorporating the I-MOS technology within new and existing designs. These tools would selectively replace certain transistors in the layout with I-MOS transistors, perform circuit-level optimizations, and help in seamlessly porting existing chip designs to take advantage of the I-MOS technology. A general flow chart showing how companies would use this technology is shown in Fig. 4. All the procedural steps shown in Fig. 4 are completely automated so that it is extremely easy to exploit the benefits of this technology.

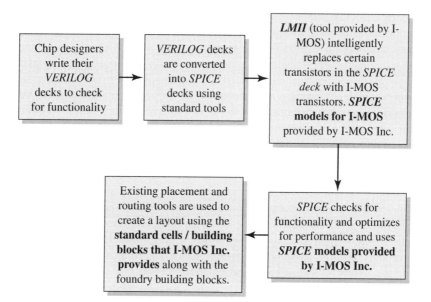

FIGURE 4   Schematic showing how semiconductor companies would use the I-MOS technology for the products.

VIII. Research and Development

I-MOS believes that its future success will depend critically on the continued development and introduction of new, scaled, and improved transistor technologies in accordance with or faster than Moore's law. To this end, company engineers are involved in developing new versions of the I-MOS technology that will allow higher performance, lower power, and higher levels of integration. In addition, we plan to have a highly-qualified technical advisory board comprised of professors and industry stalwarts from Stanford University, MIT, and SUN Microsystems, and we believe that this multidisciplinary expertise of our team of scientists and engineers will continue to advance our technological leadership and market.

I-MOS would intitally develop its transistor technology in one foundry (e.g., TSMC, TI, UMC) before expanding to satisfy the needs of customers in different foundries. The company plans to do extensive research and development on the scalability and the reliability of the transistor, gauging the power and performance in circuit frameworks and then optimizing the transistor for maximum performance. This would be carried out after extensive presentations and negotiations with the technology group in the foundries. We anticipate that the I-MOS process would require minimal process changes over the standard baseline strained-Si CMOS process in the foundry. After developing the technology and doing extensive reliability testing by March 2005, we plan to sign on beta customers (e.g., Xilinx, Nvidia) who use the foundry, that we are working with, for their own silicon processing needs. These beta customers would be invaluable in providing direct feedback between technology development and client needs. After having demonstrated our value proposition to the above customers, we then anticipate that we will be able to turn some of them into permanent customers who would need the I-MOS technology to satisfy the emerging needs for low power and higher performance. We believe that I-MOS is strategically positioned to address the evolving and highly demanding needs of these markets.

IX. Intellectual Property

The I-MOS technology was originally developed as part of a Defense Research Project (DARPA) at Stanford, and the patent for the I-MOS transistor is owned jointly by Stanford University, I-MOS CTO (Kailash Gopalakrishnan), and Prof. James D. Plummer (Dean of the Stanford School of Engineering). Currently, the Office of Technology Licensing at Stanford University holds exclusive licensing rights to this technology. I-MOS believes that we can license this technology on an exclusive basis from Stanford University. Recent conversations with OTL have justified this expectation, and there are many examples of this licensing structure including T-RAM Inc., Pixim Inc., and Google.

As an IP licensing company, we believe that our patents, copyrights, mask work rights, trademarks, and trade secret laws are critical for our success and to that end the company will have an active program to protect its proprietary technology through the various types of filings mentioned above. In addition, the company would attempt to protect its trade secrets and other proprietary information through agreements with licensees and foundries, proprietary information agreements with employees and consultants, and other security measures. We estimate that we may have to file up to two U.S. patents a month and would have about a $500,000 or more in legal expenses in the initial years of development including the initial setup charges. We also seek to protect our software, documentation, and other written materials under trade secret and copyright laws.

X. Competition

I-MOS's competitors include integrated device manufacturers (IDMs) such as Intel and IBM, fabless semiconductor companies such as Sun and Transmeta, and foundries such as TSMC and UMC. Of these competitors, Intel, IBM, and Sun pose the largest competitive threat. These three companies have a combined annual R and D budget of over $10 billion and devote a significant portion of it to creating technologies that will make semiconductors dissipate less power, run faster, and cost less.

Specifically, the problem of reducing static leakage has long been recognized by the semiconductor industry. However, it has been perceived as an impossible problem to solve. Some of these companies claim that any potential solution is limited by the basic principles of thermodynamics itself. Therefore, many of these companies have focused their efforts on circuit workarounds, each with some degree of success.

The following is a summary of the primary competitors and a description of the competing solutions that each is pursuing:

- **Intel Corporation**—Intel has been researching "sleep transistor" technology in hopes of solving the issue of static leakage current. With this solution, the power supply itself is switched off when the transistor is not being used. Sleep transistor technology can provide a thousand-fold reduction in leakage power. The primary advantage of this technique is that it does not affect the basic transistor design and functionality but rather focuses on the circuit core instead. However, the technique is very hard to implement due to the fact that it is very difficult to determine which circuits are computing and which are not computing on a microprocessor. Hence, this kind of approach may have to be applied to larger blocks, which reduces the efficiency of the process in reducing static power. In addition, the design complexity goes up as well, which translates directly into an area and cost penalty. Also there is a loss of performance because the circuits which are switched off need a finite time to switch back on. It should also be noted that it may actually not be possible to switch the power off from certain circuits such as SRAM and DRAM because these memories lose information when the power supply is switched back on.

- **International Business Machines**—IBM has been researching silicon-on-insulator (SOI) which offers lower capacitance and, hence, increased speed. SOI is a technique that gains in both dynamic power and performance by reducing the diffusion capacitance of transistors but in its native form, it does not affect static power. This technology is compatible with I-MOS, meaning that a wafer utilizing both SOI and I-MOS will experience performance improvements that are a summation of the improvements that would be realized from each technology independently.

■ **Advanced Micro Devices**—AMD, in conjunction with IBM, has been developing the Fin Field Effect Transistor (FINFET). This transistor utilizes a thin vertical "fin" to help control leakage of current through the transistor when it is in the "off" stage.[4] The FINFET is the only device solution that allows CMOS to scale to the end of the perceived roadmap. FINFET utilizes the same device physics as conventional CMOS but is nonplanar and hence require different fabrication steps. Since FINFET is based on CMOS physics, it is still limited to 60mV/dec. The Intel DST (Depleted Substrate Solution) is also limited by the same principles to 60 mV /dec.

XI. Financials

Projected Income Statement

	Year 1 Aug-2004	Year 2 Aug-2005	Year 3 Aug-2006	Year 4 Aug-2007	Year 5 Aug-2008
Revenue	—	—	8,370,000	18,825,000	37,260,000
Cost of Revenue	—	—	—	—	—
Gross Profit	—	—	8,370,000	18,825,000	37,260,000
OPERATING EXPENSES					
Sales	—	674,000	1,988,900	3,105,094	3,651,840
Marketing	—	108,000	150,480	159,509	169,079
Engineering	4,029,907	6,884,609	10,749,401	15,464,000	18,501,490
General & Admin.	877,000	1,137,000	1,155,540	1,175,192	1,196,024
Total Oper. Exp.	4,906,907	8,803,609	14,044,321	19,903,795	23,518,433
Total Non-Op. Inc. (Exp.)	—	—	—	—	—
Pre-Tax Profits	(4,906,907)	(8,803,609)	(5,674,321)	(1,078,795)	13,741,567
Taxes	—	—	—	—	—[5]
Net Income	(4,906,907)	(8,803,609)	(5,674,321)	(1,078,795)	13,741,567
Net Margin	n/a	n/a	−68%	−6%	36.9%

Key Revenue Drivers

Licensing Structure, Average Selling Price, and Volume.   We model revenues based on a licensing structure that consists of (1) an average, one-time $1.5M license fee for access to the I-MOS patent portfolio for a 5-year term, and (2) an

---

[4] Advanced Micro Device website.

[5] Although we expect to generate positive profits in Year 5, we do not anticipate paying taxes on the profit since the cumulative losses carried over from previous years exceed the expected Year 5 profit. As a reality check, our net margin in Year 5 would be 22.9 percent, consistent with the net profit margins of ARM and Rambus near 25 percent.

average royalty fee of 3 percent of the average selling price of all chips incorporating the I-MOS technology. We believe these are reasonable assumptions
given the licensing structure of similar semiconductor IP licensing companies
(see table in Appendix). We model an average price per chip at $3.50 based on
ARM's average royalty per chip ($0.07) divided by the average royalty percent-
age per chip (2%). Since ARM licenses IP to many industries including mobile
phones, storage, and consumer electronics, we believe $3.50 represents a rea-
sonable "average" selling price per chip for us.[6]

Timing and Frequency of Customer Acquisition.   We do not anticipate receiving royalty revenue from our first customers until Q1 of Year 3, when the
technology will have been developed for high scalability in commercial applications and the sales team will have had 3–6 months to land our first customer. The
timing and frequency of customer acquisition reflects our strategy of first convincing fabs to integrate our technology into their processes and then having
them walk us into their customers. We project multiple fabless customers relative to fabs because we expect each fab to have more than one customer. Supporting fabs is very costly and requires 8–10 dedicated engineers per fab. A table
of our customer acquisition plan is included in the Appendix.

Key Expense Drivers

Headcount.   The chart below summarizes headcount growth for the first five
years.

	Year 1	Year 2	Year 3	Year 4	Year 5
Sales	–	5	9	13	13
Marketing	–	1	1	1	1
Engineering	10	30	44	56	68
Genl. & Admin.	3	3	3	3	3
Total	13	39	57	73	85

Sales.   We anticipate hiring our first salesperson in Year 2 once technical risk
has been sufficiently removed, which is also 1 year prior to when we expect our
first customer to ship. We then plan on hiring one salesperson every 3 months
(one a quarter) thereafter.

Engineering.   Based on conversations with the founder of a comparable early
stage semiconductor company, we have estimated major development costs of:

- A mask set per fab per year at $1.5M. Even after signing a customer,
  developing mask sets are essential to work on the next generation of chip
  and I-MOS technology.
- Wafer costs of $1M in the first year, increasing thereafter. In the first
  month we estimate needing 30 wafers at a cost of $2500 per wafer and

---

[6] UBS Warburg, Global Equity Research Report, *At the Core of IT,* May 2002.

allocated a 2 percent increase in the cost per month to reflect rising costs of development over time.

- Leased testing equipment and software at $1.2M a year.
- Beginning in the second year, we plan to ramp development aggressively and hire two engineers a month for nine months, followed by hiring one additional engineer a month thereafter to support continued technology development.

Legal Expenses.   See Section VIII "Intellectual Property" for a breakdown of these expenses.

Proposed Company Offering:

- Series A Financing—$6.5 million by September 2003
- Series B Financing—$8.5 million by September 2004
- Series C Financing—$9.0 million by September 2005

Total Anticipated Capital Requirements = $24.0 million

Series A: I-MOS is currently seeking $6.5 million in Series A financing by September 2003. The key milestone we intend to achieve with the capital in this round is:

- Advancing the I-MOS technology from its current state to demonstrate proof of concept in simple devices and simple circuits that consist of less than 100,000 transistors.

The majority of this capital will be spent on technology development. We anticipate that the proceeds from this round will last us 15 months until November 2004 and will enable us to grow to 13 employees: 10 engineers and three G&A (two founders, one office manager).

Series B: By the end of September 2004 (month 13), we intend on raising $8.5 million in Series B financing. In case securing the Series B financing takes longer than expected, we have built in a three-month cushion from our Series A financing. The milestones we intend to achieve with the capital in this round are:

- Demonstrating the I-MOS technology in bigger circuits and optimizing the technology for power and/or performance.
- Demonstrate reliability and verifying customer specifications.
- Sign first customer.

The majority of this capital will be used to triple the number of engineers and recruit our first salespeople. We anticipate that the proceeds from this round will carry us through month 27 (November 2005) and will enable us to grow to 39 employees.

Series C: By the end of September 2005 (month 25), we intend on raising $9.0 million in Series C financing. In the cvent that securing Series C financing takes longer than expected, we have built in a three-month cushion from our Series B financing. The milestones we intend to achieve with the capital in this round are:

- Revenue from first fab and fabless customer.
- Cash flow breakeven and profitability.

The majority of this capital will be used to finance the expansion of the organization as we ramp up our sales and engineering teams in anticipation of rapid customer growth. We anticipate that this will be our last private round of financing and that the proceeds from this round will carry us through to cash flow breakeven and profitability in Q4 of Year 4.

Capitalization and Investor Return

In forecasting the future capitalization structure of the company and investor return, we assume that the company goes public at the end of Year 5 and offers 20 percent of its shares to the public. Assuming we meet our projections and value the company using the current industry median price to earnings multiple of 33, then I-MOS would be valued at the end of year 5 at $13.7 net income * 33 = $450 MM (see Appendix for public company comparison metrics).

Capitalization Structure

Premoney Valuation	$6.5	$35.0	$100.0	
Invested Capital	$6.5	$8.5	$9.0	
Postmoney Valuation	$13.0	$43.5	$109.0	$450.0

Stakeholders	Founding	Post Financing Ownership Levels			
		Series A	Series B	Series C	Exit
Founders	100.0%	35.0%	27.0%	24.3%	19.5%
Employees		15.0%	15.0%	15.0%	12.0%
Series A Preferred		50.0%	38.5%	34.8%	27.8%
Series B Preferred			19.5%	17.6%	14.1%
Series C Preferred				8.3%	6.6%
Public Market					20.0%
Total Ownership	100.0%	100.0%	100.0%	100.0%	100.0%

Projected Investor Return

	Series A	Series B	Series C	Total
Years until Exit	5	4	3	
Invested Capital	$6.5	$8.5	$9.0	$24.0
Invested Capital Value upon Exit	$125.2	$63.5	$29.7	$218.4
IRR	80.7%	65.3%	48.9%	72.1%

XII. Management

Adam Wegel

Chief Executive Officer

Adam Wegel comes from Delphi Automotive Systems with over six years of varied experience in operations, manufacturing engineering, product development, finance, and sales and marketing. Most recently at Delphi, Adam developed a corporate strategy for selling telematics to the commercial fleet vehicle industry.

Adam is a Master in Business Administration candidate (March 2003) at the Stanford Graduate School of Business and a Master of Science in Mechanical Engineering candidate (March 2003) at the Stanford School of Engineering. Adam received in Bachelor of Science degree in Mechanical Engineering from North Carolina State University.

Kailash Gopalakrishnan

Chief Technology Officer

Kailash Gopalakrishnan comes from T-RAM, where he worked on various device and circuit issues for a novel memory product. At T-RAM, Kailash also invented two modified memory cells that were later patented. These key inventions helped the company solve the static power dissipation problem in their memory cells.

Kailash is a Stanford Graduate Fellow in the Semiconductor Devices Group at Stanford University, where he is pursuing his PhD in Electrical Engineering (expected September 2003). Kailash and his advisor, Dean James D. Plummer, have filed co-patents for various aspects of I-MOS technology.

Rajit Marwah

Chief Financial Officer

Rajit Marwah most recently comes from Echelon Corporation, where he researched worldwide vendors, technology, competition, international frequency regulations, and solution costs for the emerging low power, low data rate, medium range wireless data market (ZigBee, UltraWideband) for use in automatic meter reading (AMR) and other applications. Prior to Echelon, Rajit worked as an Associate for two summers at TL Ventures, a $1.4 billion early stage technology venture capital firm, assisting partners in due diligence and structuring deals.

Rajit is a Master of Science in Management Science and Engineering and Bachelor of Arts in Economics candidate (June 2003) at Stanford University. Rajit is also a Vice President of BASES, Stanford's largest entrepreneurship organization, with over 4,500 members.

Tod Sacerdoti

Chief Marketing Officer

Tod Sacerdoti comes from Robertson Stephens, where he was a Corporate Finance Investment Banker focused on the Communication Infrastructure and Internet sectors. Working closely with research analysts, he managed two deals and developed strategic road maps, business model positioning and road show presentations for private technology companies. He was also the founder and CEO of DK Entertainment, a successful event production and marketing company.

Tod is a Master of Business Administration candidate (June 2003) at the Stanford Graduate School of Business. He received his Bachelor of Arts in Economics at Yale University, where he was President of Sigma Alpha Epsilon fraternity.

XIII. Risks and Mitigations

Risk	Mitigation

Market Risk

The primary market risk is that we are unable to establish licensing contracts for our IP. While there are examples of very successful semiconductor IP companies, the number of such companies is fairly limited.

If we are unable to secure a licensing deal within six months of meeting our second milestone (proving the technology), we will consider taking a product approach and selling a relatively simple semiconductor product such as DRAM, SRAM, and/or FPGA.

Technology/Product Risk

The primary technology risk is that I-MOS technology may not scale. Currently, the technology has been proven at the micron level. For the technology to be commercially feasible it must scale to submicron dimensions. In addition, to realize full power savings and performance increases, strained silicon must be utilized. The technology may not perform as well in strained silicon as simulations indicate.

Currently, extensive simulations and modeling have been run which validate I-MOS technology at submicron scale in strained silicon. To mitigate this risk, we plan to run I-MOS devices at a submicron scale in strained silicon as early as possible in the development process. Currently, we anticipate being able to complete this testing in nine months (September 2003).

Another major technology and market risk is that the semiconductor industry may not move toward a strained silicon based technology at all. Strained silicon is imperative in order to realize all the benefits and advantages of the I-MOS technology.

IBM, which pioneered strained silicon, has committed to introducing strained silicon in mass production by 2004 in its foundries. If we sense that the other major foundries are not moving in the direction of implementing strained silicon, we will shift our business strategy from a IP licensing business model to being a more product-oriented (SRAM or DRAM) company.

Team Risk

The I-MOS management team has not been previously involved with a start-up, nor do they have significant management experience.

The management team will reduce this risk by hiring an experienced CEO after the first milestone is achieved (feasibility of I-MOS technology is proven). Furthermore, the management team will select a board of directors based not only on their relevant backgrounds but also on their willingness to be active mentors for the management team.

Financing Risk

A significant financial risk is obtaining the capital required to launch a semiconductor IP company. Since there are relatively few successful semiconductor IP companies and since the level of required investment is high, we may experience difficulty in convincing the venture capital community to invest.

To reduce this risk, we will continue to establish relationships with potential customers early on in the development process. If we are unable to obtain the required financing, we may reconsider our licensing approach and investigate producing a relatively simple product, e.g., DRAM, SRAM, FPGA, to generate revenues sooner.

Appendix
Figure A1   Projected Cash Flow Statement

	Year 1 Aug-2004	Year 2 Aug-2005	Year 3 Aug-2006	Year 4 Aug-2007	Year 5 Aug-2008
BEGINNING CASH BALANCE	—	1,419,593	772,984	3,859,663	2,572,868
Sources of Cash					
Net Income (Loss)	(4,906,907)	(8,803,609)	(5,674,321)	(1,078,795)	13,741,567
Incr. (Decr.) in A/P	—	—	—	—	—
Incr. (Decr.) in Accrued Exp.	—	—	—	—	—
Incr. (Decr.) in S.T. Notes Payable	—	—	—	—	—
Incr. (Decr.) in Other ST Liabilities	—	—	—	—	—
Incr. (Decr.) in LT Notes Payable	—	—	—	—	—
Incr. (Decr.) in Other LT Liabilities	—	—	—	—	—
Subtotal	(4,906,907)	(8,803,609)	(5,674,321)	(1,078,795)	13,741,567
Uses of Cash					
Incr. (Decr.) in A/R	—	—	—	—	—
Incr. (Decr.) in Inventory	—	—	—	—	—
Incr. (Decr.) in Other Current Assets	—	—	—	—	—
Incr. (Decr.) in Fixed Assets	(173,500)	(343,000)	(239,000)	(208,000)	(168,000)
Incr. (Decr.) in Other LT Assets	—	—	—	—	—
Subtotal	(173,500)	(343,000)	(239,000)	(208,000)	(168,000)
Operational Cash Flow	(5,080,407)	(9,146,609)	(5,913,321)	(1,286,795)	13,573,567
Financing Activity					
Preferred Stock	6,500,000	8,500,000	9,000,000	—	—
Common Stock	—	—	—	—	—
Subtotal	6,500,000	8,500,000	9,000,000	—	—
Net Cash Flow	1,419,593	(646,609)	3,086,679	(1,286,795)	13,573,567
ENDING CASH BALANCE	1,419,593	772,984	3,859,663	2,572,868	16,146,435

Figure A2 Projected Balance Sheet

	Year 1 Aug-2004	Year 2 Aug-2005	Year 3 Aug-2006	Year 4 Aug-2007	Year 5 Aug-2008
BALANCE SHEET					
Cash	1,419,593	772,984	3,859,663	2,572,868	16,146,435
Accounts Receivable	—	—	—	—	—
Inventory	—	—	—	—	—
Other Current Assets	—	—	—	—	—
Subtotal	1,419,593	772,984	3,859,663	2,572,868	16,146,435
Fixed Assets	173,500	516,500	755,500	963,500	1,131,500
Other LT Assets	—	—	—	—	—
Total Long-Term Assets	173,500	516,500	755,500	963,500	1,131,500
TOTAL ASSETS	1,593,093	1,289,484	4,615,163	3,536,368	17,277,935
Accounts Payable	—	—	—	—	—
Accrued Expenses	—	—	—	—	—
ST Notes Payable	—	—	—	—	—
Other Current Liab.	—	—	—	—	—
Subtotal	—	—	—	—	—
LT Notes Payable	—	—	—	—	—
Other LT Liabilities	—	—	—	—	—
Subtotal	—	—	—	—	—
Pfd. A Stock	6,500,000	6,500,000	6,500,000	6,500,000	6,500,000
Pfd. B Stock	—	8,500,000	8,500,000	8,500,000	8,500,000
Pfd C Stock	—	—	9,000,000	9,000,000	9,000,000
Prior Yr. Ret. Earnings	—	(4,906,907)	(13,710,516)	(19,384,837)	(20,463,632)
YTD Earnings	(4,906,907)	(8,803,609)	(5,674,321)	(1,078,795)	13,741,567
Total Equity	1,593,093	1,289,484	4,615,163	3,536,368	17,277,935
TOTAL LIAB. & EQUITY	1,593,093	1,289,484	4,615,163	3,536,368	17,277,935

Figure A3 IP Semiconductor Licensing Structures

Company	Initial License Fee	Avg Royalty Rate	% of ASP
ARC	0.2+	0.51	8.3
ARM	5	0.07	1–3
MIPS	0.2–5.0	0.5	5–10
Parthus	<1.0	0.4–1.0	3–5

Figure A4 Customer Acquisition Plan

	Yr 1	Yr 2	Yr 3	Yr 4	Yr 5
Total Fab Customers	0	0	1	2	3
Total Chip Customers w/ own fab	0	0	1	2	3
Total Fabless Chip Customers	0	0	3	8	12
Total Customers	0	0	5	12	18

Figure A5 Industry Comparables

Company	Market Cap ($MM)	Trailing Twelve Months				
		Sales ($MM)	Net Income ($MM)	Net Margin	Sales Multiple	Earnings Multiple
ARM	1,100	251	58.9	23.5%	4.4	18.7
Rambus	867.8	96.6	24.7	25.6%	9.0	35.1
MIPS	96.5	44.7	–12.3	–27.5%	2.2	–7.8
Synopsys						
TTP Communications						
Virage Logic	278.2	45.6	0.23	0.5%	6.1	1209.6
Mentor Graphics						
Parthus Technologies	137.7	42.5	–32.5	–76.5%	3.2	–4.2
Artisan Components	314.4	38	2.1	5.5%	8.3	149.7
DSP Group	433.8	126.2	13.1	10.4%	3.4	33.1
MEDIAN	314	46	2	5.5%	4.4	33.1

Figure 6 Projected Operational Cash Flow By Quarter

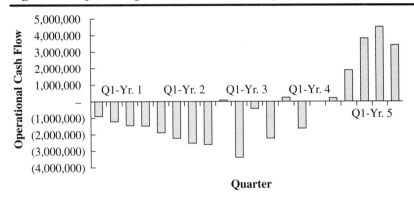

Cases

DANGER, INC.
Powering the Next Generation of Mobile Technology

In the first quarter of 2003, the U.S. economy was in the deepest recession in recent memory. Over the last three years, the collapse of the Internet bubble resulted in an extremely risk averse start-up environment. Corporate scandals such as Enron and WorldCom had caused tremendous mistrust of top executives. The technology sector was performing poorly due to cuts in spending for computing and communications by both enterprises and consumers. New York City had suffered a terrorist attack in September of 2001 that resulted in the collapse of the World Trade Center towers, and in March of 2003, the United States was at war with Iraq. In the midst of this adversity, the high technology community in Silicon Valley and around the world was struggling to get back on its feet.

The wireless industry had been especially volatile, with some analysts predicting doom and gloom, and others promising that a boom in wireless data was just around the corner. For example, the exorbitant costs of 3G licensing auctions in Europe and slow adoption rates of wireless technologies such as Wireless Application Protocol (WAP) made industry observers worry that the wireless industry had been over-hyped. On the other hand, statistics such as the steady increase of cell phone penetration rates worldwide and the tremendous success of short messaging service (SMS) were positive signs of continual growth in this space. For example, SMS text messaging, a form of short, near real-time messaging, had become a $20 billion market in Europe, and some European wireless carriers were generating 15 to 20 % of their total revenues from this crude data service. That gave the leaders of Danger confidence that integrated solutions like theirs would have a bright future. In the midst of this market and technological uncertainty, Hank Nothhaft, CEO of Danger, Inc. was in his office on University Avenue in Palo Alto, preparing for a meeting of Danger's leadership team.

Hank had taken the helm of Danger, Inc. in October of 2002. The mission was to bring the company to the next level—to become a profitable business. Danger, Inc. was building "a complete solution that enabled wireless carriers to

This case was prepared by Sei Wei Ong, Graduate Student at Stanford University School of Engineering, and Thomas J. Kosnik, Consulting Professor, Stanford School of Engineering, with guidance and support from Wong Poh Kam, Associate Professor, National University of Singapore as the basis for class discussion rather than to illustrate either effective or ineffective handling of an administrative situation. It was made possible in part through financial support from the National University of Singapore

offer innovative and affordable voice and data products to consumers over next-generation networks."[1] Hank was impressed by what the company's founders, Andy Rubin, Joe Britt and Matt Hershenson, had created. In a span of just two years, Danger had become one of the hottest start-ups in Silicon Valley. The founding team had successfully implemented a first-generation integrated solution, won numerous awards from the industry for their innovativeness, received amazing consumer reviews for their handheld product, and had signed their first major customer, T-Mobile.

Every day since his arrival, Hank had witnessed the extraordinary energy that permeated the environment of this fast-paced start-up. The team at Danger, Inc. was exceptionally talented and committed. In February 2003, Danger had reached a major milestone by securing US$35 million in Series D financing, bringing total funding to US$77 million. This included strategic investments from T-Ventures and Orange Ventures, the venture arms of two of the world's major wireless carriers. The founding team had been frugal in spending its cash, and as a result, had almost US$50 million in funds to finance future operations.

Hank and his teammates knew that to unleash the company's full potential and reach "financial critical mass," Danger needed to build a portfolio of strategic relationships that would help bring its technology to mainstream markets. The leadership team needed to build alliances with manufacturers to ensure continuous hardware innovation and to reduce the costs of handheld devices that were compatible with Danger's software solutions. They also had to convince more carriers to adopt Danger's software solution as a platform to offer wireless services to their subscribers. Partnerships with other technology and content providers would also help to bring more capabilities to the Danger platform.

Hank left his office and moved through the labyrinth of Danger's back hallways to the conference room that had been set up for the leadership team meeting. Assembling around the old, second-hand conference table were some of the best and brightest entrepreneurial minds in the Valley: Andy Rubin, Matt Hershenson and Joe Britt—Danger's co-founders; John Arledge—Vice President of Business Development; James Isaacs—Vice President of Worldwide Sales and Alliances and Les Hamilton—Senior Vice President of Worldwide Operations and Manufacturing. The biographies of Danger's founders and selected executives are provided in Exhibit 1. Note all exhibits are placed at the end of the case.

Danger's leadership team was assembling to discuss three key issues related to their portfolio of business partnerships. Which wireless carriers were the leading candidates for Danger to pursue to increase their market penetration in the U.S. and around the world? Which manufacturers were the most attractive potential partners to create Danger-compatible handheld devices? How could Danger develop trustworthy partnerships with companies that were often in fierce competition with one another?

---

[1] Danger Product Overview from Danger Press Kit, January 2003.

Wireless Industry Framework

In many developed countries, a wide variety of wireless products and services was available to end-users. In 2003, mobile phones were so popular and affordable in developed nations that they had become a standard accessory for urban living. On the other hand, high-tech companies continued to introduce cutting edge wireless innovations such as UltraWideBand Technolgy (UWB) that were still very early in their technology adoption life cycles. The industry had become very large and complex, with players ranging from start-ups to multi-national companies that formed different parts of the value chain. Exhibit 2 shows a pictorial representation of five wireless segments, which are described below.[2]

The first segment consisted of the *wireless component providers.* They competed by producing innovations in product categories such as baseband components, radio frequency (RF) components antennas, etc. which allowed devices to communicate using a variety of wireless protocols. Texas Instruments and Qualcomm were the two market leaders in providing baseband components. The major players in the RF components market were Motorola, Philips and Infineon. Other companies were positioned to provide specialized chipsets. For example, Intersil was the leading provider for 802.11 chipsets. Due to the rapid advancements of technology and the wide variety of standards, this segment was a breeding ground for start-ups such as Atheros Communications, Cambridge Silicon Radio and ArrayComm.

The second segment was the *network equipment suppliers* who made the physical infrastructure to make wireless networks function. They designed and manufactured the radio base stations, the receivers, the switches and other components of the network. Ericsson dominated the wireless infrastructure market with a market share of about 34%. Other key players in this market were Nokia, Motorola and Lucent. Traditionally, the network equipment manufacturers had sold primarily to carriers who built extensive wireless networks to serve enterprises and consumers. However, with the increased popularity of Wireless Local Area Networks (WLANs), network equipment suppliers had started to market their products to enterprises and consumers directly. From 2000–2002, a number of startups had also entered the network equipment segment with technological innovations to carve out a piece of the pie for themselves.

The third segment was the *end-terminal manufacturers.* This segment was divided into five sub-segments—laptop manufacturers, mobile phone manufacturers, Personal Digital Assistant (PDA) manufacturers, automotive telematics, and wireless-enabled appliances. Mobile phones were traditionally the end-terminals of a wireless transmission and were the key product in this space. Although laptops and PDAs had become increasingly popular over the last five years, most of them had to be plugged into a fixed Ethernet connection for network access. The increasing importance of mobility and connectivity suggested

---

[2] For more information on the wireless industry and a detailed explanation of the framework, refer to "The Wireless Industry in 2002." STVP-2002-004

that these two sub-segments could potentially become key growth areas. The automotive telematics sub-segment was still nascent in 2002, and automobile companies had started implementing navigational equipment in their vehicles, using the Global Positioning System (GPS). This was still a very expensive option and thus a relatively rare occurrence. The final segment, wireless-enabled appliances, also was a very new segment. This space included printers or digital cameras that were fitted with a wireless receiver module, thus allowing the equipment to receive wireless data, usually by using the Bluetooth protocol.

The fourth segment consisted of the *carriers,* who were *mobile service providers.* The players in this space were companies that licensed the spectrum from the government, bought infrastructure from the network equipment manufacturers, deployed the infrastructure, and provided wireless services to individual consumers or corporate users. They also served as the distribution channel for mobile phone manufacturers. Examples of mobile service providers in the United States included AT&T, Sprint, Verizon, and Cingular Wireless.

The final segment, *wireless software providers,* was divided into three segments based on the customers they served. These segments were: 1) carrier-focused, 2) enterprise-focused, and 3) consumer-focused. This subdivision further highlighted the complexity of this space, because of the different services that each segment needed. The carriers needed software applications that could improve the network management and provide value-added services to their customers. Enterprise-focused companies were trying to provide networking software that could help to extend company operations over wireless networks or business software that could leverage the wireless capabilities that corporations had set up. Finally, consumer-focused companies provided mobile applications for individuals, such as mobile gaming, content aggregation, scheduling, and other applications.

In this wireless framework, Danger fits best into the wireless software provider segment, under the subdivision—carrier-focused. This was because Danger provided an end-to-end software solution to wireless carriers that would enable them to provide value-added services on high-speed data networks, such as General Packet Radio Services (GPRS). These services were delivered through a hiptop-enabled device.

One major issue for both the handset/device manufactures and the software segment in 2003 was what operating system (O/S) would prevail. The leading contenders included: 1) Symbian, a java based OS controlled by an industry consortium, that many observers felt was really controlled by Nokia; 2) Microsoft Windows CE, .NET and other device-oriented software from Microsoft; 3) other Java-based O/S's; and 4) Palm O/S and its derivatives.

Hank Nothhaft appreciated the implications of the wireless operating systems wars for Danger: "This is very important and very strategic to Danger's prospects... Danger doesn't neatly fit into a box since we are an end-to-end client server application, but as a result we do provide the equivalent of an O/S. This is a major point of contention and opportunity for us, because Nokia and Microsoft are viewed as competitors by most participants in the wireless industry." While

Danger was careful not to position its solution as an operating system, it might be viewed by wireless carriers and handset manufacturers as a potential defense against being dominated and commoditized by either Microsoft or Nokia.

Many industry analysts mistakenly put Danger, Inc. in the same category as Personal Digital Assistant (PDA) manufacturers such as Palm or HandSpring. In the wireless framework, that would be in the end-terminal manufacturers' segment. However, for many of these PDA manufacturers, the key innovations were primarily in the hardware design; more powerful handhelds, more features and capabilities, better form factor, etc. Although hardware design was an important component of Danger's value proposition, most of its innovations occurred in the software applications on the back-end server. This server powered and supported the software applications that efficiently provided data for the hiptop-enabled device.

A closer look at Danger's business model also revealed that Danger was distinctly different from other PDA manufacturers. Although Danger played a key role in the hardware design for the hiptop® handheld, they made 0% margins on the device, and chose to sell the handheld at cost. This was different from traditional PDA manufacturers, who made 20–40% margins on their devices. Although Danger made some revenue from one-time license fees on its operating system, its primary revenue source came from licensing the back-end server software to carriers and the sale of premium software products to end-users.

Company Background

Danger, Inc. was founded by Andy Rubin, Joe Britt and Matt Hershenson in Sunnyvale, California, on Dec. 23, 1999. Although they were from different parts of the United States, (Andy—Rochester, New York; Joe—Lumberton, North Carolina; Matt—Pittsburgh, Pennsylvania) they were drawn to California in the 1980s to join Apple Computer, a pioneer in the microcomputer revolution. Since then, they had moved on to work at some of the most technologically innovative companies in Silicon Valley, such as General Magic, Catapult Entertainment, 3DO, and Web TV. Through these experiences, they had become a part of the close-knit community of technology innovators that made Silicon Valley a fertile breeding ground for start-ups. Exhibit 3 provides a timeline of the industry experience of the three founders and selected key executives at Danger.

The impetus for starting Danger occurred shortly after Microsoft's acquisition of WebTV. Andy and Joe were early employees of WebTV and had enjoyed working in the start-up environment. After the acquisition, WebTV was moved from bustling downtown Palo Alto to a dreary industrial park in Mountain View. Relaxed, casual, open cubicles were replaced by walled offices and long, quiet hallways. This transition changed the culture of the company significantly and as Andy described it, "I would drag myself into Microsoft each morning and reply to email. I really didn't know what I was doing there."

It was not long before the founders decided that it was time to move on to their next venture. Each of the founders had complementary domain expertise in a particular area—Andy in communications, Joe in software and Matt in hardware. From their past experiences in network computing, the founders knew that they needed each of these disciplines to build a great networking product. They also believed that the "thin client" model[3] was well suited for consumer products, because it made things much simpler for consumers to use. Andy described the first meeting that marked the beginnings of Danger, Inc:

> In this brainstorming session, we all chimed in and started Danger based on some of those ideas. Of course, the product evolved with time, but the fundamental ideas remained the same. My input was that I needed something to remind me to do things, Joe's was to have a connected device, and Matt's was that '...it's got to be free. It's got to be cheap.'

The company name—Danger, Inc. had a serendipitous beginning. One of Andy's hobbies was robotics and he had actually helped built a robot for the Tech Museum in San Jose. To pursue his robotics projects, Andy leased some retail space off California Ave. in Palo Alto and named his company "Robots that Kill." After experimenting with different ideas for a domain name for the company website, he settled with Danger.com which he acquired for $6,000 from a person in Vermont. Since Andy already owned the domain, the founders decided to use it, and worry about the "real" company name later on.

The seed funding of $1.7 million for Danger, Inc. came mostly from friends and family. Since this was soon after the Microsoft acquisition of WebTV, many of the founders' friends had money from Web TV stock, and were willing to support their venture. Because of their strong credentials and the easy availability of capital, the founders were able to raise money without a formal business plan. Through the first year after its inception, Danger was primarily a technology-driven company, and had made some significant technological progress. Andy described that period:

> The technology was compelling and there was something in the business model, but we didn't have the full story. There were missing pieces on the business development, sales, and partnerships aspects of the company. We could have gotten funding, but not at a valuation that would make sense for us. So we kept going on seed money, taking $100,000 chunks from our friends. There were a couple of close calls, but we managed to survive through those times.

Although Danger's product had been through several iterations, the concept of using the client-server computing model remained central in its evolution. The founders believed that by using back-end servers to provide most of the computing capabilities, they could build an affordable high performance device on the client-end. The first prototype that they built on the client-end was a NanoPDA called the Peanut that could be used to store important personal in-

---

[3] In server-client applications, a thin client is a client that is designed to be especially small so that the bulk of the data processing occurs on the server.

formation. It could be connected to the Internet, had a small LCD screen, a couple of buttons, and a wheel for easy scrolling. The device could be manufactured for as little as $10. The second iteration of the product included some experimentation with one-way transmission technology that used FM sideband data transmission, by broadcasting data from radio stations. This additional functionality increased the device cost to about $16. Through their conversations with potential investors, friends and mentors, the Danger team continued to build more functionality into the device. This included calendaring functions, information storage, basic web surfing and receiving email wirelessly. Things started to take shape when they started to experiment with two-way wireless transmissions, because this technology enabled a host of interactive applications. Andy described that moment:

> At General Magic we had basically developed the same thing years ago. The Motorola Envoy was one of the first wireless PDAs in the market that ran on the Mobitex network. For me, it was too close. I didn't want to start another company that was so similar to the company that I had previously left. So we started thinking of things that we could do beyond that.

The founders realized that the next generation data networks created an opportunity for Danger to take wireless applications to the next level. RIM's wireless PDAs operated on Mobitex that had transmission speeds of only 8 Kilobit per second (kbps). Next-generation data networks included 2.5G technologies such as GPRS, which promised transmission speeds of up to 144 kbps. Future 3G technologies boasted 2 Megabit per second transmission speeds. At that time, GPRS was still in its infancy and had not been deployed by any of the carriers. Danger had to develop a product based on a standard that was still evolving. Through a close technical partnership with Xircom, one of the key developers of GPRS radio technology, Danger successfully implemented one of the first GPRS-enabled devices in the US market—the hiptop® device.[4] This high-speed wireless communicator was "always-connected" to Danger's back-end server, which formed the backbone of the full Danger product.

In October of 2002, Hank Nothhaft joined Danger, Inc. as chairman and chief executive officer. Hank had been a successful serial entrepreneur in the telecommunications industry over the past 30 years, and had gained deep insights about how carriers operated in the United States and around the world. His experience was particularly important to Danger because the carriers owned the next generation data networks that provided the wireless connectivity for the Danger platform. As a result, Danger needed to form close strategic relationships with the wireless carriers to successfully roll out their product to the mainstream market.

Hank also attracted three of his former business colleagues, James Isaacs, Les Hamilton and George Carr, to Danger. They had worked together at Concentric Network Corporation and had nurtured it from a small start-up to a multi-

---

[4] Xircom was acquired by Intel Corporation in 2001 for US$748 million.

million dollar operation, executed a successful IPO in 1997 which was then acquired by XO Communications in 2000. Their business experience in working with wireless carriers complemented the technological expertise of the original founders of Danger. Nancy Hilker also joined Danger as its chief financial officer during this time period. With this leadership team in place, Danger, Inc. was ready to move into its next phase of development.

Danger's Business Model

Danger identified eight different functions that needed to be performed to provide consumers with end-to-end wireless service: 1) hardware design for the handheld device; 2) client side software and user interface; 3) developing and running a scalable back-end service; 4) manufacturing; 5) consumer marketing (e.g., mass media advertising and brand building); 6) consumer billing; 7) customer support; and 8) product distribution. As a start-up with limited resources, Danger elected not to engage in the capital-intensive functions such as manufacturing or consumer marketing. They focused on their core competencies and partnered with other companies to deliver the full end-to-end solution. Danger's leadership team believed that their core expertise was their technical know-how. As such, it was natural for them to focus on the technical pieces of the solution, which were the hardware design, client side software, and the back-end service. The back-end service was a series of software applications software and developmental tools for both in-house use as well as third party developers. This was Danger's most important intellectual property asset.

In the words of John Arledge, VP of Business Development, Danger "chose to use a hardware-driven service business model." This meant that most of the revenue came primarily from developing software and running the back-end service. They charged their carrier partners a licensing fee for each subscriber that used Danger's back-end service, and that became a recurring revenue stream through the lifetime of that subscriber. The carrier could choose to either host the service at their internal data centers or outsource the service to Danger. There was also a small revenue stream from licensing the software on the client. However, Danger chose to make no money on the hardware device itself. John Arledge explained:

> The hardware is a very important part of what we are doing. Without the hardware, without the slick user interface and industrial design, we would not have nearly the success that we have today. So hardware design, to an extent, is a core expertise of ours, but we have chosen not to make it a profit center. It is a cost center. It is a cost of doing business. And we use it to get into the door, and to sell carriers and end-users the concept.

By combining a zero hardware margin business model with their technology, Danger enabled carriers to offer their end-users a wireless handheld device at a much more affordable price compared with solutions from other companies such as Palm, HandSpring, Sony, Nokia and RIM. The T-Mobile Sidekick, the first example of a Danger hiptop® device, had a recommended retail price of

$199 in October 2002. The functionality of the T-Mobile Sidekick could be compared with other wireless communicators such as the Treo 180 and the Palm Tungsten W that were retailing at a price point of at least $399. Exhibit 4 shows a comparison of the different wireless communicators that were available in the market in early 2003. Exhibit 5 shows some of the wireless service plans that were offered by the carriers for these communicators.

Apart from choosing to sell the handheld device at cost, Danger's product architecture was another key factor that allowed the company to make an equivalent device at half the price. Since most of the processing was done on Danger's software residing on back-end servers, the components on the handheld could be kept to a minimum without compromising performance. Suppliers of Personal Digital Assistants (PDAs) and Smart phones, on the other hand, needed to continually innovate their hardware. They often packed new products with expensive components to power compelling features that would encourage end-users to upgrade their old handheld devices. With the Danger hiptop device, innovations could be introduced in its back-end service that would improve the performance and the functionality of the client-side handheld. Hank Nothhaft explained:

> If you were to buy a Motorola phone, you would get the same set of features no matter which carrier you buy the phone from. With us, the carrier gets to decide on the set of features that they want based on the demographics of their target customer. So, we didn't choose to go after the 18-34 year old consumer market. Our first customer, T-mobile decided that they wanted to go after that market segment, and asked the company to create a specific set of services that would appeal to them. This included e-mail, instant messaging, web browsing, calendaring, SMS, digital camera, and then later on, we added a phone function. We can do this efficiently, because most of the work or the heavy lifting is done in the back-end in the data centre.

This business model and pricing strategy perfectly aligned Danger's business objectives with those of the wireless carriers. Danger's zero margin hardware pricing strategy and low-cost licensing fees for their software for handheld devices allowed carriers to reduce their customer acquisition costs. Carriers incurred lower costs when they offered a Danger handset at a discount to entice a customer to sign up for wireless data services. Furthermore, carriers were charged a monthly licensing fee based purely on the number of subscribers that were signed up to the Danger service. This "success-based pricing" gave both Danger and the carrier strong incentives to improve the quality of the data services that they offered, in order to improve customer retention, increase average revenue per user (ARPU), and increase the lifetime value of each customer.

Danger Product Overview[5]

Danger provided an end-to-end solution that enabled wireless carriers to offer next generation wireless applications to their customers. These applications used

---

[5] This section is adapted from Danger Product Overview found in the Danger press kit.

high-speed, always-on wireless data networks based on the GPRS/GSM standard that were slowly being rolled out in 2003 in different countries worldwide. Danger's solution consisted of three key components, 1) the hiptop Device Design; 2) the hiptop Development Platform; 3) The hiptop Service Delivery Engine (SDE). This is further illustrated in Exhibit 6.

Hiptop® Device Design

Danger's hiptop device was a wireless handheld that was able to take full advantage of high-speed wireless data networks. As a component of its solution for wireless operators, Danger offered the hiptop Device Design to carriers with a set of features and applications that could be further customized for their specific target demographic. Each customized hiptop device design was also branded by the wireless carriers, e.g. the T-Mobile Sidekick. The hiptop device was an "always-on" handheld that was connected to the GSM/GPRS network, enabling end-users with real-time access to email, instant messaging and the Internet simultaneously. The hiptop also incorporated a full-featured phone, personal information management (PIM), entertainment applications, and a camera accessory. Exhibit 7 shows some snapshots of the application programs that were available on the hiptop device.

Some of key applications that were available on the hiptop device design were:

- **Mobile Phone**—Users could send and receive phone calls, store phone numbers, speed dial, customize ring tones, voice mail, call log. The phone could be used by placing it against one's ear or by using a hands-free headset. Users could send and receive messages from friends whose phones supported Short Message Service (SMS) text messaging.
- **Email**—Email could be sent and received with PDF, Word, and image attachments for up to three external POP accounts.
- **Instant Messaging**—Users could use their existing instant messaging screen names to chat with their friends. Multiple Instant Messaging programs (including AOL) were supported by the hiptop device.
- **Web Browsing**—Built-in browser enabled access to the Web with native HTML pages. Site text and graphics were optimized for viewing on the device.
- **PIM**—Included a full-featured personal information manager including Calendar, Address Book, Notes, and To Do. All PIM functionality was fully integrated with messaging and phone applications.
- **Snapshot Gallery**—Stored and managed low-resolution snapshots that were taken with the digital camera accessory. These images could be instantly emailed to family and friends.
- **Games**—A selection of games that used quality sound, vibration and color scroll wheel to provide multimedia entertainment.

The hiptop device was the most visible component to the market. This was partly because the hiptop device could be purchased at carrier store outlets and be carried around in the streets. Furthermore, the hiptop device had been featured on numerous consumer technology magazines and television programs and had garnered excellent reviews. Exhibit 8 shows the reviews given by the media.

The hiptop design had also won numerous industry awards. One of these awards was TechTV's 2002 Best of Consumer Electronics Show (CES), which honored the most outstanding new consumer technology product each year. Danger's hiptop won the best product in the "PDA, Handheld and Mobile Wireless" category when the device debuted at this show in January 2002. It also won the Best Industrial Design at WIRED Magazine's Rave Awards, and Best Products of 2002 (PDA Category) by Handheld Media Group. While these numerous awards generated positive press and word of mouth about the company, they also had the potential to confuse Danger's future customers—wireless carriers—about Danger's business model and potential value to them as a software and services provider. They also had the potential to confuse potential allies—branded hardware manufacturers—who might view Danger as a possible competitor.

Hiptop Development Platform

The hiptop Development Platform enabled third-party application developers to use standard development tools such as Metroworks CodeWarrior-J™ and Microsoft Visual J++™ to write mobile applications for the hiptop device, which used the proprietary hiptop Operating System (HTOS). The hiptop device itself did not run the Java program directly. Instead, a piece of software running in the backend service took the class files coming out of a Java compiler and modified it so that it could be sent to the device. This conversion and compression technology made the program more compact and efficient on the hiptop device. At the same time, it allowed third-party application developers to use the normal Java 2 Micro Edition (J2ME) environment that they were familiar with to write programs for the hiptop device.

Danger provided a complete Software Developers Kit (SDK) that included all the required software libraries, a resource compiler, a device simulator, sample code, and HTML-based documentation to assist developers in their development efforts. A complete set of application programming interfaces (APIs) was included to enable development of applications. The hiptop Development Platform included an integrated download management system to enable the vending capabilities of third-party applications. This would enable carriers to offer revenue-generating after-market premium services and content provided by third party software developers.

Hiptop Service Delivery Engine (SDE)

The hiptop Service Delivery Engine (SDE) was a suite of server-based software hosted as a service for wireless operators, enabling them to quickly deploy new applications over their next-generation data networks. SDE was designed to evolve with advancements in hardware, software and network standards. In a space where wireless standards were still evolving, this design reduced the risk of adopting this service and provided a "future-proof" platform for wireless content and service delivery. Key features of the SDE included:

- **Session management and load balancing**—The SDE provided individual session management, enabling the seamless transition and preservation of user settings and data between the hiptop device and the SDE as the user moved in and out of wireless network coverage. For example, with session management, if a user were engaged in an Instant Messaging conversation and lost network coverage, the hiptop device would be updated with any messages that were missed when the user re-entered a coverage area.

- **Wireless Synchronization**—With wireless synchronization, any data entered on the hiptop device was automatically synchronized instantly to the Web portal via the SDE, without the need for user intervention. Any browser-enabled PC could then access this information through the Web portal.

- **Application servers and proxies**—Application servers and proxies provided server-side processing for the applications running on hiptop devices. In early 2003, the SDE had an email server, an IM proxy, web proxy, as well as a full PIM server.

- **Content and application vending**—The SDE provided an integrated download management system, the Premium Download Manager (PDM), which enabled over-the-air (OTA) delivery of software and content to hiptop-enabled devices.

- **Flexible and modular framework**—The framework had a modular architecture with interfaces based on industry standards. This enabled carriers to integrate their existing systems and applications with the SDE software.

All three components of Danger's whole product were critical to enhance the carriers' ability to attract and retain customers.

Mobile Internet for the Mass Market

Danger believed that the Internet had become a necessary source for information, communication and entertainment in the daily lives of ordinary consumers. The Internet had billions of pages of web content that could be easily accessed when consumers logged onto the Internet. The personal computer served as an effective gateway to the Internet, but its mobility was limited. Joe Britt, CTO of Danger, explained,

> I like to use the analogy of the telephone. When telephones were only available as landlines, people knew that the value of the phone was in the content—other phone users. However, you couldn't carry the phone with you because it had a fixed connection. That was the great thing about cell phones. People got it instantly. People knew what the phone was used for, and knew the value of having it with them all the time. We applied that same school of thought to the Internet: People already knew the value of the Internet, but there were no really good consumer-focused mobile Internet access devices. Through our software and networking innovations, we have

created a compelling solution with a rich graphical interface that makes Internet content available all the time.

From their experiences, the founders knew that the thin client model was well suited for consumer products, because it made a complicated device easy to use. The successful implementation of WebTV was a clear example of this. Users could surf the web easily with a simple remote control and did not have to worry about TCP/IP addresses, POP servers, SMTP or even phone numbers. All of this was hidden because there was a backend data center that was processing the data for the user. In the case of Danger, a complicated machine such as the personal computer that was required for accessing Internet content, could be reduced to a relatively simple handheld with an intuitive user interface. The technical configurations and processing could be done in the back-end server and pushed to the device when necessary. From Hank's perspective:

> Danger's founders understood this business model very well. In a way, you could call Danger *WebTV on steroids,* or *WebTV on wireless.*

One of the key benefits of this system was that it significantly reduced the amount of bandwidth required to have the Internet available on a mobile device. If a user were to wirelessly download the actual Internet page onto a mobile device, a lot of bandwidth would be required. Following that, the device would have to process the webpage and scale it down to a size that fit its screen, and discard the majority of downloaded information. There was tremendous waste in terms of processing power, battery power, and bandwidth. By using a client-server model, where most of the processing was done on the server, the interaction between the hiptop device and the target website became much more efficient. When a user logged onto a website from the hiptop device, a small request would first be wirelessly transmitted to the back-end service. The server would then download the HTML webpage, re-format and configure its contents to suit the device, before compressing the data and transmitting it back to the hiptop device. In this way, the hiptop was able to access any webpage that was available on the Internet, yet maintain efficient bandwidth and power usage.

This model also provided a significant amount of protection against obsolescence for the consumer. During a time when Internet standards were still rapidly evolving, consumers were more careful about investing in a product that was unable to support new messaging protocols, attachment types or even new types of web content. By having most of the processing done at the data center, bug fixes and enhancements could be updated at one central location, instead of pushing software updates out to each device.

In commenting about the consumer market, Joe remarked:

> In the enterprise market, you can get away with a poor user interface; you can even have some bugs. The users view products purely as tools. In the consumer market, users take a more personal interest in the products they choose. Most successful consumer products are sexy and engaging. Consumers want to buy products that are inexpensive, fun to use, and that will impress their friends.

Competition

As mentioned earlier, industry analysts often considered Danger, Inc. to be a PDA manufacturer. This was an inaccurate categorization of the company because it did not reflect the end-to-end solution that Danger offered to the carriers. Research in Motion (RIM) and Good Technology were two companies that provided better comparisons because they also had end-to-end solutions. In addition, RIM had an existing alliance with Nextel, demonstrating their desire to move into a similar competitive space. However, their solutions were focused mainly on secure email synchronization and did not have the versatility of the Danger platform. Furthermore, their solutions were primarily targeted at enterprise users.

Some potential customers such as AT&T and Vodafone were also trying to construct their own end-to-end wireless data systems, which overlapped with Danger's product offering. For example, AT&T Wireless had 400 in-house employees focused on content development for Wireless Application Protocol (WAP). As a result, they were less willing to invest in the Danger platform because it posed as a competitive threat to their own development efforts. Other carriers could also conceivably choose this route and build their own internal systems. On the other hand, AT&T's in-house development project could also be used to convince other carriers that Danger could enable them to rapidly deploy revenue generating wireless data solutions and gain market share without incurring the cost and risk of an in-house, custom system.

There were other sources of competition that concerned the Danger team. Nokia was a potential competitor because they had extended their business model beyond hardware to include services as well. Club Nokia was an online community and loyalty program that offered WAP services to Nokia phone owners. Nokia also created a smart-phone reference design called the Nokia Series 60 that it licensed to other mobile handset manufacturers, including Matsushita and Samsung Electronics. This platform used the Symbian OS and supported Multimedia Messaging Service (MMS), JAVA applications and content downloads. Nokia could be contrasted with Samsung Electronics who was a pure hardware provider, who had licensed software from Nokia, Palm, Symbian and Microsoft. Another potential competitor was Microsoft, who had introduced a new operating system—the Microsoft Smartphone 2002, designed specifically for next generation phones. The first mobile phone that used this new operating system was the Orange SPV that was co-launched by Orange and Microsoft in November of 2002. Analysts believed that Microsoft intended to extend its software application capabilities into the services space through this new operating system.

Carrier Relationships

Danger's business model required it to attract wireless carriers to become both customers and business partners.

Danger had strong incentives to develop partnerships with the wireless carriers. In addition to wireless connectivity, carriers could also perform several of the eight functions that were described earlier in the Danger business model. Carriers had significant experience in consumer marketing/brand building, and committed a substantial budget each year for that function. Danger could ride on the coattails of their partners' multi-million dollar marketing campaigns, and avoid many of the costs and risks of big-ticket brand building. They had already implemented complex billing systems for their customers, so Danger could simply develop interfaces to those systems and let the carriers do the end-user billing. Carriers already employed thousands of Customer Support Representatives (CSRs) who answered the phones 24/7 to provide support to end-users. Carriers also had an extensive distribution system. They purchased handsets from end terminal manufacturers and distributed them, bundled with a variety of wireless service packages, through their own stores and independent retailers. In other words, the carriers were able to provide all the functions that Danger wanted to outsource except handset manufacturing. As a result, Danger's leaders saw the carriers as more than customers. They had the potential to be strategic business partners, complementing Danger's strengths, and reducing costs and risks as Danger took its new technologies to market.

Carrier Industry Background

The wireless carriers owned the next-generation data networks for consumers who wanted to use handheld devices to improve their lifestyles or become more efficient on the job. These wireless providers, especially the European carriers, had spent an estimated US$100 billion for 3G licenses in government auctions, and were expected to invest an equivalent amount to build out the 3G infrastructure. GPRS investments were significantly lower because GPRS ran on existing GSM network infrastructure.

In 2003, there were very few applications in the market that required high-speed wireless data networks. Although most industry analysts believed that wireless data was the next big opportunity for the carriers, there was still a great deal of uncertainty about how the opportunity would be realized. Danger's leadership team believed that their solution could demonstrate the full potential of high-speed wireless networks and deliver data services that were compelling to end-users. By working with Danger, a carrier could generate additional revenues from existing customers, and attract new ones, thereby providing a higher return on its network investment.

The hiptop Service Delivery Engine (SDE) allowed the carriers to characterize and control the data usage of each subscriber. Since most of the computing work was done in a central back-end server, the amount of wireless data that was transmitted to users' handsets was significantly reduced. This made it possible for carriers to estimate the maximum amount of wireless data usage per user each month. Contention for wireless bandwidth made this factor particularly important because it prevented subscribers from clogging up the entire network. Thus the hiptop Service Delivery Engine (SDE) made it possible for carriers to offer unlimited data usage to their subscribers at a fixed rate, without having to worry about slowdowns and crashes due to network congestion. Fixed prices and more reliable service availability would make wireless data more appealing to subscribers, and thereby increase their appetite for wireless applications.

The carrier business was extremely competitive and capital intensive. In addition to acquiring licenses in government auctions, carriers had to invest millions of dollars to purchase telecommunications equipment and software from their vendors before they could start offering services to their customers. Since they bought the same hardware from network equipment providers such as Cisco and Nortel, the services that they were able to offer were very similar. If two or more carriers provided the same coverage in an area, they usually wound up competing primarily on price. This was a cutthroat business, and some carriers were looking for ways to differentiate themselves.

T-Mobile: Danger's First Carrier Partner

In describing their first carrier customer, James Isaacs, VP of Worldwide Sales and Alliances, explained:

> T-Mobile was a combination of luck and fortuitous meetings, but also some amount of calculation. This was because T-Mobile was behind the leaders like Verizon and AT&T, in terms of the number of subscribers. They were willing to gamble a bit on a breakout strategy and that was what Danger represented for them. And most of the other carriers would not do what T-Mobile did, which was to commit to buying a significant number of units, which amounted to a high capital expenditure. They were basically sponsoring us into existence, putting their bets on an unknown unproven startup. Don't mistake the first big customer for your market. The other carriers are much more risk-averse, more methodical.

The financial statements for T-Mobile are shown in Exhibit 9. Danger understood that helping T-Mobile to improve its financial performance was critically important to the continued success of that partnership, and also to encourage other carriers to do business with Danger. Since the market for wireless data was still in its infancy, it was a challenge to forecast how working with Danger might actually improve T-Mobile's bottom line. However, most wireless carriers were skeptical of technological innovations, and wanted credible evidence that working with Danger would help them to increase revenues, reduce costs, and improve their profitability.

Challenges in Partnering with Carriers

The selling process to carriers was a long one. It took between 12-18 months to complete a sale with a potential carrier customer. The Danger team designed a standard selling process to help keep track of the prospects. This process is shown in Exhibit 10. Since there were only about 50 major carriers worldwide, the list of prospects was fairly straightforward. Danger had to decide which carriers that they wanted to target and then make themselves attractive to those carriers. James explained:

One of the things that we need to do is to figure out a way to reduce the up-front commitment and the perceived opportunity cost of launching with Danger. We are introducing a rapid deployment guide, where there are fewer configurable and customizable elements, and where the back-end authentication and systems are more easily hooked up to the carrier's software systems. We want to make it easy, make it comfortable, while at the same time, extracting enough of a commitment that it is worth Danger's while. We believe that Danger's services will sell well, but we really need to get over a hump so that the carriers will work with us. That's the challenge.

Danger's first challenge was that carriers were unwilling to commit to partnership agreements that might threaten their long-term competitiveness. For example, Danger realized early on that carriers were reluctant to wholesale their networks to third parties. Carriers would not agree to allow a startup like Danger to buy capacity from them, and resell it along with value added services to end-users. Andy explained:

We did think about going it alone, but it took a couple of phone calls to realize that the carriers would not be willing to wholesale their network. We could not call the wireless carriers and say that we wanted to buy the data at $1/MB and sell it for $2/MB. This was because they did not want to be dumb pipes. They wanted to have a key role in developing next generation wireless services and applications. We had to think of a way to partner with the carriers so that we were able to effectively help them with the innovation.

Second, carriers also wanted to protect the investment they had made in building their brands, and were reluctant to promote other brands that might compete for the loyalty of their customers. As a result, many carriers felt threatened if a third party offered a branded service for their end-users. They guarded their customer relationships jealously, and wanted their own brands to be associated with any innovative services provided over their networks.

To address this concern, Danger decided to remain unbranded, and allow the carriers to put their brands on Danger-enabled devices. In this way, Danger acted as a "silent partner," working behind the scenes to build the brands of its carrier partners. Incidentally, the company name "Danger" became an advantage in carrier relationships, because it clearly demonstrated that Danger had no desire to put their brand on the devices for end-users. After all, who would want to buy a wireless product marked "Danger?"

The third challenge facing Danger was the fact that some carriers demanded exclusivity. For example, Danger granted its first customer—T-Mobile, some elements of exclusivity over the hardware form factor and design in the United States for one year. This meant that end-users in the US who were interested in a hiptop-enabled device had to sign up for a wireless plan with T-Mobile. The exclusivity agreement enabled T-Mobile to differentiate its offer from its competitors, and gain market share relative to other wireless carriers during the period of exclusivity.

However, exclusivity came at a high cost to Danger, because it created difficulties when the sales team engaged with other carriers who operated in the same geography. As the market demand for hiptop-enabled devices increased, and as wireless carriers migrated to new geographic regions via mergers and acquisitions, Danger's opportunity costs and risks in offering exclusivity would increase substantially. Offering exclusivity would also automatically prevent Danger's software solution from becoming an industry standard, thereby making both Danger and its carrier partners more vulnerable to competition from other software platforms. What appeared to be a short-term advantage for a carrier could actually become a major disadvantage in the long term. For these reasons, Danger needed to formulate an effective response for situations when a carrier requested exclusivity as a condition for entering a partnership. How could Danger convince carriers that it could help them to differentiate themselves from their competitors *without* granting them exclusivity?

The fourth challenge facing Danger was the lack of uniform standards among U.S. carriers with regard to wireless protocols. As was mentioned earlier, some wireless carriers in the United States had adopted GSM/GPRS, while others supported CDMA. This meant that because Danger supported only GSM/GPRS in 2003, it could only address about 50% of the U.S. market. Because all of the European carriers had adopted GSM, Danger's solution could potentially address 100% of Europe's end users. Exhibit 5 provides examples of carriers that were supporting GSM/GPRS and CDMA in 2003.

The fifth challenge facing Danger was how to manage relationships with sworn enemies. Danger had to be both diplomatic and strategic in their interactions with the different carriers. The carriers were intensely competitive and were on "a mission to grab, steal, rob market share from one another." The executives leading one of Danger's carrier partners might be personally offended if they perceived that Danger was helping their competitors to gain market share at their expense. As a result, Danger had to be very careful in managing their portfolio of current and potential carrier customers. James gave an analogy:

> One of my friends described the wireless space is like being on a chess board with four queens. You know that if you go this way, you will get your legs sawed off, or that if you go that way, you will get run over. For a small company like us, we need to figure out how to go in between the white and black squares on the chess board, and try to ooze our way through.

From Danger's point of view, establishing partnerships with multiple carriers was necessary to survive. From some carriers' perspective, if Danger was doing business with their most direct competitors, the notion of partnership was unthinkable. They demanded fidelity as a requirement for a true partnership. Ironically, some of these same carriers had no qualms about working with Danger, and also with Danger's competitors. Most carriers were unwilling to consider making Danger's software solution their exclusive platform for wireless data, because they had to hedge their bets.

Potential Carrier Partners Worldwide

Although Danger had offered T-Mobile limited exclusivity for one year, the time had come to begin courting other wireless carriers in the United States. In addition, Danger was also pursuing partnerships with several European carriers. They were in active conversations with major players such as Vodafone, Orange, O2, Telefonica, etc. Some of the carriers wanted specific software applications such as multimedia messaging service (MMS) that the technical team was in the process of developing. Danger was hoping to launch a European carrier by the summer of 2003. Danger had been less active in marketing to carriers in Asia, primarily because it did not yet have a large enough sales force to effectively cover that region. They were considering the possibility of expanding their sales operations in Asia in the future.

Exhibit 11 shows the financial highlights for a representative sample of wireless carriers who some industry observers believed had the potential to become Danger's partners in the long term. It was not obvious how many of the companies in Exhibit 11 Danger could enter into partnerships with, because some were fierce competitors with others. Was it possible for Danger to manage trustworthy partnerships with two carriers that were bitter enemies?

Manufacturing Relationships

Developing partnerships to manufacture hiptop-enabled handheld devices was another critical element of Danger's business model. Manufacturing was extremely capital intensive, and it was unrealistic for a start-up to perform it in-house. Danger was exploring partnerships with three different types of manufacturers: 1) Contract Manufacturers (CMs), 2) Original Design Manufacturers (ODMs) and 3) Branded Original Equipment Manufacturers (Branded OEMs). Exhibit 12 gives a breakdown of the manufacturers and the vendors in this space.

Contract Manufacturers (CMs) provided the least number of services of the three categories. They were focused on one basic function, which was to assemble hardware. With a reference design furnished by their customer, their job was to order the components from their suppliers, to assemble the hardware in their factories, and to ship the final product to their client. They undertook no design risk and no warranty risk. Materials risk and inventory risk were usually underwritten by the CM's customers. In good years, CMs had gross margins of 3% or less, and needed to generate high volume contracts to remain profitable. Examples of CMs in the market were Solectron, Benchmark and Jabil Circuit.

Original Design Manufacturers (ODMs) offered more services than the CMs. They had the capabilities to do design work, and thus assumed some design risk and warranty risk. Many Taiwanese CMs were moving into the ODM category because of the fear that contract manufacturing would become completely commoditized and drive down profit margins even further. In good years, ODMs usually had gross margins of 15%. Examples of ODMs included Quanta, Flextronics and HTC.

Branded Original Equipment Manufacturers (Branded OEMs) provided even more services than the ODMs. Apart from assembly and design, they also provided marketing and logistics services. They maintained close relationships with retailers and had the ability to perform reverse logistic processes where they collected broken handsets, repaired them, and returned them to the channel. A Branded OEM also provided value because of its brand name, which could be appealing to the end-user. Most of these companies also managed a complex inventory system and thus assumed a certain amount of inventory risk. They usually had gross margins of 30%. Examples of Branded OEMs were Nokia, Samsung and Sony-Ericsson.

Exhibit 13 shows a pictorial representation of how Danger and the carriers would interact with manufacturers using each of these models.

Danger wanted to have relationships with each of these types of manufacturers to best meet the needs of different carrier customers. For some carriers such as T-Mobile who were very flexible but very price-sensitive, a CM that provided few services at a low price would be most attractive. Other carriers such as AT&T had a selected list of approved vendors, such as Nokia or Motorola, and wanted more services, and thus they preferred to work with a branded OEM. John described this situation:

> We don't want to push any of our partners out of their comfort zone. We realized that every time we try to change a company's momentum to fit some other business model, even if it may suit us in the short term, ultimately, it will regress back to their standard business model, and we get into trouble. So, the idea is to keep everybody in his comfort zone. So if AT&T doesn't want to buy from a CM, let's not force a CM into AT&T's model, let's go get an OEM that they are familiar with.

In the first quarter of 2003, Danger's business development team was considering three different partnership models with the manufacturers. In the CM model, Danger would enter into a contract manufacturing relationship with a CM such as Solectron who would produce hiptop devices for the company. Following that, Danger would then sell the handhelds and provide the back-end services to the carrier customer. There were two disadvantages of this model for Danger. First, it meant that Danger had to assume hardware and inventory risk. Second, Danger had to enter into a complex handset relationship with the carrier.

In the ODM model, Danger would license the hardware reference design it had created to an ODM and provide back-end services to the carrier customer. The ODM and the carrier would then enter into a direct relationship to manufacture handsets to the carrier's specifications. This relationship was less costly and risky for Danger than working with a CM, because Danger only needed to provide consulting services to the ODM and back-end services to the carriers. Assuming that the carrier was willing to deal with the ODM, they would issue a standard contract to the ODM, and treat them like another handset vendor, like Sony-Ericsson or Nokia. This model might be less attractive to an ODM such as Flextronics because they would have to deal with the intricacies of retail, such as marketing co-op dollars, seed stock, inventory turns, price protection, etc. that they were typically isolated from when they manufactured handsets for branded OEMs.

In the branded OEM model, Danger's relationship with the manufacturer would be similar to the ODM model. However, in this case, a branded OEM such as Samsung would already have an established handset relationship with the carrier, with all of the intricate terms and conditions worked out. The hiptop device would simply become another stock-keeping unit (SKU) in the purchase order from the carrier. Some branded OEMs such as Sony-Ericsson did not have any manufacturing facilities and differentiated themselves purely by the design capabilities. In this case, Danger would provide the reference design to a CM who would then sell the units to the branded OEM. Another possibility was to have the branded OEM subcontract the manufacturing to one of their CMs. This model allowed each player to maintain any existing relationships, thereby saving time, reducing costs, and reducing implementation risk for all parties.

To attract ODMs or branded OEMs to manufacture hiptop-enabled devices, Danger decided that they would license the hardware reference design royalty free to the manufacturers. This would encourage the major manufacturers to consider Danger as a potential partner, especially if they believed that the hiptop device could become a profitable line item in their product portfolio. The manufacturers could also create derivatives of the basic design, such as adding an MP3 player, built-in cameras, different form factors, etc. Danger would then engage with the manufacturer in a design-focused partnership where it could help develop software to support new hardware features and capabilities. Matt explained:

> The great thing about our architecture is that it makes minimum demands on the hardware. Even though we have a captivating hardware design, our unique point of value is not about advancing the hardware on all fronts. Although hardware design is not trivial, there are a lot of companies that understand it well, so instead of competing against them, why don't we simply turn them into our partners by licensing them the reference design?

Decision Time

As Danger's leadership team assembled in a company conference room, three issues related to developing their portfolio of partnerships were on everyone's mind. First, which wireless carriers were the leading candidates for Danger to pursue to increase their market penetration in the U.S. and around the world? Danger had to walk a fine line to ensure they could deepen and expand their existing carrier relationships while also adding new carriers.

Second, which manufacturers were the most attractive potential partners to build handheld devices compatible with Danger's hardware design and software architecture? In particular, which Branded OEMs could best serve Danger's potential wireless carrier partners? Could partnering with a particular branded handset manufacturer make it more difficult to attract some wireless carrier partners?

Third, how could Danger develop trustworthy partnerships with companies that were often in fierce competition with one another? As a small startup, Danger had a great deal to gain if it could earn the trust of bitter rivals in the wireless carrier market. They also had a great deal to lose if signing up a few carriers caused others to align with Danger's competition.

Back in Danger's conference room, it was time to decide.

Hank: Our first launch with T-Mobile has been very successful and the sales of T-Mobile Sidekicks are exceeding expectations. As we had predicted, there is significant demand for a low-cost high-performance wireless data device. However, more work needs to be done. It's time to negotiate partnership agreements with other carriers.

Andy: I agree. We need to convince the carriers that we have a service that appeals not only to the early adopters but also to the mass market. Both Danger and our carrier partners will benefit from getting as many hiptop devices out into the market as possible. To maintain a ready supply of hiptop devices, we need to partner with several large manufacturers. In addition, branded OEMs will also introduce hardware innovations to the hiptop product line that will appeal to a wider variety of consumers. Our two most important partners are wireless carriers and handset manufacturers. We need to have a clear game plan for how to engage them going forward.

James: The carrier space is extremely competitive and so we need to be very careful in our interactions with them. We don't want to incur the wrath of a potential partner or sign too many exclusive deals that could shut us out from future opportunities. The list of prospects is pretty straightforward. There are only 6 carriers in the US that have a national footprint. Given that we already have an alliance with T-Mobile, which of the other U.S. carriers should we target as our highest priority? The other two regions that have a significant number of wireless users are in Europe and in Asia. What approach should we take in those regions? Which specific carriers should we target in each region? I have my own ideas on priorities, but I'd like to hear from the rest of you.

John: With regard to the manufacturers, we already have deals with a CM and an ODM, but not a Branded OEM. What do we need to do to secure a contract with a Branded OEM? Nokia is the clear market leader in the Branded OEM's space, and a deal with them would definitely give our company a lot of exposure. However, Nokia has a competing platform and seems to have interest in extending their business to wireless service. Perhaps a Branded OEM like Sony-Ericsson that is focused primarily on differentiating based on innovative hardware design might be a better potential partner for Danger.

Matt: Some manufacturers like Samsung have both design and manufacturing capabilities. A partnership with them will be reasonably straightforward.

On the other hand, they are completely agnostic to software and have done deals with everyone, including Palm, Symbian, Windows CE and Nokia's Series 60. They might be willing to work with us, but it might be quite difficult to cut a deal that gives us preferential treatment.

Joe: Many of the Branded OEMs already have existing relationships with the wireless carriers, and so we have to take that into account when analyzing potential partnerships. These OEM—carrier relationships are very sensitive and we need to be aware of the implications that a Danger relationship with one partner might have on other relationships. In a market where there is such fierce competition, we need to think of ways to develop trust and collaboration with all of our prospective partners. We don't want to make it difficult for them to partner with us.

Hank: We have identified some critical factors that will affect our future partnerships. Now let's put together a game plan for building Danger's portfolio of partnerships.

EXHIBIT 1 Biographies of selected key executives in Danger, Inc.

Hank Nothhaft, Chairman and Chief Executive Officer

As Chairman and Chief Executive Officer, Hank Nothhaft is responsible for managing all functional areas of the company—including sales, marketing, business development, finance, and engineering.

Hank has been a pathfinder in the telecommunications and networking industries, growing pioneering startups, such as Concentric Network Corp., DSC Communications and GTE Telenet Communications (now Sprint), into mature industry leaders. Hank most recently was the chairman, president and chief executive officer of startup SmartPipes, Inc., a leading provider of policy-based software that simplifies and automates the design, deployment and management of IP VPN networks, firewalls and remote-client software.

Before joining SmartPipes, Hank served as President and CEO of Concentric Network Corporation from 1995 to 2000, where during his tenure he raised $380 million in equity financing, created $2 billion in shareholder value, and drove the company's rapid growth through strategic distribution agreements with major telecommunications and technology companies such as SBC Communications and Microsoft.

Prior to Concentric, Hank held several roles at DSC Communications Corporation, including Senior Vice President of Marketing, and Group President, in addition to serving as a Member of the Board of Directors. During his tenure, Hank helped grow DSC from a $10 million startup to a $650 million business that was eventually sold to Alcatel for $4 billion.

Hank also held the roles of Vice President of Marketing and Vice President of Sales for GTE Telenet Communications Corporation (now Sprint), which pioneered x.25 networks, the initial public data networking protocol.

Hank has an M.B.A. from George Washington University and a B.S. degree from the U.S. Naval Academy, and he is a former officer in the U.S. Marine Corps.

Andy Rubin, President and Chief Strategy Officer

As President and Chief Strategy Officer, Andy Rubin brings more than a decade of experience in consumer products and technology to Danger. Andy is the visionary behind the corporate strategy and manages Danger's product and technology roadmap.

Before co-founding Danger, Andy led the communications engineering team at WebTV Networks. He was instrumental in building and shipping WebTV, the first interactive television-based Internet service, which was acquired by Microsoft in 1995.

Prior to WebTV, Andy designed and implemented the communications capabilities for General Magic's handheld devices. In 1993, Andy led the effort to ship the Motorola Envoy, one of the first wireless PDAs (personal digital assistants). Andy has also held various leadership roles focused on communications technology with Apple Computer, where he and his team shipped the world's first host-based software modem.

In his spare time, Andy enjoys tinkering with his mobile robot. Some of his robotic creations can be found wandering the galleries at the Tech Museum of Innovation in San Jose, California.

Andy holds a B.S. in Computer Science from Syracuse University.

John Arledge, Vice President of Business Development

As Vice President of Business Development for Danger, John Arledge is responsible for establishing and maintaining relationships with content partners, wireless carriers, consumer electronics manufacturers and technology providers.

Prior to joining Danger, John was director of business development at TiVo, where he managed the company's relationships with pay television service operators, consumer electronics manufacturers, and technology suppliers. While at TiVo he signed major agreements with Philips, Sony, Thomson Multimedia, Quantum and Liberate.

Before TiVo, John was responsible for product marketing at OpenTV, Inc. overseeing the OpenTV operating system, which now resides in 11 million set-tops. He also secured and maintained key partnerships with corporations such as IBM, LSI Logic and Pace. John has also held marketing positions with Pacific Bell in the broadband networking group and with Nissan North America in the product-planning group for the Nissan Truck line.

John holds an M.B.A. from the Stanford Graduate School of Business and a B.A. in Political Science from the University of California, Los Angeles.

Joe Britt, Chief Technology Officer, Senior Vice President of Software

As Chief Technology Officer and Senior Vice President of Software for Danger, Joe Britt is responsible for the intellectual property and software aspects of Danger's technology. Joe brings more than 13 years of experience building consumer products to Danger. His specialty is designing system software for consumer electronic devices.

Prior to co-founding Danger, Joe spent over four years at WebTV Networks, which was acquired by Microsoft in 1995. As the first non-founding employee, he was responsible for the architecture and creation of the system software used in the WebTV set-top boxes. Joe was involved in the design of every hardware product shipped by WebTV and has been awarded nine patents as a result of his work.

Before WebTV, Joe worked at Catapult Entertainment, since acquired by Hearme. Joe was part of the team that created the Xband Video Game Network that enabled multi-player gaming over the Internet. Joe contributed to the system software as well as the technology required to enable video games for network play. Before Catapult, Joe worked at the 3DO Company, contributing to the design of a game console powered by a PowerPC CPU.

Joe started his Silicon Valley career at Apple Computer at age 20, working in the RISC (Reduced Instruction Set Computer) Products Group. Joe was a core member of the ROM (Read Only Memory) development team for the first generation PowerPC-based Macintosh.

Joe holds a B.S. in Computer Engineering from North Carolina State University.

Les Hamilton, Senior Vice President of Worldwide Operations and Manufacturing

As Senior Vice President of Worldwide Operations and Manufacturing, Les Hamilton has global responsibility for Danger's data centers, and he will oversee Danger's outsourced manufacturing processes.

Les has more than thirty years of experience in building and managing complex information services and network operations, working for companies such as Infonet Services Corp., TRW Information Services (now Experion), and Lockheed. Before joining Danger, Les was Senior Vice President of Network Services for GX Networks UK Ltd, a spin-off of XO Communications Europe. At GX Networks, Les was responsible for operations, including information systems, network and data center operations, security, provisioning, systems development and customer care.

Prior to assuming his position at GX Networks, Les was Senior Vice President of Network and Systems for Europe for XO Communications. Prior to the merger of Nextlink and Concentric Network Corp., which resulted in the formation of XO Communications, Les was the Senior Vice President of Engineering and Operations for Concentric.

Previously, Les Hamilton managed network services and engineering operations for Infonet Service Corporation, and served as Director of Customer and Technology Services for TRW Information Services (now Experion). Les began his career at British Aerospace Corporation as a technical programmer analyst.

Les Hamilton holds a bachelor of science degree in mechanical engineering from Teesside University in the United Kingdom and an M.B.A. from the Peter F. Drucker graduate school in Claremont, California.

Matt Hershenson, Senior Vice President of Hardware and Operations

As Senior Vice President of Hardware and Operations, Matt Hershenson brings a wealth of experience to Danger. He is responsible for the day-to-day operations of the company, as well as Danger's hardware engineering, development and design efforts.

Before co-founding Danger, Matt managed the hardware group at Mainbrace Corporation, a Windows CE systems integrator. Prior to Mainbrace Corporation, Matt served in various roles at Philips Electronics. During his tenure with the Philips Mobile Computing Group, he was responsible for the hardware of the Velo-1 handheld PC, one of the first Windows CE devices. While at Philips Semiconductors, Matt served as a systems architect, where he played an integral role in the design of numerous consumer handheld devices, including the Sharp Mobilon, Philips Nino, and the Compaq C-series.

Before Philips, Matt was a hardware engineer with Catapult Entertainment, since acquired by Hearme. He was part of the team that turned the vision of multi-player gaming over the Internet into a reality. He handled all aspects of product development and design for the XBAND Video Game Modem.

Matt also played a key role in the product development and design of the Apple Powerbook 150, then Apple's most affordable PowerBook computer. Matt also co-founded MOTO Development Group, a product design consultancy firm specializing in product development. MOTO aided in the design of many technical products, such as remote controls for Apple Interactive TV.

James Isaacs, Vice President of Worldwide Sales and Alliances

As Vice President of Worldwide Sales and Alliances, James Isaacs is responsible for sales to wireless operators and for building a world-class organization for long-term support of Danger's wireless operator customers. James is also responsible for establishing alliances that will further Danger's market penetration and growth.

Before joining Danger, James served as a Vice President at XO Communications, Inc., formerly Concentric Network Corporation. At XO and Concentric, James led product management for the company's hosting and application services, where he was the key driver in the growth of the private-label distribution arrangements with carriers and large partners for shared hosting, a software-based services platform. While with Concentric, James led several of the company's mergers and acquisitions and strategic partnerships, and developed the company's product management effort.

Previously, James was with Apple Computer, Inc. While at Apple, he managed marketing for the company's online services division and served in a business strategy role for the Apple U.S.A. Sales and Marketing division.

James has an M.B.A. degree from the Haas School of Business at U.C. Berkeley and a B.A. degree from Stanford University.

Source: Danger website.

EXHIBIT 2 Pictorial Representation of Wireless Segments

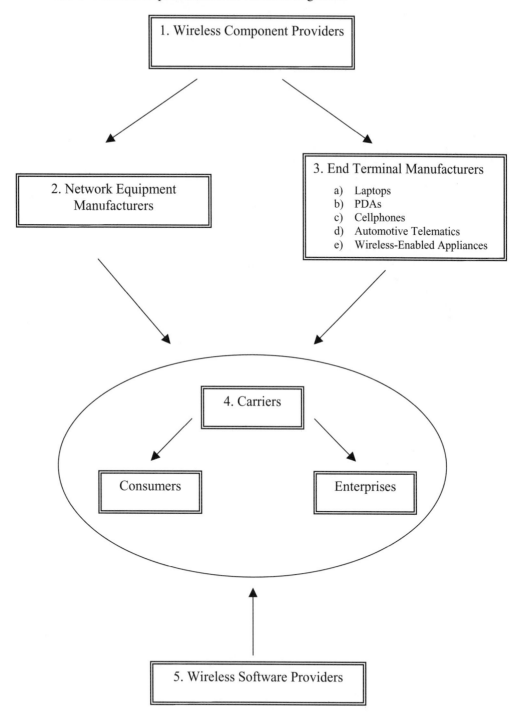

EXHIBIT 3 Industry Experience of key executives at Danger, Inc.

	1975	1980	1985	1990	1995	2000	
Andy Rubin		Carl Zeiss	Apple Computers		General Magic	WebTV	Danger
Joe Britt				Apple Computers	3DO / Cata-pult	WebTV	Danger
Matt Hershenson				Apple Computers	Catapult	Philips / Main-Brace	Danger
John Arledge						OpenTV / TiVo	Danger
Hank Nothhaft	Prior Experience[1]	GTE Telenet	DSC Communications	David Systems	Concentric Network/ XO Communications		SmartPipes / Danger
James Isaacs				Apple Computers		Concentric Network/ XO Communications	Danger
Les Hamilton		British Aerospace	TRW Information Services	Infonet Services		Concentric Network/ XO Communications	Danger

Note: 1) Hank Nothhaft was an officer in the U.S. Marine Corps. between 1966 and 1970. He was a marketing and sales executive in the 70s.

2) The Exhibit is arranged according to the starting date at Danger for each executive.

EXHIBIT 4 Comparison of Wireless Handhelds available in Market

Handheld Device	Microsoft Siemens SX56	Nokia 9290 Communicator	Handspring TREO 300	Samsung SPH-1330	Audiovox Thera	RIM 5810	Palm Tungsten W	T-Mobile Sidekick
Price								
w/o service	$650	$600	$700	NA	NA	NA	$550	NA
w service	$450	$400	$400	$400	$800	$200	$500	$200
Processor	Intel Strong ARM 32 bit	Intel Strong ARM 32 bit	Motorola DragonBall 33	Motorola DragonBall 33	Intel Strong ARM206 MHz	Custom	Motorola DragonBall 33	NA
Operating System	Microsoft PocketPC 2002	Symbian	Palm OS	Palm OS	Microsoft PocketPC	Blackberry	Palm OS	Danger
Memory								
RAM	32MB	56 MB	16 MB	8 MB	32 MB	1 MB (SRAM)	16 MB	16 MB
ROM	32MB	NA	NA	NA	32 MB	8 MB (Flash)	NA	NA
Display								
Colors	4096	4096	4096	256	65 K	None	65 K	None
Resolution	240x320	640 x 200	320 x 240	160 x 240	240x320	160x160	320x320	240x160
Battery Life								
Talk	3.5 hours	10 hours	3 hours	4 hours	1.5 hours	3 hours	NA	3 hours
Standby	150 hours	240 hours	150 hours	100 hours	8 hours	10 days	NA	60 hours
Size (inches)	4.6 x 3.1 x 0.7	6.2 X2.2x 1.0	4.4x2.8x0.8	4.9x2.3x0.8	5.0x3.0x0.8	4.6x3.1x0.7	4.8x3.1x0.7	2.6x4.5x1.1
Weight (ounces)	4.7	8.6	5.7	6.0	7	4.7	6.5	6.0
Network	GSM/GPRS	GSM	CDMA 1xRTT	CDMA 1xRTT	CDMA 1xRTT	GSM/GPRS	GPRS	GPRS
Carrier	AT&T	Cingular/ T-Mobile	Sprint PCS	Sprint PCS	Sprint PCS/ Verizon	T-Mobile/ AT&T	AT&T	T-Mobile

Sources: Company Websites accessed on 3/18/03

EXHIBIT 5 Sample Wireless Data Service Plans for Wireless Handhelds

Handheld	Carrier	Network	Plan Name	Voice Plan	Data Plan	Cost /month
T-Mobile Sidekick	T-Mobile	GPRS	Sidekick $39.99 plan	200 Anytime Min. 1000 Weekend Min 35c/min overage	Unlimited	$39.99
T-Mobile Sidekick	T-Mobile	GPRS	Sidekick $59.99 plan	500 Anytime Min Unlimited Nights and Weekend 35c/min overage	Unlimited	$59.99
Handspring Treo 300	Sprint PCS	CDMA 1xRTT	PCS Free & Clear with Vision	300 Anytime 1000 Night and Weekend	Unlimited	$45
Handspring Treo 300	Sprint PCS	CDMA 1xRTT	PCS Free & Clear with Vision	500 Anytime Unlimited Night and Weekend	Unlimited	$55
Handspring Treo 270	Cingular	GPRS	Wireless Internet Express	None	Up to 7MB Overage: $20/MB	$29.99
Samsung SPH I330	Sprint PCS	CDMA 1xRTT	PCS Vision	None	Up to 20MB Overage: $0.002/KB	$40
Audiovox Thera	Verizon Wireless	CDMA 1xRTT	Express Network	None	Up to 10MB Overage: $0.008/KB	$35
Palm Tungsten W	AT&T Wireless	GPRS	Mobile Internet Plan	None	Up to 8MB Overage: $0.006/KB	$19.99
Palm Tungsten W	AT&T Wireless	GPRS	Mobile Internet Plan	None	20MB data Overage: $0.002/KB	$39.99
BlackBerry 5180	AT&T Wireless	GPRS	BlackBerry Access Plan	None	Up to 4MB Overage: $0.0048/KB	$34.99
BlackBerry 5180	AT&T Wireless	GPRS	BlackBerry Access Plan	None	Unlimited	$44.99
Blackberry 6510	Nextel	Xxxx				
Motorola	Nextel	xxxx				

Source: Carrier Websites.

EXHIBIT 6 Danger's Complete Solution

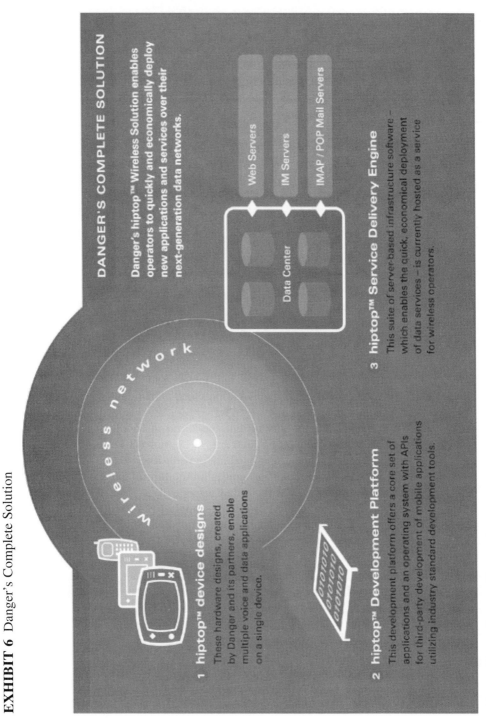

Source: Danger Company Resources

EXHIBIT 7 Screenshots of the T-Mobile Sidekick

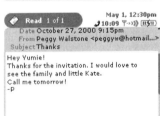

Source: Danger Company Resources.

EXHIBIT 8 Media Highlights from Danger's Press Kit

"Danger has come up with a product that can teach a few things to those making handhelds for the corporate set…At such reasonable prices, Sidekick may have what it takes to make the wireless handheld a true mass-market product."

– Steve Wildstrom, Business Week, September 16, 2002

"But the big winner could be T-Mobile's plan for Sidekick…That kind of no-surprises package is just the sort of thing that might finally drag ordinary consumers into the mobile data world."

– Forbes Magazine, Stephen Manes, September 16, 2002

"Three of the Sidekick's attributes are particularly admirable: a no-nonsense service fee of $40 a month…; an intuitive user interface that's ideal for people who are all thumbs; and a server-based storage system that speeds mail handling and prevents data loss, even if the Sidekick itself is destroyed in an X-Games accident."

– Fortune Magazine, Peter Lewis, October 14, 2002

"Sidekick represents a departure from the widening field of handhelds targeted at the corporate crowd. Instead, T-Mobile is training its sights on a mass market of 18-to-34-year-olds, not that those of us with a few more rings around our trunks might not crave the device as well."

– USA Today, Ed Baig, September 18, 2002

"….the Sidekick is a fantastic success, thanks to the most thoughtfully conceived hardware and software since the original Palm Pilot…service price is right, too."

– The New York Times, David Pogue, October 3, 2002

"The Sidekick is a well-designed, very usable gadget that looks like nothing else on the market and that is packed with clever features…As an e-mail, Web and instant-messaging device, I found the Sidekick highly usable and effective, even addictive... the Sidekick is a true breakthrough….."

– The Wall Street Journal, Walter Mossberg, August 7, 2002

Source: Danger company resources.

EXHIBIT 9 Financial Statements for T-Mobile USA. (Figures in thousands of U.S. Dollars)

Operating Income Statement	Year ended Dec 31, 2002
Revenues:	
Post pay revenues	3,629,052
Prepaid revenues	386,871
Roaming revenues	229,312
Equipment sales	690,174
Affiliate and other revenues	68,436
Total revenues	**5,003,845**
Operating expenses:	
Network costs	913,509
Cost of equipment sales	1,078,582
General and administrative	1,029,906
Customer acquisition	1,657,722
Depreciation and amortization	1,079,965
Impairment charges	15,628,000
Stock-based compensation	14,852
Total operating expenses	**21,402,536**
Operating loss	(16,398,691)
Other expenses, net	(501,143)
Income tax benefit	272,489
Net loss	**(16,627,345)**
Adjusted EBITDA (loss) (2)	324,126
Cash flows provided by (used in):	
Operating activities	(254,647)
Investing activities	(2,015,835)
Financing activities	2,306,992
Other data:	
Population covered by spectrum licenses	248,338,000
Population covered by network	198,945,000
Customers:	
Post pay	7,689,400
Prepaid	1,002,500

Balance Sheet	Year ended Dec 31, 2002
ASSETS	
Current assets:	
Cash and cash equivalents	36,510
Accounts receivable,	911,726
Inventory	299,237
Other current assets	226,070
Total current assets	**1,473,543**
Property and equipment	4,427,115
Goodwill	9,868,082
Spectrum licenses	9,951,288
Other intangible assets	380,047
Investments in and advances to unconsolidated affiliates	885,470
Other assets and investments	135,666
Total Assets	**27,121,211**
LIABILITIES AND SHAREHOLDERS EQUITY	
Current liabilities:	
Accounts payable	336,440
Accrued liabilities	856,281
Construction accounts payable	344,953
Deferred revenue	157,190
Total current liabilities	**1,694,864**
Long-term debt	1,274,638
Long-term notes payable to affiliates	7,041,944
Deferred tax liabilities	3,259,452
Other long-term liabilities	98,861
Total long-term liabilities	**11,674,895**
Minority interest in equity of consolidated subsidiaries	8,480
Voting preferred stock	5,000,000
Shareholders equity:	
Common stock	26,851,821
Deferred stock compensation	(21,039)
Accumulated other comprehensive income (loss)	(264)
Accumulated deficit	(18,087,546)
Total shareholders equity	**8,742,972**
Total Liabilities and Shareholders Equity	**27,121,211**

Note 2 to Operating Income Statement: Adjusted EBITDA represents operating income (loss) before depreciation, amortization, impairment charges and non-cash stock-based compensation. Adjusted EBITDA should not be construed as an alternative to operating income (loss) as determined in accordance with GAAP, as an alternative to cash flows from operating activities as determined in accordance with GAAP, or as a measure of liquidity.

Source: T-Mobile 10K Report filed 03/11/2003

EXHIBIT 9 Consolidated Statements of Cash Flows for T-Mobile, USA. (in thousands)

	Year ended Dec 31, 2002
Operating activities:	
Net loss	$ (16,627,345)
Adjustments to reconcile net loss to net cash used in operating activities:	
Depreciation and amortization	1,079,965
Impairment charges	15,628,000
Income tax benefit	(272,489)
Amortization of debt discount and premium, net	15,310
Equity in net losses of unconsolidated affiliates	170,903
Stock-based compensation	14,852
Allowance for bad debts	5,457
Other, net	(19,617)
Changes in operating assets & liabilities, net of effects of purchase accounting:	
Accounts receivable	(310,180)
Inventory	(144,845)
Other current assets	(77,028)
Accounts payable	123,486
Accrued liabilities	158,884
Net cash used in operating activities	**(254,647)**
Investing activities:	
Purchases of property and equipment	(1,749,084)
Acquisitions of wireless properties, net of cash acquired	(80,147)
Sales (purchases) of short-term investments, net	—
Investments in and advances to unconsolidated affiliates, net	(383,093)
Refund (payment) of deposits held by FCC	195,956
Other, net	533
Net cash provided by (used in) investing activities	**(2,015,835)**
Financing activities:	
Net proceeds from issuance of common and preferred stock	—
Long-term debt borrowings	—
Long-term debt repayments	(651,312)
Long-term debt borrowings from affiliates, net	2,908,007
Outstanding checks in excess of bank balance	92,596
Cash entitlements on conversion of preferred stock of consolidated subsidiary	—
Deferred financing costs	—
Other, net	(42,299)
Net cash provided by (used in) financing activities	**2,306,992**
Change in cash and cash equivalents	**36,510**

Adjusted EBITDA represents operating income before depreciation, amortization, impairment charges and non-cash stock-based compensation. We believe Adjusted EBITDA provides meaningful additional information regarding our operating results, our ability to service our long-term debt and other fixed obligations and to fund our continued growth. Adjusted EBITDA is considered by many financial analysts to be a meaningful indicator of an entity's ability to meet its future financial obligations. Growth in Adjusted EBITDA is considered to be an indicator of the potential for future profitability, especially in a capital-intensive industry such as wireless telecommunications.

Source: T-Mobile 10K Report filed 03/11/2003

EXHIBIT 10 Danger Selling Process to Carriers

	Carrier 1	Carrier 2	Carrier 3
1. Account Qualification			
1A Preliminary sales activities	X	X	X
1B Internal champion identified	X	X	X
1C Executive sponsor in place	X	X	X
1D Product manager assigned	X	X	X
2. Marketing Evaluation			
2A Marketing evaluation units sent (monochrome)	X		X
2B Consumer and/or business product offer teams briefed	X		X
2C Product offer (or handsets) team prepares volume forecasts	X		X
2D 3rd-party manufacturer introduction	X		X
2F IM vendor introduction	X		
2G Business case prepared	X		
2H Product committee or business unit approval			
3. Technical Evaluation			
3A Technical evaluation units sent	X		
3B Terminal due diligence	X		
3C Infrastructure and integration due diligence	X		
3D Terminal signoff	X		
3E Infrastructure and integration signoff	X		
4. Commercial Negotiations			
4A Carrier identifies individual to drive commercial terms	X	X	
4B Commercial discussion underway	X		
4C Project plan drafted with carrier deployment team	X		
4D 3rd-party mfr LOI in place or no outstanding issues	X		
4E Carrier has IM solution in place or no outstanding issues	X		
4F LOI or term sheet signed	X		
5. Contract Negotiations			
5A Draft contract issued to carrier	X		
5B Contract discussion underway	X		
5C Customization requirements – client			
5D Customization requirements – service			
5E Contract in place with 3rd-party manufacturer			
5F Danger/carrier contract signed			

Source: Adapted from Danger Inc. internal records.

EXHIBIT 11 Financial Data for a Representative Sample of Wireless Carriers (US$)

North American Carriers	AT&T Wireless (Redmond, WA)	Sprint PCS (Overland Park, KS)	Nextel (Reston, VA)	Verizon WS (New York, NY)	Cingular (Atlanta, GA)
Total Revenue (millions)	12,544	8,579	6,560	16,011	13,216
EBIDTA (millions)	2,859	1,513	1,900	5,783	4,505
Net Income after tax (millions)	(306)	(1,268)	(745)	1,300	1,669
# of mobile subscribers ('000s)	18,047	13,555	8,667	29,398	21,596
Mobile ARPU	63	61	71	48	52
Mobile Minutes of Use	382	506	563	262	308
2000-2001 Revenue Growth	34%	58%	32%	NA	NA
2000-2001 Net Income Growth	-155%	-33%	80%	NA	NA

European Carriers	Orange (Paris, France)	mm02 (Slough, UK)	Vodafone (Berkshire, UK)	Telecom Italia Mobile (Turin, Italy)	Telefonica Moviles (Madrid, Spain)
Total Revenue (millions)	15,992	6,696	35,775	10,878	8,916
EBIDTA (millions)	3,485	678	12,451	5,046	3,534
Net Income after tax (millions)	(938)	(1,132)	5,478	1,007	1,111
# of mobile subscribers ('000s)	37,951	17,194	83,381	23,950	16,793
Mobile ARPU	36	49	NA	30	33
Mobile Minutes of Use	NA	NA	NA	NA	NA
2000-2001 Revenue Growth	25%	11%	52%	9%	14%
2000-2001 Net Income Growth	-33%	-12%	80%	-48%	54%

Asian Carriers	NTT DoCoMo (Tokyo, Japan)	China Mobile (Beijing, China)	SK Telecom (Seoul, S. Korea)	TCC Taiwan (Taipei, Taiwan)	China Unicom (Beijing, China)
Total Revenue (millions)	34,211	12,140	6,016	1,401	3,557
EBIDTA (millions)	14,011	7,100	3,157	615	1,635
Net Income after tax (millions)	8	3,390	899	481	539
# of mobile subscribers ('000s)	40,783	90,566	15,179	6,589	27,030
Mobile ARPU	70	18	NA	21	11
Mobile Minutes of Use	178	NA	NA	NA	NA
2000-2001 Revenue Growth	NA	54%	14%	8%	24%
2000-2001 Net Income Growth	NA	55%	-100%	18%	39%

Source: Salomon Smith Barney *Global Telecoms Outlook*, Jan 1 2003.

Note: Deutsche Telekom, one of the largest carriers in Europe, is not included in this exhibit because Danger already had a relationship with T-Mobile, a U.S. subsidiary of Deutsche Telekom.

EXHIBIT 12 Global Handset Outsourcing

Handset OEM	Tier 1 Partner	Tier 2 Partner	OEM units (millions)	OEM Market Share	Out-sourced[1] (%)	CM[2]	ODM[3]
Nokia	Elcoteq	Celestica, Jabil, Solectron, Telson, Hon Hai	165	35.1%	<20%	<20%	0%
Motorola	Flextronics, BenQ, Pantech	Solectron, Celestica, Compal, Telson	65	13.8%	<60%	<10%	50%
Samsung			65	13.8%	0%	0%	0%
Sony-Ericsson	Flextronics	Elcoteq, Arima, LiteOn	29	6.2%	100%	40%	60%
Siemens	Flextronics, Sanmina-SCI	Quanta	27	5.7%	>60%	30%	30%
LG Elec			25	5.3%	0%		
Panasonic		Celestica	16	3.4%	<25%	<5%	20%
NEC		Celestica, BenQ, Arima	14	2.9%	<20%	<5%	15%
Mitsubishi		Solectron	11	2.3%	<10%	<10%	0%
Kyocera	Solectron		10	2.1%	<30%	<30%	0%
Alcatel	Flextronics	GVC	4	0.9%	100%	40%	60%
TCL (China)			12	2.6%	35%		35%
Others			28	6.0%	80%	10%	70%
Asian			22	4.7%			
Non-Asian			6	1.3%			
Total			470	100%	25%-26%	13%	19%

Source: Morgan Stanley Research, Asia/Pacific Handsets, Feb. 25, 2003.

Note: 1) This refers to the percentage of OEM units that were outsourced to Tier 1 and Tier 2 partners.

2) This refers to the percentage of OEM units that were outsourced to Contract Manufacturers.

3) This refers to the percentage of OEM units that were outsourced to Original Design Manufacturers.

EXHIBIT 13 Manufacturing Relationship Models Available to Danger, Inc. (Source: Danger)

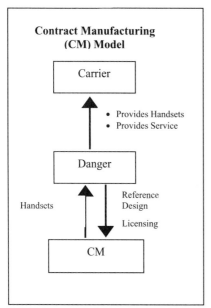

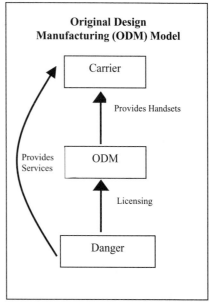

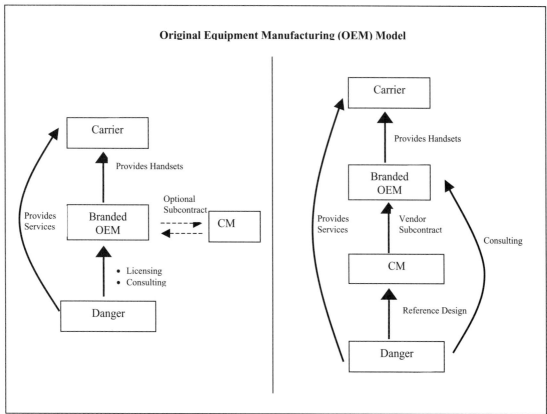

BIODIESEL INCORPORATED

Joshua Maxwell shut down his laptop and looked out the window. From the second floor of the Graduate School of Management's new building, he could see a number of cars driving on the nearby freeway and sitting in the adjacent parking lot.

Josh was in his last term of the full-time MBA program at UC Davis. He would soon be graduating and entering a new chapter of his life. While he had the luxury of having several management-level job offers from which to choose, he was unsure whether he wanted to follow such a traditional route. There was one opportunity in particular that had recently come across his path which gave him pause.

Background

The previous term, Josh had been enrolled in Professor Dorf's class on Business and Sustainability. While the class was offered at the GSM, it was open to the entire university. In this class, he met Hannah Long, who was in her final year of her undergraduate studies in Agricultural Economics, and Matthew Hammond, who was a senior in the Mechanical Engineering department.

The three began working on a class project, which would ultimately turn into a formidable business opportunity. The impetus for their collaboration began with a lecture-discussion regarding the challenges and opportunities in the emerging renewable energy industry.

The Challenge

Dependence on energy is a worldwide reality. Energy powers the machines and equipment around us in order to make life more convenient and efficient. In our everyday lives, energy is synonymous with the forms that it can assume. The major generation sources, of course, petroleum, coal, natural gas and nuclear are non-renewable resources and have detrimental effects on the environment. In our daily lives, the two most common forms of this energy are liquid fuel (refined from petroleum) and electricity.[1]

Increasingly, developed and developing countries alike are consuming liquid fuel for the purposes of mobility, food production, and the facilitation of trade. All of these functions essentially provide a substitute for human effort. Due to the widespread consumption of petrol-based liquid fuel, an incredibly large global infrastructure and set of surrounding institutions have grown around the support of such consumption. The petrochemical fuel industry manifests itself in the form of oil fields and reserves, pipelines, transport ships, and fueling stations.

Prepared by MBA candidate Benjamin Finkelor; Assistance from MBA candidate Sonja Yates and Paul Yu-Yang under the supervision of Professor Richard C. Dorf, Graduate School of Management of UC Davis

[1] Technically speaking, liquid fuel is a form of energy, while electricity is considered a carrier of energy. For the purposes of this proposal, the distinction is not significant.

The way energy is used worldwide is not sustainable. It is well-documented that the use of these fuels is depleting the world's natural resource reserves, harming communities in terms of health and displacement, and polluting the air and water in local environments. The drilling, refining, and transporting of oil leads to spills on land and in oceans, and when petrol-based fuels are used to power machines and automobiles, the air is polluted with greenhouse gases and particulate matter such as carbon dioxide, carbon monoxide, sulfur, and nitrous oxide emissions.

In spite of the drawbacks, the current energy industry is committed to the continuation of these ways, primarily because of considerable assets and investment in the existing form of infrastructure.

The challenge, which became clear to the team from class discussion and further brainstorming, is to find a form of fuel or technology that can mitigate the current negative affects on the environment of petrol-based fuel while utilizing the existing infrastructure. The urgency of this challenge is heightened by the astounding projected growth in the global population and per-capita consumption of liquid fuels.

The Concept

Matthew's coursework in engineering coupled with a bit of networking with fellow engineers suggested the emerging technology of biodiesel as a possible solution to this challenge. As the group explored the environmental benefits and the viability of the diesel fuel substitute, the three began to realize the potential of the biodiesel market.

Biodiesel is a vegetable- and/or animal-based product that serves as a substitute for traditional diesel fuel. Although its chemical composition is dissimilar from the petrol-based diesel, biodiesel will still work in diesel engines built in and after 1996 with no modification. For engines made before that time, modifications can be made to allow for the use of biodiesel fuel. The choice of biodiesel as a product of biomass is an intentional one. Producing a product that can be utilized by the existing infrastructure and social patterns of use[2] increases the likelihood of its adoption. "Entrepreneurs must locate their ideas within the set of existing understandings and actions that constitute the institutional environment yet set their innovations apart from what already exists."[3] This economic viability is coupled with a significant potential to the environment: biodiesel showcases an innovation that is a step in the right direction for air quality.

Biodiesel's greatest promise to sustainability as a renewable energy source is its lower emissions over conventional diesel. Compared to traditional diesel, biodiesel achieves significant reductions in harmful emissions. Additionally, the

---

[2] Yates, JoAnne, "The Structuring of Early Computer Use in Life Insurance," *Journal of Design History,* 12(1999): pp. 5–22.

[3] Hargadon, A., and Y. Douglas, "When Innovations Meet Institutions," *Administrative Science Quarterly,* 46(2001): 476.

ozone-forming impact of biodiesel is nearly half of that of petroleum fuel. Further benefits can be counted when looking at lifecycle effects. If biodiesel is obtained using soybeans as an example, the amount of CO_2 taken up by soybeans and released upon burning the fuel, is a near zero sum balance. Contrast this with petroleum products where release of CO_2 is unidirectional into the atmosphere.

Because biodiesel is biodegradable and dependent on organic material as opposed to fossil fuels, the energy source is considered renewable. Production of biodiesel begins with feedstock, preferably in the form of oils or fats. Oils can be processed from oleic varieties of plants such as soy, canula, sunflower and safflower. Fats can come directly from grease such as and tallow/lard and recycled cooking grease from restaurants. The oils or fats are mixed with alcohol and a catalyst in a process that forms esters resulting in biodiesel defined as mono-alkyl esters of long chain fatty acids and glycerin.

Ultimately, the large-scale production of biodiesel would generate a dramatic impact on the economic value of the feedstocks involved. For example, according to one study, if biodiesel demand over the next ten years were to increase to 200 million gallons, a commensurate amount of soy oil would be required and net average farm income would increase by $300 million per year. A bushel of soybeans would increase by an average of 17 cents over the ten-year period.[4] The potential economic benefit to farmers seems considerable.

Even with such economies of scale, however, the wholesale price of 100 percent biodiesel would rarely be lower, and therefore cost-competitive, with traditional diesel fuel. Barring some crisis that would drive up the price of crude oil or reduce the capacity of diesel refineries, the current regulatory structure and assets devoted to petrol-diesel will more often than not yield a lower price with petrol-based diesel. Biodiesel as a fuel additive however, does provide a cost-competitive potential. Studies have shown that splash-mixing even 1 percent biodiesel with traditional diesel "can increase the lubricity of petroleum diesel by up to 65 percent."[5] This is not to mention the sulfur- and other emissions-reducing benefits that splash-mixing provides. As more consumer and regulatory pressure is placed on traditional diesel users, biodiesel producers will be able to charge the premium necessary to offset higher relative costs. Markets for 100 percent biodiesel will grow as well in such specialty markets as the marine industry, railroads, electricity generators, and even agriculture.

Biodiesel Incorporated

Josh, Hannah, and Matthew presented a compelling business case for their final class project: Biodiesel Incorporated. This new venture would enlist and develop a series of local producer's cooperatives in an effort to capitalize on the emerging biodiesel market as described in the following:

---

[4] National Biodiesel Board, "Benefits of Biodiesel," www.biodiesel.org.

[5] "Biodiesel Carries New Weight in Premium Diesel Market." *Biodiesel Bulletin,* Sept. 2002. http://www.biodiesel.org/news/bulletin/2003/080403.pdf

- Members would grow feedstock crops and gather crop residues with high fat content.

- Capital equipment costs would be shared and spread over membership; Oils would be extracted from the collected biomass and biodiesel would be produced using these oils.

- Biodiesel Incorporated would distribute the biodiesel locally using the existing petroleum-based infrastructure.

Advantages of the Cooperative Business Form   The cooperative model has been successfully used to allow small farmers to maintain a competitive edge against the larger corporate farming organizations. "Today, there are more than 4,000 agricultural cooperatives in the U.S., with a total net income of nearly $2 billion and net business volume of more than $89 billion."[6] A coop is owned and controlled by the members, with self-reliance and self-help being key characteristics—ideal for the implementation of emerging and disruptive innovations such as biodiesel.

Biodiesel Incorporated will:

- Utilize the collective purchase power of the coop to obtain necessary capital intensive equipment and gain economies of scale.

- Increase negotiating power
  - Stabilizes crop prices and biodiesel output
  - Gain access to higher-volume contracts

- Serve to unite rural communities and preserve agricultural economy

Biodiesel Incorporated offers the unique service of both the bargaining and manufacturing of biodiesel on behalf of its farmer members. It will serve to control the production of agricultural products (i.e., the biomass feedstock), the price and terms set for members' production, and price and terms for biodiesel output.

Questions

1. What are the key factors in determining if this is a viable business opportunity for Josh, Hannah, and Matthew?
2. What market drivers should they research and be aware of?
3. What are the flaws in the current business strategy?
4. What type of financing should they use if they choose to go forward with this?
5. What types of distribution channels should they go into?
6. How can they improve their chances for success?
7. What is the next step?

---

[6] http://co-operatives.ucdavis.edu/Agricultural Co-operatives.htm

YAHOO! 1995

"I guess, three and a half years ago, if we were looking to start a business and make a lot of money, we wouldn't have done this."
—Jerry Yang, 1997

It was April of 1995—a key decision point for Jerry Yang and David Filo. These two Stanford School of Engineering graduate students were the founders of Yahoo!, the most popular Internet search site on the World Wide Web. Yang and Filo had decided that they could transform their Internet hobby into a viable business. While trying to decide between several different financing and partnering options that were available to them, they attended a meeting with Michael Moritz, a partner at Sequoia Capital. Sequoia, one of the leading venture capital firms in Silicon Valley, had been discussing the possibility of investing in Yahoo!.

Michael Moritz leaned forward in his chair. As he looked across the conference table at Jerry and Dave, he laid out Sequoia's offer to fund Yahoo!:

> As you know, we have been working together on this for some time now. We have done a lot of hard work and research to come up with a fair value for Yahoo!, and we have decided on a $4 million valuation. We at Sequoia Capital are prepared to offer you $1 million in venture funding in exchange for a 25 percent share in your company. We think that with our help, you have a real chance to make Yahoo! something special. Our first order of business will be to help you assemble a complete management team, after which we should be able to really start helping you to develop and manage your site's vast amount of content.
>
> Right now, the biggest risk that you guys run is ***not*** making a decision. You ***have*** to make a decision, because if you don't, someone else is going to

run you over. You might get run over by Netscape. You might get run over by AOL. You might get run over by one of these other venture-backed start-ups. It is imperative that you make a decision now if you are going to survive. To help you make a decision, I am going to give you a deadline: tomorrow. If you don't want to do business with Sequoia, that's OK. I'll be disappointed, but that's OK. But you are going to have to call me by 10 A.M. tomorrow morning to tell me yes or no.[1]

Yang and Filo gazed around the Sequoia conference room and noticed the many posters of companies such as Cisco, Oracle, and Apple that were hung from the walls—all success stories from past Sequoia investments. They wondered if Yahoo!'s poster would someday join that group. The two were excited at the possibilities; however, they still had some decisions to make. There were several other financing options available, and they were still not sure if they wanted to accept Sequoia's funding. Yang responded:

> That sounds like a pretty fair offer, Mike. Let us talk this over tonight, and we will get back to you by tomorrow after we weigh all of our options. However you have to realize that we're still grad students, and we don't even usually wake up by 10 A.M., so can you give us until noon?

Yahoo!

Yahoo! was an Internet site that provided a hierarchically organized list of links to sites on the World Wide Web. It offered a way for the general public to easily navigate and explore the Web. Users could click through multiple topic and category headings until they found a list of direct links to Web sites related to their interest. In addition, Yahoo! offered a central place where people could go to just to see what was out there. This made it easy for people with little previous exposure to the Web to start searching through Yahoo!'s lists of links, often just to see if they could find something of interest. In a little over a year since its inception, it had become one of the most heavily visited sites on the Web.

But Yang and Filo believed Yahoo! had the potential to be much more that a way for Web surfers to find what they were looking for. In 1995, John Taysom, a marketing vice president of marketing of Reuters, a London-based provider of news and financial data, called Jerry Yang to explore the idea of a Yahoo!-Reuters partnership. It seemed to Taysom that affiliating with Yahoo! could help Reuters to build a distribution network on the Web.

> ". . . The first thing Jerry said to me," Taysom remembers, "was 'if you hadn't called me, I would have called you.'" Jerry *got* the news feed vision. He had been thinking about it for months. He further surprised Taysom by informing him that as far as he was concerned, Yahoo! was "not just a directory but a *media property*."[2]

---

[1] Moritz, Michael, personal interview, November 10, 1998.

[2] Reid, Rob, (1997) *Architects of the Web,* John Wiley and Sons, New York (p. 253).

Yang further believed that:

> "Primarily we're a brand. We're trying to promote the brand and build the product so that it has reliability, pizzazz, and credibility. The focus of the business deals we are doing right now is not on revenues but on our brand."[3]

Dave and Jerry at Stanford

David Filo, a native of Moss Bluff, Louisiana, attended Tulane University's undergraduate program in computer engineering. In 1988, Filo finished his undergraduate work and enrolled in Stanford's master's program in electrical engineering. Completing his master's degree, he opted to stay at Stanford and try for his PhD in electrical engineering. Extremely competent in the technical arena, Filo had been described by many as a quiet and reserved individual.

Jerry Yang was a Taiwanese native who moved to California at the age of 10. Yang was raised by his widowed mother and grew up in San Jose with his younger brother, Ken. Yang was a member of the Stanford class of 1990 and completed both his bachelor's and master's degrees in electrical engineering. Yang also opted to stay at Stanford for a PhD in electrical engineering. Also technically competent, Yang was considered much more outgoing than Filo.

Yang and Filo met each other in the electrical engineering department at Stanford; Filo was Yang's teaching assistant for one of his classes. They also both worked in the same design automation software research group. They became close friends while teaching at the Stanford campus in Kyoto, Japan. Upon returning to the Stanford campus, they moved into adjacent cubicles in the same trailer to conduct their graduate research. They both enjoyed working together, as their individual personalities perfectly complemented each other, forming a unique combination.

Their office was not much to look at, but it served as a place for them to work on their research as well as a place from which they could run their website. "The launching pad (for Yahoo!) was an oxygen-depleted, double-wide trailer, stocked by the university with computer workstations and by the students with life's necessities... that prompted a friend to call the scene 'a cockroach's picture of Christmas'."[4] Michael Moritz remembered his early visits to Jerry and Dave's cube:

> "With the shades drawn tight, the Sun servers generating a ferocious amount of heat, the answering machine going on and off every couple of minutes, golf clubs stashed against the walls, pizza cartons on the floor, and unwashed clothes strewn around . . . it was every mother's idea of the bedroom she wished her sons never had."[5]

---

[3] Yang, Jerry, (1995) interview in *Red Herring,* (October), (online back issue, p. 9).

[4] Stross, Randall E., (1998) "How Yahoo! Won the Search War," *Fortune,* http://www.pathfinder.com.fortune/1998/980302/yah.html., (p. 2).

[5] Reid, Rob, (1997) *Architects of the Web,* John Wiley and Sons, New York (p. 254).

Mosaic and the World Wide Web

In 1993, the University of Illinois-Urbana Champagne's National Center for Supercomputing Applications (NCSA) revolutionized the growth and popularity of the World Wide Web by introducing a Web browser they had developed called Mosaic. Mosaic made the Web "an ideal distribution vehicle for all kinds of information in the professional and academic circles in which it was known."[6] It provided an easy-to-use graphical interface that allowed users to travel from site to site simply by clicking on specified links. This led to the widespread practice of surfing the Web, as people spent hours trying to find new and interesting sites. This easy-to-use browser for navigating the Internet was estimated to have 2 million users worldwide in just over one year.

Creating Jerry's Guide to the World Wide Web

With Mosaic's introduction in late 1993, Filo and Yang, along with thousands of other students, began devoting large amounts of time to surfing the Web and exploring the vast content available. As they discovered interesting sites, they made bookmarks of the sites. The Mosaic Web browser had an option to store a bookmark list of your favorite sites. This feature allowed users to return directly to a page that they had visited, without having to navigate through several different links. As the popularity of the Web quickly increased, so did the total number of sites created, which in turn led to an increase in the number of interesting sites that Filo and Yang wanted to bookmark. Eventually, their personal list of favorite Web sites grew large and unwieldy, due to the fact that the earliest versions of Mosaic were unable to sort bookmarks in any convenient manner.

To address this problem, Filo and Yang wrote software using Tcl/TK and Perl scripts that allowed them to group their bookmarks into subject areas. They named their list of sites "Jerry's Guide to the World Wide Web" and developed a Web interface for their list. People from all over the world started sending Jerry and Dave e-mail, saying how much they appreciated the effort. Yang explained: "We just wanted to avoid doing our dissertations."[7]

The two set out to cover the entire Web. They tried to visit and categorize at least 1,000 sites a day. When a subject category grew too large, subcategories were created, and then sub-subcategories. The hierarchy made it easy for even novices to find websites quickly. "Jerry's Guide" was a labor of love—lots of labor, since no software program could evaluate and categorize sites. Filo persuaded Yang to resist the engineer's first impulse to try to automate the process. "No technology could beat human filtering," Filo argued.[8]

---

[6] Reid, Rob, (1997) *Architects of the Web,* John Wiley and Sons, New York (p. 11).

[7] Stross, Randall E., (1988), "How Yahoo! Won the Search War," *Fortune,* http://www.pathfinder.com.fortune/1998/980302/yah.html., (p. 2).

[8] Stross, Randall E., (1998), "How Yahoo! Won the Search War," *Fortune,* http://www.pathfinder.com.fortune/1998/980302/yah.html., (p. 3).

Though engineers, Yang and Filo had a great sense of what real people wanted. Consider their choice of name. Jerry hated "Jerry's Guide," so he and Filo opted for "Yahoo!," a memorable parody of the tech community's obsession with acronyms (this one stood for "Yet Another Hierarchical Officious Oracle"). Why the exclamation point? Said Yang: "Pure marketing hype."[9]

Yahoo!'s Growing Popularity

At first, Yahoo! was only accessible by the two engineering students. Eventually, they created a Web interface that allowed other people access to their guide. As knowledge of Yahoo!'s existence spread by word of mouth and e-mail, more people began using their site, and Yahoo!'s network resource requirements increased exponentially. Stanford provided them with sufficient bandwidth to the Internet, but bottlenecks came from limitations in the number of TCP/IP connections that could be made to the two students' workstations.[10] Additionally, the time required to maintain the site was becoming unmanageable, as Yang and Filo found themselves continually updating their Web site with new links. Classes and research fell behind as Yang and Filo devoted more and more time to their ever-expanding hobby.

Competing Services

A number of businesses already existed in the Internet search space. While none offered the same service that Yahoo! did, these companies could definitely provide potential competition to any new business that Yahoo! would start. Among the competitors were Architext, soon to be renamed Excite, Webcrawler at the University of Washington, Lycos at Carnegie Mellon, the World Wide Web Worm, and Infoseek, founded by Steven Kirsh. AOL and Microsoft in 1995 represented larger competitors who could enter the market either by building their own capability or acquiring one of the other start-ups.

Yahoo!'s human-crafted hierarchical approach to organizing the information for intuitive searches was a key component of its value proposition. Rob Reid, a Venture Capitalist with 21st Century Internet Venture Partners, explained how this made Yahoo! unique among Internet search providers.

> "The Yahoo! hierarchy is a handcrafted tool in that all of its . . . categories were designated by people, not computers. The sites that they link to are likewise deliberately chosen, not assigned by software algorithms. In this, Yahoo! is a very labor intensive product. But it is also a guide with human discretion and judgment built into it—and this can at times make it almost uncannily effective. . . .

---

[9] Stross, Randall E. (1998), "How Yahoo! Won the Search War," *Fortune,* http://www.pathfinder.com.fortune/1998/980302/yah.html., (p. 3).

[10] Holt, Mark, and Marc Sacoolas, (1995), "Chief Yahoos: David Filo and Jerry Yang," *Mark & Marc Interviews,* May, http://www.sun.com/950523/yahoostory.html.

This is the essence of Yahoo!'s uniqueness and (let's say it) genius. It isn't especially interesting to point to information that many people are known to find interesting. *TV Guide* does this. So do phone books, and countless Web sites that cater to well-defined interest groups. . . . But Yahoo! is able to build intuitive paths that might be singularly, or even temporarily important to the people seeking it. And it does this in a way that no other service has truly replicated."[11]

However, if Yahoo!, as a business, was to survive and flourish in the face of increasingly well-funded competition, it would quickly need to find some outside capital.

Leaving Stanford and Starting the Business

Yang and Filo had been in Silicon Valley long enough to realize that what they really wanted to do was to start their own business. They split much of their free time between their Internet hobby and sitting around thinking up possible business ideas.

"A considerable period of time passed before it occurred to them that the most promising idea was sitting under their noses, and some of the credit for their eventual illumination belongs to their PhD adviser, Giovanni De Micheli. Toward the end of 1994, De Micheli noted that inquiries to Yahoo were rising at an alarming rate. In a single month, the number of hits jumped from thousands to hundreds of thousands daily. With their workstations maxed out, and the university's computer system beginning to feel the load, De Micheli told them that they would have to move their hobby off campus if they wanted to keep it going."[12]

By fall of 1994, the two received over 2 million hits a day on their site. It was then that Jerry and Dave commenced the search for outside backing to help them continue to build up Yahoo!, but with only modest hopes. Yang thought they might be able to bootstrap a workable system, using personal savings to buy a computer and negotiating the use of a network and a Web server in return for thank-you banners. Unexpected overtures from AOL and Netscape caused them to raise their sights, although both companies wanted to turn Filo and Yang into employees.

If they were going to abandon their academic careers (as they soon did, six months shy of their doctorates), they reasoned that they should hold out for some control. Filo and Yang had three main potential options to explore: 1) sell Yahoo! outright; 2) partner with a corporate sponsor; 3) start an independent business using venture capital financing.

---

[11] Reid, Rob, (1997) *Architects of the Web,* John Wiley and Sons, New York (pp. 243–244).

[12] Lardner, James, (1998), "Yahoo! Rising," *U.S. News,* May 18, http://www.usnews.com.usnews/issue/80518/18yaho.html., (p. 3).

The Search for Funding

Looking to receive funding and create a credible business out of Yahoo!, Filo and Yang began preliminary discussions with potential partners in October 1994. One of the first people who contacted them was John Taysom, a vice-president of marketing at Reuters, the London-based media service. Taysom was interested in integrating Reuters' news service into Yahoo!'s Web pages. Yahoo! would gain the advantage of being able to provide news services from a well-known source, while Reuters would be able to begin developing its own presence on the Internet. Unfortunately, since Yahoo! did not generate revenues, it was in a poor negotiating position. Talks between the two were cordial, but they also progressed very slowly.

Yahoo! also talked to Randy Adams, founder of the Internet Shopping Network (ISN), a company that styled itself as "the first online retailer in the world." ISN, funded by Draper Fisher Jurvetson, was one of the first venture funded Internet companies. It had recently been purchased by the Home Shopping Network, in order to expand its possible exposure. ISN was interested in being a host site for Yahoo!, offering them the chance to finally generate some revenue. However, there were also definite possible disadvantages that came from being associated with a shopping network.

Another company that approached Yahoo! was Netscape Communications Corporation. Founded in April 1994 by Jim Clark, who also founded Silicon Graphics, and Marc Andressen, who created the NCSA Mosaic browser with a team of other UIUC students and staff, Netscape was a hot private company developing an improved browser based on the old Mosaic technology. Andressen contacted Yang and Filo over e-mail and, in Yang's words said, "Well, I heard you guys were looking for some space. Why don't you come on into the Netscape network? We'll host you for free and you can give us some recognition for it."[13] This was a fortuitous contact that allowed Yahoo! to move itself off of Stanford's campus. By early 1995, Yahoo! was running on four Netscape workstations.

Soon after, Netscape offered to purchase Yahoo! outright in exchange for Netscape stock. The advantage of this option was that Netscape was already planning its initial public offering and had tremendous publicity and momentum behind it. Coupled with high profile founders and backers like Clark, James Barksdale, former president and CEO of AT&T Wireless Services, and the venture capital firm Kleiner Perkins Caufield & Byers, this offer was a potentially lucrative one for the two Yahoo! founders. Additionally, Netscape's company culture was more in tune with what the two students were looking for, in comparison to some of the more established market players.

---

[13] Ubois, Jeff, (1996), "One Thing Leads to Another," *Internet World,* January, http://www.Internetworld.com/print/monthly/1996/01/yahoo.html, (p. 1).

Corporate Partnerships

Yahoo! was also feeling tremendous pressure to partner or accept corporate sponsorship from other large content companies and online service providers like America Online (AOL), Prodigy, and Compuserve. These companies offered the carrot of money, stock, and/or possible management positions. They argued that if Yahoo! did not partner with them, as large players they could develop their own competing services that would cause Yahoo! to fail. One potential disadvantage with corporate funding was the potential taint that came with such sponsorship. Yahoo! had started as a grass-roots effort, free of commercialization. A second disadvantage was the lack of control that the two Yahoo! founders would have over their creation. "Building Yahoo! was fun, particularly without adult supervision. (Dave) and Jerry were also worried that selling to AOL would have 'most likely killed' Yahoo! in the end."[14]

With partner discussions beginning to heat up, Yang requested help from Tim Brady, a friend and second-year Harvard Business School student. As a class project, Brady generated a business plan for Yahoo! during the 1994–1995 Christmas vacation. (See the Appendix for excerpts of the business plan circa 1995.)

With Brady's business plan in hand, Filo and Yang began to approach different venture capital firms on nearby Sand Hill Road. Venture capital firms brought experience, valuable contacts in the Silicon Valley, and most importantly, money. However, they also required substantial ownership in return for their services. One venture firm that the Yahoo! founders approached was Kleiner Perkins Caufield & Byers. KPCB had an excellent reputation as one of the most prestigious VC firms in the Silicon Valley, and their list of successful investments included Sun Microsystems and Netscape. KPCB showed a definite interest in Yahoo!; however, Vinod Khosla of KPCB and Geoffrey Yang of Institutional Venture Partners had just invested $0.5M in Architext (later renamed Excite), another company started by Stanford engineering students that was developing a search-and-retrieval text engine. Architext was receiving increased press coverage, with a March 1995 *Red Herring* magazine spotlighting the company and its venture capital partners. KPCB proposed to fund Yahoo!, but only if they agreed to merge with Architext.

Sequoia Capital

Another venture capital firm that Yahoo! approached was Sequoia Capital. It was during partnership discussions with Adams at the Internet Shopping Network that Yang and Filo were first introduced to Michael Moritz, a partner at Sequoia Capital. Moritz went to visit Jerry and Dave, who were at the time still operating out of their tiny Stanford trailer. Said Yang, "The first time we sat down with Sequoia, Mike (Moritz) asked, 'So, how much are you going to

---

[14] Reid, Rob, (1997) *Architects of the Web,* John Wiley and Sons, New York (p. 256).

charge subscribers?' Dave and I looked at each other and said, 'Well, it's going to be a long conversation."[15] Fortunately, Moritz, who came from a journalistic background at *Time* was flexible in his thinking. Some of the major advantages that Moritz brought to the negotiating table were his contacts with publications and knowledge about how to manage content. Moritz talked about the roots of Sequoia's interest in working with Yang and Filo. "I think we are always enamored with people that seem to be on to something, even if they can't define that something. They had a real passion and a real spark."[16]

Sequoia Capital had a long tradition of success in the venture capital market, citing that the total market capitalization for Sequoia backed companies exceeded that of any other venture capital firm. Sequoia's trademark *modus operandi* was funding successful companies using only a small amount of capital. Its list of successful investments included Apple Computer, Oracle, Electronic Arts, Cisco Systems, Atari, and LSI Logic. Said Moritz, Sequoia preferred "to start wicked infernos with a single match rather than 10 million gallons of kerosene."[17]

In February 1995, Filo and Yang were weighing a number of possibilities and in no hurry to accept any of them, when Michael Moritz made them an offer. Sequoia Capital would fund Yahoo! for $1 million and would help them to assemble a top management team. In return, Sequoia would receive a 25 percent share of the company. Additionally, Moritz gave them only 24 hours to accept the deal before it was pulled off the table. "I felt a need to deliver them from the agony of indecision," claimed Moritz. With the deadline quickly approaching, Yang and Filo sat down to weigh their options. The decisions that they made that night would determine the direction of their careers as well and the future of Yahoo!

The Decision

Sitting in their tiny office on the Stanford campus, Jerry and Dave shared a late-night pepperoni and mushroom pizza as they explored their options and tried to come to a decision. It was already getting pretty late, and they only had until noon the next day to make their decision.

Yang took a bite from his pizza as he looked over the terms sheet that Sequoia had given them.

> We have some pretty tough decisions to make, and Michael has really forced the issue now with this 24-hour deadline. As I see it, we have a couple of options. The first is to accept Sequoia's offer and launch Yahoo! as our own company. We would be giving up a significant percentage of Yahoo!, but we really need the money if we are going to survive. Moritz and the rest of the

---

[15] Yang, Jerry, in interview "Found You on Yahoo!" *Red Herring,* October 1995 (p. 3).

[16] Moritz, Michael, personal interview, November 10, 1998.

[17] Perkins, Anthony, *Red Herring,* June 1996.

resources at Sequoia could also prove to be invaluable as we try to assemble the rest of our management team.

Our second option is to accept corporate sponsorship. This would allow us to get the funding we need and still retain 100 percent ownership of Yahoo!. However, I am worried about selling out to corporate America. We were fortunate to be able to develop our site in an educational setting as a noncommercial free site. I am afraid if we accept the corporate sponsor ship, it will taint Yahoo!'s image.

Finally, we could agree to merge with an existing corporation. The word is that Netscape is pretty close to their IPO, and Architext has some really big time investors behind it. If we merge with Netscape or Architext in exchange for stock options, it could mean a lot of money for us in the next couple of years.

Filo got up from his seat and kicked aside some of the empty pizza boxes that had started to accumulate. He walked over to Yahoo!'s tiny office window and stared at Stanford's Hoover Tower, which was barely visible in the distance.

It's true that we could make some money if we sell to Netscape or Architext, but we would have to give up primary control of Yahoo! if we did. We would never know what we could have done if we would have maintained control of the site ourselves.

There is also a fourth option you forgot to mention. I'm excited by Sequoia's offer, but I'm wondering if maybe we are giving up too much of our company. A fourth option could be to not decide tonight and look for better terms with another VC firm. I know Michael said that we should decide quickly, but I would hate to give up 25 percent of our company, only to find out in a week that another firm would have offered us $3 million for the same percentage. I know that time is really important, and we like working with Michael Moritz. On the other hand, I don't want to be regretting our decision two months from now.

As they grappled with the alternatives facing them, Filo and Yang began to envision life outside of the Stanford trailer in which Yahoo! was born. It was well past 2 A.M., and they had to make a decision in less that ten hours. What should they do?

Questions

1. What makes Yahoo! an attractive opportunity (and not just a good idea)?
2. How will Yahoo! make money?
3. Identify the major risks in each of these categories: technology, market, team, and financial. Rank order them.
4. What are the advantages and disadvantages of each of the funding options they could pursue?

EXHIBIT 1 Yahoo! Founders and Potential Investor

Jerry Yang

Jerry Yang was a Taiwanese native who was raised in San Jose, California. He co-created the Yahoo! online guide in April of 1994. Jerry took a leave of absence from Stanford University's electrical engineering PhD program after earning both his BS and MS degrees in electrical engineering from Stanford University.

David Filo

David Filo, a native from Moss Bluff, Louisiana, co-created the Yahoo! online guide in April 1994 and took a leave of absence from Stanford University's electrical engineering PhD program in April 1995 to co-found Yahoo!, Inc. Filo received a BS degree in computer engineering from Tulane University and a MS degree in electrical engineering from Stanford University.

Michael Moritz, Partner, Sequoia Capital

Moritz was a general partner at Sequoia Capital since 1988 and focused on information technology investments. Moritz served as a director of Flextronics International and Global Village Communication, as well as several private companies. Between 1979 and 1984, Moritz was employed in a variety of positions by *Time,* Inc. Moritz had an MA degree in history from Oxford University and an MBA from the Wharton School.

Appendix Selected Excerpts from the Yahoo! Business Plan.

Yahoo!'s first business plan was developed by Tim Brady as part of a course project at the Harvard Business School. The plan was continuing to evolve during discussions between Jerry Yang and David Filo at Yahoo! and Michael Moritz of Sequoia Capital. The company has provided excerpts of this business plan that are not proprietary for this case.

The case writers thank Mr. J. J. Healy, Director of Corporate Development, and others at Yahoo! for their efforts in providing this original archival information to enhance the learning experience of future entrepreneurs.

Business Strategy

Yahoo's goal is to remain the most popular and widely used guide to information on the Internet. The Internet is in a period of market development characterized by extremely high rates of both user traffic growth and entry of new companies focused on various products and services. By virtue of its early entry, Yahoo has developed its current position as the leader in this segment. Yahoo's ability to expand its position and develop long-term, sustainable advantages will depend on a number of things. Some of these relate to its current position and others relate to its future strategy.

Today, Yahoo! solves the main problem facing all Internet users. It is next to impossible for users, faced with millions of pieces of information scattered globally on the Internet, to easily find that what is relevant to them without a guide like Yahoo! Not only is the amount of information huge, it is expanding almost exponentially.

All enhancements to Yahoo! will be governed by the goal of making useful information easy to find for individuals.

We believe that Yahoo's enormous following has been generated by the following:

■ Yahoo was first to create a fast, comprehensive and enjoyable guide to the Internet, and in so doing, built a strong brand early and created momentum.

■ The unique interest area based structure of Yahoo! makes it an easier and more enjoyable way for the user to find relevant information than the classic search engine approach where key words and phrases are used as the starting point.

■ Through its editorial efforts, Yahoo! has continually built a guide which is noticeably better than its competition through a combination of comprehensiveness and high quality.

The company will focus on the directory and the guide business and generate revenue from advertising and sponsorship.

Yahoo!'s strategy is to:

■ **Continue to build user traffic and brand strength** on the primary server site through product enhancements and extensions as well as through an aggressive marketing communications program.

■ **Develop and integrate the leading technology** required to maintain a leadership position. Underlying the extremely appealing guide is Yahoo!'s scaleable core technology in search engine, database structure, and communication software. These core technologies are relevant to the user's experience to the extent that it enables Yahoo! customers' access to a broader array of high quality information in an intuitive way, faster than any competitors product. Yahoo! is discussing a full license to advanced web-wide search engine technologies, web-wide index data, and crawler services with Open Text of Waterloo, Canada, Yahoo! will be the first guide with a seamless integrated directory/web-wide search product. The proposed agreement with Open Text also includes ongoing joint development of advanced search and database technologies leveraging the strengths of both companies. All jointly developed products will be distributed by Yahoo! allowing the company to continue to introduce advanced features on a regular and aggressive basis.

■ **Extend the reach to a broader audience** through establishment of contractual relationships with Internet access providers such as MSN, America Online, and Compuserve and very popular Web sites.

- **Extend the reach and appeal to international users** through partnerships with international access providers who can operate foreign mirror sites for Yahoo and add localization in the form of foreign language, local advertisers, and local content.
- **Retain the users ("readership") of Yahoo!** through constant enhancements to the content and interface of the guide.
- **Rapidly extend the product line** by introducing regional guides, vertical market guides, and more importantly, individually personalizeable guides. Our intention is to be the first to market in all of most of these categories and outrun our competition by constantly "changing the competitive rules and targets." Our introduction of personalized guides will be a first in the market and will leverage core technology owned both internally as well as through our license with Open Text.

Market Analysis

The Internet, whose roots trace back almost 20 years, is experiencing a period of incredibly rapid growth in the area of online access base and user population. According to IDC and a recent report by Montgomery Securities, there are approximately 40 million users of the Internet, a majority using it only, for email. However, it is estimated that about 8 million people have access to the Internet and World Wide Web. Most of these access the Web from the workplace because of the availability of high bandwidth hardware and communications ports there. It is expected that over the next two to four years as higher bandwidth modems, home-based ISDN lines and cable modems are adopted, that both the growth and penetration of Web access into the home will increase dramatically. IDC estimates that by 2000, 40 percent of the homes and 70 percent of all businesses in the U.S. will have access to the Internet. In the Western European and Japan markets, the comparable penetration rates might be as high as 25 percent and 40 percent respectively. If this holds true, there will be as many as 200 million users on the Internet and Web by the year 2000.

Market Segmentation and Development

We believe that between now and the year 2000 there will be three principal user groups driving the growth of the Web:

- Large businesses using the Internet for both internal wide area information management and communication as well as intrabusiness communication and commerce.
- Small home based businesses using it for retrieval of information relevant to the business as well as for vendor communication and commerce.
- The individual user/consumer using it initially to find and access information which is relevant to their personal entertainment and learning and later to make purchases of products and services.

We also believe that the evolution of the Internet will include three stages of market development:

- Availability and proliferation of enabling technology.
- Establishment of widespread access and communication services.
- Widespread distribution of high value content.

We are currently in the first stage of market development consisting primarily of infrastructure building and including rapid growth in the adoption and sale of computer, network, and communication products and entering into the second stage involving the initial establishment of "access" service based businesses.

Internet Market Size

Estimates of the amount of current and projected revenue for Internet related business vary. However, primary research conducted by both Montgomery Securities as well as Goldman Sachs indicate that the total served market for Internet hardware, software, and services will total approximately $1B in 1995, up from approximately $300M in 1994. Projections are that these categories might grow to a total of $10B by the year 2000. Several research firms including Forrester and Alex Brown & Sons have estimated the revenues to be produced by Web-based advertising at approximately $20M in 1995, $200M in 1996, and over $2B by the year 2000.

Market Trends

During the current, rapidly expanding stages of market and industry development, the following trends are clear:

- There is large scale adoption of enabling technology in the areas of network hardware and software, as well as communication hardware and software. The World Wide Web with its inherent support of multimedia begs for the adoption of higher and higher bandwidth platform and communication hardware and software.
- Telecommunication companies and newly entering Internet access providers are rushing to put in place basic "hook-ups" in high bandwidth form.
- The price for high-speed computer and communication "port" hardware and software of adequate bandwidth to support acceptable levels of transport and display is still somewhat high. Partly for this reason, the adoption of fully capable ports onto the Web is still principally occurring at businesses.
- With the availability of 28.8K baud modems, ISDN lines and high performance/low price personal computers, home adoption of Internet access is on the rise and slated to have extremely high growth over the next five years. Adoption of cable modems could accelerate this trend.

- Formerly closed network online services such as America Online, Compuserve, and Prodigy are now offering Internet access and opening up their services. Other companies such as Microsoft as well as divisions of MCI, AT&T, and others are attempting to put in place Internet online services in which a range of programming content is presented.
- Companies such as Yahoo! which provide means to navigate the Web are growing rapidly as measured by amount of end user traffic.
- These high traffic sites already provide a high volume platform for delivering electronic advertising.

During this stage, and sustainably for all stages to come, there is one fundamental need which users have: The location of meaningful information easily and quickly on this large and exponentially growing source called the Internet.

Competition

Yahoo! intends to effectively beat any emerging competitors by:

- Establishing broader distribution earlier than any other competitor in order to maintain the Yahoo! guide as the most widely used in its class.
- Broadening the product line faster than the competition through the introduction of vertical market focused guides and personalizeable editions of the guide.
- Staying ahead of the competition with regular core product updates which continue to make it faster, easier to use, and more effective.
- Delivering high quality audiences and compelling results to advertisers.

Risks

The main risks facing Yahoo! are:

- *The ability to increase traffic and enhance the Yahoo! brand.* Management believes it can achieve both these goals.
- *Ability to introduce key new products faster and better than the competition.* We believe that our current core technologies and platform will allow us to do this if supplemented by funded expansion of product development and marketing functions.
- *Ability to develop an international presence and leading brand internationally before the competition.* At the present time, Yahoo! is being pursued by a number of very high visibility and capable international affiliates. The funded addition of limited marketing and business development resources will allow us to respond to these opportunities in a timely way.
- *The introduction of competitive products internally developed by access providers.* While there is no assurance that this will not happen, we have secured relationships with several of the leading providers already in

which the Yahoo! product is featured and are in advanced discussions with others. We believe that many of the access providers already respect Yahoo!'s strong brand, comprehensive guide and focus and are concluding that they will not be inclined to reinvent this late in lieu of a mutually favorable affiliate business relationship with Yahoo!.

- *Ability to scale our support of both the traffic through our main site as well as mirror sites of our affiliates.* If the demands of traffic outgrow the bandwidth of servers we install, then response rates might go down and lead to customer dissatisfaction. Yahoo! has successfully scaled and operated its server site. We believe we will be able to support the needed growth.

- *That the growth of the Internet industry as a whole slows significantly, or that the adoption of the Web as a significant platform for advertising does not grow as projected.* These are both out of Yahoo!'s control. However, the company believes that the industry is in a secure phase of adoption which should fuel growth.

Yahoo!'s sustainable advantages

The Internet is in a period of market development characterized by extremely high rates of both user traffic growth and entry of new companies focused on various products and services. By virtue of its early entry, Yahoo! has developed its current position as the leader in its segment. Yahoo!'s ability to sustain and grow its position will depend on a number of things. Some of these relate to its current core advantages and others relate to future execution of its strategy.

At present, Yahoo!'s core strategic advantages include:

- *It's strong brand.* The company executed early and well with its unique, context focused, quick and intuitive guide and benefited from the widespread adoption of the Yahoo! product. The guide is the standard in the world of Web navigation.

- *Yahoo!'s scalable core technology in search engine, database structure, and communication software.* These core technologies are relevant to the user's experience to the extent that it enables the Yahoo! customer's access to a broader array of high quality information in an intuitive way, faster than any competitor's product.

GLOBAL WIRELESS VENTURES

In November of 2002, Bjorn Magnusson, Theresa Byers, and Kai Leung Ping, the cofounders of Global Wireless Ventures, had to select the best location to establish the new worldwide headquarters of their new company. Global Wireless Ventures was a start-up that was developing a killer application for mobile business professionals, and it had the potential to revive the wireless Internet services marketplace. Wireless Internet services were in the doldrums, due to the worldwide recession and the uncertainty about the future in the wake of terrorist attacks on the World Trade Center and the Pentagon in September of 2001.

The founders of Global Wireless Ventures were three friends and colleagues who had worked together for four years at NTT DoCoMo, helping to create a phenomenally successful market for wireless services in Japan. Bjorn Magnusson had graduated with a PhD from the department of TeleInformatics at the Royal Institute of Technology (KTH) in Kista, Sweden. Theresa Byers had earned an undergraduate and master's degree in computer science at Stanford. Kai Leung Ping had completed an undergraduate degree in electrical engineering at National University of Singapore and an MBA at the Harvard Business School. All three had learned Japanese in their undergraduate years and had taken jobs in Japan with NTT DoCoMo in 1998. They wanted to be at the forefront of the wireless Internet services revolution and believed that it would become a major global market. They had learned a lot in their years with NTT DoCoMo and in 2001 decided that they were ready to launch a new company of their own. Working nights and weekends, they developed a prototype of a new application for mobile business professionals. Following that, they demonstrated their prototype to a small circle of venture capitalists who were considered to be the world leaders in funding wireless Internet ventures.

Global Wireless Ventures had received multiple term sheets from leading venture capitalists (VCs) in three high-tech regions: Stockholm/Kista in Sweden, Singapore, and Silicon Valley. The VCs were convinced that Global Wireless Ventures had enormous potential. They offered copious quantities of "smart money," with reasonable terms and only one catch. The founders had to locate

This case was prepared by Sei Wei Ong, a graduate student at Stanford University School of Engineering; Thomas J. Kosnik, consulting professor, Stanford School of Engineering; and Lena Ramfelt, assistant professor at Royal Institute of Technology (KTH), with the guidance and support of Wong Poh Kam, associate professor, National University of Singapore, as the basis for class discussion rather than to illustrate either effective or ineffective handling of an administrative situation. It was made possible in part through financial support from the National University of Singapore.

the headquarters of their new company in the home region of the VCs whose term sheets they selected. After careful analysis of the VCs offers, the founders were convinced that none of the firms or the term sheets had a clear advantage over the others. Previously, the cofounders had also visited all three locations and believed that they could be happy living in any of the three locations. Eventually, they would need operations in all three regions. The issue facing them in November 2002 was: What place offered the best context to launch the company and to serve as the home base for Global Wireless Ventures?

Overview of the Wireless Industry

In 2002, the wireless industry had become very large and complex. The number of wireless subscribers had quadrupled from less than 200 million in 1997 to 770 million by the end of 2001. There were many players, ranging from start-ups to multinational companies, that formed different parts of the value chain. To describe this industry better, a framework was developed to help understand the different segments of the market. The five segments of the framework are described briefly below, and Exhibit 1 shows a pictorial representation of the wireless segments.[1]

The first segment consisted of the *wireless component providers.* They competed by producing innovations in components such as baseband components, radio frequency (RF) components antennas, etc., which allowed devices to communicate using a variety of wireless protocols. Texas Instruments and Qualcomm were the two market leaders in providing baseband components. The major players in the RF components market were Motorola, Philips, and Infineon. Other companies were positioned to provide specialized chipsets. For example, Intersil and Agere were the leading providers for 802.11 chipsets. Due to the rapid advancements of technology and the huge variety of standards, this segment was a key breeding ground for many start-ups such as Atheros Communications, Cambridge Silicon Radio, and ArrayComm.

The second segment was the *network equipment suppliers* that made the physical infrastructure that was needed to make wireless networks function. They designed and manufactured the radio base stations, the receivers, the switches, and other components of the network. Ericsson dominated the wireless infrastructure market with a market share of about 34 percent. Other key players in this market were Nokia, Motorola, and Lucent. In 2001, the valuations of these companies suffered a significant decline. This was a result of the drastic capital expenditure reductions of their customers—mobile service providers. However, with the pressure to upgrade their networks to support 3G, analysts believed that it would not be long before this segment would pick up again. Looking ahead at the 3G contracts that had already been signed, Nokia had increased

---

[1] For more information on the wireless industry and a detailed explanation of the framework, refer to "Wireless Industry in 2002" at http://edcorner.stanford.edu.

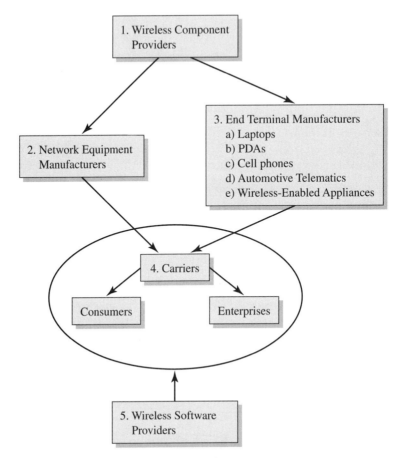

EXHIBIT 1 Pictorial Representation of Wireless Segments

its market share and was currently competing head-to-head with Motorola. Traditionally, the network equipment manufacturers had sold primarily to carriers who build extensive wireless networks for the enterprises and the consumers. However, with the increased popularity of wireless local area networks (WLANs), players in this segment had started to target enterprises and consumers directly. Due to the evolving marketplace, this segment was also rich with start-ups, such as NetGear or FHP Wireless, who were trying to carve out a piece of the pie for themselves.

The third segment was the *end-terminal manufacturers*. This segment was divided into five subsegments—laptop manufacturers, mobile phone manufacturers, personal digital assistant (PDA) manufacturers, automotive telematics, and wireless-enabled appliances. Mobile phones were traditionally the end terminals of a wireless transmission and were the key product in this space. Although laptops and PDAs had become increasingly popular over the last five years, most of them were not wirelessly connected and had to be plugged into a fixed Ethernet connection for network access. The increasing importance of mobility and connectivity suggested that these two subsegments could potentially become key growth areas. Although voice transmission would continue to be the key application for mobile phones, data transmission was the key usage for the laptops and PDAs. The automotive telematics subsegment was still nascent in 2002, and automobile companies had started implementing navigational equipment in their vehicles using global positioning system (GPS). This was still a very expensive option and thus a relatively rare occurrence. The final segment, wireless-enabled appliances was also a very new segment. This space included printers or digital cameras that were fitted with a wireless receiver module, thus allowing the equipment to receive wireless data, usually using the Bluetooth protocol.

The fourth segment consisted of the *carriers* who were mobile service providers. The players in this space were companies that licensed the spectrum from the government, bought infrastructure from the network equipment manufacturers, deployed the infrastructure, and provided wireless services to individual consumers or corporate users. They also served as the distribution channel for mobile phone manufacturers. Examples of mobile service providers in the United States included AT&T, Sprint, Verizon, and Cingular Wireless.

The final segment, *wireless software providers,* was divided into three segments depending on their customer types. These segments were 1) carrier-focused, 2) enterprise-focused, 3) consumer-focused. This subdivision further highlighted the complexity of this space because of the different services that each segment needed. The carriers needed software applications that could improve the network management and to provide value-added services to their customers. Enterprise-focused companies were trying to provide software that could help to extend company operations over wireless networks or business software that could leverage the wireless capabilities that corporations had set up. Finally, consumer-focused companies provided mobile applications for individuals, and this included mobile gaming, content aggregators, scheduling, and other software. This segment was probably the newest in the entire wireless space, and many start-ups were rising up to create software that could use the wireless networks of the future.

A Comparison of the Different Wireless Clusters

There were many hotbeds for wireless clusters all over the world in 2002. Some of the most innovative centers were in Tokyo (Japan), Kista (Sweden), and Helsinki (Finland). This was partly due to the dominance of wireless giants such as Ericsson, Nokia, and NTT DoCoMo, the presence of sophisticated wireless users, and the high penetration rates of wireless technology in these regions. Other technology clusters such as Silicon Valley in the United States and Israel had also developed wireless clusters that meshed intricately with the existing information technology clusters. There were also some important regions such as Singapore, Hong Kong, and Taiwan that were less renowned as innovative

EXHIBIT 2 Worldwide Mobile Subscriber Forecast by Region (Thousands)

	2000	2001	2002	2003	2004	CAGR 01-04
NA/SA	182,300	222,400	247,897	277,535	300,355	10.5%
Europe	270,200	311,802	338,993	358,218	371,556	6.0%
Asia Pacific	231,900	304,200	356,413	384,235	414,813	10.9%
MEA	43,900	55,000	65,000	73,400	80,100	13.4%
ROW	3,400	5,000	7,000	8,000	8,500	19.3%
Total	731,700	898,402	1,015,302	1,101,388	1,175,324	9.4%
Growth		22.8%	13.0%	8.5%	6.7%	

Source: CIBC World Markets, Global Mobile, CIA *2001 Fact Book*

centers for developing breakthrough technologies. However, they had a high wireless penetration rate and were early adopters of wireless technology. Exhibits 2 and 3 divide the world into five regions: America, Europe, Asia Pacific, the Middle East and Africa (MEA), and the rest of world (ROW). Although the penetration rates for Asia Pacific were relatively low, it had one of the highest in absolute numbers of mobile subscribers because of the high penetration rates of advanced countries such as Japan, Taiwan, Korea, and Singapore. Exhibits 4, 5, and 6 further break down the data into the individual countries.

EXHIBIT 3 Mobile Subscriber Penetration Rate by Region

	2000	2001	2002	2003	2004
North America	37.4%	46.2%	51.0%	55.9%	59.8%
South America	12.7%	15.6%	18.0%	20.2%	22.3%
Western Europe	62.8%	71.1%	75.6%	78.3%	80.1%
Eastern Europe	8.3%	11.4%	14.4%	17.1%	19.0%
Asia Pacific	7.0%	9.1%	10.5%	11.5%	12.4%
MEA	5.1%	6.3%	7.5%	8.4%	9.3%

Source: CIBC World Markets, Global Mobile, CIA *2001 Fact Book*

EXHIBIT 4 North American Subscriber Penetration Rate by Country
             (Population in Millions)

	Population	2000	2001	2002	2003	2004
U.S.	278.0	41.5%	47.9%	53.1%	58.0%	62.0%
Canada	31.6	25.9%	31.2%	33.0%	37.0%	40.0%

Source: CIBC World Markets, Global Mobile, CIA *2001 Fact Book*

EXHIBIT 5 Western European Subscriber Penetration Rate by Country
(Population in Millions)

	Population	2000	2001	2002	2003	2004
Austria	8.1	77.4%	78.6%	79.1%	79.4%	79.5%
Belgium	10.2	54.4%	70.5%	75.0%	80.0%	81.0%
Denmark	5.3	64.9%	76.0%	81.0%	82.0%	83.0%
Finland	5.2	78.8%	79.8%	80.3%	80.3%	80.4%
France	58.9	49.3%	61.5%	66.5%	71.5%	75.0%
Germany	83	58.3%	67.0%	75.0%	79.0%	79.5%
Greece	10.7	55.4%	69.5%	79.5%	84.5%	85.5%
Ireland	3.6	63.4%	73.4%	78.4%	80.4%	81.4%
Italy	56.8	74.7%	76.7%	78.7%	80.7%	82.7%
Netherlands	15.8	66.8%	76.8%	79.7%	80.7%	81.2%
Norway	3.1	74.6%	76.6%	78.6%	80.6%	82.6%
Portugal	9.9	63.0%	73.0%	78.0%	83.0%	84.0%
Spain	39.2	62.0%	70.0%	74.5%	75.5%	78.5%
Sweden	8.9	73.8%	75.8%	77.8%	79.8%	81.8%
Switzerland	7.3	63.1%	73.1%	83.1%	84.1%	85.1%
U.K.	59	67.8%	77.0%	78.0%	79.0%	81.0%

Source: CIBC World Markets, Global Mobile, CIA *2001 Fact Book*

EXHIBIT 6 Asia Pacific Subscriber Penetration Rate by Country
(Population in Millions)

	Population	2000	2001	2002	2003	2004
Australia	18.8	53.3%	68.0%	75.0%	75.0%	75.0%
China	1,266.0	6.8%	11.0%	13.4%	15.0%	16.5%
Hong Kong	7.1	74.6%	79.0%	80.0%	80.0%	80.0%
India	1,014.0	0.0%	0.4%	0.5%	0.6%	0.7%
Indonesia	224.8	0.0%	2.3%	3.0%	4.0%	5.0%
Japan	126.9	45.7%	52.2%	58.0%	62.0%	65.0%
South Korea	47.3	56.6%	58.5%	59.5%	62.0%	63.0%
Malaysia	21.4	26.7%	31.0%	34.0%	37.0%	40.0%
Philippines	79.3	7.7%	10.8%	13.0%	15.0%	17.0%
Singapore	3.2	76.4%	85.0%	85.0%	85.0%	85.0%
Taiwan	22.1	79.7%	90.0%	90.0%	90.0%	90.0%
Thailand	60.6	6.1%	9.4%	12.0%	15.0%	18.0%

Source: CIBC World Markets, Global Mobile, CIA *2001 Fact Book*

To further describe these regions better, four different axes were used to analyze the development of these wireless clusters.

The first axis was to compare consumer-orientated technologies versus enterprise-orientated technologies. The wireless industry in Asia and Europe seemed to be focused in providing services to the consumer, whereas the wireless industry in the United States seemed to be more focused on enterprises. Many analysts believed that cultural differences might be one of the major reasons for the deep penetration rates of wireless cell phones in Asia compared to the United States. In technologically advanced Asian countries such as Singapore, Hong Kong, or Japan, almost every teenager had a cell phone. The "coolness" factor was measured by the sophistication of electronic gadgets owned. This could range from the latest Sony Playstation to electronic pets that were in fashion in the late 1990s. Furthermore, a greater proportion of the population in Asia and Europe commuted to work using public transportation and thus found the cell phone critical in making "dead time" useful. On the other hand, most of the U.S. population commuted to work by cars, and thus cell phones were less convenient. As a result, wireless technologies in Asia and Europe appeared to be tailored more toward the consumer market. In the United States, companies had extensive IT departments that implemented enterprise-wide software, such as supply chain management (SCM), customer relationship management (CRM), and enterprise resource planning (ERP) software. These technologies had become highly integrated into the enterprise and were key components to the daily operations of the company. Many companies wanted to extend these applications wirelessly so that their employees were able to leverage these resources even when they were out in the field. U.S. companies were also more aggressive in adopting new technologies and integrating it into their organization than other countries. As a result, wireless technologies in the United States tended to be orientated more toward the enterprise.[2]

The second axis was to compare telecommunications-oriented versus computing communications-orientated technologies. Although the Sweden-Finland region and Silicon Valley were all considered highly innovative wireless centers, Scandinavia came from a telecommunications perspective, whereas Silicon Valley came from a computing communications perspective. The presence of telecommunication giants such as Nokia and Ericsson were the driving force behind the growth of the wireless cluster in Scandinavia. Since these companies were primarily telecommunication equipment manufacturers, the innovation tended to be targeted toward the needs of these giants. Silicon Valley, on the other hand, did not have wireless gorillas that threw their weight around. However, in Silicon Valley, the presence of other clusters, such as the microelectronics cluster and the information technology cluster, fostered an environment that tackled the wireless problem from a computing angle.

The third and fourth axes were somewhat related. The third axis was the speed of wireless innovation, and the fourth was the speed of adoption of wireless technology. In wireless clusters such as Silicon Valley, there was tremendous techno-

---

[2] For more information, look at *The Economist,* "The Internet Untethered," 13 October 2001.

logical innovation in the wireless space. However, the deployments of these technologies were still found to be lacking. Regions such as Singapore and Hong Kong did not have much innovative *development* of new wireless technologies, but they had been the most aggressive in *deployment* of wireless technology. These included the latest Nokia cell phones and Cisco 802.11b access points. Sweden and Finland had high levels of technology innovation as well as deployment. The low penetration rates in Silicon Valley were primarily due to the scarcity of spectrum as well as the fragmented wireless space that had retarded the deployment of wireless technologies. The low innovation rates in Singapore and Hong Kong were a result of the scarcity of wireless experts and technological innovators. Scandinavia and Japan did not suffer from either of these problems and thus were able to have deep innovation centers as well as extensive deployments.

The founders of Global Wireless Ventures had to analyze the three different regions—Silicon Valley, Sweden, and Singapore. Exhibits 7, 8, and 9 show pictorial representations of each of these wireless regions using the wireless framework that we developed earlier. The size of each component represents the importance and strength of its influence. Each of the exhibits is divided into two parts. The top half represents the internal forces within the wireless cluster, and the bottom half represents the external forces that were influencing the clusters.

Silicon Valley Wireless Cluster   Silicon Valley was known as the high-tech entrepreneurship center of the world. Over the last 30 years, some of the most successful high-tech multinational enterprises such as Intel, Cisco, and Oracle were founded in the valley. However, in terms of wireless innovation, different segments of the value chain were at different levels of maturity. This could be seen in Exhibit 7 that shows that segments 2 and 4 were still very weak, whereas there was more innovation in segments 1, 3, and 5. As mentioned earlier in the text, the slow development of network equipment manufacturers (segment 2) and the carriers (segment 4) could partly be attributed the scarcity of spectrum in the United States and the lack of a dominant wireless standard. Some industry observers felt that the government should have played a more active role in trying to develop these segments. If the government had intervened to free up the spectrum earlier or stipulated one particular standard, the wireless industry in the United States would not be two to three years behind other developed nations in Europe and Asia.

A strong influence to the rapid growth of the wireless software providers (segment 5) was the presence of the information technology cluster in this region. Experienced software engineers were able to switch from developing software for desktop computers to mobile devices. Similarly, the well-developed microelectronics cluster spurred the growth of the wireless component providers (segment 1). For the end-terminal manufacturers (segment 3), there were already incumbent laptop and PDA manufacturers in the region. Thus, it was relatively easy for them to add a mobile component to their existing devices. Although wireless innovation was a relatively recent phenomenon in the valley, there were already related technologies strongly rooted in the valley that helped the swift development of these segments.

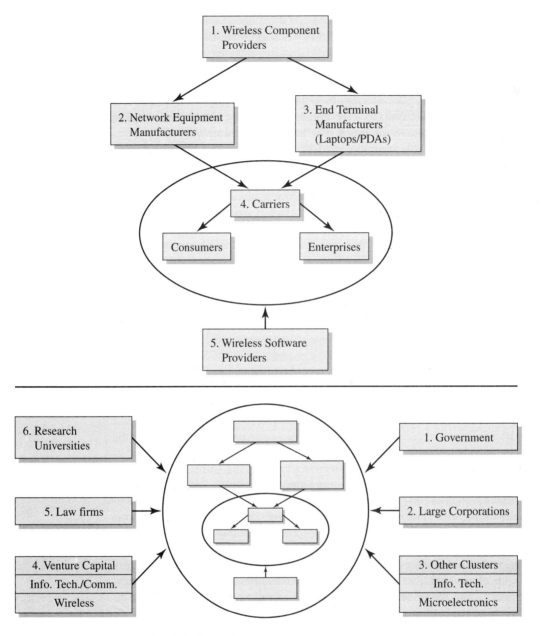

EXHIBIT 7 · Silicon Valley Wireless Cluster

Silicon Valley also had a strong start-up infrastructure that helped to support the growth of start-ups in this region. Specialized law firms such as Wilson Sonsini and Venture Law Group were experienced in helping start-ups formulate term sheets and to give appropriate legal advice to help the inexperienced entrepreneurs with their ventures. Other important elements of the start-up puzzle in-

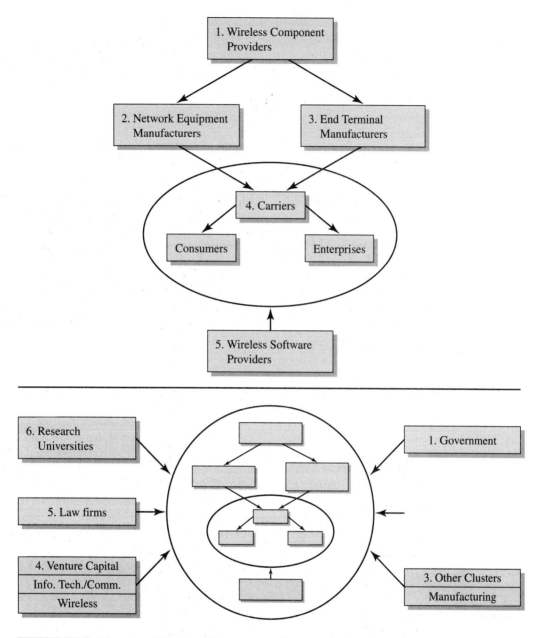

EXHIBIT 8 Singapore Wireless Cluster

cluded the presence of certified public accountant (CPA) firms and investment banking firms that provided professional services that helped the growth of start-ups. There was also an easy access to research and skilled talent from world-renowned universities such as Stanford and Berkeley. The mature venture capital industry formed another crucial component in the valley. Although the

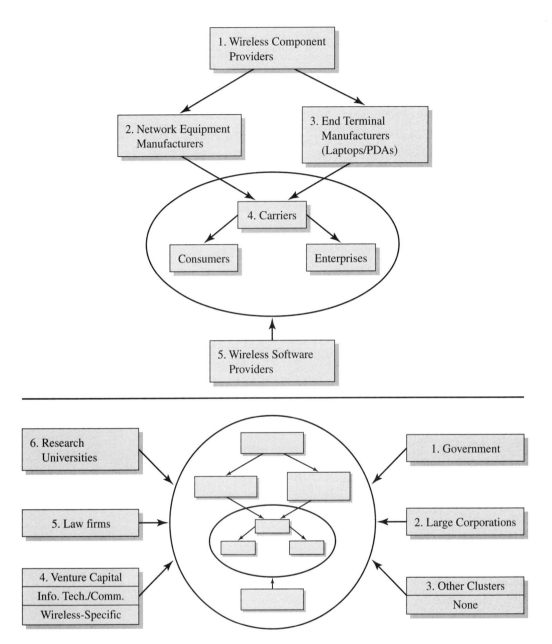

EXHIBIT 9 Sweden Wireless Cluster

majority of VC firms were still generic, there was a recent trend of VC firms such as Ignition Ventures that targeted wireless start-ups specifically. This strong network infrastructure that supported the wireless start-up industry was vital to the rapid development of this cluster.

Another interesting aspect of the Silicon Valley wireless cluster was the symbiotic relationship between start-ups and the large technology corporations that were in the valley. Large corporations like Cisco were looking to acquire start-ups with technology that had a good strategic fit with their portfolios. On the other hand, some corporations were spinning out divisions in the company so that they could focus on their core competencies. One example would be Hewlett Packard spinning out Agilent. This continuous interaction between corporations and start-ups created a healthy environment for the growth of start-ups, strategic alliances, potential exit strategies, and strong distribution channels.

Singapore Wireless Cluster   The wireless cluster in Singapore was not as innovative as that of Silicon Valley and Sweden. As can be seen from Exhibit 8, the only two segments with significant growth were in the carriers (segment 4) and the wireless software providers (segment 5). This was partly due to the shortage of technical and entrepreneurial talent in the country and partly due to the strong presence of multinational companies that had contributed significantly to the growth of the country in its short history of 37 years. Although there was not much domestic technological innovation, Singapore was one of the leading adopters of wireless technology worldwide.

With an estimated wireless penetration rate of 85 percent in 2002, Singapore had a very sophisticated wireless community. This had led to many wireless companies choosing Singapore as one of the first target markets for their products. For example, Handspring chose to launch the new Treo Communicator in Singapore ahead of Europe and United States. In fact, Singapore was the second country worldwide to retail this product, a couple of days behind Hong Kong. Aether Systems also chose to partner with a Singapore wireless software provider (segment 5), EON Infotech, to provide wireless and mobile computing solutions for the mobile sales force to financial, government and enterprise markets in Southeast Asia. With a population of only about 4 million people, some companies saw Singapore as an ideal test bed for wireless technologies. For example, Citibank chose Singapore as the first country to offer its new wireless service, Citibank Alerts. This service enabled customers to track their bank accounts and receive the latest news and stock prices through their mobile phones. After successful trials in Singapore, Citibank had continued to roll out this service in other parts of Asia and also in the United States.

Apart from being an ideal test site, Singapore was also often seen as the springboard for companies who were interested in the capturing the Asian market. Since its independence in 1965, many multinational companies have flocked to Singapore to set up their Asia headquarters. The state-of-the-art infrastructure, effective communications networks, and a stable government made Singapore the ideal home base for many companies. Furthermore, Singapore had one of the most competitive economies of the world. According to the International Institute for Management Development (IMD), Singapore was ranked the fifth most competitive economy in 2002. However, from 1998-2001, it was second only to the United States. Exhibit 11 shows the *World Competitive Yearbook* rankings over the last five years.

Many wireless companies were very interested in Asia for two primary reasons. First, Asia contained some of the most advanced wireless users such as Korea, Hong Kong, Taiwan, and Japan. Second, it also contained more than half of the world population, especially the untapped markets of China and India. Many analysts believed that wireless technology would leapfrog wireline technology in the less developed countries, as they were easier and less expensive to deploy. Global entrepreneurs could not afford to ignore this market when considering their start-up strategies.

Within the wireless cluster, the carriers (segment 4) in Singapore were the most developed. SingTel, the leading mobile service provider in Singapore, was one of the most advanced carriers worldwide. Although Singapore only had a population of about 4 million, SingTel had moved beyond the domestic market and had more than 16 million subscribers from the Australasian region. The presence of a strong regional mobile service provider in Singapore was one of the key driving forces to the deep wireless penetration rates in Singapore. The strong connections between SingTel and the government also made it possible for them to deploy nationwide projects more readily.

Another important incentive for these companies to deploy their products in Singapore was the active involvement of the Singapore government in promoting these technologies. The InfoComm Development Authority (IDA) had been the primary driving force in promoting the growth of the wireless industry in Singapore. In October of 2000, IDA launched its "Wired with Wireless" Program that was aimed at developing Singapore into a living lab for wireless development in Asia. One of the interesting projects under this scheme was the Call for Collaboration (CFCs), an open invitation to the wireless industry, both local and overseas, to propose innovative wireless projects in Singapore. Since its inception in May of 2001, five CFCs had been issued that focused on different parts of the wireless value chain, such as mobile payment, wireless Java solutions, etc. CFCs would also be awarded funding to help them deploy their projects. It was too early to tell if these projects would be successful, but they had received significant attention from industry leaders such as Nokia and Sun Microsystems.[3] In addition, Singapore had also made great strides as a regional center for higher education. Besides three excellent local universities—the National University of Singapore, Nanyang Technological University, and the Singapore Management University—they had also attracted leading universities including MIT, Johns Hopkins, and University of Pennsylvania's Wharton School to set up Asian centers in Singapore. The presence of these research institutes would attract technical and professional talent to Singapore and thus nurture the next generation of technologists.

Sweden Wireless Cluster   Of the three regions under consideration by Global Wireless Ventures, Sweden was the most developed cluster in terms of development and also deployment. Exhibit 9 shows that all five segments had a certain

---

[3] For more information, www.ida.gov.sg.

level of maturity. However, the network equipment manufacturers segment was by far the most developed segment in this wireless cluster. This was primarily due to the presence of Ericsson, the gorilla in this space with more than 30 percent of the global market share for network infrastructure. The Swedish wireless cluster was centered in the Stockholm metropolitan area with the largest concentration of wireless companies located in Kista, north of Stockholm.

It was not possible to discuss the Sweden wireless cluster without Ericsson. Ericsson was one of the first companies to move into Kista when it was established in the 1970s and had continued to serve as the core of the wireless cluster in Kista. Ericsson had also played an important role in standards definition as well as the development of software to complement wireless networks and devices. This rich technological environment had led to the development of many wireless start-ups from all the segments in the wireless framework.

Unlike Silicon Valley, where there were a multitude of large corporations that continuously interacted with the start-ups, there was only one major player in the Sweden cluster, Ericsson. As a result, many of the start-ups were developing technology that was compatible with Ericsson systems. Although Ericsson had been the driving force behind wireless innovation in Stockholm, some analysts were arguing that Ericsson's presence was actually starting to stifle innovation because start-ups had to pander to the strategic goals of just one company. As a result, a small but increasing number of wireless innovators were choosing to leave Stockholm for other clusters such as Silicon Valley. On the other hand, many other global industry players were starting wireless research centers in Kista to develop technology that could complement their existing products. Some of these companies included Oracle, Apple, Microsoft, Intel, and Compaq. Exhibit 10 shows a more comprehensive list of the companies that were involved in this cluster and the research areas that they were focusing on. The presence of these big companies and the hundreds of wireless start-up companies in this region had led Stockholm to be known as the "Wireless Valley."

One of the key players in the wireless sector in Sweden was the Royal Institute of Technology (Kungliga Tekniska Högskolan). The university had been the primary source of technical talent and research that fed into Stockholm as well as Kista Science Park, educating one-third of Sweden's engineering talent. In 2000, the university opened an Information and Communication Technology (ICT) University in Kista with the goal of becoming the largest ICT university in Europe. They targeted to have more than 12,000 students by 2010. The venture capital industry in Sweden had also been rapidly developing over the last five years. Many of these VC firms were investing specifically in the wireless industry.

Decision Time

In early 2002, Bjorn Magnusson, Theresa Byers, and Kai Leung Ping reviewed the information about the three regions that they had collected from their Web research and discussions with entrepreneurs, venture capitalists, and other industry professionals. As cofounders, they had to select the best location to establish the worldwide headquarters of their new company.

EXHIBIT 10 Companies with Centers in Kista Science Park

Company	Description
Microsoft, U.S.	Mobile Internet—joint venture with Ericsson
Intel, U.S.	Wireless competence center and e-business solution center
Nortel, Canada	R and D in mobile communication
Motorola, U.S.	Development center for wireless applications and services
IBM, U.S.	Wireless Internet Center
Oracle, U.S.	Wireless research and a joint project with Ericsson and Telia
Hewlett-Packard, U.S.	Wireless research and a joint project with Ericsson and Telia
Andersen Consulting, U.S.	Global center for WAP-applications and services
RSA Security, U.S.	Development of secure wireless communications
Compaq, U.S.	Wireless Competence Center and an eCommerce Knowledge Center
Nokia, Finland	R and D in mobile communication infrastructure
Cambridge Technology Partners, CTP, U.S.	Global Wireless Competence Center
Siemens, Germany	R and D-center for mobile applications
Sybase, U.S.	Test center for mobile business applications and a strategic alliance with Ericsson around Mobile Banking Solution
Sun Microsystems, U.S.	Wireless Center of Excellence
CapGemini Ernst&Young, France/U.S., and Cisco, U.S.	Joint competence center for third-generation mobile systems

Source: Invest in Sweden Agency, 2001 presentation.

Unfortunately, things were not cut and dried. In addition to the information about each region, they had to think about how the choice of headquarters location might affect them personally in the months and years ahead. For example, all three were currently single but looked forward to finding the right person some day, getting married, and having a family. All three also had parents and brothers and sisters still living in their respective birthplaces of Sweden, Singapore, and Silicon Valley.

They knew that the decision of where to start would have unpredictable implications for both the company in its life cycle and each of them in their personal lives. They wondered how to integrate both the macro and micro issues in an effective way to make the best decision.

Theresa: I think we should come to a decision about the headquarters of our new start-up. The VCs are fascinated both with our technology and our business plan, so we are in a strong position to choose who we want to support us. I don't have a strong preference for any particular VC, so I think we should make our decisions based on the needs of the company and our preferences as cofounders.

EXHIBIT 11 IMD World Competitiveness Scoreboard 2002
(Ranking as of April 2002)

Country	Rankings				
	2002	**2001**	**2000**	**1999**	**1998**
USA	1	1	1	1	1
Finland	2	3	4	5	6
Luxembourg	3	4	6	3	3
Netherlands	4	5	3	4	4
Singapore	5	2	2	2	2
Denmark	6	15	13	9	10
Switzerland	7	10	7	7	9
Canada	8	9	8	10	8
Hong Kong	9	6	12	6	5
Ireland	10	7	5	8	7
Sweden	11	8	14	14	16
Iceland	12	13	9	13	18
Austria	13	14	15	18	24
Australia	14	11	10	11	12
Germany	15	12	11	12	15
U.K.	16	19	16	19	13
Norway	17	20	17	16	11
Belgium	18	17	19	21	23
New Zealand	19	21	18	17	17
Chile	20	24	25	25	27

Bjorn: I agree with Theresa. The choice of our home base will be very important because it reflects the identity of the company. We need to be careful of the kind of message that we want to project to our customers and partners. I would vote strongly to base our start-up in Sweden, because it has both a strong culture for wireless innovation and a wide base of sophisticated wireless users. Besides, I know the professors at KTH where I earned my PhD, and I think we can use my relationships with them to get access to new technological developments from their labs and to recruit their best students as new employees.

Kai Leung: Although I agree that Sweden is a good place to base our start-up, I'm a little wary about the strong presence of Ericsson and its influences in that region. We want to project ourselves as an independent global company. Besides, I am particularly attracted by the financial incentives that the

Singapore government is giving to wireless start-ups, the growth in world-class research institutes, and the technical talent. But most importantly, I believe that Singapore is the perfect springboard into Asia, especially the untapped markets of China and India. On a personal level, I would also like to get back to Singapore. I've been away from home for many years, and I need to return home to take care of my aging parents. I know this is a personal request, but it is an issue that I'm currently facing.

Theresa: I understand your situation, Kai Leung. And I think we should definitely take our personal lives into consideration when making this decision. We can't put our whole heart into this start-up if we are worried about issues elsewhere. For the moment, let me play the devil's advocate for Silicon Valley. With its high concentration of high-tech companies, Silicon Valley could provide us with a strong customer base for our initial product launch. This is particularly important, because our initial product targets the mobile business professionals. If these companies endorse our product, that could be a tremendous boost to our product reputation. I agree with Bjorn that access to a research university will be valuable, and I think that we can use my network as a Stanford alumna to connect with faculty and students with wireless software expertise at Stanford University.

Kai Leung: There seems to be so many different issues that we need to think about. How should we go about making the decision? The VCs want to know our decision by early next week.

Questions

1. What factors should Global Wireless Ventures consider in comparing the entrepreneurial context in Sweden, Singapore, and Silicon Valley?
2. Conduct some research about Kista, Sweden, Singapore, and Silicon Valley, using libraries and the Internet. Based on this research, what are the assets, liabilities, and risks in the context of each region?
3. What assets (besides smart money) are most critical for a start-up like Global Wireless Ventures?
4. What institutions are likely to help them gain access to those assets in each region?

JON HIRSCHTICK'S NEW VENTURE

In August, 1994, 12 months after Jon Hirschtick left a great job to found a new venture in the software industry, SolidWorks, the deal was looking good. The seed capital discussions had shifted into high gear as soon as Michael Payne joined the SolidWorks team. After working on the deal for nine months, Axel Bichara, the Atlas Venture vice president originating the project, finally got a syndicate excited about it: Atlas Venture, North Bridge Venture Capital Partners, and Burr, Egan, Deleage & Co. presented an offer sheet to SolidWorks two weeks after Michael was on board.

This process was particularly interesting because Jon and Axel had worked together for most of the past eight years. They met at MIT in 1986 and co-founded Premise, Inc., a computer aided design (CAD) software company, in 1987. After Premise was bought by Computervision, they joined that team as managers. Now, they sat on opposite sides of the table for Axel's first deal as the lead venture capitalist.

Jon and the other founders thought the valuation and terms were fair, but the post-money* equity issue was unresolved. They had to decide how much money to raise. Did they want enough capital to support SolidWorks until it achieved a positive cash flow, or should they take less money and attempt to increase the entrepreneurial team's postmoney equity?

If they took less money now, they could raise funds later, when SolidWorks might have a higher valuation. But they would be gambling on the success of the development team and the investment climate. If their product was in beta testing with high customer acceptance, raising more money would probably be fast and fun, but if they hit any development snags, the process could take a lot of time and yield a poor result.

Jon Hirschtick: 1962–1987

Jon grew up in Chicago in an entrepreneurial family. He fondly remembers helping with his father's part-time business by traveling to stamp collectors' shows across the Midwest. In high school, he was self-employed as a magician.

The entrepreneurial impulse continued during his undergraduate years. Jon recalls the blackjack team he played with at MIT:

> We raised money to get started. At the same time, we developed a probabilistic system for winning at blackjack. The results were amazing! We tripled our money in the first six months, doubled it during the next six months, and doubled it again in the next six months. We produced a 900

This case was written by Dan D'Heilly and Tricia Jaekle under the direction of Professor William Bygrave. Funding provided by the Ewing Marion Kauffman Foundation and the Frederic C. Hamilton Chair for Free Enterprise Studies. © Copyright Babson College, 1995. All rights reserved.

---

*Post-money valuation: the value of a company's equity after additional money is invested.

percent annualized return. I learned a useful lesson: you really can know more than the next guy and make money by applying that knowledge. We tackled blackjack because people thought it was unbeatable; we studied it, and we won. The same principle applies to entrepreneurship. Opportunities often exist where popular opinion holds that they don't.

Jon's introduction to CAD came from a college internship with Computervision during the summer of 1981. Computervision was one of the most successful start-up companies to emerge during the 1970s. By the early 1980s, it dominated the CAD market.

After earning a master's degree in mechanical engineering (M.E.) at MIT, Jon managed the MIT CAD laboratory. He supervised student employees, coordinated research projects, and conducted tours for visitors.

Axel Bichara: 1963–1987

Axel was born in Berlin and attended a French high school. In 1986, while studying at the Technical University of Berlin for a master's degree in mechanical engineering, he won a scholarship to MIT. Axel had worked in a CAD research lab in Germany, so he selected the CAD laboratory for his work-study assignment at MIT.

Early CAD Software

CAD software traces its roots to 1969, when computers were first used by engineers to automate the production of drawings. CAD was used by architects, engineers, designers, and other planners to create various types of drawings and blueprints. Any company that designed and manufactured products (e.g., Ford, Sony, Black & Decker) was a prospective CAD software customer.

An Entrepreneurship Class: January 1987

Visitors to the MIT CAD lab often complained about problems that Jon knew he could solve. He enrolled in an entrepreneurship class to write a business plan for a CAD start-up company, Premise, Inc. Jon described the decision to quit his job and start a company:

> I once heard Mitch Kapor* use a game show metaphor to describe the entrepreneurial impulse. He said, "Part of the entrepreneurial instinct is to push the button before you know the answer and hope it will come to you before the buzzer." That's what happened for us: we didn't know how to start a company, or how to fund it, but Premise got rolling, and we came up with answers before we ran out of time.

Jon and Axel were surprised and delighted to find each other in the entrepreneurship class. They had worked together for the past month on a project at

---

*Mitchell Kapor founded Lotus Development Corporation.

the CAD lab, and they decided to become partners in the first class session. Axel recalled:

> It was a coincidence that we enrolled in the same class, but it was clear that we should work together. Jon had had the idea for a couple of months, and we started work on the product and the business plan immediately.

Axel took the master's exam at MIT in October 1987 and at Technical University of Berlin in July 1988. He was still a student at both universities when he and Jon started Premise. Axel graduated with highest honors from both institutions.

Premise, Inc.: 1987–1991

Premise went from concept to business plan to venture capitalist-backed start-up in less than six months. As Axel remembered:

> The class deadline for the business plan was May 14. On June 1, we had our first meeting with venture capitalists, and by June 22, we had a handshake deal with Harvard Management Company for $1.5 million. We actually received an advance that week. It was much easier than it should have been, but the story's 100 percent true.

In the first quarter of 1989, Premise raised its second round of capital. Harvard Management and Kleiner Perkins Caufield & Byers combined to finance the product launch. The product shipped in May to very positive industry reviews, but sales were slow. Premise's software didn't solve a large mass-market problem. As Jon later recalled:

> I've seen successful companies get started without talent, time, or money—but I've never seen a successful company without a market. Premise targeted a small market. I had a professor who said it all, "The only necessary and sufficient condition for a business is customers."

By the end of 1990, the partners had decided that the best way to harvest Premise was an industry buyout. They hired a Minneapolis investment banking firm to find a buyer. Wessels, Arnold & Henderson was considered one of the elite investment banking firms serving the CAD industry. Premise attracted top-level service providers because of the prestige of its venture capitalist partners. Jon explained:

> Several bankers wanted to do the deal, and a big reason was because they wanted to work with our venture capitalists. We had top venture capitalists, and that opened all kinds of doors. This is often under-appreciated. I believe in shopping for venture capital partners.

Wessels, Arnold & Henderson were as good as their reputation. As Axel recalled:

> We sold Premise to Computervision on 7 March 1991. Computervision bought us for our proprietary technology and engineering team. It was a good deal for both companies.

Computervision: 1991–1993

As part of the purchase agreement, Jon and Axel joined the management team at Computervision. They managed the integration of Premise's development team and product line for one year before Axel left to study business in Europe. Jon stayed on after Axel's departure.

Revenues for the Premise team's products grew 200 percent between 1991 and 1993, and perhaps as important as direct revenue, their technology was incorporated into some of Computervision's high-end products. In January 1993, Jon was promoted to director of product definition for another CAD product. He stayed in this position for eight months. After two years at Computervision, he was ready for new horizons. He resigned effective August 23, 1993. (See Exhibit 1 for excerpts from his letter of resignation.)

After a holiday in the Caribbean, Jon purchased new computer equipment, called business friends and associates, and began working on a business plan. He

---

EXHIBIT 1 Excerpts from Jon Hirschtick's Letter of Resignation from Computervision (CV)

This is my explanation for wanting to leave CV. . . . The other day you asked me whether I was leaving because I was unhappy, or whether I really want to start another company. <u>I strongly believe that it is because I really want to work on another entrepreneurial venture.</u>* I want to try to build another company that achieves business value. . . .

I am interested in leaving CV to pursue another entrepreneurial opportunity because I seek to:

1. <u>Be a part of business strategy decisions</u>. I want to attend board meetings and create business plans, as I did at Premise.

2. <u>Select, recruit, lead, and motivate a team of outstanding people</u>. I believe that one of my strengths is the ability to select great people and form strong teams.

3. <u>Represent a company with customers, press, investors, and analysts</u>. I enjoy the challenge of selling and presenting to these groups.

4. <u>Work on multidisciplinary problems: market analysis, strategy, product, funding, distribution, and marketing</u>. I am good at cross-functional problem-solving and deal-making.

5. <u>Work in a fast-moving environment</u>. I like to be in a place where decisions can be made quickly, and individuals (not just me) are empowered to use their own judgment.

6. <u>Work in a customer-driven and market-driven organization</u>. I find technology and computer architecture interesting only as they directly relate to winning business. I want to focus on building products customers want to buy.

7. <u>Have significant equity-based incentives</u>. I thrive on calculated risks with large potential rewards.

8. <u>Be recognized for having built business success</u>. I measure "business success" by sales, profitability, and company valuation; I want to directly impact business success. Recognition will follow. I admit that this ego-need plays a part in my decision.

Summary
I've decided I want to work on an entrepreneurial venture. . . . This is more a function of what I do best than any problems at CV. . . . I don't have any delusions about an entrepreneurial company being any easier. I know first-hand that start-up companies have at least as many obstacles as large established companies—but they are the obstacles I want.

---

*Underlines in original.

didn't have a clear product idea, but his market research suggested that the time was ripe for a new CAD start-up.

CAD Software Market in the 1990s

By the 1990s, the hottest CAD software performed a function called solid modeling. Solid modeling produced three-dimensional computer objects that resembled the products being built in almost every detail. It was primarily used for designing manufacturing tools and parts. Solid modeling was SolidWorks' focus. The key benefits driving the boom in solid modeling were:

1. Relatively inexpensive CAD prototypes could be accurate enough to replace costly (labor, materials, tooling, etc.) physical prototypes.
2. The elimination of physical prototypes dramatically improved time-to-market.
3. More prototypes could be created and tested, so product quality was improved.

However, not all CAD software could manage solid modeling well enough to effectively replace physical prototypes.

Most vendors offered CAD software based on computer technology from the 1970s and 1980s. IBM, Computervision, Intergraph, and other traditional market leaders were losing market share because solid modeling required software architecture that worked poorly on older systems.

As one of the industry's newest competitors, Parametric Technology Corporation (PTC) was setting new benchmarks for state-of-the-art solid modeling software. (It was an eight-year-old company in 1994.) CAD was a mature and fragmented industry with many competitors, but PTC thrived because other companies tried to make older technology perform solid modeling functions.

Worldwide mechanical CAD software revenues were projected at $1.8 billion for 1995, with IBM expected to lead the category with sales of $388 million. PTC was growing over 50 percent annually and had the second highest sales, with $305 million in projected revenue. Industry analysts predicted 3 percent to 5 percent revenue growth per year, with annual unit volume projected to grow at 15 percent. The downward pressure on prices was squeezing margins, so many stock analysts thought that the market was becoming unattractive. However, PTC traded at a P/E between 21 and 40 in 1994.

Axel after Computervision: 1992–1994

After five years in the United States, Axel decided to attend an MBA program in Europe. From his experiences at Premise and Computervision, he had become intrigued with the art and science of business management, and he was ready for a geographic change.

INSEAD was his choice. Located in Fontainebleau, an hour south of Paris, INSEAD was considered one of the top three business schools in Europe. The application process included two alumni interviews, and one of Axel's interviewers was Christopher Spray, the founder of Atlas Venture's Boston office. Atlas Venture was a venture capital firm with offices in Europe and the United States. It had $250 million under management in 1994.

Since Axel had a three-month break before INSEAD started, Chris asked him to consult on a couple of Atlas Venture's projects. Axel found he enjoyed evaluating business proposals "from the other side of the table." He graduated in June 1993 and joined the Boston office of Atlas Venture as a vice president with responsibility for developing high-tech deals.

Axel reflected on the relationship between business school training and venture capital practice:

> I was qualified to become a venture capitalist because of my technical and entrepreneurial background; business school just rounded out my skills. You do not need a bunch of MBA courses to be a successful venture capitalist. Take finance, for example, I learned everything I needed from the core course. People without entrepreneurial experience who want to be venture capitalists should take as many entrepreneurship courses as possible.

Jon Founds SolidWorks: 1993–1994

Jon's business plan focused on CAD opportunities. He explained:

> I knew that this big market was going through major changes, with more changes to come. From an entrepreneur's perspective, I saw the right conditions for giving birth to a new business. I also knew I had the technical skills, industry credibility, and vision needed to make it happen. This was a pretty rare situation.

SolidWorks' product vision evolved slowly from Jon's personal research and from discussions with friends. He was careful to avoid using research that Computervision might claim as proprietary. He was concerned about legal issues, because he would be designing software similar to what Computervision was trying to produce. Axel explained:

> Both Computervision and SolidWorks wanted to produce a quality solid modeling product. Solid modeling technology was still too difficult to learn and use. Only PTC's solid modeling software really worked well enough. The rest made nice drawings but could not replace physical prototypes for testing purposes.

There were only 50,000 licensed solid modeling terminals in the United States, and most of them belonged to PTC, but there were over 500,000 CAD terminals. There were two main reasons PTC did not have a larger market: (1) its products required very powerful computers, and (2) it took up to nine months of daily use to become proficient with PTC software. Solidworks' goal was to create solid modeling software that was easier to learn and modeled real-world parts on less specialized hardware (see Figure 1).

FIGURE 1 Competitive Positioning Grid

	Computer aided drafting and add-ons	Production solid modeling
Low-end:		
Windows	Autodesk	SolidWorks
~$5K per station	Bentley	
VAR channel	CADKEY	
High-end:		
UNIX	Applicon	PTC
~$20K per station	CADAM	
Direct sales		

This vision was not unique in the industry. Many CAD companies were developing solid modeling software, and the low-end market was wide open. SolidWorks's major advantage was its ability to use recent advances in software architecture and new hardware platforms—it wasn't tied to antique technology. Attracting talented developers was the top priority in this leading-edge strategy.

Team Building

Jon's wife, Melissa, enthusiastically supported his decision to resign from Computervision. Jon explained:

> Some spouses couldn't deal with a husband who quits a secure job to start a new company. Melissa never gave me a hard time about being an entrepreneur.

Jon described his priorities in October 1993, when he decided to launch SolidWorks:

> I knew I needed three things: good people, a good business plan, and a good proof-of-concept.* I needed a talented team that could set new industry benchmarks, but there was no way I could get those people without a persuasive prototype demonstration.
>
> The venture capitalists wanted a solid business plan, but that wouldn't be enough. They wanted a strong team. I needed fundable people who were also CAD masters. Venture capitalists couldn't understand most complex

---

Proof-of-concept is a term that refers to a computer program designed to illustrate a proposed project. Also referred to as a *prototype,* it is used for demonstration purposes, and it is limited but functional in ideal circumstances.

technologies well enough to be confident a high-tech business plan was really sound, so they looked at the team and placed their bets largely on that basis. If the proof-of-concept attracted the team, then the team and the business plan would attract the money. I needed a team that could create the vision and make venture capitalists believe it was real.

Jon worked on finding the team and developing the proof-of-concept concurrently; but the proof-of-concept was his first priority. He worked on it daily. In his search for cofounders, Jon talked to dozens of people; he even posted a notice on the Internet, but "none of those guys worked out."

Recruiting posed another dilemma—how to get people to work full time without pay, while the company retained the right to their output? He resolved this problem by creating consulting agreements that gave SolidWorks ownership of employees' work and made salaries payable at the time of funding. As it turned out, this arrangement only lasted nine months. Jon described his approach to recruiting:

> I always paid for the meal when I talked with someone about SolidWorks. I wanted them to feel confident about it, and that meant that I had to act with confidence. The deal I offered was: no salary, buy your own computer, work out of your house, and we're going to build a great company. I'd done it before, so people signed on.

Axel described Jon's management style as, "visionary, he's a talented motivator, and a strong leader."

Robert Zuffante: CAD Engineer/Consultant

A major development in 1993 was the addition of super-star consultant Bob Zuffante as manager of proof-of-concept development. Jon needed time to write the business plan and recruit his team. He had been working on the prototype for over a month when Bob took over development. Jon recalled the situation:

> I hadn't seen him since we were students together at MIT, but when I thought about the skills I needed, my mental Rolodex came up with his name. I always thought about working with him again. We talked in late November, and about a month later, he began work on the prototype.

Bob knew Jon and Axel from MIT, where he earned a master's degree in mechanical engineering. He had worked in the CAD industry for over 10 years and had managed a successful consulting business. His arrival at SolidWorks allowed Jon to focus on other pressing issues.

Scott Harris: CAD Marketer

Scott Harris worked at Computervision for 11 years, where he managed development and marketing activities. Most notably, Scott was the founder and manager of Computervision's product design and definition group. He also managed the 11-person solid modeling development group and acted as technical liaison between Computervision's customers and R and D engineers.

Scott was let go by Computervision during a large-scale layoff. He was skeptical when Jon first told him about the SolidWorks vision, but he became a believer after seeing a proof-of-concept demonstration. Scott stopped looking for a job and started working full time for SolidWorks almost immediately. Scott was impressed, "The prototype was the embodiment of a lot of the things I was thinking about. This was the way solid modeling should perform."

Scott started with SolidWorks about six weeks after Bob signed on. He became involved in the marketing sections of the business plan and in the product definition process. He ran focus groups, conducted demonstrations for potential customers, and analyzed the purchasing process. He kept the development team focused on customer needs—how did customers really use CAD software, and what did they need that current products lacked?

The Business Plan

When Bob came on in January, Jon turned to the business plan with a passion. The plan went through a number of versions as Jon and his advisors wrestled with key issues such as positioning, competitive strategy, and functionality. By the end of March, the plan was polished enough for Jon to show it to venture capitalists. Axel recalled:

> Jon and I decided that the business plan was ready to show in April, so I scheduled a presentation at Atlas. Jon gave the presentation to Barry [Fidelman, Atlas general partner] and myself—market, team, and concept. Overall, Barry was encouraging, but not excited. He thought Jon's story was not crisp enough; he was looking for money to take on some very large companies, and the CAD market was not that attractive. It was a rocky start.

Initial Financing Attempts

In addition to negotiating with Atlas Venture, Jon met with other venture capital firms and rewrote the business plan several times. Axel described the rationale behind this process:

> If you talk to too many people and you do not make a good impression, it will be much harder to get funding, because the word on the street will be, "this deal will not fly." Meet with four or five venture capitalists at most, then revise the plan if you are not getting the right response. After each major revision, show it again to the lead venture partner.

While there were promising discussions with several venture capitalists, Atlas did not want to be the sole investor, and SolidWorks did not win support from other venture capitalists during the spring or summer.

Jon was contacted by an established CAD software company in May 1994. It wanted to acquire SolidWorks—essentially the development team and the prototype. The proposal was attractive; it included signing bonuses and stock. Scott recalled his excitement:

> This was a big shot in the arm. It meant that other industry insiders respected our vision and talent enough to put up their money and take the risk. This was like an cold bucket of Gatorade on a hot day.

Jon stopped seeking venture capital for about a month while he considered the buyout offer. If the offer was a boost to morale, the way the team rejected it was even more meaningful. Jon talked to each person (several other programmers had joined during the spring), and they were unanimous in wanting to continue toward their original goal. Affirming their commitment reinvigorated the team.

Turning Point: Michael Payne, CAD Company Founder

The most significant advance that summer began with a due diligence meeting set up by Atlas Venture. Atlas wanted the SolidWorks team to meet its agent, Michael Payne, who had recently resigned from PTC. Michael had cofounded PTC, the number one company in CAD software. He was one of the most influential people in the industry.

Michael had grown up in London. He earned his bachelor's degree in electrical engineering from Southampton University and his master's degree in solid-state physics from the University of London. He came to the United States and worked many years for RCA designing computer chips. Michael continued his education at Pace University, where he earned an MBA. His senior CAD development experience began in the 1970s, when he ran the CAD/CAM design lab at Prime Computer. He was subsequently recruited by Sam Geisberg, the visionary behind PTC. Michael recalled their first meeting in 1986: "Sam had some kind of crazy prototype, and I said, 'Hey, we can do something with that. This is what we should be working on.'"

PTC was founded in 1986 with Michael as vice president of development, and within five years the company had created a new set of CAD industry benchmarks. For FY 1993, PTC sales were $163 million, it earned a pretax profit margin over 40 percent, and it reached a market capitalization* of $1.9 billion. Michael's reputation as a development manager was outstanding. Remarkably, PTC had never missed a new product release date, and it released products every six months. This was considered a near-impossible feat in software development. He left PTC in April 1994 during a management dispute, about two months before the due diligence meeting with SolidWorks.

Jon had never met Michael but knew by reputation that he was a tough character. The SolidWorks team was worried about two possibilities: that Michael would say they were on the wrong track, or that he might take their ideas back to PTC. Jon recalled the meeting:

---

*Market capitalization is the value of the company established by the selling price of the stock times the number of shares outstanding.

Bob and I were on one side of the table and Michael and Axel on the other. I decided to gamble on a dramatic entrance. Before we told him anything about SolidWorks, I asked Michael to show his cards. I asked him to tell us what he thought were the greatest opportunities in the CAD market. Michael mentioned many of the things we were targeting. I couldn't imagine a better way to start the meeting.

We presented our plan and prototype. Michael asked us a lot of tough, confrontational questions. Afterwards, he told Atlas Venture, "These guys have a chance." Coming from him, that was high praise.

The due diligence meeting was also the beginning of a dialogue between Michael and Jon about joining SolidWorks. Over the next couple of months, Michael decided to join the team. Jon described the synergy between them:

> You almost couldn't ask for two people with more different styles, but we got along well because we were united in our philosophy and vision. We found that our stylistic differences were assets; they created more options for solving problems.

Michael talked about his motivation for joining the SolidWorks team:

> I couldn't go work for a big company because I didn't have any patience for petty politics. A start-up was my only option. The larger the company, the more focused it would be on internal issues rather than on making a product that customers would buy. Customers don't care about technique, they care about the benefits of the technology.

Jon focused on CAD features that he knew customers wanted, and he had a prototype demonstrating that he could do it. It was also quicker and easier than what was on the market. Being able to develop it was another matter. They still had to build it. Implementation, that's where he would be useful. He told them, "Give me whatever title you want; I just want to run development."

Team Adjustments

Michael's arrival created an imbalance in the SolidWorks team, and it took time to sort it out. In fact, Michael didn't join the team until the last week in August. Jon described his thoughts about team cohesion:

> When I decided to start SolidWorks, I had three goals:
>
> 1) work with great people,
> 2) realize the vision of a new generation of software, and
> 3) make a lot of money.

We didn't go looking for Michael Payne, but when he came along, it was an easy decision. It can be hard to bring in strong players, but if those are your three goals, the decision falls out of the analysis rather naturally.

Bob and I had to give up the reins in some areas so Michael could come on board. We weren't looking for a top development manager because we thought we already had two. The change took some getting used to, but it was clearly the right thing to do.

Jon focused on team building, and Michael became the development manager. There were still big talent gaps, especially in sales and finance, but those positions could be filled when they were closer to the product launch. Michael was satisfied, "We didn't have a vast team, but you don't start out with a vast team, and we had a terrific nucleus."

September 1994

Atlas arranged for Jon to talk with venture firms interested in joining the investment syndicate. The team met with Jon Flint of Burr, Egan, Deleage & Company, and Rich D'Amore of North Bridge Venture Partners. After completing their due diligence investigations, both firms joined the syndicate. Jon Hirschtick recalled the situation:

I was pleased that Jon Flint and Rich D'Amore decided to invest. I had met Jon many years earlier and thought very well of him. Rich also impressed me as a very knowledgeable investor. Both had excellent reputations and I looked forward to having them join our board.

An offer sheet was presented to SolidWorks two weeks after Michael officially joined the SolidWorks management team. Now the team had to decide how much money they really wanted. Michael's last venture, PTC, only needed one round of capital, and this team wanted to go for one round, too. SolidWorks' monthly cash burn rate was projected to average about $250,000 and they planned to launch the product in a year, so they needed $3 million for development. Sales and marketing would also need money; they decided that $1 million should be enough to take them through the product launch to generating positive cash flow. To that total, they added a $500,000 safety margin. SolidWorks asked Atlas to put together an offer sheet based on raising $4.5 million.

SolidWorks received the offer sheet during the first week of September. It gave a $2.5 million pre-money valuation with a 15 percent post-money stock option pool.* (For SolidWorks' business plan projections, see Exhibit 2.) These terms were fairly typical for a first round deal, but the SolidWorks team didn't like what happened to their post-money equity when they ran the numbers.

Questions:

1. Why has this deal attracted venture capital?
2. Can the founders optimize their personal financial returns and simultaneously ensure that SolidWorks has sufficient capital to optimize its chance of succeeding? What factors should the founders consider?

---

*The pool of company stock reserved for rewarding employees in the future.

3. How can the venture capitalists optimize their return? What factors should they consider?
4. After you have answered questions 2 and 3, structure a deal that will serve the best interests of the founders, the company, and the venture capital firms.

EXHIBIT 2 Business Plan Projections

	1994	1995	1996	1997	1998
Revenue	$ —	$ 175,000	$ 3,010,000	$ 8,225,000	$ 17,115,000
Cost of sales	$ —	$ 31,500	$ 541,800	$ 1,480,500	$ 3,080,700
Sales and marketing	$ 71,919	$ 765,920	$ 1,930,000	$ 3,030,000	$ 5,822,500
R and D	$ 605,544	$ 1,126,208	$ 1,350,000	$ 1,500,000	$ 2,050,000
G and A	$ 185,954	$ 445,175	$ 650,000	$ 800,000	$ 1,050,000
Total expenses	$ 863,417	$ 2,368,803	$ 4,471,800	$ 6,810,500	$ 12,003,200
Operating income	$ (863,417)	$ (2,193,803)	$ (1,461,800)	$ 1,414,500	$ 5,111,800
Margin analysis					
Cost of Sales		18.0%	18.0%	18.0%	18.0%
Gross Profit		82.0%	82.0%	82.0%	82.0%
Sales and marketing		437.7%	64.1%	36.8%	34.0%
R and D		643.5%	44.9%	18.2%	12.0%
G and A		254.4%	21.6%	9.7%	6.1%
Operating income		−1253.6%	−48.6%	17.2%	29.9%

ARTEMIS IMAGES

Christine Nazarenus tried to retain her optimism. Thirteen had always been a lucky number for her, but Friday, the thirteenth of July, 2001, had the earmarks of being the unluckiest day of her life. She was more than disappointed. She was shattered. Yet she knew that she had hard facts, not just gut feel, that offering images and products on the World Wide Web was the wave of the future. She was sure that the management team she had put together had the creativity and skills to turn her vision into reality. Managing her own company had seemed the obvious solution, but she hadn't counted on how overwhelming the start-up process would be. Now, two years later, she was trying to figure out what went wrong and if the company could survive.

It had been so clear on day one. Archived photographs and images had tremendous value if they could be efficiently digitized and catalogued. Sports promoters and publishers had stores of archived information, most of it inaccessible to those who wanted it. Owners and fans represented only part of the untapped markets that the Internet and digital technology could serve. She had conceived a simple business model: digitize documents using the latest technology, tag them with easy-to-read labels, and link them to search engines for easy retrieval and widespread use. But over the ensuing months, so many factors affected the look, feel and substance of the company that Artemis Images would become.

So many things seemed outside her control that she wondered how she could have been so sure of herself back in February of 1999. Enthusiastically, Chris had approached a number of friends and acquaintances to help in the formation of a new "dot.com" company that seemed a sure bet. Frank Costanzo, a former colleague from Applied Graphics Technologies (AGT), shared Chris's enthusiasm, as did long-time friend George Dickert. George, in turn, contacted Greg Hughes, who was enrolled in a Business Planning course. Grateful for the opportunity to help launch a real company, Greg took the idea and honed it as part of a class assignment. The plan was a confirmation of Chris's confidence in the venture. But as she looked over the original plan, she knew there was a lot of work yet to do. Greg understood the business idea, but he didn't understand the work involved to actually run a business. George and Frank understood digital technology and project management, but, like Chris, had never launched, much less worked for, a start-up company. Chris knew that she had the technology and

© 2002 by Joseph R. Bell, University of Northern Colorado, and Joan Winn, University of Denver. Published in *Entrepreneurship Theory & Practice,* **28**(2) Winter 2003. The authors wish to thank Chris Nazarenus and the staff of Artemis Images for their cooperation in the preparation of this case. Special thanks to Herbert Sherman, Southampton College, Long Island University, and Dan Rowley, University of Northern Colorado. This case is intended to stimulate class discussion rather than to illustrate the effective or ineffective handling of a managerial situation. *All events and individuals in this case are real, but some names may have been disguised.*

talent she needed and felt confident that the four friends could construct a business model that would put Artemis ahead of the current image providers. Greg's business plan looked like the perfect vehicle to appeal to investors for the funds they needed to proceed.

The Business Idea

In 1999, Chris had been working for three years as VP-Sales out of the Colorado office of AGT, a media management company that provided digital imaging management and archiving services for some of the largest publishers and advertisers in the world. AGT had sent Chris to Indianapolis to present a content management technology solution to the Indianapolis Motor Speedway Corporation (IMSC) as it prepared marketing materials for the 2001 Indy 500. IMSC is the host of the 80-plus-year-old Indy 500, the largest single-day sporting event in the world, NASCAR's Brickyard 400, the second-largest single-day sporting event in the world, and other events staged at the track. Chris's original assignment was a clear one: IMSC needed to protect its archive of photographs, many of which had begun to decay with age. The archive included 5 million to 7 million photographs and dynamically rich multimedia formats of video, audio, and in-car camera footage.

Chris discovered that the photo archives at IMSC were deluged with requests (personally or via letters) from fans requesting images. She was amazed that a relatively unknown archive had generated nearly $500,000 in revenues in 1999 alone. Further discussions with IMSC researchers revealed that requests often took up to two weeks to research and resulted in a sale of only $60 to $100. However, IMSC was not in a position, strategically or financially, to acquire a system to digitize and preserve these archives. Not willing to leave the opportunity on the table, Chris asked herself, "What is the value of these assets for e-commerce and retail opportunities?" Without a doubt, IMSC and some of her other clients (Conde Nast, BBC, National Motor Museum) would be prime customers for digitization and content management of their collections.

Chris knew that selling photos on the Internet could generate substantial revenue. She conceived of a business model where the system would be financed through revenue-sharing, rather than the standard model where the organization paid for the system up front. IMSC was interested in this arrangement, but it was outside the normal business practices of AGT. AGT wanted to sell systems, not give them away. They couldn't see the value of managing other organizations' content.

As Chris told the story, her visit to the archives at IMSC was her *Jerry Maguire* experience. In the movie, Jerry is sitting on the bed when everything suddenly becomes clear and now he must pursue his dream. Like Jerry, Chris believed so passionately that her idea would bear fruit that when AGT turned down Chris's request for the third time, she quit her job to start Artemis Images on her own.

Building a Team

When AGT was not interested in Chris's idea of on-site digitization and sale of IMSC's photo archives, Chris was not willing to walk away from what she saw as a gold mine. She contacted her friends and colleagues from AGT. Swept up in the dot.com mania, Chris named her company "e-Catalyst." e-Catalyst was incorporated as an S-corporation on May 3, 1999, by a team of four people: Christine Nazarenus, George Dickert, Frank Costanzo, and Greg Hughes. (See Exhibit 1 for profiles of these partners.) Expecting that they would each contribute equally, each partner was given a 25 percent interest in the company. Chris fully expected them to work as a team, so no formal titles were assigned, largely as a statement to investors that key additions to the team might be needed and welcomed. As another appeal to potential investors—and to broaden the team's expertise—Chris and George put together a roster of experts with content management, systems and technology experience as their first advisory board. Greg's professor and several local business professionals agreed to serve on the board of advisors, along with an Indy 500 winning driver-turned-entrepreneur, and Krista Elliott Riley, president of Elliott Riley, the marketing and public relations agency that represented Indy 500 and Le Mans Sports Car teams and drivers. Chris felt confident that her team had the expertise she needed to launch a truly world-class company.

EXHIBIT 1 Artemis Images Management Team 1999–2000

Christine Nazarenus, 34, was formerly Vice President of National Accounts for AGT, one of the top three content management system providers in the world securing million dollar deals for this $500 million company. She is an expert in creating digital workflow strategies and has designed and implemented content management solutions for some of the largest corporations in the world including Sears, Conde Nast, Spiegel, Vio, State Farm, and Pillsbury. Ms. Nazarenus has extensive general management experience and has managed a division of over 100 people. Chris holds a BA in Communications from the University of Puget Sound.

George Dickert, 32, most recently worked as a project manager for the Hibbert Group, a marketing materials distribution company. He has experience with e-commerce, Web-enabled fulfillment, domestic and international shipping, call centers and CD-ROM. He has overseen the implementation of a million-dollar account, has managed over $20 million in sales, and has worked with large companies including Hitachi, Motorola, ON Semiconductor, and Lucent Technologies. Mr. Dickert has an MBA from the University of Colorado. George and Christine have been friends since high school.

Frank Costanzo, 40, is currently a Senior Vice President at Petersons.com. Petersons.com has consistently been ranked as one of the top 100 sites worldwide. Mr. Costanzo is an expert in content management technology and strategy and was previously a Vice President at AGT. Mr. Costanzo has done in-depth business analysis and created on-site service solutions in the content management industry. He has worked on content management solutions for the world's top corporations including General Motors, Hasbro, Bristol-Meyers Squibb, and Sears.

Greg Hughes, 32, is currently a senior sales executive with one of the largest commercial printers in the world. Mr. Hughes has 10 years' sales experience and has sold million-dollar projects to companies like US West, AT&T, R. R. Donnelly, and MCI. His functional expertise includes financial and operational analysis, strategic marketing, fulfillment strategies and the evaluation of start to finish marketing campaigns. Mr. Hughes has an MBA from the University of Colorado.

Chris and George quit their jobs and took the challenge of building a company seriously. They contacted one of the Rocky Mountain region's oldest and most respected law firms for legal advice. They worked with two lawyers, one who specialized in representing Internet companies as general counsel and one who specialized in intellectual property rights. With leads from her many contacts at AGT, Chris contacted venture capitalists to raise money for the hardware, software licensing, and personnel costs of launching the business.

The dot.com bust of 2000 did not make things easy. Not wanting to look like "yet another dot.com" in search of money to throw to the wind, Chris and her team changed their name to Artemis Images. Artemis, the Greek goddess of the hunt, had been the name of Chris's first horse as well as her first company, Artemis Graphics Greeting Cards, her first entrepreneurial dabble at the age of 16. Chris had always been enthralled with beautiful images.

Artemis Images's Niche

In her work at AGT, Chris had observed that many organizations had vast stores of intellectual property (photos, videos, sounds and text), valuable assets often underutilized because they exist in analog form and may deteriorate over time. Chris's vision was to preserve and enable the past using digital technology and the transportability of the World Wide Web. Chris envisioned a company that would create a digitized collection of image, audio and video content that she could sell to companies interested in turning their intellectual property into a source of revenue.

Publishers and sports promoters were among the many organizations with large collections of archived photos and videos. Companies like Boeing, General Motors, and IMSC are in the business of producing planes, cars, or sporting events, not selling memorabilia. However, airplane, car, and sports fans are a ready market for photos of their favorite vehicle or videos of their favorite sports event.

Proper storage and categorization of archived photos and videos is complex and expensive. In 2000, the two common solutions were to sell the assets outright or to set up an in-house division devoted to managing and marketing them. Most organizations were unwilling to sell their assets, as they represented their priceless brand and heritage. Purchasing software and hiring specialized personnel to digitize and properly archive their assets was a costly proposition that lay beyond the core competence of most companies. Chris's work with AGT convinced her that there were literally thousands of companies with millions of assets that would be interested in a company that would digitize and manage their photo and video archives.

Chris understood a company's resistance to selling its archives, and the high cost of obtaining and scanning select images for sale. However, she also understood the value to an organization of having its entire inventory digitized, thus creating a permanent history for the organization. She proposed a revenue sharing model whereby Artemis Images would digitize a client's archives but would not take ownership. Instead, her company would secure exclusive license to the archive, with 85 percent of all revenue retained by Artemis Images and 15 percent paid to the archive owner. She expected that the presence of viewable archives on the Artemis Images website would lure buyers to the site for subsequent purchases.

The original business model was a "B2C" (business-to-consumer) model. Starting with the IMSC contract, Artemis Images would work with IMSC to promote the Indy 500 and draw the Indy race fans to the Artemis Images website. Photos of the current-year Indy 500 participants—and historical photos including past Indy participants, winners, entertainers, celebrities (e.g., Arnold Palmer on the Indy golf course)—would be added to IMSC's archived images and sold for $20 to $150 apiece to loyal fans. A customer could review a variety of photo options on the Artemis website, then select and order a high-resolution image. The order would be secured through the Web with a credit card, the image transferred to the fulfillment provider, and a hard copy mailed to the eager recipient. The website was sure to generate revenue easier than IMSC's traditional sales model of the past.

Having established the model with IMSC content in the auto racing market, Chris and George built the business plan around obvious market possibilities that might appeal to a wider range of consumers and create a comprehensive resource for stock photography. Since the Artemis Images team had prior business dealings with two of the three largest publishers in the world, publishing was the obvious target for future contracts. Future markets would be chosen similarly, where the Artemis team had established relationships. These markets would be able to build on the archive already created and would bring both consumer-oriented content and saleable stock images. Greg made a list of examples of some industries and the content that they owned:

- Sports: images of wrestling, soccer, basketball, bodybuilding, football, extreme sports
- Entertainment: recording artists, the art from their CDs, movie stars, pictures of events, pictures from movie sets
- Museums: paintings, images of sculpture, photos, events
- Corporations: images of food, fishing, planes, trains, automobiles
- Government: coins, stamps, galaxies, satellite imaging

As Chris and George worked with Greg to put together the business plan, they began to see other revenue-generating opportunities for their virtual-archive company. Customers going to IMSC or any other Artemis client's website would be linked to Artemis Images's website for purchase of photos or videos. Customer satisfaction with image sales would provide opportunities to sell merchandise targeted to specific markets and to syndicate content to other websites. For motor sports, obvious merchandise opportunities would include T-shirts, hats, and model cars. For landscapes, it might be travel packages or hiking gear. Corporate customers might be interested in software, design services, or office supplies. Unique content on Artemis Images's website could be used to draw traffic to other companies' sites. Chris and her team planned to license the content on an annual basis to these sites, creating reach and revenues for Artemis Images.

Another potential market for Artemis Images lay in the unrealized value of the billions of images kept by consumers worldwide in their closets and drawers. These images were treasured family heirlooms which typically sat unprotected and underutilized. Consumers could offer their photographs for sale or simply pay for digitization services for their own use. If just 10% of the U.S. population were to allow Artemis Images to digitize their archive and half of these people ordered just one 8"×10" print, Artemis Images could create a list of 25 million consumers and generate revenues of approximately $250 million. Because images suffer no language barriers, the worldwide reach of the Internet and the popularity of photography suggested potential revenues in the billions.

Working together on the business plan, the Artemis team brainstormed ways they could attract customers to the Artemis Images site by providing unique content and customer experiences. A study by Forrester Research analyzed the key factors driving repeat site visits and found that high-quality content was cited by 75 percent of consumers as the number one reason they would return to a site. The Artemis team wanted to create a community of loyal customers through additional unique content created by the customers themselves. This would include the critical chats and bulletin boards that are the cornerstone of any community-building program. Artemis Images could continuously monitor this portion of the site to add new fan experiences to keep the experience "fresh." Communities would be developed based on customer interests.

As the company gained clients and rights to sell their archived photos and videos, Artemis would move toward a "B2B" (business-to-business) model. Chris and George knew marketing managers at *National Geographic,* CMG World Wide, the BBC, Haymarket Publishing (includes the Formula 1 archive), Conde Nast, and International Publishing Corporation. These large publishers controlled and solicited a wide range of subject matter (fashion, nature, travel, hobbies, etc.) yet often had little idea of what existed in their own archives or had difficulty in getting access to it. Finding new images was usually an expensive and time-consuming proposition. Artemis Images could provide the solution. For example, Conde Nast (publisher of *Vogue, Bon Appetit, Conde Nast Traveler, House & Garden,* and *Vanity Fair*) might like a photo for its travel magazine from the *National Geographic* archives. They would be willing to pay top dollar for classic stock images, given the number of viewers who would see the image. Price-per-image was typically calculated on circulation volume, much like royalty fees on copyrighted materials. Similarly, advertising agencies use hundreds of images in customer mockups. For example, an agency may desire an image of a Pacific island. If Artemis Images held the rights to Conde Nast and *National Geographic,* there might be hundreds of Pacific island photos from which to choose. As with the B2C concept, a copy of the image would be transferred through the Web with a credit card or on account, if adequate bandwidth were available (only low-resolution images would be available to view initially), or via overnight mail in hard copy or on disk.

The transition from B2C to B2B seemed a logical progression, one that would amass a large inventory of saleable prints and, at the same time, draw in larger per-unit sales. The basic business model was the same. Artemis would archive photos and videos that could be sold to other companies for publication and promotion brochures. Chris and George expected that this model could be replicated for other vertical markets including other sports, nature, entertainment, and education.

While the refocus on the B2B market seemed a surer long-term revenue stream for the company, both B2B and B2C were losing favor with the investing community. Chris and George refocused the business plan as an application service provider (ASP). With the ASP designation, Artemis Images could position itself as a software company, generating revenue from the licensing of its software processes. In 2000, ASPs were still in favor with investors.

Artemis Images's revenue would come from three streams: (1) sales of images to businesses and consumers, (2) syndication of content, and (3) sales of merchandise. Projected sales were expected to exceed $100 million within the first four years, with breakeven occurring in year three. (See Exhibits 2, 3, and 4 for projected volume and revenues.)

To implement this strategy, Artemis Images, Inc. needed an initial investment of $500,000 to begin operations, hire the team, and sign four additional content agreements. A second round of $1.5 million and a third round of $3 million to $8 million (depending on number of contracts) were planned, to scale the concept to 28 archives and over $100 million in assets by 2004. (See Exhibit 5 for funding and ownership plan.)

The Content Management Industry

According to GISTICS, the trade organization for digital asset management, the content management market (including the labor, software, hardware, and physical assets necessary to manage the billions of digital images) was projected to be a $2 trillion market worldwide in the year 2000 (1999 Market Report). Content could include images, video, text and sound. Artemis Images intended to pursue two subsets of the content management market. The first was the existing stock photo market, a business-to-business market where rights to images were sold for limited use in publications such as magazines, books, and websites. Deutsche Bank's Alex Brown estimated this to be a $1.5 billion market in 2000. Corbis, one of the two major competitors in the digital imaging industry, estimated it to be a $5 billion market by 2000.

Commercially produced images were also in demand by consumers. Industry insiders believed that this market was poised for explosive growth in 2000, as Web-enabled technology facilitated display and transmission of images directly from their owners to individual consumers. The archives from the Indianapolis Motor Speedway was an example of this business-to-consumer model.

EXHIBIT 2 Anticipated Sales Volume and On-Site Operations

Volumes 2001

	Jan-01	Feb-01	Mar-01	Apr-01	May-01	Jun-01	Jul-01	Aug-01	Sep-01	Oct-01	Nov-01	Dec-01	Total
Consumer Photos	0	0	7,500	7,500	27,000	9,000	9,000	22,500	18,000	9,000	4,500	22,500	136,500
Stock Photos	0	0	0	3,750	4,500	5,250	6,750	7,500	9,000	9,750	10,500	11,250	68,250
Subtotal	0	0	7,500	11,250	31,500	14,250	15,750	30,000	27,000	18,750	15,000	33,750	204,750
Licensing Deals	0	0	0	0	0	1	1	2	4	5	6	7	26
Merchandise Orders	0	0	6,000	6,000	21,600	7,200	7,200	18,000	14,400	7,200	3,600	18,000	109,200

Volumes 2002

	Jan-02	Feb-02	Mar-02	Apr-02	May-02	Jun-02	Jul-02	Aug-02	Sep-02	Oct-02	Nov-02	Dec-02	Total
Consumer Photos	12,000	15,000	15,000	15,000	54,000	18,000	18,000	45,000	36,000	18,000	9,000	45,000	300,000
Stock Photos	6,000	7,500	9,000	10,500	12,000	15,000	15,000	15,000	15,000	15,000	15,000	15,000	150,000
Subtotal	18,000	22,500	24,000	25,500	66,000	33,000	33,000	60,000	51,000	33,000	24,000	60,000	450,000
Licensing Deals	8	9	10	11	12	13	14	15	16	16	16	16	156
Merchandise Orders	9,600	12,000	12,000	12,000	43,200	14,400	14,400	36,000	28,800	14,400	7,200	36,000	240,000

Volumes 2003

	Jan-03	Feb-03	Mar-03	Apr-03	May-03	Jun-03	Jul-03	Aug-03	Sep-03	Oct-03	Nov-03	Dec-03	Total
Consumer Photos	24,800	31,000	31,000	31,000	111,600	37,200	37,200	93,000	74,400	37,200	18,600	93,000	620,000
Stock Photos	17,000	21,250	25,500	29,750	34,000	42,500	42,500	42,500	42,500	42,500	42,500	42,500	425,000
Subtotal	41,800	52,250	56,500	60,750	145,600	79,700	79,700	135,500	116,900	79,700	61,100	135,500	1,045,000
Licensing Deals	16	16	16	16	16	16	16	16	16	16	16	16	192
Merchandise Orders	15,600	19,500	19,500	19,500	70,200	23,400	23,400	58,500	46,800	23,400	11,700	58,500	390,000

ONSITE OPERATIONS (by quarters)

Year	2000	2001				2002				2003			
Quarter	Qtr 4	Qtr 1	Qtr 2	Qtr 3	Qtr 4	Qtr 1	Qtr 2	Qtr 3	Qtr 4	Qtr 1	Qtr 2	Qtr 3	Qtr 4
Onsites (cumulative)	1	4	4	7	10	13	13	16	19	22	25	28	28

Source: e-Catalyst Business Plan, February 28, 2000.

EXHIBIT 3 Projected Monthly Revenue Stream

Revenues 2001

	Jan-01	Feb-01	Mar-01	Apr-01	May-01	Jun-01	Jul-01	Aug-01	Sep-01	Oct-01	Nov-01	Dec-01	Total
Consumer Photos	$ 0	$ 0	$ 149,925	$ 149,925	$ 539,730	$ 179,910	$ 179,910	$ 449,775	$ 359,820	$ 179,910	$ 89,955	$ 449,775	$ 2,728,635
Stock Photos	$ 0	$ 0	$ 0	$ 562,500	$ 675,000	$ 787,500	$1,012,500	$ 1,125,000	$ 1,350,000	$ 1,462,500	$1,575,000	$ 1,687,500	$ 10,237,500
Subtotal	$ 0	$ 0	$ 149,925	$ 712,425	$ 1,214,730	$ 967,410	$1,192,410	$ 1,574,775	$ 1,709,820	$ 1,642,410	$1,664,955	$ 2,137,275	$ 12,966,135
Syndication	$ 0	$ 0	$ 0	$ 0	$ 0	$ 8,333	$ 16,667	$ 33,333	$ 66,667	$ 108,333	$ 158,333	$ 216,667	$ 608,333
Merchandise	$ 0	$ 0	$ 45,000	$ 45,000	$ 162,000	$ 54,000	$ 54,000	$ 135,000	$ 108,000	$ 54,000	$ 27,000	$ 135,000	$ 819,000
Total	$ 0	$ 0	$ 194,925	$ 757,425	$ 1,376,730	$1,029,743	$1,263,077	$ 1,743,108	$ 1,884,487	$ 1,804,743	$1,850,288	$ 2,488,942	$ 14,393,468

Revenues 2002

	Jan-02	Feb-02	Mar-02	Apr-02	May-02	Jun-02	Jul-02	Aug-02	Sep-02	Oct-02	Nov-02	Dec-02	Total
Consumer Photos	$ 239,880	$ 299,850	$ 299,850	$ 299,850	$ 1,079,460	$ 359,820	$ 359,820	$ 899,550	$ 719,640	$ 359,820	$ 179,910	$ 899,550	$ 5,997,000
Stock Photos	$ 900,000	$1,125,000	$1,350,000	$1,575,000	$ 1,800,000	$2,250,000	$2,250,000	$ 2,250,000	$ 2,250,000	$ 2,250,000	$2,250,000	$ 2,250,000	$ 22,500,000
Subtotal	$1,139,880	$1,424,850	$1,649,850	$1,874,850	$ 2,879,460	$2,609,820	$2,609,820	$ 3,149,550	$ 2,969,640	$ 2,609,820	$2,429,910	$ 3,149,550	$ 28,497,000
Syndication	$ 283,333	$ 358,333	$ 441,667	$ 533,333	$ 633,333	$ 741,667	$ 858,333	$ 983,333	$ 1,116,667	$ 1,250,000	$1,383,333	$ 1,516,667	$ 10,100,000
Merchandise	$ 72,000	$ 90,000	$ 90,000	$ 90,000	$ 324,000	$ 108,000	$ 108,000	$ 270,000	$ 216,000	$ 108,000	$ 54,000	$ 270,000	$ 1,800,000
Total	$1,495,213	$1,873,183	$2,181,517	$2,498,183	$ 3,836,793	$3,459,487	$3,576,153	$ 4,402,883	$ 4,302,307	$ 3,967,820	$3,867,243	$ 4,936,217	$ 40,397,000

Revenues 2003

	Jan-03	Feb-03	Mar-03	Apr-03	May-03	Jun-03	Jul-03	Aug-03	Sep-03	Oct-03	Nov-03	Dec-03	Total
Consumer Photos	$ 495,752	$ 619,690	$ 619,690	$ 619,690	$ 2,230,884	$ 743,628	$ 743,628	$ 1,859,070	$ 1,487,256	$ 743,628	$ 371,814	$ 1,859,070	$ 12,393,80
Stock Photos	$2,550,000	$3,187,500	$3,825,000	$4,462,500	$ 5,100,000	$6,375,000	$6,375,000	$ 6,375,000	$ 6,375,000	$ 6,375,000	$6,375,000	$ 6,375,000	$ 63,750,000
Subtotal	$3,045,752	$3,807,190	$4,444,690	$5,082,190	$ 7,330,884	$7,118,628	$7,118,628	$ 8,234,070	$ 7,862,256	$ 7,118,628	$6,746,814	$ 8,234,070	$ 76,143,800
Syndication	$1,650,000	$1,783,333	$1,916,667	$2,050,000	$ 2,183,333	$2,316,667	$2,450,000	$ 2,583,333	$ 2,716,667	$ 2,850,000	$2,983,333	$ 3,116,667	$ 28,600,000
Merchandise	$ 117,000	$ 146,250	$ 146,250	$ 146,250	$ 526,500	$ 175,500	$ 175,500	$ 438,750	$ 351,000	$ 175,500	$ 87,750	$ 438,750	$ 2,925,000
Total	$4,812,752	$5,736,773	$6,507,607	$7,278,440	$10,040,717	$9,610,795	$9,744,128	$11,256,153	$10,929,923	$10,144,128	$9,817,897	$11,789,487	$107,668,800

Source: e-Catalyst Business Plan, February 28, 2000.

EXHIBIT 4 Pro Forma Financial Summary 2000

Summary Profit and Loss Statement

	2000	2001	2002	2003	Total
Revenues	$ 0	$14,393,468	$40,397,000	$107,668,800	$162,459,268
Cost of sales	$ 0	$ 5,186,454	$11,398,800	$ 30,457,520	$ 47,042,774
Gross profit	$ 0	$ 9,207,014	$28,998,200	$ 77,211,280	$115,416,494
Operations	$ 439,847	$13,623,571	$27,109,143	$ 47,078,657	$ 88,251,217
Net income before tax	($ 439,847)	($ 4,416,556)	$ 1,889,057	$ 30,132,623	$ 27,165,277
Taxes (38%)	$ 0	$ 0	$ 0	$ 10,322,805	$ 10,322,805
Net income	($ 439,847)	($ 4,416,556)	$ 1,889,057	$ 19,809,818	$ 16,842,472

Summary Balance Sheet

Assets	2000	2001	2002	2003
Cash and equivalents	$ 428,020	$ 4,490,768	$ 4,958,270	$21,508,477
Accounts receivable	$ 0	$ 2,488,942	$ 4,936,217	$11,789,487
Inventories	$ 0	$ 0	$ 0	$ 0
Prepaid expenses	$ 0	$ 0	$ 0	$ 0
Depreciable assets	$ 0	$ 0	$ 0	$ 0
Other depreciable assets	$ 0	$ 0	$ 0	$ 0
Depreciation	$ 0	$ 0	$ 0	$ 0
Net depreciable assets	$ 0	$ 0	$ 0	$ 0
Total assets	$ 428,020	$ 6,979,710	$ 9,894,487	$33,297,964

Liabilities and capital				
Accounts payable	$ 367,867	$ 1,836,113	$ 2,861,833	$ 5,589,379
Accrued income taxes	$ 0	$ 0	$ 0	$ 866,113
Accrued payroll taxes	$ 0	$ 0	$ 0	$ 0
Total liabilities	$ 367,867	$ 1,836,113	$ 2,861,833	$ 6,455,492
Capital contribution	$ 500,000	$10,000,000	$10,000,000	$10,000,000
Stockholders' equit	$ 0	$ 0	$ 0	$ 0
Retained earnings	($ 439,847)	($ 4,856,403)	($ 2,967,346)	$16,842,472
Net capital	$ 60,153	$ 5,143,597	$ 7,032,654	$26,842,472
Total liabilities and capital	$ 428,020	$ 6,979,710	$ 9,894,487	$33,297,964

Source: e-Catalyst Business Plan, February 28, 2000.

•

EXHIBIT 5 Artemis Images Original Funding Plan

Projected Plan	Round 1	Round 2	Round 3	Round 4	Exit
Financing assumptions:					
2003 Revenues	$110,000,000				
2003 EBITDA	$ 30,000,000				
2003 Revenue growth rate	40%				
2003 Valuation	$440,000,000				
Valuation/revenue	4				
Valuation/EBITDA	14.67				
Round 1 Financing	$ 500,000				
Round 2 Financing	$ 1,500,000				
Round 3 Financing	$ 3,000,000				
Round 4 Financing	$ 5,000,000				
	Round 1 Oct-00	Round 2 1-Jan	Round 3 1-Mar	Round 4 1-Jun	Exit 3-Dec
Number of shares outstanding					
Total number of shares outstanding prior to financing	6,000,000	7,200,000	9,000,000	11,250,000	11,250,000
Shares issues this round	1,200,000	1,800,000	2,250,000	1,406,250	1,406,250
Total number shares outstanding after financing	7,200,000	9,000,000	11,250,000	12,656,250	12,656,250
Valuations					
Premoney valuation	$2,500,000	$6,000,000	$12,000,000	$40,000,000	$440,000,000
Amount of financing	$ 500,000	$1,500,000	$ 3,000,000	$ 5,000,000	0
Postmoney valuation	$3,000,000	$7,500,000	$15,000,000	$45,000,000	$440,000,000
Price per share	$0.42	$0.83	$1.33	$3.56	$34.77
Resulting ownership					
Founders	83.33%	66.67%	53.33%	47.41%	47.41%
Round 1 investors	16.67%	13.33%	10.67%	9.48%	9.48%
Round 2 investors	0.00%	20.00%	16.00%	14.22%	14.22%
Round 3 investors	0.00%	0.00%	20.00%	17.78%	17.78%
Round 4 investors	0.00%	0.00%	0.00%	11.11%	11.11%
Total	100.00%	100.00%	100.00%	100.00%	100.00%
Value of ownership					
Founders	$2,500,000	$5,000,000	$ 8,000,000	$21,333,333	$208,592,593
Round 1 investors	$ 500,000	$1,000,000	$ 1,600,000	$ 4,266,667	$ 41,718,519
Round 2 investors	$ 0	$1,500,000	$ 2,400,000	$ 6,400,000	$ 62,577,778
Round 3 investors	$ 0	$ 0	$ 3,000,000	$ 8,000,000	$ 78,222,222
Round 4 investors	$ 0	$ 0	$ 0	$ 5,000,000	$ 48,888,889
Total	$3,000,000	$7,500,000	$15,000,000	$40,000,000	$440,000,000
Payback to investors	**Round 1**	**Round 2**	**Round 3**	**Round 4**	
Holding period (years)	3.25	3	2.75	2.5	
Times money back	83.44	41.72	26.07	9.78	
Internal rate of return (IRR)	290%	247%	227%	149%	

Source: e-Catalyst Business Plan, February 28, 2000.

Historically, consumers who bought from the archive had to visit the museum at IMSC or write a letter to the staff. Retrieval and fulfillment of images then required a manual search of a physical inventory, a process which could take as long as two weeks. Web-based digitization and search engines would reduce the search time and personnel needed for order fulfillment and allow customers the convenience of selecting products and placing orders on-line. The *Daily Mirror,* a newspaper in London, had displayed its archived images on its own website and had generated over $30,000 in sales to consumers in its first month of availability. IMG, a sports marketing group, placed a value of $10 million on the IMSC contract.

Competition

There were a variety of stock and consumer photo sites ranging from those that served only the business-to-business stock photo market to amateur photographers posting their pictures. Most sites did not offer a "community," the Internet vehicle for consumer comments and discussion, a powerful search engine, and ways to repurpose the content (e-greeting cards, prints, photo mugs, calendars, etc.). In addition, the archives available in digital form were limited because other content providers worked from the virtual world to the physical world versus the Artemis Images model of working from the physical world to the virtual world. Competitors had problems with integrated digital workflows and knowing where the original asset resided due to the distributed nature of their archives. They scanned images on demand, which severely limited the content available to be searched on their websites.

Chris and Greg evaluated the five major competitors for their business plan:

www.corbis.com: Owned by Bill Gates with an archive of over 65 million images, only 650,000 were available on the Web to be accessed by consumers for Web distribution (e-greeting cards, screen savers, etc.). Only 350,000 images were available to be purchased as prints. The site was well designed and the search features were good, but there was no community on the site. The niche Corbis pursued was outright ownership of archives and scanning on demand. Corbis had recently acquired the Louvre archive, for a reported purchase price of over $30 million.

www.getty-images.com: An archive of over 70 million images. In 1999, this site was only a source to link to their other wholly owned subsidiaries, including art.com. There were no search capabilities, no community. This website functioned only as a brochure for the company. Like Corbis, Getty was focused on owning content and then scanning on demand.

www.art.com: A good site in design and navigation, this site was a wholly owned subsidiary of Getty and was positioned as the consumer window to a portion of the Getty archive. Similar to Corbis, customers were able to buy prints, send e-greeting cards, etc. Despite the breadth of the Getty archive, this site had a limited number of digitized images available.

www.mediaexchange.com: Strictly a stock photo site targeted toward news sources, the site was largely reliant on text. It was difficult to navigate and had an unattractive graphical user interface.

www.thepicturecollection.com: Strictly a stock site offering the *Time* photo archive, this site was well designed with good search capabilities. Searches yielded not only a thumbnail image but a display of the attached locator tags, or metadata.

www.ditto.com: The world's leading visual search engine, ditto.com enabled people to navigate the Web through pictures. The premise was two-fold: deliver highly relevant thumbnail images and link to relevant Web sites underlying these images. By 2000, they had developed the largest searchable index of visual content on the Internet.

Exhibit 6 shows a comparison of Artemis Images to the two major players in the stock photography market, Getty and Corbis. This table illustrates only

EXHIBIT 6 Anticipated Sales Volume Comparisons

Stock Photo Market						
	Artemis Images	**Artemis Images**	**Artemis Images**	**Artemis Images**	**Getty**	**Corbis**
	Indy Archive 2000*	**2000**	**2001**	**2002**	**1999**	**1999**
Archive size	5,000,000	5,000,000	50,000,000	95,000,000	70,000,000	65,000,000
Cumulative number of images digitized	345,600	345,600	6,796,800	21,542,400	1,200,000	2,100,000
% digitized**	7%	7%	14%	23%	1.71%	3.2%
# of image sales needed to hit revenue target	0	0	151,484	623,493	1,646,667	666,666
% of archive that must be sold to hit revenue target***	0	0	0.30%	0.16%	2.35%	1.00%
Revenues****	0	0	$22,722,600	$93,523,950	$247,000,000	$100,000,000
Revenue per image in archive	0	0	$0.45	$0.98	$ 3.53	$ 1.54
Revenue per digitized image	0	0	$3.25	$4.30	$205.83	$47.62

*Artemis Images had already secured an exclusive content agreement from the Indianapolis Motor Speedway Corporation.
**Estimates based on scanning 1,920 images a day per scanner, 2 scanners per archive. As scanning technologies improve, the throughput numbers were expected to go up.
***The percentage of the Artemis Images archive that needed to be sold to hit revenues projections varied between 0.03% and 0.22%, as compared to an actual 2.35% for Getty and to 0.6% for Corbis.
****The Artemis Images revenue numbers were based on selling a certain number of images at $150 per image; $150 was the minimum average price paid for stock photographs. Corbis was privately held; this figure was an estimate.

revenues from stock photo sales and does not include potential revenue from consumer sales, merchandise, advertising or other potential revenue sources.

According to its marketing director, Corbis intended to digitize its entire archive, and was in the process of converting analog images into digital images, with 63 million images yet to be converted. While Getty and Corbis were established players in the content industry, they were just recently feeling the effects of e-commerce:

- In 1999, Corbis generated 80 percent of its revenues from the Web versus none in 1996.
- Getty's e-commerce sales were up 160 percent between 1998 and 1999.
- 34 percent of Getty's 1999 revenues came from e-commerce versus 17 percent in 1998.

Strategy

Artemis Images intended to provide digitization and archive management by employing a professional staff who would work within each client-company's organization, rather than in an off-site facility of its own. Chris's model was to provide digitized archive services in exchange for (1) exclusive rights to market the content on the Internet, (2) merchandising rights, and (3) promotion of Artemis Images's URL, effectively co-branding Artemis Images with each client-partner. Chris envisioned a software process that would be owned or licensed by Artemis, and which could be used for digitizing different archive media, such as photos, videos, and text.

Chris and George expected Artemis Images to partner with existing sellers of stock photography and trade digitizing services for promotion through their sales channels. Artemis Images would pursue these relationships with traditional sales and marketing techniques. Sales people would call on the major players and targeted direct mail, trade magazine advertising and PR would be used to reach the huge audience of smaller players. In addition, content partners were expected to become customers, as they were all users of stock photography.

As Artemis Images gained clients, the company would have access to some of the finest and most desirable content in the world. Chris knew that the workflow expertise of the management team would put them in a good position to provide better quality more consistently than either Corbis or Getty. This same expertise would allow Artemis to have a much larger digital selection, with a website design that would be easily navigable for customers to find what they needed.

Using on-site equipment, the client's content would be digitized, annotated (by attaching digital information tags, or metadata) and uploaded to the corporate hub site. Metadata would allow the content to be located by the search engine and thus viewed by the consumer. For example, a photo of Eddie Cheever winning the Indy 500 would have tags like Indy 500, Eddie Cheever, win photo, 1998, etc. Therefore, a customer going to the website and searching for "Eddie Cheever" would find this specific photo, along with the hundreds of other photos associated with him. The Artemis corporate database was intended to serve as the repository for search and retrieval from the website.

The traditional content management strategy forced organizations to purchase technology and expertise. Artemis Images's model intended to alleviate this burden by exchanging technology and expertise for exclusive web distribution rights and a share of revenues. The operational strategy was to create an infrastructure based on installing and operating digital asset management systems at their customers' facilities to create a global digital archive of images, video, sound and text. This would serve to lock Artemis Images into long-term relationships with these organizations and ensure that Artemis Images would have both the historical and the most up-to-date content. Artemis Images would own and operate the content management technology, with all other operational needs outsourced including Web development, Web hosting, consumer data collection, and warehousing and fulfillment of merchandise (printing and mailing posters or prints). Artemis Images would scan thousands of images per day, driving down the cost per image to less than $2.00, versus the Corbis and Getty model of scan-on-demand, where the cost per image was approximately $40.00. The equipment needed for both the content management and photo production would be leased to minimize start-up costs and ensure greater flexibility in the system's configuration.

The original plan was to purchase and install software and hardware at their main office in Denver, Colorado, contract with a Web development partner, and set up the first on-site facility at Indianapolis Motor Speedway Corporation. The Denver facility would serve as a development lab, to create a standard set of metadata to be used by all of their partners' content. This consistency of annotation information was intended to allow for consistent search and retrieval of content. Artemis Images's goal was to build a world-class infrastructure to handle content management, consumer data collection, and e-commerce. This infrastructure would allow them to amass a large content and transaction volume by expanding to other market segments. Developing their own structure would ensure standardization of content and reduced implementation time. Outreach for news coverage and the development of community features would be negotiated concurrently. The time line in Exhibit 7 illustrates the Artemis Images development plan.

Financial Projections

Revenues were expected to come from four primary sources:

Consumer photos: IMSC's archive sold approximately 53,000 photos in 1999 to a market limited to consumers who visited the archive or wrote to its staff. Artemis Images based its projected sales on an average of 15,000 images sold per archive in 2001, increasing to 20,000 images per archive in 2003. Price: $19.99 (8"×10").

Stock photos: Stock photos ranged in price from $150 to $100,000, depending on the uniqueness of the photo. Competitors Getty and Corbis, two of the leaders in this market, sold 2.35 percent and 0.6 percent of their archive, respectively. Based on an average selling price of $150, Getty

Three Phases of Development

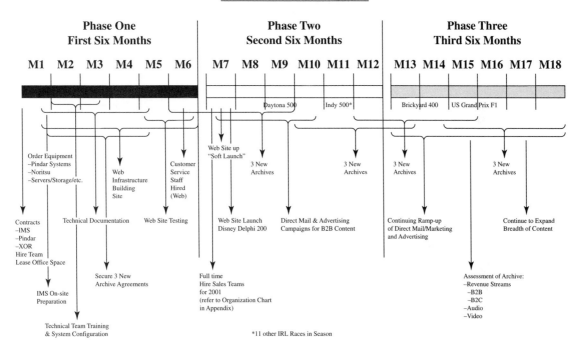

Phase One �█████████▶

This phase was intended to take Artemis Images from initial funding to operationally being ready to sell images and take orders. The three main components included establishing the on-site facility at IMSC, construction and testing of the website and establishing the fulfillment operations. Phase one assumed money was in the bank.

Phase Two ▭▭▭▭

This phase assumed that three additional archives had been secured and implemented, at least one of which would include breadth of content. Focus would be sales and ramping up revenues. B2B and B2C marketing strategies were to be executed and evaluated. Toward the end of Phase two three more archives would be secured.

Phase Three ▭▭▭▭

Phase three continued to build more archives and breadth of content. Marketing and sales would continue to be core focus for revenue development. Audio and video content assessed based on the state of market and technology (e.g., bandwidths) and a decision would be made on timing to enter this market.

EXHIBIT 7 Artemis Images Development Time Line

Source: e-Catalyst Business Plan, February 28, 2000.

generated approximately $6.00 in revenue for each image in its archive; Corbis generated approximately $1.85. Artemis Images constructed financial projections based on sales of 0.30 percent of its archive in 2001 and 0.16 percent of its archive in 2002. Artemis Images's margin was based on a return of $0.20 per image in its archive for 2001, increasing to $0.60 per image in 2003.

Syndication: The team's dot.com experience led them to believe that websites with exclusive content were able to syndicate their content to other websites. They anticipated that Artemis Images would generate revenues of $100,000 per year from each contract for content supplied as marketing tools on websites. Existing companies with strong content had been able to negotiate five new agreements per week for potential annual revenues of $5 million.

Merchandise: According to America Online/ Roper Starch Worldwide, approximately 30 percent of Internet users regularly make purchases. Artemis Images used a more conservative assumption that only 1 percent of unique visitors would make a purchase. Estimates of the average purchase online varied widely, ranging from Wharton's estimate of $86.13 to eMarketers's estimate of $219. The Artemis Images team viewed $50 per purchase as a conservative figure.

Chris and George felt confident that Artemis Images would be able to reach the revenue projections for number of photos sold. IMSC's archive had sold approximately 53,000 photos in 1999, an increase of 33 percent over 1998. These sales had been generated solely by consumers who had visited the archive in person, estimated at 1 million people. In other words, one out of every 28 possible consumers actually purchased an image. Chris and George assumed that if even one out of 160 unique visitors to the website purchased a photo, the Artemis website would generate 42 percent more than IMSC's 1999 figures (see Exhibit 6 for projected sales volume). Chris and George believed that this projection was reasonable in light of the fact that IMSC did not market its archive and significant publicity and advertising would accompany Artemis Images's handling of the archive. As breadth of content and reach of the Web increased, 2002 revenues should easily be double those of 2001.

Since the team previously had configured and sold content management systems, they were familiar with the costs associated with this process, including both equipment and personnel. They carefully conducted research to stay abreast of recent improvements in technology and intended to be on the lookout for cost reductions and process improvements.

The Launch: Problems from the Start

Chris dove into the Artemis Images project with a vengeance. Having secured a five-year contract for exclusive rights and access to the IMSC archive, she found a dependable technician who was eager to relocate to Indianapolis to start the scanning and digitizing process. A reputable, independent photo lab agreed to handle printing and order fulfillment. Chris's visit to the Indy 500 in May 2000 was a wonderful networking opportunity. She met executives from large companies and got leads for investors and clients. She secured an agreement with a Web design company to build the Artemis Images site, careful to retain ownership of the design. She contacted over 100 potential venture capitalists and angel investors.

Personally, she was on a roll. Financially, she was rapidly going into debt. Frank and Greg, legal owners of the company, had long since contributed ideas, contacts, or legwork to the Artemis Images launch. While confident that his work on the business plan would appeal to investors, Greg viewed the start-up company as a risk to which he was unwilling to commit. Likewise, Frank decided to hold onto his job at Petersons.com, a unit of Thompson Learning, until the first round of investor funding had been secured. Frank continued to offer advice, but he had a wife and two preschool-age children to support.

Each meeting with a potential funder resulted in a suggestion on how to make the business more attractive for investment. Sometimes they helped, sometimes they just added to Chris's and George's frustration. Beating the bushes for money over two years was exhausting, to say the least. The lack of funds impacted the look and feel of the business and severely strained relationships among the founding partners. Heated discussions ensued as to the roles that each was expected to play, the reallocation of equity ownership in the company, and the immediate cash needed to maintain the Indianapolis apartment and pay the scanning technician and Web developers, not to mention out-of-pocket expenses needed to manage and market the business.

Chris and George appealed to their families for help. George's father contributed $5,000. Chris's mother tapped into her retirement, mostly to pay Chris's mortgage and to fund Chris's trips to potential clients and investors in London, New York, and Boston. By May 2001, Chris's mother's contribution had exceeded $200,000. A $50,000 loan from a supportive racing enthusiast provided the impetus for Artemis Images to reorganize as a C-corporation. All four original partners had stock in the new company, but Chris held the majority share (66 percent), George held 30 percent, and Frank and Greg's shares were each reduced to 2 percent. Financial projections were revised downward (see Exhibit 8).

The site was officially launched on May 18, 2001. It was beautiful. Chris held her breath as she put in her credit card late that evening when the site went live. The shopping cart failed and the order could not be processed. Chris knew she was in trouble.

The Crash

From the first, the website had problems. The Web development contract stipulated that the website for the Indy 500 would go live by May 8, 2001, to coincide with the month-long series of events held at the Indianapolis Motor Speedway

EXHIBIT 8 Revised Pro Forma Financial Summary 2001

Summary Profit and Loss Statement

	2001	2002	2003	2004	Total
Revenues	$ 5,312*	$373,779	$2,294,116	$4,735,400	$7,408,607
Cost of sales	$ 1,700	$ 43,368	$ 265,312	$ 564,480	$ 874,860
Gross profit	$ 3,612	$330,411	$2,028,804	$4,170,920	$6,533,747
Operations	$52,499	$328,550	$1,235,363	$2,035,430	$3,651,842
Net income before tax	($48,887)	$ 1,861	$ 793,441	$2,135,490	$2,881,905
Taxes (38%)	$ 0	$ 0	$ 283,638	$ 811,486	$1,095,124
Net income	($48,887)	$ 1,861	$ 509,803	$1,324,004	$1,786,781

Summary Balance Sheet

Assets	2001	2002	2003	2004
Cash and equivalents	$45,113	$ 78,260	$ 675.347	$2,615,573
Accounts receivable	$ 0	$ 13,610	$ 222,950	$ 462,200
Inventories	$ 0	$ 0	$ 0	$ 0
Prepaid expenses	$ 0	$ 0	$ 0	$ 0
Depreciable assets	$ 0	$ 0	$ 0	$ 0
Other depreciable assets	$ 0	$ 0	$ 0	$ 0
Depreciation	$ 0	$ 0	$ 0	$ 0
Net depreciable assets	$ 0	$ 0	$ 0	$ 0
Total assets	$45,113	$ 40,574	$ 898,297	$3,077,773

Liabilities and capital				
Accounts payable	$ 4,000	$ 12,355	$ 61,882	$ 105,868
Accrued income taxes	$ 0	$ 0	$ 283,638	$1,095,124
Accrued payroll taxes	$ 0	$ 0	$ 0	$ 0
Total liabilities	$ 4,000	$ 12,355	$ 345,520	$1,200,992
Capital contribution	$90,000†	$ 90,000	$ 90,000	$ 90,000
Stockholders' equity	$ 0	$ 0	$ 0	$ 0
Retained earnings	($48,887)	($ 61,781)	$ 462,777	$1,786,781
Net capital	$41,113	$ 28,219	$ 552,777	$1,876,781
Total liabilities and capital	$45,113	$ 40,574	$ 898,287	$3,077,773

Notes: *Approximately two-thirds of these transactions were executed by Artemis staff and friends to test the website.
† Chris's mother's contribution to her daughter for mortgage and living expenses is not included.

leading up to the Indy 500 on May 27. However, the Web development took longer than anticipated, and the site was first operational on May 18. Having neglected to test the Web interface properly, serious failures were encountered when the site was activated. The site went down for 24 hours, only to face similar problems throughout the following week, again shutting down on May 27. More technical difficulties delayed the reactivation of the site until May 31, after the Indy racing series had ended.

Throughout June, consumer traffic was far less than originally anticipated. The site was not easily navigable. The shopping cart didn't work. Yet the Web builder demanded more money. Fearful of a possible lawsuit, investors stayed away. The crash of the dot.coms added kindling to the woodpile. Chris and George started to rethink their original business model. They were held hostage, as they owned no tangible assets.

Website tracking data indicated that between May and July there had been at least $40,000 worth of attempted purchases. Chris read through hundreds of angry e-mails, and tried manually to process orders. Orders which were successfully executed resulted in spotty fulfillment. Many photos ordered were never shipped, were duplicated, or were incorrectly billed. At the same time, she tried to negotiate with the software developers' demand for payment and keep alive a $250,000 investment prospect.

On July 9, 2001, the Web development company threatened an all-or-nothing settlement. They wanted payment in full for the balance of the contract even though the sites didn't work. Absent full payment, they would shut down the sites within the week. The investor offered to put up 80 percent of the balance owed on the full contract to acquire the code to fix it. The company refused. On Friday, July 13, Chris had to tell IMSC that in less than 48 hours the sites would be shut down. The investor took his $250,000 elsewhere.

On Tuesday, July 17, Chris called an emergency meeting with George. George had had enough. The stress was affecting his health, his relationships, and his lifestyle. He believed that his family had already contributed more money than he had a right to ask. He was putting in long hours with no money to show for his efforts. His girlfriend had been putting pressure on George to quit for some time. Now he had run out of reasons to stay.

Chris was devastated. How could she face the people in Indianapolis? It was hard for her to come to grips with having let them down. Having put so much of herself into this venture, she wasn't sure she could let go. At the same time, she wasn't sure how to go on.

Chris reflected, "At one time, I defined success by my title, my salary, and my possessions. Working for AGT, I had it all. I started Artemis Images because I really cared about IMSC and making the Indy motorsports images available to its fans. Now, I realize that there is a profound satisfaction in building a company. I can see my future so clearly, but living day to day now is so hard. And I'm still enthralled with beautiful images."

RADCO© ELECTRONICS

Background

The RAD Company was founded 15 years ago by Anton Biggs, Alexandra Beller and Albert Sells. The two knew each other as students from the University of California, Davis. Biggs graduated with a degree in Agricultural Horticulture and Beller with a degree in Computer Science. Sells graduated with degrees in Chemical Engineering and Business Administration from UCLA. During their undergraduate days, the two friends Biggs and Sells conceived of a method of probing agricultural fields with inexpensive chemical sensors in order to detect moisture, mineral content, acidity, and fertilization levels. They realized that a large number of sensor measurements would require extensive computer processing and hence sought the advice of computer experts at the university. They were referred to Beller by a member of the agriculture faculty. Beller came up with an approach that looked promising but would require considerable software development and testing time.

Biggs realized that if they could put it all together, they would have a technique which could be of considerable value to farmers. Their concept would allow selective fertilization and irrigation based on soil requirements, thus reducing costs, and similar techniques could be extended to detect crop diseases at an early stage. Biggs proposed that he, Sells, and Beller form a company to develop the concept. They agreed that the company should be run as a for-profit organization cooperating with the University of California, Davis. Profits would be generated from fees charged on federal and state contracts as well as from sales to commercial customers. Initially, they would use all profits to reinvest in growth of the business. They decided to call their organization the RAD Company, "RAD" standing for "Research Associates of Davis." Since all three founders came from well-to-do families, they were able to convince their respective families to provide "forgivable loans" of $60,000 each. These loans would be paid back at an interest rate of 10% if the company was successful, but forgiven if the company went broke.

Starting in Beller's apartment, Beller went to work on developing the computer data-analysis program while Biggs and Sells started a market research effort. Initial interest from large agricultural firms and from several state agricultural agencies was high, but potential customers wanted to see some experimental results before utilizing the technology. Within 15 months Beller achieved spectacular theoretical results on her computer model, and a year later the results were confirmed by field tests, which were conducted in cooperation with a UCD agricultural field station. Technical papers on the field tests were published in agricultural journals in order to establish the young company's reputation in the field. However, the three founders agreed to keep Beller's computer software developments as trade secrets in order to maintain a competitive advantage.

©Professor Jerome J. Suran, Graduate School of Management, University of California, Davis.

The author gratefully acknowledges the editorial contributions of Professor Richard C. Dorf.

So far, the RAD Company's key inventions were the sensors used in the field and the mathematical models which allowed the sensor information to be analyzed and displayed graphically. The major advantage of the sensors was their surprisingly low cost and the ease with which they could be distributed and monitored in the fields. This allowed them to be used in very large numbers, to be distributed by air drops, and to be monitored from the air. Because of their low cost, the sensors could be considered expendable from crop to crop. Since the sensors were organically based, they were designed to degrade without polluting the fields. Unfortunately, the sensor technology had been developed at UC Davis and could not be protected by company patents or as trade secrets.

Three years after the formation of the RAD Company, now known as RADCO, it started a spectacular growth. Both federal and state research contracts were successfully solicited, and agricultural firms began to fund the company to apply its technology to their operations. Furthermore, it became obvious that the basic inventions had applicability to more than the agricultural sector. The same techniques, especially the software, could be used for mineral explorations, pollution detection, precious-metal prospecting, and even oceanographic exploration. Specialists with management skills were hired to head company departments in each of these market areas. Donna Poulos, a civil engineering graduate of UCD, was given the goal of developing the pollution-monitoring business. Robert Ellis, who was hired from Exxon, was asked to develop the energy market, which primarily involved oil exploration. Jill Mills, a geologist by training, was hired to head the minerals department and to develop the business of precious-metal prospecting. Ira Frank, an oceanographer hired from the Scripps Institute, was given the goal of applying the pollution-monitoring and resource-prospecting technologies to markets which involved the seas and oceans, including fisheries. Finis Morris, another UCD agricultural engineering graduate, was hired to manage the burgeoning agricultural market opportunities.

The company sales grew at a compound annual rate of almost 80 percent, requiring the founders to develop more formal organization and planning procedures. They hired an accounting graduate from the Wharton School of the University of Pennsylvania, Constance Braun, who was named Vice-President of Finance. A graduate of UCD's Graduate School of Management, Alan Rausch, was hired as Vice-President of Administration. The three founders agreed to form an Executive Committee consisting of the three of them, which would collectively determine the strategic direction of the company. Day-to-day decisions would be made by the "line" department managers, although the Executive Committee never relinquished sign-off authority on the purchase of equipment, space, or any items costing more than $1,000. Further, the Executive Committee retained its authority to approve or disapprove any proposals for customer contracts. Beller took the title Executive Vice-President of Research and Development, and would concentrate her activities on new-product development. Sells became Executive Vice-President of Marketing and concentrated his time on expanding current customer sales as well as penetrating new markets. Biggs was elected President of the Company by the Executive Committee and increasingly became "Mr. Outside," focusing on public relations and lobbying activities in both Sacramento and Washington. The three founders agreed that all decisions by the Executive Committee would be made by a majority vote.

The Current Situation

Organization   RADCO has grown over the past six years to a company of about $100 million in sales and over 800 people in size. Its organization, shown in Exhibit 1, consists of five "line" departments headed by general managers and oriented to the specific markets they serve. Biggs, Sells, and Beller have been strong believers in tight central control of the company. During its early entrepreneurial stage, when the company was small and everyone interacted with everyone else on almost a daily basis, centralized control was easy. But as the company grew to its present size and business diversity, informal communication channels were no longer effective. The company began to develop the usual symptoms of growth: more time in formal meetings and in the preparation of reports and presentations by the vice-presidents and general managers to the Executive Committee. However, a more severe problem was the lack of speed in responding to market opportunities. Department managers complained of being unable to locate a majority of the Executive Committee whenever decisions needed to be made about submitting proposals or purchasing new equipment. Jill Mills and Ira Frank suggested that proposal sign-off and purchasing authority be decentralized to their level, or at least to the level of Alan Rausch and Constance Braun.

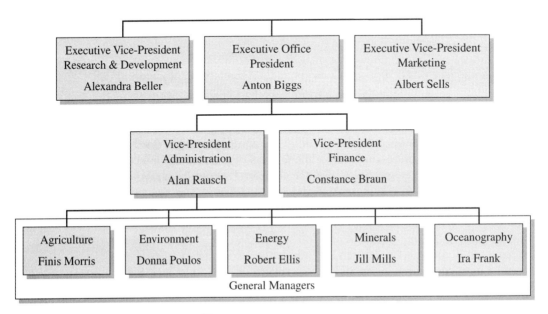

EXHIBIT 1 RADCO Organization Chart

Another organizational problem, brought to the attention of the Executive Committee by Donna Poulos, was a problem of "functional imbalance." When asked what she meant by that, she answered: "When I was hired, the company was still small and we had informal but satisfactory channels of communication to everyone else. So whenever I needed to hire new people as we expanded, I put the word out and got some excellent suggestions, especially from friends at the university. We hired without much regard to affirmative action or diversity considerations. But soon Al (Albert Sells) began to pressure us to conform with government-contract requirements, and recruiting became a much more formal and legalistic process. In place of our old boys, or old girls, networks, we had to advertise our positions in a diverse set of publications to meet legal requirements. There is no human resources group of experts to implement this function for us. As a matter of fact, Al's group sort of took human resources over by default, because they were the ones concerned with compliance to government regulations in our proposal efforts. So we became reactive rather than proactive in our hiring practices, which occasionally led to placing the wrong people in critical positions. Unfortunately, it is almost impossible to rectify these errors without risking major lawsuits. We need an operation which can better plan our human resource needs and give us more time and care in the important hiring process."

Constance Braun supported Donna in her arguments for adding a human resource manager to the company structure. But she also added her concerns for stronger financial representation in the Executive Committee (ExCom). She felt that although marketing and technology expertise was well represented in ExCom, many important financial issues were not adequately addressed during critical discussions and decision making. According to Braun: "Frequently I have to go back to ExCom and point out things which they overlooked, or perhaps didn't have sufficient information about. This often leads to delays in implementing important decisions, or worse, in implementing the wrong decisions if I do not catch the errors quickly enough. Many of these areas are highly technical, like tax law or cash-flow ramifications. Someone with my knowledge and experience is really needed on ExCom."

In addition to these problems, many managers were beginning to feel uncomfortable with conflict-of-interest issues. One of these concerned compensation. Initially, all profits made on contracts to government agencies or sales to the commercial market were used in the company to offset expenses in new ventures and research. As the company became more successful, the three founders decided to take some of the profits as compensation for themselves in the form of bonuses. According to Anton Biggs: "For several years the three of us worked our tails off to establish this company. We risked our parents' money, ultimately our own inheritance, on the success of an 'iffy' venture. Now that we made it, we're entitled to be compensated for the sweat equity we put into this company."

A few years after the founders established bonuses for themselves based on company profits, they decided to extend profit-sharing to the other manager levels. However, the decision on how much of the profit would be distributed to the three levels (Executive Office, vice presidents and general managers) as incentive bonuses was made by the Executive Committee on a year-to-year basis. Many of the managers felt that this procedure constituted a rather blatant conflict-of-interest. The Executive Committee never established an explicit formula for how profit-sharing was to be decided.

Another delicate conflict-of-interest issue developed on the personal side. Shortly after Donna Poulos joined the company, she and Albert Sells began dating. Before long, the relationship turned deep enough for the two of them to decide to live together. Recently, Sells confided to Biggs that the two of them would not get married until they decided to have children, but would continue to live together. When the company was much smaller, no one regarded this as a problem. However, as the company grew, many of the managers began to wonder if Poulos didn't have an unfair edge in influencing Executive Committee decisions. Biggs had brought the issue up with Sells, but Sells's position was that "it was nobody else's business who I lived with or dated." Biggs and Beller had private discussions about "the affair," but were on opposite sides of the issue. According to Biggs: "Managers need to be concerned about appearances as well as substance. Any private action which reflects on the company, or which results in organizational problems within the company, is of concern to the company. Furthermore, if that affair should end badly, and Al (Sells) makes decisions which Donna (Poulos) doesn't like, she can sue us for sexual harassment." But Beller had other views. She said: "When the company started, we developed close friendships with many of our business associates. Some of these people have continued to be our close friends as the company grew. We need not have a sexual relationship with someone to be influenced by that person. So where does company prying into personal affairs stop? So far, I have not seen one shred of evidence that either Al or Donna have made business decisions which are influenced by their personal relationship. Until we have such evidence, it's none of our business what they do on their own time. One can develop all kinds of what-if problems about any of us. We should not manage the business on hypothetical situations, but on reality." Ironically, since Biggs and Beller take opposite positions on the company's options in dealing with "the affair," Sells has the swing vote in the Executive Committee.

Financial control and strategy   The Executive Committee has maintained financial and strategic control of RADCO since its inception. Both financial and strategic control is exercised through the budgeting process. Biggs explains the management concept this way:

> We (ExCom) view our primary role as business strategists. We implement strategy through resource allocation. Our role is that of a portfolio manager, extracting funds from resource generators (cash cows) and using those funds to grow our ventures (question marks). (See Exhibit 2, Strategic Positioning.) Revenue producers that are in growing markets (stars) are expected to break even, approximately (generate revenues about equal to their true costs), effectively funding their own growth when possible. If the total business needs additional funds, our first option is to retain our profits in the

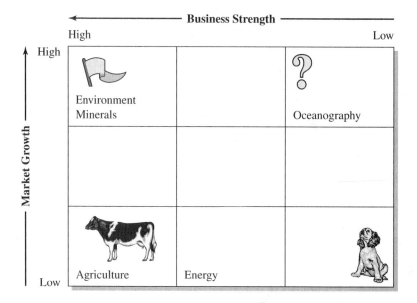

EXHIBIT 2 RADCO Strategic Positioning

business. If that's not enough, we negotiate bank loans. Ali (Alexandra Beller) sees to it that our venture and star operations, in particular, are supported in product innovation by adequate research and development (R and D). Beller's operation is centralized, and her costs are billed out to the operating components as part of our general overhead. Since Ali's lab operations get a significant part of their revenues in the form of government R and D contracts, a good part of her organization is self-supporting. In order to maintain efficiency, we have centralized space and equipment management. Alan (Rausch) has all line units (departments) reporting to him and his job is to make sure that each of the units has sufficient equipment and space to do their jobs, but no more space or equipment than is needed. A lot of our expensive stuff is shared between units. We have a centralized supply and shop area as well, all under Alan's control. So far, we rent all of our space and major equipment, and we subcontract the manufacturing of our sensors, but Connie (Constance Braun) is pushing to change that in order to take fuller advantage of depreciation write-offs. In any event, all space rentals, equipment rentals, equipment depreciation charges, and general supplies are part of our centralized overhead. ExCom signs off on all space and equipment purchases, as well as major expenditures for supplies. This way we make sure that our resource allocations are consistent with our positioning strategies and that we operate as efficiently as possible."

Anton Biggs prepared the positioning matrix shown in Exhibit 2. He believes that both the Agriculture and Energy Departments are resource generators (revenues in excess of costs). Biggs also claims that the two star operations, the Environment and Minerals Departments, are approximately self-liquidating (revenues about equal to costs). The question-mark position is held by the Oceanography Department, which is considerably underliquidated (costs exceed revenues), but according to Biggs, that department is moving rapidly toward the star position as revenues continue to increase.

All three members of the Executive Committee are opposed to starting a sensor manufacturing facility within RADCO. The reason is that the university holds the patents on the sensors and is unwilling to grant RADCO an exclusive patent license. Without patent protection, the RADCO leadership is unwilling to risk the expenditure required to start a manufacturing operation. The two university professors who invented the basic sensor configuration started their own manufacturing company and reached an agreement with the university, which holds the sensor patents, on royalties to be paid to the university. Of course, these royalties are factored into the price which RADCO pays for the sensors. Beller's laboratory has been trying to circumvent the sensor patents with new innovative approaches, but so far has been unsuccessful in coming up with an invention that is cost-competitive. While RADCO's Executive Committee feels comfortable with the supply of sensors, the line managers are uneasy about depending on a single source. As Beller summarizes the situation, "The issue is one of opportunity cost. If we invest in an in-house manufacturing facility, we'll have less money for market development and for research. I'd rather use this money to bet on the future than to provide a security blanket for the present."

Budgeting The budget process starts in the Finance Operation, managed by Constance Braun, during the last quarter of each fiscal year. Finance provides the ExCom with two estimates for the following year: total company expenses and available billing time. Billing time is the number of person-hours which are available to directly charge customer jobs. Unbillable hours include time that is spent on vacations, holidays, personal or family illness, management, conferences, trade meetings (to keep track of the latest technology and competitor activities), employee education, etc. A substantial and critical part of unbillable time is used in marketing activities by personnel in the line departments. Specialists in those departments must actually write the proposals and make the presentations to prospective clients. This time is usually not paid for directly by the client. Provision in the budget of unbillable hours for proposal preparation is critical to a department's marketing success. However, fewer billable hours for the company translates to higher overhead rates, which results in increased costs to customers. ExCom must make the decision on how high the overhead rate can be before the company becomes uncompetitive in its markets. Once that decision is made, the required billable hours can be calculated. The billable hours are then subdivided and allocated to the various departments in accordance with their location in the positioning matrix (Exhibit 2). For example, "cows" are assigned proportionately more billing hours than "question marks," since the latter must spend much more time in proposal and other marketing activities.

Constance Braun has been pushing for a major overhaul of this internal budgeting process. She has questioned the accuracy of the strategic positioning matrix (Exhibit 2) developed by Biggs and has claimed that the line managers are demotivated to exercise tight expense controls. Braun has made the following statement:

"Currently, budgeting is done in the ExCom by assigning billing hours to the various departments on the basis of their strategic position in the Company, and to Ali's (Alexandra Beller) Research and Development Department on the basis of her estimate of customer-funded R and D contracts. But all expenses are centralized, and these centralized expenses are allocated out to the departments as an overhead adder to their billed salaries (employee salaries charged directly to customer jobs). A major part of the overhead is unbillable time, of which a substantial portion is spent on marketing activities in the departments. So the department managers fight as hard as they can during the budgeting process to keep their billable hours to a minimum. Notice that this budgeting process encourages higher costs, not lower costs! Compounding this perverse incentive is the fact that a department's use of space, equipment and supplies is not directly charged to that department. These costs are incorporated into the company's general overhead charge. Hence, there is no incentive to minimize these costs at the department-manager level. Those departments which are least efficient are, in effect, subsidized by the more efficient operations!"

During the times the company has conducted inventory audits, Braun has tried to estimate space and equipment used by each of the departments. She claims that if the costs of this space and equipment were assigned to each of the departments according to actual use, the Oceanography and Environment Departments would be resource generators (revenues exceeding costs), the Energy Department would be a big resource user (costs exceeding revenues), and the Agricultural Department would be about at breakeven (costs about equal to revenues). This runs counter to Biggs' definition of the way resources should be generated and used, relative to the strategic positioning matrix of Exhibit 2. "If I am right," says Braun, "our real resource generators (cows) are the Oceanography and Pollution Departments, while our supposed cows may even be resource users! The only way we'll find out for sure is if we make the effort to calculate, and then allocate, the true costs of each department directly to that department's budget."

Alan Rausch, however, defends the current budgeting process. According to Rausch: "Connie (Constance Braun) is proposing to decentralize cost control to the level of department managers in order to incentivize them to become more cost conscious. All of us want our managers to be more frugal, but Connie's method will result in short-range expediency at the expense of long-range productivity. In other words, the argument is one between strategic motivation and tactical motivation. I am afraid that if department managers are allowed to trade off billable hours against costs, they will be reluctant to buy equipment which is expensive in the short run but productive over the long run. They may be tempted to keep obsolete equipment for longer periods and thus jeopardize their ability to perform customer jobs at competitively high quality and speed. Under the current centralized system, all space and equipment requests come to me first. I make sure that we standardize as much as possible to take advantage of quantity discounts, but I also make sure that new equipment will add to our quality and productivity. About as often as I turn down requests for 'gold-plating' and 'bells and whistles,' I suggest adding features and upgrading to more productive equipment. Thus, the current budgeting system and centralized cost control ensures a balance between short-range cost savings and long-range quality and productivity improvements.

"Furthermore, if we accept Connie's theory of decentralization, why shouldn't we decentralize all the way and make the departments look like independent companies? I have already had department managers suggest that we use department, rather than company, overhead rates in charging customers. This would give our intrinsically low-overhead departments a cost advantage in their markets, while giving the high-overhead departments a real imperative to lower their operating costs. We can even motivate the department managers to make more profit by offering them the opportunity to retain a portion of their earnings in their own departments! Decisions on how much to invest in further growth and how much to take out in incentive compensation could then be left to each department manager. Going this far would, in effect, decentralize strategic management. Our company would lose coherence, and we would no longer be able to exercise portfolio management. Nevertheless, this level of decentralization would be the logical next-step of implementing Connie's suggestion. By the way, we can still test the accuracy of Al's (Anton Biggs) positioning matrix by using Connie's calculated department costs to determine if our cows are really delivering resources to the company. This is important information for the Ex-Com to have, but we can do this without giving up management's strategic control of the company."

The Beller Committee

The Executive Committee is quite concerned about the controversies which have developed during its rapid growth. ExCom regards the lack of resolution of these issues as hurting morale at all levels and compromising the ability to manage the company in an integrated way. Consequently, ExCom has appointed Alexandra Beller to head a study committee composed of Alan Rausch (vice-president, administration), Constance Braun (vice-president, finance), Ira Frank (manager, oceanography), Finis Morris (manager, agriculture) and Jill Mills (manager, minerals). The committee's objectives are to determine the organizational and financial issues which are of concern to company management, prioritize these issues from the most urgent to the least urgent, and then present recommendations to ExCom on how to deal with the four most urgent issues. Beller has been asked to chair the committee but not to participate in its substantive discussions or in voting. She will supply secretarial and other resources to the committee as needed. Her appointment as chair of the committee is to emphasize the study's importance and to help set the meeting agendas so that the committee does not stray from the objectives. As a member of ExCom, Beller will have ample opportunity to discuss her views and cast her votes within that governing body.

Anton Biggs has been pleased with the growth of the firm, but is concerned with its future. After six years, the firm seems to need to regularize its management and human relations practices and policies. However, Biggs is unsure of the next step. What is a good course of action for this rapidly growing company? Biggs arranged for a review by Constance Braun, who has prepared an analysis of the issue (see Exhibit 3). He also is worried about the conflict-of-interest issues underlying the company turmoil. He is unsure what to do next. How can they keep the entrepreneurial spirit and yet regularize management policies at RADCO? He hopes the solution will be provided by the Beller Committee.

EXHIBIT 3 Memo from Constance Braun

AN EXAMPLE OF THE FINANCIAL CONTROL AND STRATEGY ISSUE

For illustration purposes, suppose that RADCO consists of three operating departments, A, B, and C, of equal size. An administrative department, under Vice-President Alan Rausch, is established to manage all fixed and variable expenses in a centralized mode of control. The Executive Committee of RADCO has determined that the only direct charges to customer accounts will consist of salaries and benefits, with the rest of the company expenses charged on the basis of a central billing rate of 200 percent. For example, if a RADCO employee spends four hours on a customer project, and the employee earns $1,200.00 of salary and benefits in that time, the customer bill may be calculated as:

$$
\begin{array}{ll}
\text{Salary and benefit charges} \dots\dots\dots\dots\dots\dots & \$1,200.00 \\
\text{Overhead @ 200 percent} \dots\dots\dots\dots\dots\dots & 2,400.00 \\
\text{Total charge} \dots\dots\dots\dots\dots\dots\dots\dots\dots & \$3,600.00
\end{array}
$$

An operating budget for the company would appear as follows.

ExCom's centralized budget	A	B	C	Admin	Total
Salaries and benefits	100	100	100	10	310
Fixed expenses	0	0	0	200	200
Variable expenses	0	0	0	90	90
TOTAL EXPENSES	100	100	100	300	600
Billable salaries and benefits	60	75	65	0	200
CUSTOMER BILLINGS*	180	225	195	0	600
Cost Liquidation	80	125	95	(300)	0

*Billing rate = 200 percent

The *billable* salaries and benefits are determined by the Executive Committee based upon their assumptions of how the operating departments are strategically positioned. Department

(continued on next page)

EXHIBIT 3 Memo from Constance Braun (continued)

B is considered a "cow" and hence is budgeted at a higher applied rate (ratio of billable salaries and benefits to total salaries and benefits) than the others. Department A is designated a question mark, requiring high marketing expenses to build a presence in its designated market, and hence is budgeted to bill fewer hours. Department C is considered a star, required to bill more hours than A but not as much as B. These allocations of billable hours, which translate to billable salaries and benefits in the budget table, determine the billings to customers. The total customer billings are calculated as:

$$\text{Customer billings} = (\text{billable salaries and benefits}) * (1 + \text{billing rate})$$

The "bottom line" (cost liquidation) is the difference between customer billings and total expenses.

I, Vice-President of Finance, have taken issue with this budget. My claim is that if expenses are allocated to Departments A, B, and C according to their actual use, the strategic positioning assumptions would be proven incorrect. I also believe that decentralizing expense control would motivate the department managers to exercise better cost controls. Accordingly, if the costs were allocated out to the departments as they are actually incurred, the budget would look as follows.

Braun's decentralized budget	A	B	C	Admin	Total
Salaries and benefits	100	100	100	10	310
Fixed expenses	50	80	0	0	200
Variable expenses	10	50	30	0	90
TOTAL EXPENSES	160	230	200	0	600
Billable salaries and benefits	60	75	65	0	200
CUSTOMER BILLINGS*	180	225	195	0	600
Cost Liquidation	20	(5)	(5)	(10)	0

*Billing rate = 200 percent

In my budget, all fixed and variable expenses are charged to the departments according to actual or inferred use. Only management salaries and benefits are retained as a centralized expense in the administration column. My budget shows Department B using many more resources, proportionately, than the other departments. As a result, Department B's expenses are higher than its income from customer billings. Instead of performing like a "cow" in the strategic positioning matrix it is more like a "dog!" Furthermore, Department A is performing more like a "star" than a "question mark." This leads me to question whether or not effective cost controls are being implemented in the management of Department B and whether Department A is spending sufficient resources to follow a market growth strategy.

Note that in both ExCom's cost-centralized budget and my cost-decentralized budget, the total company's figures are the same. While the centralized versus decentralized budgets show quite a difference at the department levels, these differences "wash" at the company level.

Information Sources on the Internet

Note: The prefix "www" is omitted on all addresses below except where noted.

D1 General Information on Entrepreneurship

Wall Street Journal	startupjournal.com
Fast Company	fastcompany.com
Service Corps of Retired Executives (SCORE)	score.org
National Dialogue on Entrepreneurship	publicforuminstitute.org/nde
Kauffman Foundation	kauffman.org
Entreworld	entreworld.org
Entrepreneur America	entrepreneur-america.com
Global Entrepreneurship Monitor	gemconsortium.org
Always On	alwayson-network.com
Inc.	inc.com

D2 Student Organizations

Collegiate Entrepreneurs Organization	c-e-o.org
Students in Free Enterprise	sife.org
National Foundation for Teaching Entrepreneurship	nfte.com

D3 Venture Forums and Women's Networks

MIT Enterprise Forum	http://web.mit.edu/entforum
Venture clubs	venturea.com/clubs2.htm
Venture forums	ventureforums.org
Forum for Women Entrepreneurs	fwe.org
Women entrepreneurs	springboardenterprises.org

D4 Entrepreneurship Education

American Society of Engineering Education	asee.org
National Collegiate Investors and Innovators Alliance	nciia.org
U.S. Association for Small Business and Entrepreneurship	usasbe.org
National Network for Technology Entrepreneurship and Commercialization	n2tec.org
Academy of Management	aomonline.org
National Consortium of Entrepreneurship Centers	nationalconsortium.org
Stanford Technology Ventures Program (STVP)	http://stvp.stanford.edu
Georgia Institute of Technology	entrepreneurship.gatech.edu
Columbia University	gsb.Columbia.edu/entprog
MIT Entrepreneurship Center	http://entrepreneurship.mit.edu
Centre for Scientific Enterprise London	cselondon.com
National University of Singapore	enterprise.nus.edu.sg/nec
Penn State University	smeal.psu.edu/fcfe
University of Maryland	rhsmith.umd.edu/dingman
STVP Educators Corner (Teaching Resources)	http://edcorner.stanford.edu
Harvard Business Online (Teaching Resources)	hbsp.harvard.edu
European Case Clearing House (Teaching Resources)	ecch.cranfield.ac.uk
Eweb (Teaching Resources)	eweb.slu.edu
Center for Venture Education (Kauffman Fellows Program)	kauffmanfellows.org

D5 Ideas, Brainstorming and Creativity

Idea Fisher	ideafisher.com
Inspiration Software	inspiration.com
Problem solving	whynot.net

D6 Business Plans

Sample plans	bplans.com
National Business Incubator Association	nbia.org
Start-up plans and forms	startupbiz.com
Business plans	bizplanet.com

D7 Industry Information, Technology Trends and Simulations

Lexis Nexis	lexisnexis.com
Balanced Scorecard Collaborative	bscol.com
MIT Technology Review	technologyreview.com
Silicon Valley	siliconvalley.com
Life sciences information	bio.com
Life sciences information	biospace.com
Technology news	worldtechnews.com
Larta	larta.org
Business game	hotshotbusiness.com
Scenarios	gbn.org

D8 Tax, Legal and Intellectual Capital Information

Nolo	nolo.com
Smart Money	smartmoney.com/tax
Intellectual property	patentcafe.com
U.S. Patent and Trademark Office	uspto.gov
Inventors Association	uiausa.org
Patent search	delphion.com
U.S. Copyright Office	loc.gov/copyright
IBM Intellectual Property and Licensing	patents.ibm.com
World Intellectual Property Organization	wipo.org

D9 Marketing and Global Business

International Trade Administration	ita.doc.gov
National economies	cia.gov
World Bank	worldbank.org
Example of global website	Holland.com
Small Business Exporter Association	sbea.org
Global trade	globetrade.com
Marketing research resources	researchinfo.com
Links to industry web sites	industrylink.com
Thomas Register	thomasregister.com
American Demographics	demographics.com
Nielsen Media Research	nielsenmedia.com
Direct Marketing Association	the-dma.org
Direct marketing	dmnews.com
Surveys	surveymonkey.com

D10 Franchising and Family Businesses

Franchising information	franchise.com
Franchising opportunities	franchiseopportunities.com
Family business	fambiz.com

D11 United States Government Sources

Securities and Exchange Commission	sec.gov
SEC filings and forms	sec.gov/edgar.shtml
Small Business Administration	sba.gov
Department of Commerce	doc.gov
Patent and Trademark Office	uspto.gov
Census Bureau	census.gov

D12 General Information on Funding Sources

Funding directory	businessfinance.com
Capital Venue	capitalvenue.com
Venture capital	vcbuzz.com
Startup Nation	startupnation.com
Venture Reporter	venturereporter.net
National Venture Capital Association	nvca.org
Venture Economics	ventureeconomics.com
VentureWire	venturewire.com
VentureOne	ventureone.com
Venture Blog	ventureblog.com

D13 Angel Investor Groups

Silicon Valley	bandangels.com
Nevada and Northern California	sierraangels.com
Chicago	prairieangels.com
Arizona	arizonaangels.com
North Carolina	tignc.com
Southern California	techcoastangels.com
Washington state	allianceofangels.com
United States	gatheringofangels.com
Pasadena, California	pasadenaangels.com
Pennsylvania	ppig.com
Ohio	c-cap.net
Kansas City	kcep.com
Boston	hcangels.com
Boston–MIT	tcnmit.org

Toronto, Canada   tvg.org
Angel Capital Association   angelcapitalassociation.org

D14 Venture Capital Firms

Accel   accel.com
Advent International   adventinternational.com
Apax   apax.com
Atlas   atlasventure.com
Austin Ventures   austinventures.com
Ben Franklin   benfranklin.org
Benchmark Capital   benchmark.com
Carlyle   thecarlylegroup.com
Charles River Ventures   crv.com
Draper Fisher Jurvetson   drapervc.com
Financial Technology Ventures   ftventures.com
Flywheel Ventures   flywheelventures.com
Garage Technology Ventures   garage.com
Kleiner Perkins   kpcb.com
Matrix Partners   matrixpartners.com
Mayfield   mayfield.com
Mobius   mobiusvc.com
Mohr Davidow Ventures   mdv.com
New Enterprise Associates   nea.com
North Bridge   nbvp.com
Norwest   norwestvp.com
Sequoia   sequoiacap.com
Sutter Hill   shv.com
Ventana Global   ventanaglobal.com
Venrock   venrock.com

D15 Corporate Venture Capital

Chevron Texaco, San Ramon, California   chevrontexaco.com/
                                                        technology/ventures

Dell Ventures, Austin, Texas   dell.com
Eastman Ventures, Kingsport, Tennessee eastmanventures.com
Intel Capital, Santa Clara, California intel.com/capital
Motorola Ventures, Chicago, Illinois   motorola.com/ventures
Nokia Venture Partners, Menlo Park,
   California   nokiaventurepartners.com
Intrapreneurs   intrapreneur.com

D16 Social Innovation and Responsibility

General information	enterprisingnonprofits.org
USF Institute	inom.org
"Win-Win" Partners	winwinpartner.com
Nonprofits	nonprofits.org
Investors Circle	investorscircle.net
Social Enterprise Alliance	se-alliance.org
Silicon Valley Social Venture Fund	svz.org
Venture Philanthropy Partners	venturephilanthropypartners.org
Ashoka	ashoka.org
Center for Social Innovation	gsb.Stanford.edu/csi
Ethics Resource Center	ethics.org
Social responsibility	bsr.org

absorptive capacity   A firm's ability to assimilate new knowledge for the production of innovations. The more related knowledge a firm has, the easier it is for it to assimilate the new knowledge.

acquisition   When one firm purchases another, the acquired company gives up its independence and the surviving firm assumes all assets and liabilities.

adaptive enterprise   An enterprise that changes its strategy or business model, as the conditions of the marketplace require.

advertising   Public messages sent via any media that are intended to attract and influence consumers.

advertising revenue model   Selling firm, usually media companies, provides space or time for advertisements and collects revenues for each use.

affiliate revenue model   A model based on steering business to an affiliate firm and receiving a referral fee or percentage of revenues.

alliance   See *partnership*.

angels   Wealthy individuals, usually experienced entrepreneurs, who invest in business start-ups in exchange for equity in the new ventures.

architectural innovation   A change in how components of a product are linked together while core design concepts are left untouched.

asset   Something of monetary value that is owned by a firm or an individual.

balanced scorecard   A strategy formulation device as well as a report of performance.

bankruptcy   A voluntary or involuntary state declared by a court in which an individual or company is unable to meet its financial obligations.

barriers to entry   Whatever keeps a firm from entering an industry or market.

base case   The calculation of cash flows based on a set of assumptions that portray outcomes that are most likely to happen.

best customer   One who values your brand, buys it regularly whether your product is on sale or not, tells his or her friends about your product, and will not readily switch to a competitor.

Black-Scholes equation   An exact formula for the price of a call option. The formula requires five variables: the risk-free interest rate, the variance of the underlying stock, the exercise price, the price of the underlying stock, and the time to expiration.

board of advisors   A group constituted to provide advice and contacts to a venture. The members have extensive skills and knowledge and provide good advice.

board of directors   A group composed of key officers of a corporation and outside members responsible for the general oversight of the affairs of the entity.

bond   Debt security issued by a firm.

book value   The net worth (equity) of the firm, calculated by total assets minus intangible assets (patents, goodwill) and liabilities.

bootstrap financing   Launching a start-up with modest funds from the entrepreneurial team, friends, and family.

brand   A combination of name, sign, or symbol that identifies the goods sold by a firm.

brand equity   The brand assets linked to a brand's name and symbol that add value to a product.

breakeven   The point at which the total sales equals the total costs.

breakeven analysis   A means of determining the quantity that has to be sold at a given price so that revenues will equal cost.

burn rate   Defined as cash in minus cash out on a monthly basis.

business design   A design that incorporates the venture's selection of customers, its offerings, the tasks it will do itself and those it will outsource, and how it will capture profits.

business format franchise   Involves the provision of a complete method including a license for the trade name and logo, the products and methods, the form

649

of the physical facility, the strategy, and the purchasing system.

business method patent   A type of a utility patent that involves the classification of a process.

business model   A set of planned assumptions about how a firm will create value for all its stakeholders; the resulting outcome of the business design process.

business plan   A document that describes the opportunity, product, context, strategy, team, required resources, financial return, and harvest of a business venture.

buy-sell agreements   Contracts between associates that set the terms and conditions by which one or more of the associates can buy out one or more of the other associates.

cannibalization   The act of introducing products that compete with a firm's existing product line.

capacity   The ability to act or do something. A firm has processes and assets that need to be expanded as the firm grows its sales volume.

cash flow   The sum of retained earnings plus the depreciation provision made by a firm. The cash coming into a firm minus the cash going out over a predetermined time period.

C-corporation   A business that provides limited liability, unlimited life, the ability to accept investments from other corporations, and the ability to merge with other corporations.

certain   An outcome resulting from an action in that it will definitely happen.

challenge   A call to respond to a difficult task and the commitment to undertake the required enterprise.

champion   An executive or leader in the parent company who advocates or provides support and resources as well as protection of the venture when parent company routines are breached. The champion helps describe and defend the venture and secure the necessary resources.

chasm   A large gap between visionaries and pragmatists in the adoption process.

cluster   A geographic concentration of interconnected companies in a particular field. Clusters can include companies, suppliers, trade associations, financial institutions, and universities active in a field.

collaborative structure   Primarily consists of teams with few underlying functional departments. In a collaborative structure, the operating unit is a team that may consist of five to 10 members.

common stock   Evidence of ownership share.

competence   Skill or aptitude; ability to perform an activity. This skill or aptitude is core (i.e., a *core competence*) only if it provides customer value, differentiates a firm from its competitors, and is extendable to other products.

competitive advantage   A firm's distinctive factors that give it a superior or favorable position in relation to its competitors.

competitive intelligence   The process of legally gathering data about competitors.

complement   A product that improves or perfects another product.

complementors   Companies that sell complements to another enterprise's product offerings.

concept summary   A simple statement of the problem the new venture is solving and how the new venture will act to solve it.

conjoint analysis   A quantitative measure of the relative importance of one attribute as opposed to another.

convergence   The coming together or merging of several technologies or industries thought to be different or separate.

copyright   An exclusive right granted by the federal government to the owner to publish and sell literary, musical, or other artistic materials. A copyright is honored for 50 years after the death of the author.

core competencies   The unique skills and capabilities of a firm.

corporate culture   The basic style of a company and how people work with each other.

corporate new venture (or **corporate venture**)   A new venture started by an existing corporation for the purpose of initiating and building an important new business unit or organization, solely owned subsidiary, or spin-off as a new public company.

corporate venture capital   An initiative by a corporation to invest in either young firms outside the corporation or units formerly part of the corporation. These are often organized as corporate subsidiaries, not as limited partnerships.

corporation   A legal entity separate from its owners. A body of owners granted a charter to act as a separate entity distinct from its owners.

creative destruction   The creation of new industrial structures and companies and the destruction of older structures.

creativity   The ability to use the imagination to develop new ideas, new things, or new solutions.

customer relationship management   A set of conversations that consist of 1) economic exchanges, 2) the product offering that is the subject of the exchange, 3) the space in which the exchange takes place, and 4) the context of the exchange with the customers.

customization   Provision of a product designed to meet a user's preferences.

debt capital   Money that a business has borrowed and must repay in a specified time with interest.

design   The activity leading to the arrangement of concrete details that embodies a new product idea or concept.

design patents   Grants of exclusive right of use for new original, ornamental, and nonobvious designs for articles of manufacture for a period of 14 years.

diffusion of innovations   The process by which innovations spread through a population of potential adopters.

diffusion period   The time required to move from 10 percent to 90 percent of the potential adopters.

discontinuity   A change or interruption of an industrial structure or methods.

discount rate   The rate (r) at which future earnings or cash flow is discounted because of the time value of money.

disruptive application   A new product that establishes an entirely new category and dominates that category.

distributor   A firm that takes title to finished goods and then sells them.

dominant design   A design whose major components and underlying core concepts do not vary substantially from one product model to the other and that commands a high percentage of the market share for the product.

dynamic capitalism   The process of wealth creation characterized by the dynamics of new, creative firms forming and growing and old, large firms declining and failing.

dynamic disequilibrium   The constant change of factors in an economy.

e-commerce   Digitally enabled commercial transactions between and among organizations and individuals.

economic capital   The value of an economy and the associated standard of living.

economics   The study of humans in the ordinary business of life. Economics can also be defined as the study of how society manages its scarce resources.

economies of scale   The concept that larger volumes sold reduce per-unit costs.

economies of scope   Economies obtained by sharing of resources by multiple products or business units.

efficient market   A market where prices reflect all available information.

elevator pitch   A short version of the venture story that quickly demonstrates that the entrepreneurs know their business and can communicate it effectively.

emergent industries   Newly created or re-created industries formed by product, customer, or context changes.

emotional intelligence   A bundle of four psychological capabilities that leaders exhibit: self-awareness, self-management, social awareness, and relationship management.

entrepreneur   1) A person who undertakes an enterprise or business with the chance of profit or loss (or success or failure); or 2) a person or group that engages in the initiation and growth of a purposeful enterprise for the production of goods and services.

entrepreneurial capital   A combination of entrepreneurial competence and entrepreneurial commitment.

entrepreneurial commitment   A dedication of the time and energy necessary to bring the enterprise to initiation and fruition.

entrepreneurial competence   The ability 1) to recognize and envision taking advantage of opportunity and 2) to access and manage the necessary resources to take advantage of the opportunity.

entrepreneurial intensity   The degree of commitment of the entrepreneurial team to the growth of the firm.

entrepreneurship   Focused on the identification and exploitation of previously unexploited opportunities.

equity capital   The investment by a person in ownership through purchase of the stock of the firm.

ergonomics   The science of making a physical task easier and less stressful to accomplish.

ethics   A set of moral principles for good human behavior. Ethics provides the rules for conducting activities in a manner acceptable to society.

execution   A discipline for meshing strategy with reality, aligning the firm's people with goals, and achieving the results promised.

exit strategy   The way entrepreneurs or investors get their money out of a venture.

experience curve   Systematic production cost reductions that occur over the life of a product.

experience curve pricing   Aggressive pricing designed to increase volume and help the firm realize experience curve economies.

family-owned business   A firm that includes two or more members of a family who hold control of the firm.

financial capital   Financial assets such as money, bonds, securities, land, patents, and trademarks.

financial structure   Mix of debt and equity used to finance a business.

first-mover advantage   Gain accruing to the first to enter a market.

flexibility   A measure of a firm's ability to react to a customer's needs quickly.

flow-through entities   Firms where all profits flow to the owners free of prior taxation.

focus group   A small group of people who are brought together to discuss a product or service.

founders   The people responsible for starting a firm, usually all the members of the initial leadership team.

franchise   A legal arrangement in which the owner of a business format has licensed it to an individual or a local firm.

franchisee   An individual or a local firm that receives the right through a contract to use the franchisor's business format, brand, and logo in a specific geographic region.

franchisor   The organization that owns and operates a firm that controls the business format and its associated trademarks and logo.

global   A strategy that emphasizes worldwide creation of new products, sales and marketing.

globalization   The integration of markets, nation-states, and technologies enabling people and companies to offer and sell their products in any country in the world.

growing industries   Industries that exhibit moderate revenue growth and have moderate stability and uncertainty.

harvest plan   A plan that defines how and when the owners and investors expect to realize or attain an actual cash return on their investment.

high growth   Characterizes a business corporation that aims to build an important new business and requires a significant initial investment to start up.

horizontal merger   A merger between firms that make and sell similar products in a similar market.

hybrid model   This model, sometimes called "bricks and clicks," utilizes the best of the Internet as well as other channels. A hybrid model can extend a company's reach to new market segments as well as globally.

implementation   Putting into effect the elements of the business plan.

increasing returns   When the marginal benefits of a good or of an activity are growing with the total quantity of the good or the activity consumed or produced.

independent venture   A new venture that is not owned or controlled by an established corporation. An independent venture is typically unconstrained in its choice of a potential opportunity yet is usually constrained by limited resources.

industry   A group of firms producing products that are close substitutes for each other and serve the same customers.

initial public offering   The first public equity issue of stock by a company.

innovation   Invention that has produced economic value in the marketplace. It is the commercialization of new technology.

installed base profit model   One of the most powerful profit models; the supplier builds a large installed base of users who then buy the supplier's consumable products.

integrity   Truthfulness, wholeness, and soundness; the consistency of our words and our actions or our character and our conduct.

intellectual capital   The sum of knowledge assets of an organization.

intellectual property   The valuable intangible property owned by persons or companies.

international   A strategy that aims to create value by transferring products and capabilities from the home market to other nations using export or licensing arrangements.

internet   A worldwide network of computer networks linking businesses, organizations, and individuals.

intrapreneur   An entrepreneur working within the confines of a corporation.

IPO   See *initial public offering.*

joint venture   A short-lived partnership with each partner sharing in the costs and rewards of the project; common in research, investment banking, and the health care industry.

just-in-time   A method that focuses on reducing unnecessary inventory and removing non-value-added activities by receiving items only when needed.

knowledge   The awareness and possession of information, facts, ideas, truths, and principles in an area of expertise.

knowledge management   The practice of collecting, organizing, and disseminating the intellectual

knowledge of a firm for the purpose of enhancing its competitive advantages.

layout   The arrangement of a facility to provide a productive workplace. This can be accomplished by aligning the form of the space with its use or function.

leadership   The ability to create change or transform organizations. A real measure of leadership is the ability to acquire needed new skills as the situation changes.

lead users   People who have an advanced understanding of a product and are experts in its use.

lean systems   Operations systems that are designed to create efficient processes by using a total systems perspective.

learning organization   A firm that captures, generates, shares, acts, and uses its corporate experiences to improve and adapt.

license   A grant to another firm to make use of the rights of the licensor's intellectual property.

licensing   Occurs when a firm (the licensor) grants the right to produce its product, use its production processes, or use its brand name or trademark to another firm (the licensee). In return for giving the licensee these rights, the licensor collects a royalty fee on every unit the licensee sells.

local   A strategy focusing all efforts locally (or regionally) since that is the venture's pathway to a competitive advantage.

logistics   The organization of moving, storing, and tracking parts, materials, and equipment. Logistic systems usually are based on electronic networks such as a supply-chain intranet.

loyalty   A measure of a customer's commitment to a company's product or product line.

management   A set of processes such as planning, budgeting, organizing, staffing, and controlling that keep the organization running well.

marketing   A set of activities with the objective of securing, serving, and retaining customers for the product offerings of the firm. Marketing is getting the right message to the right customer segment via the appropriate media and methods.

marketing objectives statement   A clear description of the key objectives of the marketing program.

marketing plan   A written document serving as a section of the business plan and containing the necessary steps required to achieve the marketing objective.

market potential   A prospective of the maximum sales under expected conditions.

market research   The process of gathering the information that serves as the basis for a sound marketing plan.

market segment   A group with similar needs or wants and may include geographical location, purchasing power, and buying attitudes.

market segmentation   The division of the market into segments that have different buying needs, wants, and habits.

mature industries   Industries that have slow revenue growth, high stability, and intense competitiveness.

merger   The combining together of two companies.

metanational   A company that possesses three core capabilities: 1) being the first to identify and capture new knowledge emerging all over the world; 2) mobilizing this globally scattered knowledge to out-innovate competitors; and 3) turning this innovation into value by producing, marketing, and delivering efficiently on a global scale.

module   An independent, interchangeable unit that can be combined with others to form a larger system.

moral principles   Tenets concerned with goodness (or badness) of human behavior and usually provided as rules and standards of human behavior.

multidomestic   A strategy that calls for a presence in more than one nation as resources permit.

natural capital   Those features of nature, such as minerals, fuels, energy, biological yield, or pollution absorption capacity, that are directly or indirectly utilized or potentially utilizable in human social and economic systems.

negotiation   A decision-making process among interdependent parties who do not share identical preferences.

net present value (NPV)   The present value of the future cash flow of a venture discounted at an appropriate rate (r).

network economies   Observed effects in industries where a network of complementary products is a determinant of demand.

new venture team   A small group of individuals who possess expertise, management, and leadership skills in the requisite areas.

new venture valuation rule   Uses the projected sales, profit, and cash flow in a target year (N), the projected earning growth rate, (g) for five years after year N to calculate value of a firm.

niche business   A firm that seeks to exploit a limited opportunity or market to provide the entrepreneurs with independence and a slow-growth buildup of the business.

nonprofit organization   A corporation or a member association initiated to serve a social or charitable purpose.

oligopoly   An industry characterized by just a few seller firms.

on-time speed   A measure of lead-time, on-time delivery, and product development speed.

operation   A series of actions.

operations management   The supervising, monitoring, and coordinating of the activities of a firm carried out along the value chain. Operations management deals with processes that produce goods and services.

opportunity   A timely and favorable juncture of circumstances providing a good chance for a successful venture or progress; an auspicious chance of an action occurring at a favorable time.

opportunity cost   The value (cost) of the foregone action.

option   The right to purchase an asset at some future date and at a predetermined price.

organic growth   Growth enabled by internally generated funds.

organic organizations   Organizations that are flexible and effectively adapt to change.

organizational culture   The bundle of values, norms, and rituals that are shared by people in an organization and govern the way they interact with each other and with other stakeholders.

organizational design   The design of an organization in terms of its leadership and management arrangements, its selection, training, and compensation of its talent (people), its shared values and culture, and its structure and style.

organizational norms   The guidelines and expectations that impose appropriate kinds of behavior for members of the organization.

organizational rituals   The rites, ceremonies, and observances that serve to bind together members of the organization.

organizational values   The beliefs and ideas of an organization about what goals should be pursued and what behavior standards should be used to achieve these goals. Values include entrepreneurship, creativity, honesty and openness.

outsourcing   Purchasing services or goods from suppliers rather than doing or producing them within the firm.

partnership   Business association of two or more people or firms who agree to cooperate with one another to achieve mutually compatible goals that would be difficult for each to accomplish alone. There are two types of partnerships: the general partnership and the limited partnership.

patent   A grant by the U.S. government to an inventor giving exclusive rights to an invention or process for 18 years. A U.S. patent does not always grant rights in foreign countries.

PE ratio   The ratio of the price of a stock to the company's earnings.

personalization   The provision of content specific to a user's preferences and interests.

pessimistic case   When the outcome of the calculation of cash flows, based on a set of assumptions, is less than expected.

place   The channels for distribution of the product and, when appropriate, the physical location of the stores.

plant patent   A grant of exclusive right of use for a term of 17 years for certain new varieties of plants that have been asexually reproduced.

portal   A place or gate of access to a company's offerings.

positioning   The act of designing the product offering and image to occupy a distinctive place in the target customer's mind.

post-money valuation   The valuation accorded a company after investment by venture capitalists or angels.

preferred stock   Stock with preferences or claims on dividends and assets before common stock owners.

pre-money valuation   The value accorded a company before investment from venture capitalists or angels.

pricing policies   Methods for setting prices for various customer categories and volume discount plans.

private placements   The sale of stocks or bonds to wealthy individuals, pension funds, insurance companies, or other investors without a public offering or any oversight from the Securities and Exchange Commission.

process   Any activity or set of activities that takes one or more inputs, transforms and adds value to them, and provides one or more outputs.

product   The item or service that serves the needs of the customer.

product distribution franchise   A license to sell specific products under a manufacturer's trademark and brand.

productivity   The quantity of goods and services produced from the sum of all inputs, such as hours worked and fuels used.

product offering   Communicates the key values of the product and describes the benefits to the customer.

product platform   A set of modules and interfaces that form a common architecture from which a stream of derivative products can be efficiently developed and produced.

product sales model   Sales of a product in units to a customer. Selling items for a price.

profit   The net return after subtracting the costs from the revenues.

profit margin   The ratio of profit divided by sales revenues.

profit model   The mechanism a firm uses to reap profits from its revenues.

pro forma   Provided in advance of actual data. Pro forma statements are forecasts of financial outcomes.

promotion   The communication of an initial product message using public relations, advertising, and sales methods to attract customers.

proprietary   That which is owned, such as a patent, formula, brand name, or trademark associated with the product or service, and not usable by another without permission.

prototype   A model that has the essential features of the proposed product or service but remains open to modification.

public offering   The sale of a company's shares of stock to the public by the company or its major stockholders.

quality   A measure of a product that usually includes performance and reliability.

radical innovation   An important new development that leads to a new industry or way of operating.

rapid prototyping   The fast development of a useful prototype that can be used for collaborative review and modification.

real option   The right to invest in (or purchase) a real asset (the start-up firm) at a future date.

relational coordination   Describes how people act as well as how they see themselves in relationship to one another.

reliability   A measure of how long a product performs before it fails.

regional   See *local.*

regret   The amount of loss that a person can tolerate.

relational coordination   Describes how a firm's people act as well as how they see themselves in relationship to one another.

resilience   The ability to recover quickly from setbacks. It is a skill that can be learned and increased.

restricted stock   Stock issued in an employee's name and reserved for his or her purchase at a specified price after a period of time.

return on capital   The ratio of profit to the total invested capital of a firm.

return on equity   The ratio of net income divided by owner's equity.

return on investment   The ratio of net income divided by invested capital.

revenue model   Describes how the firm will generate revenue.

revenues   A firm's revenues are its sales in dollars expressed after deducting for all returns, rebates, and discounts.

risk   The chance or possibility of loss, which could pertain to finance, physicality, or reputation. Risk is a measure of the potential variability that will be experienced in the future.

robust product   A product that is relatively insensitive to aging, deterioration, component variations, and environmental conditions.

sales forecast   An estimate of the amount of sales to be achieved under a set of assumed conditions within a specified period of time.

scalability   The extent to which a firm can grow in various dimensions to provide more service.

scale of a firm   The extent of the activity of a firm as described by its size. The scale of a firm's activity can be described by its revenues, units sold, or some other measure of size.

scenario   An imagined sequence of possible events or outcomes, sometimes called a mental model.

scope of a firm   The range of products offered or distribution channels utilized.

S-corporation   A firm that has elected to be taxed as a partnership under the subchapter S provision of the Internal Revenue Code.

second-mover strategy   Following the first-mover and learning from their mistakes.

secrecy   The state of concealment or the state of being concealed or maintained as a secret.

seed capital   The first funds used to launch the new firm.

self-organizing organization   Teams of individuals that benefit from the diversity of the individuals and the robustness of their network of interactions.

selling   The transfer of products from one person or entity to another through an exchange mechanism.

six forces model   A model for evaluating the competitive forces in an industry: 1) firm rivalry, 2) threat of entry by new competitors, 3) threat of substitute products, 4) bargaining power of customers, 5) bargaining power of complementors, and 6) bargaining power of suppliers.

small business   A firm with fewer than 50 employees operating as a sole proprietorship, a partnership, or a corporation owned by a few people.

social capital   The accumulation of active connections among people in a network. Social capital refers to the resources available in and through personal and organizational networks.

social entrepreneur   A person or team that acts to form a new venture in response to an opportunity to deliver social benefits while satisfying environmental and economic values.

sole proprietorship   A firm owned by only one person and operated for his or her profit.

sources of innovation   For new ventures, these include universities, research laboratories, and independent inventors.

spin-off   An organization that is first established within an existing company and then sent off on its own.

staging   The provision of capital to entrepreneurs in multiple installments, with each financing conditional on meeting particular business targets. This helps ensure that the money is not squandered on unprofitable projects.

stock   The owner's shares of the corporation.

stock options   An offer in a plan under which employees can purchase shares of the company at a fixed price. Stock options take on value once the market price of a company's stock exceeds the exercise price. A stock option gives employees the right to buy the company's stock in the future at a preset price.

story   A narrative of factual or imagined events. The story tells the goal, the challenge, and the response of the new firm.

strategic control   The process used by firms to monitor their activities and evaluate the efficiency and performance of these activities and to take corrective action to improve performance, if necessary.

strategic learning   A cyclical process of adaptive learning using four steps: learn, focus, align, and execute.

strategy   A plan or road map of the actions that a firm or organization will take to achieve its mission and goals.

subscription revenue model   A type of business that offers content or a membership to its customers or members and charges a fee permitting access to the information or participation for a certain period of time.

sunk costs   A cost that has already incurred cannot be affected by any present or future decisions. In other words, funds and time invested on a new venture are already spent, regardless of any action taken today or later.

supply-chain management   A firm's processes and those of its suppliers that enable the flow of materials, resources, and information to meet customer demand.

sustainable competitive advantage   A competitive advantage that can be maintained over a period of time.

switching costs   The costs to the customer to switch from the product of an incumbent company to the product of the new entrant.

synergy   The increased effectiveness and achievement produced as a result of the combined action of two or more firms.

talent   The people, often called employees, of an organization.

team   A small number of people with complementary capabilities and skills who are committed to a common objective, goals, and tasks for which they hold themselves mutually accountable.

technology   Devices, artifacts, processes, tools, methods, and materials applied to industrial and commercial purposes.

"telework"   All kinds of remote work from home, satellite offices, and on the road.

term sheet   A summary of the principal conditions for a proposed investment by a venture capital firm in a company.

theory of a business   How a firm comprehends its total activities, resources, and relationships.

throughput   The amount of units processed within a given time.

tipping point   The moment of critical mass or threshold that results in a jump in adoption of a product or service.

trademark   Any distinctive word, name, symbol, slogan, shape, sound, or logo a firm uses to designate its product.

trade-name franchise   A franchise name that primarily involves a brand name, such as Western Auto or ACE Hardware.

trade secret   An intellectual asset protected by confidentiality, nondisclosure, and assignment of inventions agreements as well as physical barriers such as safes and limited access.

transaction fee revenue model   A type of business that provides a transaction source or activity for a fee.

transnational   A strategy resting on a flow of product offerings created in any one of the countries of operation and transferred between countries.

triple bottom line   The three factors of a product or business: economic, environmental, and social equity.

trust   A firm belief in the reliability or truth of a person or an organization.

uncertain   An outcome resulting from an action in that the outcome is not known or is likely to be variable.

unique selling proposition   A short version of a firm's value proposition often used as a slogan or summary phrase to explain the key benefits of the firm's offering versus that of a key competitor.

usability   A measure of the quality of a user's experience when interacting with a product.

utility patents   Rights of exclusive use issued for the protection of new, useful, nonobvious, and adequately specified processes, machines, and manufacturing processes for a period of 17 years.

valuation rule   The algorithm by which an investor, such as an angel or venture capitalist, assigns a monetary value to a new venture.

value   The worth, importance, or usefulness to the customer. In business terms, value is the worth in monetary terms of the social and economic benefits a customer pays for a product or service.

value added   The value of a process output minus the value of its inputs.

value chain   The sequence of steps or subprocesses that a firm uses to produce its product or service.

value proposition   Summarizes the values offered to the customer.

value web   Consists of the extended enterprise within a network of interrelated stakeholders that create, sustain, and enhance its value-creating capacity. It is usually based on an Internet infrastructure to manage operations dispersed in many firms.

venture capital   A source of funds for new ventures that is managed by investment professionals on behalf of the investors in the venture capital fund.

venture capitalists   Professional managers of investment funds.

versioning   The creation of multiple versions of a products and selling their modified versions to different market segments at different prices.

vertical integration   The extension of a firm's activities into adjacent stages of productions (i.e., those providing the firm's inputs or those that purchase the firm's outputs).

vertical merger   The merger of two firms at different places on the value chain.

viral marketing   Building knowledge of a product through word of month.

virtual organization   A venture that manages a set of partners and suppliers linked by the Internet, fax, and telephone to provide a source or product.

vision   An informed and forward-looking statement of purpose in response to an opportunity.

working capital   The amount of funds available to support a firm's normal operations, such as unexpected or out-of-the-ordinary, one-time-only expenses. Working capital is a firm's current assets minus its current liabilities.

Page numbers followed by f indicate figures, t tables, respectively.